Expanding the Horizons of Manufacturing: Towards Wide Integration, Smart Systems and Tools

Expanding the Horizons of Manufacturing: Towards Wide Integration, Smart Systems and Tools

Editors

Luis Puigjaner
Antonio Espuña Camarasa
Edrisi Muñoz Mata
Elisabet Capón García

MDPI • Basel • Beijing • Wuhan • Barcelona • Belgrade • Manchester • Tokyo • Cluj • Tianjin

Editors
Luis Puigjaner
Center for Process and Environment Engineering (CEPIMA), Chemical Engineering Department, EEBE-c, Universitat Politècnica de Catalunya (UPC), Eduard Maristany 10-14, Ed. I-5, 08019 Barcelona, Spain

Antonio Espuña Camarasa
Department of Chemical Engineering, Universitat Politècnica de Catalunya (UPC), E-08019 Barcelona, Spain

Edrisi Muñoz Mata
Chemical Engineering Department, EEBE, Universitat Politècnica de Catalunya, Av. Eduard Maristany, 10-14, 08019 Barcelona, Spain

Elisabet Capón García
ABB Switzerland Ltd., 5400 Baden, Switzerland

Editorial Office
MDPI
St. Alban-Anlage 66
4052 Basel, Switzerland

This is a reprint of articles from the Special Issue published online in the open access journal *Processes* (ISSN 2227-9717) (available at: https://www.mdpi.com/journal/processes/special_issues/manufacturing_horizons).

For citation purposes, cite each article independently as indicated on the article page online and as indicated below:

LastName, A.A.; LastName, B.B.; LastName, C.C. Article Title. *Journal Name* **Year**, *Volume Number*, Page Range.

ISBN 978-3-0365-4361-1 (Hbk)
ISBN 978-3-0365-4362-8 (PDF)

Contents

Preface to "Expanding the Horizons of Manufacturing: Towards Wide Integration, Smart Systems and Tools"

Over the last twenty years, most companies and researchers have tended to employ a company-centric view of the supply chain. From this perspective, the supply chain (SC) is perceived to consisting of the enterprise as a central entity, potentially together with some peripheral partners—typically, first-tier suppliers and customers. These views involve integrating production and logistics planning across the enterprise, value-chain management, global network planning, and investment appraisal. Research on the "**extended**" **supply chain**—where the view is much broader, e.g., **encompassing the suppliers' suppliers and the customers' customers—is far less prevalent**. This is almost certainly due to **(i) the relative infancy of the discipline, and the fact that considerable benefits can be achieved simply using company-centric views of the supply chain; and (ii) a wariness of supply-chain "partners" and a lack of data sharing.**

The aim of creating this research topic is to a) achieve enterprise-wide modeling and optimization (EWMO) through the development and application of integrated modeling, simulation, and optimization methodologies, and b) create computer-aided tools for reliable and sustainable improvement opportunities within the entire manufacturing network (raw materials, production plants, distribution, retailers, and customers) and its components. This integrated approach incorporates information from local basic control and supervisory modules to the scheduling/planning formulation. This makes it possible to dynamically react to incidents occurring in network components at the appropriate decision-making level, reducing the amount of resources required, emitting less waste, and facilitating better responsiveness to the changing market requirements and operational variations. This reduces the cost, waste, energy consumption, and environmental impacts, increasing benefits.

Moreover, the increase in globalization has significantly increased the scale and complexity of current businesses. Businesses have become global networks of multiple business units and functions. Operational functions include R&D, production networks (continuous, discontinuous, and discrete), and supply networks. These functions are supported by financial planning and marketing strategy functions. Furthermore, businesses are also subject to internal and external uncertainties. Internal uncertainties include the unpredictable success rate of R&D projects; the technological risks involved; and disruptions to production, such as production failures and unforeseen stoppages. External uncertainties include the fluctuating cost of raw materials and products (unless they are subject to monopoly fluctuations in the exchange rate), and uncertainties in market size and demand due to competition and macroeconomic factors. Businesses control their operations through their decisions regarding capital expenditure, company finances, growth strategies, and operations. Strategic decisions on capital spending and planning include the technology used, the choice of R&D projects, and decisions regarding infrastructure and supply-chain management (SCM). Financial decisions are made by identifying the assets and liabilities required from the working capital for larger projects and operations, assessing, and protecting the company from, change risk. Examples of tactical production decisions include planning activities in plants which run on a discontinuous basis to respond to an anticipated demand, making decisions about the sources of energy used following market prices, and increasing production capacity in response to the demand-related pressures.

The current solutions in the field of process systems engineering (PSE) and operations research tend to only consider subsets of such decisions, even though a business must act as a cohesive body in which its various functions are, to a certain extent, coordinated. Therefore, from the perspective of a company, the overall performance will be suboptimum if strategic and tactical decisions are made independently, as has been the case to date. However, it is significantly more complex for a company to make decisions that involve its overall interests than it is to make decisions about specific functions. This explains why research with integral modeling that reflects the overall operation of companies is scarce.

The PSE community faces an increasing number of challenges; to address these challenges, enterprise and SCM remain subjects of significant interest because they offer multiple opportunities. It is anticipated that further progress in this area will bring a unique opportunity to demonstrate the potential of the PSE approach to enhance a company's value. As previously mentioned, one of the critical components of SCM and EWMO is decision-making coordination and integration at all levels. Most recent contributions offer models that separately address problems arising in the three standard supply chain (SC) hierarchical decision levels (i.e., strategic, tactical aggregate planning, and short-term scheduling).

More recently, integrating new technologies through semantic models in formal knowledge models allows for the capture and utilization of domain knowledge, human knowledge, and expert knowledge of comprehensive intelligent management. Otherwise, the development of advanced technologies and tools, such as cyber-physical systems, the Internet of Things, the Industrial Internet of Things, Artificial Intelligence, Big Data, Cloud Computing, Blockchain, etc., have directed the attention of manufacturing enterprises toward intelligent manufacturing systems. This Special Issue also invites contributions from these advanced areas.

In summary, this Special Issue addresses the following concepts:

- **Development of advanced mathematical models** and **methodologies** for the integrated approach of:
    - **The network design problem**, such as the location of the plant, warehouses, and distribution centers, capacity and technology selection, etc.;
    - **The supply chain planning problem**, including distribution planning, inventory control, and product demand forecasting;
    - **The integration of production, financial and environmental aspects, risk, and uncertainty.**

The expected models will tackle a **multi-objective** view of achieving the necessary trade-off between often contradictory benefits in terms of economic and environmental benefits, customer satisfaction, and increased response to dynamic market changes:

- **Development of detailed production scheduling at the plant level** for batch, continuous, and discrete manufacturing for **online scheduling** implemented in practice under real-time variations and uncertainty;
- **Integration of the tracking system of network dynamics** within the holistic decision-making model (e.g., by enclosing a model predictive control framework), thus facilitating equipment capacity handling similarly at strategic and operational levels and enabling an adequate response to incidents **for enhanced production sustainability**;
- **Development of suitable frameworks and algorithms** for solving these problems in an efficient and integrated manner (e.g., surrogate problem decomposition, disjunctive programming, and potentially, Lagrange decomposition);

- **Development of software prototypes** for the implementation of the above methodologies and algorithms, illustrating their applicability in several **real-life industrial case studies** involving typical manufacturing/distribution networks belonging to relevant sectors in the world;
- **Development of novel frameworks** focusing on the utilization of formal knowledge models, facilitating new technology implementation, and transactional system integration;
- **Further development of smart manufacturing systems** for the transformation of manufacturing enterprises, from traditional to the intellectualized ones;
- **Development of Intelligent Systems and Intelligent Agents** focused on cooperative work between human beings and computers, enhancing the capability of human decision making and problem solving in the process engineering field.

Luis Puigjaner, Antonio Espuña Camarasa, Edrisi Muñoz Mata, and Elisabet Capón García
Editors

processes

Editorial

Expanding the Horizons of Manufacturing, towards Wide Integration, Smart System, and Tools

Luis Puigjaner [1,*], Antonio Espuña [1], Edrisi Muñoz [1] and Elisabet Capón-García [2]

1 Department of Chemical Engineering, Universitat Politècnica de Catalunya, Campus Diagonal-Besòs, 08930 Barcelona, Spain; antonio.espuna@upc.edu (A.E.); edrisi.munoz@gmail.com (E.M.)
2 ABB Switzerland Ltd., 5400 Baden, Switzerland; elisabet.capon@ch.abb.com
* Correspondence: luis.puigjaner@upc.edu

This research topic aims at enterprise-wide modeling and optimization (EWMO) through the development and application of integrated modeling, simulation and optimization methodologies, and computer-aided tools for reliable and sustainable improvement opportunities within the entire manufacturing network (raw materials, production plants, distribution, retailers, and customers) and its components. Such an integrated approach incorporates information from the local basic control and supervisory modules into the scheduling/planning formulation, making it possible to react dynamically to incidents occurring in the network components at the appropriate decision-making level.

A wide-integrated solution should allow enhanced coordination and cooperation between network components by avoiding competition, eventually leading to local optima and inefficiency associated with inconsistent isolated decisions at different levels. Such a wide-integrated solution approach would provide new structural alternatives, more effective management policies, more economical design options. Moreover, the solution obtained can work in practice requiring fewer resources, emitting less waste, and allowing for better responsiveness to changing market requirements and operational variations, thus reducing cost, waste, energy consumption, environmental impact, and increased benefits.

More recently, the exploitation of new technology integration, such as through semantic models in formal knowledge models, allows capturing and utilizing domain knowledge, human knowledge, and expert knowledge towards comprehensive intelligent management. Otherwise, the development of advanced technologies and tools such as cyber-physical systems, the Internet of Things, the Industrial Internet of Things, artificial intelligence, big data, cloud computing, and blockchain, have captured the attention of manufacturing enterprises toward intelligent manufacturing systems. This Special Issue also calls for contributions from these advanced areas.

In summary, we look for articles addressing (but not limited to) the following concepts:

- the development of advanced mathematical models and methodologies for the integrated approach;
- the network design problem, such as the location of the plant, warehouses, and distribution centers, and capacity and technology selection;
- the supply chain planning problem, including distribution planning, inventory control, and product demand forecasting;
- the integration of production, financial and environmental aspects, risk, and uncertainty.

The expected models will tackle a multi-objective view of achieving the necessary trade-off between often contradictory benefits in terms of economic, environmental, customer satisfaction, and increased response to dynamic market changes:

- the development of detailed production scheduling at the plant level for batch, continuous and discrete manufacturing for online scheduling implemented in practice under real-time variations and uncertainty;

Citation: Puigjaner, L.; Espuña, A.; Muñoz, E.; Capón-García, E. Expanding the Horizons of Manufacturing, towards Wide Integration, Smart System, and Tools. *Processes* **2022**, *10*, 772. https://doi.org/10.3390/pr10040772

Received: 10 March 2022
Accepted: 12 April 2022
Published: 14 April 2022

- the integration of the tracking system of network dynamics within the holistic decision-making model (e.g., by enclosing a model predictive control framework), thus facilitating equipment capacity handling similarly at strategic and operational levels and enabling adequate response to incidents for enhanced production sustainability;
- the development of suitable frameworks and algorithms for solving these problems in an efficient and integrated manner (e.g., surrogate problem decomposition, disjunctive programming, Lagrange decomposition);
- the development of software prototypes for the implementation of the above methodologies and algorithms, illustrating their applicability in several real-life industrial case studies involving typical manufacturing/distribution networks belonging to relevant sectors in the world;
- the development of novel frameworks focusing on the utilization of formal knowledge models, facilitating new technologies implementation, and transactional system integration;
- the further development of intelligent manufacturing systems for the transformation of manufacturing enterprises, from the traditional to the intellectualized;
- the development of intelligent systems and intelligent agents focused on cooperative work between human beings and computers, enhancing the capability of human decision-making and problem solutions in the process engineering field.

In the following, you will find selected contributions (original research, reviews, opinions, and perspectives) regarding this research topic. They bring novel solution approaches accompanied by rich case studies and examples of practical interest.

1. Original Research

The articles in this Special Issue examine different facets of enterprise-wide modeling and optimization through the development and application of integrated modeling, simulation and optimization methodologies, and computer-aided tools for reliable and sustainable improvement opportunities within the entire manufacturing network.

The contribution by David Bogle et al. [1] addresses the present and future situation of the high complexity inherent with batch manufacturing of many products. The intrinsic flexibility associated with batch manufacturing has been the choice of most pharma products manufacturing. As a result, it is usual to see a battery of batch reactors in the pharma industry or single batch reactors with high and complex inlet and outlet pipe connections, some of them unused, because different or new products appear in the industry portfolio. The net result is increasing the cost of the final product. Instead, this original paper offers a novel approach: The use of operational envelopes to study the trade-off between the design and operational flexibility of a tablet manufacturing process. Moreover, using an alternative adaptive sampling technique will alleviate the significant computational burden associated with the operational envelopes. Finally, a critical fluidized bed dryer case study at the heart of the continuous manufacturing of tablets supports the paradigm shift change to continuous manufacturing.

The paper by M. Ziyan Sheriff et al. [2] also examines the transition from batch to continuous processes in the pharmaceutical industry. In order to enable the quality-by-control (QbC) paradigm to move forward, this work developed and presented a moving horizon estimation-based nonlinear model predictive control (MHE-NMPC) framework to accomplish the dual requirement of accurate estimation and efficient control. The real-time implementation feasibility of the developed framework was also discussed, and the ability of the proposed framework to solve the optimization problem at each time step in a manner that enabled real-time implementation was highlighted. The practical applicability of the developed framework was corroborated through two realistic case studies that incorporated the effects of glidant to better control CQAs such as the tensile strength. Both examples demonstrated the ability of the framework to achieve reasonable control performance despite the presence of varying sources and degrees of plant model mismatch.

The work by Zixue Guo et al. [3] proposes an evaluation model that addresses the problem of fuzziness and randomness in regional logistics decarbonization, assessing its development. An evaluation index contemplates three dimensions: low-carbon logistics environment support, low-carbon logistics strength, and low-carbon logistics potential. Secondly, the evaluation indexes serve as cloud model variables, and the cloud model theory determines numerical characteristic values and cloud affiliation degrees. Finally, the entropy weight method determines the index weights and calculates the comprehensive determination degree of the research object affiliated with the logistics decarbonization level. Finally, the Beijing–Tianjin–Hebei region is the example used for empirical evidence, analyzing the development of logistics decarbonization and its temporal variability in Beijing, Tianjin, and Hebei provinces and cities. The study results show that the development of logistics decarbonization in Beijing, Tianjin, and Hebei Province has improved to different degrees from 2013–2019, but the development is uneven. Developing to 2019, the three provinces and cities of Beijing, Tianjin, and Hebei still have significant differences in terms of the economic environment, logistics industry scale, logistics industry inputs and outputs, and technical support.

In their article, Vivek Dua et al. [4] aim to introduce a method for designing multi set-point explicit controllers for nonlinear systems through recent advances in multi-parametric programming. Multi-parametric programming (mp-P) has received considerable attention from the process systems engineering community because of its unique ability to aid in the design of explicit model predictive controllers and thus shift the computational burden associated with offline control. The authors examine a case of multi-parametric nonlinear programs (mp-NLPs) that involve both endogenous uncertainties, in the form of left-hand side parameters (LHS), as well as exogenous uncertainty in the cost coefficient of the objective function (OFC), and, on the right-hand side of the constraints (RHS), uncertain parameters on the right-hand side (RHS). In engineering problems, LHS uncertainty arises from variations in model coefficients, due to parameter estimation errors or model mismatch; OFC uncertainty arises due to fluctuation in market prices or control penalties while RHS uncertainty can be due to varying system exogenous factors. The contribution of the present work is a novel framework for the design of multi set-point explicit controllers for nonlinear process systems.

As a demarcation of the past, present and future of intelligent systems, a Tri-X Intelligence (T.I.) model is proposed in this paper by Baicun Wang et al. [5] to state the mechanism, factors, and connotation of three main entities (conscious humans, physical objects, and cyber entities), including single-X intelligence, two-X integrated intelligence, and three-X complex intelligence. Every single entity shows primitive intelligence. Two-entity integration creates integrated intelligence. Three-entity fusion generates advanced intelligence. The intelligentization mechanism of artificial systems continuously converts human intelligence to machine intelligence via different channels and interfaces. With the increasing use of machine intelligence, humans will gradually play a less significant role in intelligent systems. However, human intelligence will keep influencing artificial systems in the form of software/algorithms to drive intelligent systems. Therefore, we cannot take humans out of the systems given the accelerating development of technology. The key to success is to adapt humans to new work environments, i.e., not to replace but to enhance. According to the Tri-X Intelligence (T.I.) model, humans need to think more about how to collaborate with cyber systems rather than with intelligent systems, a Tri-X Intel than training operators to work like computers. The proposed Tri-X model (e.g., Human-Cyber-Physical-System HCPS) will integrate the intelligence in the complex system with a combination of human-cyber-physical and machine subsystems.

In addition, Sujeon Baek et al. [6] prepared a testbed for conducting a pick-up operation using a vacuum gripper with a single suction cup. Using the proposed method, the air pressure in the Venturi line was automatically monitored in real-time. When a command for starting suction was provided to the gripper, a sharp decrease in the collected air pressure signals appeared at approximately 0.5 s. However, the same decline was not observed in the signal for faulty box surfaces; consequently, the suction action and the corresponding gripper operation were not performed owing to insufficient contact between the suction cup(s) and the contact surface of the object. Using the early detection results derived from the air pressure signal analysis, a prediction-based process adjustment method for the pick-up operation was proposed. Through pick-up experiments using the developed testbed, it was revealed that the z-position of the suction cup significantly affects whether an object is gripped adequately by the vacuum gripper or not. Therefore, it is possible to determine a possible error situation in advance (before the failure of the lifting operation) and provide appropriate feedback control commands so that the target operation is finished successfully without stopping machine operations.

The process, manufacturing, and service industries face many non-trivial challenges in a customized market environment, from product conception, design, development, commercialization, and delivery. Thus, industries can benefit by integrating new technologies into their day-by-day tasks gaining companies profitability. Puigjaner et al. [7] present an integrated model framework for enterprise process development activities called a "Comprehensive intelligent management architecture model for integrating new technologies for services, processes, and manufacturing who strive for finding the most efficient way towards enterprise and process intelligence." The model comprises and structures three critical systems: process, knowledge, and transactional. As a result, analytical tools belonging to process activities and transactional data systems are guided by a systematic development framework consolidated with formal knowledge models. Thus, the model improves the interaction among processes lifecycle, analytical models, transactional systems, and knowledge. Finally, a case study systematically presents an acrylic fiber production plant applying the proposed model, demonstrating how the three models described in the methodology work together to systematically achieve the desired technology application of life cycle assessment. The results conclude that the interaction between the semantics of formal knowledge models and the processes-transactional system development framework facilitates and simplifies new technology implementation along with enterprise development activities.

Compared with other fossil fuels, natural gas (N.G.) is considered a sustainable and potential energy source in the future. Being liquefied, natural gas (LNG) is 600 times smaller than the gaseous state of N.G., LNG becomes especially attractive if obtained at a competitive cost. The authors of this article, Liang Zhao et al. [8], show that modeling and optimizing the LNG terminals may also reduce energy consumption and GHG emissions. In this work, the authors propose an operational optimization model of the LNG terminal to minimize the energy consumption of boil-off gas (BOG) compressors and low pressure (L.P.) pumps. Finally, an MINLP model determines whether the pumps are running or on standby, and the number of compressor level chosen as a binary variable. The model can propose operating strategies for varied flow rates of the send-out speed, and the ambient temperature can be offered using the model. An actual case study on the LNG terminal is presented to indicate the effectiveness of the proposed approach. Finally, the optimization model provides the minimum energy consumption and the corresponding decision variables. The optimized compressor load and recirculation flow rate were 8.44 t/h and 122.58 t/h, respectively. Compared with the previous period, 26.1% of energy can be saved after optimization. About 16.21% of energy consumption can be saved annually.

In his challenging article, Heinz A. Preisig [9] presents "Reductionism and splitting application domain into disciplines and identifying the smallest required model-granules, termed "basic entity" combined with systematic construction of the basic entities, yields a systematic approach to process modeling." They do not aim toward a single modeling domain, but enabling specific application domains and object inheritances to be addressed. They start with reductionism and demonstrate how the basic entities depend on the targeted application domain. They use directed graphs to capture process models, and introduce a new concept, which they call "tokens," that enables the extension of the context beyond physical systems. The network representation is hierarchical to capture complex systems. The interacting basic entities are defined in the leave nodes of the hierarchy, making the overall model the interacting networks in the leave nodes. Multi-disciplinary and multi-scale models result in a web of networks. They identify two distinct network communication ports, namely, ports that exchange tokens and ports that transfer information of tokens in accumulators. An ontology captures the structural elements and the applicable rules and defines the syntax to establish the behavior equations. Linking the behaviors to the fundamental entities defines the alphabet of a graphical language. They use this graphic language to represent processes which have proven to be efficient and valuable. Then, a set of three examples demonstrates the power of graphical language. Finally, the Process Modelling framework (ProMo) implements an ontology-centered approach to process modeling and uses graphic vocabulary to construct process models.

In this article, Ignacio Grossmann et al. [10] address an inventory management problem for a make-to-order supply chain with inventory holding and-or manufacturing locations at each node. The lead times between nodes and production capacity limits are heterogeneous across the network. This study focuses on a single product, a multi-period centralized system in which a retailer is subject to uncertain stationary consumer demand at each time period. The authors consider two sales scenarios for unfulfilled demand: backlogging or lost sales. The daily inventory replenishment requests from immediate suppliers throughout the network are modeled and optimized using three different approaches: (1) deterministic linear programming, (2) multi-stage stochastic linear programming, and (3) reinforcement learning. The performance of the three methods is compared and contrasted in terms of profit (reward), service level, and inventory profiles throughout the supply chain. The proposed optimization strategies testing occurs in a stochastic simulation environment built upon the open-source OR-Gym Python package. The results indicate that stochastic modeling yields the most significant increase in profit of the three approaches. In contrast, reinforcement learning creates more balanced inventory policies that would potentially respond well to network disruptions. Furthermore, deterministic models perform well in determining dynamic reorder policies comparable to reinforcement learning in terms of their profitability.

In an inspiring novel article, Zeinab Shahbazi and Yung-Cheol Byun [11] bring the latest developments and experimental results on smart manufacturing. The modern industry, production, and manufacturing core is developed based on smart manufacturing (S.M.) systems and digitalization. Smart manufacturing's practical and meaningful design follows data, information, and operational technology through the blockchain, edge computing, and machine learning to develop and facilitate the smart manufacturing system. This process's proposed intelligent manufacturing system considers the integration of blockchain, edge computing, and machine learning approaches. Edge computing balances the computational workload and similarly provides a timely response for the devices. Blockchain technology utilizes the data transmission and the manufacturing system's transactions, and the machine learning approach provides advanced data analysis for a vast manufacturing dataset. Finally, the model solves the problems using a swarm intelligence-based method regarding intelligent manufacturing systems' computational environments. The experimental results present the edge computing mechanism and similarly improve the processing time of a large number of tasks in the manufacturing system.

Present increasing regulatory demands force the pharmaceutical industry to invest its available resources carefully. That is especially challenging for small and middle-sized companies. For example, computer simulation software such as FlexSim allows one to explore variations in production processes without interrupting the running process. Claus-Michael Lehr et al. [12] applied a discrete-event simulation to two approved film-coated tablet production processes in this article. The simulations were performed with FlexSim (FlexSim Deutschland, Ingenieurbüro für Simulationsdienstleistung Ralf Gruber, Kirchlengern, Germany). Process visualization required the use of Cmap Tools (Florida Institute for Human and Machine Cognition, Pensacola FL, USA), and statistical analysis used MiniTab® (Minitab GmbH, Munich, Germany). The most critical elements identified during model building were the model logic, operating schedule, and processing times.

These factors required graphically and statistically verification. In addition, employee utilization optimization required three different shift systems to be simulated, revealing the advantages of two-shift and one-and-a-half-shift systems compared to a one-shift system. Finally, without interrupting any currently running production processes, we found that changing the shift system could save 50–53% of the campaign duration and 9–14% of the labor costs. In summary, we demonstrated that FlexSim, mainly used in logistics, can also be advantageous for modeling and optimizing pharmaceutical production processes.

Tibor Krenicky et al. [13] present a study on the surface quality dependency on the selected parameters of cuts made in Hardox by abrasive water jet (AWJ). The authors applied the regression process to measured data and prepared the Ra and Rz roughness parameters equation. One set of regression equations describes the relationship of Ra and Rz on cutting parameters—pumping pressure, traverse speed, and abrasive mass flow rate. The second set of regression equations describes relationships between the declination angle in kerf as the independent variable and the Ra or the Rz parameters as dependent variables. Finally, the models can predict cutting variables to predict the surface quality parameters.

The complexity of the automated guided vehicles (AGV) system requires substantial decision-making and is challenging to solve. The authors Adrian Kampa et al. [14], use the flexible manufacturing system solution with the associated AGV transport system and discuss such systems' design and simulation issues. The initial system design optimization stage is crucial, and computer simulation enables relatively easy elaboration and testing of various manufacturing and logistics systems variants. On the other hand, excessive simplifications may appear applied at the modeling stage, making the simulation not reflect the production system properly. On the other hand, it is worth noticing that detailed modeling is very labor-intensive and requires the involvement of experienced specialists. Therefore, choosing which parameters to use in the modeling process and which metric to evaluate the model. Finally, to make the simulation more accurate and assess the system's productivity, the authors propose using overall equipment effectiveness (OEE) metrics. The results obtained from the presented simulations show that the OEE metrics may be helpful in the modeling and productivity evaluation of manufacturing and logistics systems, with the generalization of overall factory effectiveness (OFE) and overall transport effectiveness (OTE). The use of OEE factors also allows for comparison of the results obtained from different manufacturing systems. For example, many of them with OEE scores lower than 45% in the real world and a small number of world-class companies have an OEE value higher than 85%. Accordingly, the simulation results can also be helpful in analyzing the costs involved in implementing a given project and at the stage of the in-depth design of the production system.

The following work addresses the closed-loop stability problem with an application to refinery preheats trains' online cleaning schedule stability problem under fouling. Lozano Santamaría and Sandro Macchietto [15] focus on the sources of instability and ways to mitigate it. The various metrics developed to quantify schedule instability for online scheduling account for distinct aspects, such as changes in task allocation, task sequence, starting time of the task, and the earlier or later occurrence of such changes in the future scheduling horizon. Based on the proposed methods, further stability metric variations could be quickly developed (for example, ways of assigning weights to distinct contributions to a schedule change). These stability considerations can be practically and, in a rather general way, introduced in a closed-loop nonlinear model predictive control (NMPC) formulation of the optimal scheduling and control problem and solved online over a moving horizon, in terms of penalties in an economic objective or via additional constraints. The above methods demonstrated to be helpful for the online cleaning scheduling and flow control of refinery preheat trains, a challenging application with significant economic, safety, and environmental impact. A demanding industrial case study followed an illustrative, small but realistic case study. Results show that, of the three alternatives evaluated, the terminal cost penalty proved to be inefficient in this case. The other two (fixing some predictions horizon decisions and penalizing schedule changes between consecutive evaluations) improved the closed-loop schedule stability against various economic penalties. The results highlight the importance of including stability considerations in an economically oriented online scheduling problem to obtain feasible solutions for operators over long operating horizons without sacrificing the benefits of a reactive system to reject disturbances or take advantage of them. The application of the metrics developed in this manuscript is not restrictive to the specific closed-loop NMPC scheduling implementation detailed here. They are helpful to assess schedule stability in general regardless of how schedules are calculated, only relying on the existence of two consecutive evaluations or predictions of the schedule within a common period. The two successive instances may have different control horizons, scheduling horizons, or update frequency. Lastly, although this work dealt with a specific application (the optimization of refinery heat exchanger networks subject to fouling), the formulations and solution approach demonstrated here should apply to important systems, such as batch and semi-continuous processes.

Mohammed Alkahtani et al. [16] enlighten the management of the man-machine interaction as essential to achieving a competitive advantage among production firms and specifically more highlighted in the case of processing agricultural products. The authors design a non-derivative technique to integrate an algebraic approach in the agri-product based supply chain to optimize the resources and coup with variable demands through a controllable production rate. The analysis provides a platform for manufacturing managers to invest in advanced technology in agricultural supply chain management (agri-SCM), leading to a less rejection production environment for clean manufacturing. The solution methodology of the proposed model included manufacturing limitations in the integration of the objective formulations with the developed system. The authors use sensitivity analysis to evaluate sensitivity for an optimal solution to the value of uncertain parameters, providing confidence in the model's resolution. Managerial insights are beneficial to agricultural supply chain management (agri-SCM) for the agri-food processing industry, and the people with cleaner production and carbon emission prioritized policies. The authors can extend the research into a three-echelon agri-SCM model by considering the farming industry and agri-retailer. The fuzzy set theorems can deal with costs, prices, inflation, and time value uncertain factors. Finally, the authors envisage a feasible conversion of the deterministic model into probabilistic or stochastic theorems for application in real scenarios. Overall, the agri-product supply chain requires global development to make food more secure and accessible.

In the following article, Dejan Gradišar and Miha Glavan [17] consider a manufacturing problem requirement plan. This plan must satisfy the capacity needs and be available by the work order's due date. In addition to this, the program must also consider a group of work orders to produce from the same batch of raw material. In this way, the manufacturer can systematically compensate for some undesirable variations in raw material quality. In day-to-day practice, the plan management makes it challenging to maintain the plan up-to-date, even in smaller dimensions. As a result, the operator's decisions are time-consuming and prone to errors. That results in situations in which the operator must constantly make plan corrections. Finally, this paper proposes using an extended bin-packing problem formulation to solve the material planning systematically. Finally, a fundamental bin-packing problem (BPP) formulation requires an extension to include constraints such as variable bin and item sizes. For example, one can use time limitations and only a group of bins to produce one group of items. The suggested solution offers a tool for supporting the production planner's decisions. With it, they can determine how to efficiently cut the raw material to satisfy the planned work orders. Depending on the situation, the planner can choose between various model formulations. Additionally, they can optimize the leftover, tardiness, or both. Finally, we demonstrated that the proposed solution could quickly solve a problem of realistic dimensions to be of use in an industrial application. However, case-specific requirements would first need to be analyzed to prioritize the importance of leftovers and/or tardiness in real applications.

Andrzej Paszkiewicz et al. [18] propose a novel approach for integrating the distributed additive manufacturing process enabling remote designing, selecting appropriate manufacturing means, and implementing a physical production process and control at all stages. This approach was possible thanks to the development of an unprecedented framework. The authors integrated distributed and functionally different elements (Informative Technology (I.T.) and manufacturing), forming a coherent design and manufacturing system. Importantly, this framework ensures an increase in production efficiency, shortens production time, reduces costs, and increases flexibility and accessibility to the latest methods and design and manufacturing tools. In addition, they presented a mechanism that facilitates the integration of independent manufacturing environments by considering and implementing appropriate levels of maturity in the system. The implementation in a natural production environment, i.e., at Infosoftware Poland, confirms the proposed solution's validity. At present, work is in progress to integrate the rapid prototyping laboratory of the Rzeszów University of Technology. In addition, the automotive and aerospace industries can widely use the presented platform. In addition, it will facilitate cooperation between industrial clusters and academic centers to a higher degree and encourage collaboration between small enterprises and startups. Finally, from the perspective of management, the technical implementation of the presented framework allows one to adapt to the needs of globalization and facilitates the integration of distributed resources. Thus, this framework affects business, logistics, and technological processes. One of the implications of implementing such a framework is the need to develop or adapt existing workflows to the new heterogeneous and distributed work environment.

2. Review

The review by Marianthi Ierapetritou et al. [19] informs the reader of the latest development and application of emerging technologies of Industry 4.0, enabling the realization of digital twins (D.T.). D.T.s is a crucial development of the close integration of manufacturing information and physical resources that raise much attention across industries. The critical parts of a fully developed D.T. include the physical and virtual components and the interlinked data communication channels. Following the development of Internet of Things (IoT) technologies, there are many applications of D.T. in various industries, but the progress is lagging for pharmaceutical and biopharmaceutical manufacturing. This review paper summarizes the current state of D.T. in the two application scenarios, providing insights to stakeholders and highlighting possible challenges and solutions to implementing a fully integrated D.T. In pharmaceutical manufacturing, building blocks of a D.T., including process analytical technology (PAT) methods, data management systems, unit operations, flowsheet models, system analyses methods, and integration approaches, have all been developed in the last few years, but gaps in PAT accuracy, real-time model computation, model maintenance capabilities, real-time data communication, as well as concerns in data security and confidentiality, are preventing the full integration of all the components. Several insights seem appropriate to solve these challenges. First, developing new tools such as near-infrared spectroscopy (NIRS) and in-line U.V. spectroscopy, iterative optimization technologies, and different online adaptive methodologies can help resolve the existing issues in PAT methods. Second, efficient algorithms and reduced-order modeling approaches need further study for process models to reduce simulation time to achieve real-time computation. Third, adaptive modeling methods with online streaming data will be under further investigation in model maintenance. Third, to have a fully integrated and automated D.T., the information flow from the virtual component to the physical plant also must be established. Moreover, the virtual plant should be able to change system settings and control the physical plant to help achieve an optimized process within the design space. Ideally, all these components require appropriate physical and virtual security protocols.

3. Opinion

In the past few years, pharmaceutical products have evolved toward disease- and patient-specific therapeutics involving meticulous manufacturing steps. In addition, cell-based therapeutics and vaccines present high sensitivity to environmental and transport conditions, complicating supply chain logistics. Increased drug specificity and demand uncertainty add further complexity to the design and operation of robust manufacturing processes and distribution networks. As Maria M. Papathanasiou et al. [20] discuss in their paper, the pharmaceutical industry has taken significant steps toward improving existing and-or developing novel processes that promise agile, responsive, and reproducible manufacturing. Similarly, distribution networks in the pharmaceutical sector are undergoing a paradigm shift, exploring the capabilities of decentralized models. Such developments accompany digital innovation in the pharmaceutical industry that enables seamless communication between process units, production plants, and distribution nodes. As discussed earlier, process systems engineering has been at the forefront of allowing digitalization through the development of computer modeling tools. The latter can assist with real-time monitoring of critical storage conditions for sensitive pharmaceutical products with short shelf-life, thus increasing drug safety. One of the main challenges hindering the fast exploitation of Industry 4.0 principles in pharmaceutical manufacturing is a mindset change. Practitioners should embrace the benefits arising from the realization of Pharma 4.0 towards replacing paper-based systems with cloud-based servers. That will allow significantly improved agility and productivity in the operations of the pharmaceutical sector.

4. Perspective

The authors of the perspective, Krist V. Gernaey et al. [21], emphasize that despite the benefits of continuous over-batch bioprocessing, its adoption has lagged, with few exceptions. However, the batch manufacturing paradigm's dominance in the industry for reasons such as "batch processing is familiar and works very well" cannot be sustained in the long term, given the new biomanufacturing challenges. Moreover, the industry-held perception of complexity in continuous bioprocessing is becoming obsolete as more and more new technologies and solutions continually improve the situation. Several academic- and industry-led consortia are working to improve the perception regarding continuous bioprocessing by bringing the questions to the correct stakeholders who can address them. The training provided by these initiatives to the top management of the companies is playing an essential role in changing the perception and, at the same time, also creating new scientists and operators that can understand and respond to a new set of operational challenges. However, wider adoption of continuous bioprocessing will only be possible if the technical, management, and regulatory gaps are acknowledged. This paper argues that concerted efforts focusing on technology, management, and regulatory aspects are abridging them.

Funding: This research received no external funding.

Acknowledgments: The authors would like to acknowledge the support of CEPIMA (UPC) Barcelona, Spain.

Conflicts of Interest: The authors declare that the research was conducted in the absence of any commercial or financial relationships that could be construed as a potential conflict of interest.

References

1. Jiang, S.-L.; Papageorgiou, L.G.; Bogle, I.D.L.; Charitopoulos, V.M. Investigating the Trade-Off between Design and Operational Flexibility in Continuous Manufacturing of Pharmaceutical Tablets: A Case Study of the Fluid Bed Dryer. *Processes* **2022**, *10*, 454. [CrossRef]
2. Huang, Y.-S.; Sheriff, M.Z.; Bachawala, S.; Gonzalez, M.; Nagy, Z.K.; Reklaitis, G.V. Evaluation of a Combined MHE-NMPC Approach to Handle Plant-Model Mismatch in a Rotary Tablet Press. *Processes* **2021**, *9*, 1612. [CrossRef]
3. Guo, Z.; Tian, Y.; Guo, X.; He, Z. Research on Measurement and Application of China's Regional Logistics Development Level under Low Carbon Environment. *Processes* **2021**, *9*, 2273. [CrossRef]

4. Charitopoulos, V.M.; Papageorgiou, L.G.; Dua, V. Multi Set-Point Explicit Model Predictive Control for Nonlinear Process Systems. *Processes* **2021**, *9*, 1156. [CrossRef]
5. Zhao, M.; Ning, Z.; Wang, B.; Peng, C.; Li, X.; Huang, S. Understanding the Evolution and Applications of Intelligent Systems via a Tri-X Intelligence (T.I.) Model. *Processes* **2021**, *9*, 1080. [CrossRef]
6. Baek, S.; Kim, D.O. Predictive Process Adjustment by Detecting System Status of Vacuum Gripper in Real-Time during Pick-Up Operations. *Processes* **2021**, *9*, 634. [CrossRef]
7. Muñoz, E.; Capon-Garcia, E.; Muñoz, E.M.; Puigjaner, L. A Systematic Model for Process Development Activities to Support Process Intelligence. *Processes* **2021**, *9*, 600. [CrossRef]
8. Ye, Z.; Mo, X.; Zhao, L. MINLP Model for Operational Optimization of LNG Terminals. *Processes* **2021**, *9*, 599. [CrossRef]
9. Preisig, H.A. Ontology-Based Process Modelling-with Examples of Physical Topologies. *Processes* **2021**, *9*, 592. [CrossRef]
10. Perez, H.D.; Hubbs, C.D.; Li, C.; Grossmann, I.E. Algorithmic Approaches to Inventory Management Optimization. *Processes* **2021**, *9*, 102. [CrossRef]
11. Shahbazi, Z.; Byun, Y.-C. Improving Transactional Data System Based on an Edge Computing–Blockchain–Machine Learning Integrated Framework. *Processes* **2021**, *9*, 92. [CrossRef]
12. Hering, S.; Schäuble, N.; Buck, T.M.; Loretz, B.; Rillmann, T.; Stieneker, F.; Lehr, C.-M. Analysis and Optimization of Two Film-Coated Tablet Production Processes by Computer Simulation: A Case Study. *Processes* **2021**, *9*, 67. [CrossRef]
13. Krenicky, T.; Servatka, M.; Gaspar, S.; Mascenik, J. Abrasive Water Jet Cutting of Hardox Steels—Quality Investigation. *Processes* **2020**, *8*, 1652. [CrossRef]
14. Foit, K.; Gołda, G.; Kampa, A. Integration and Evaluation of Intra-Logistics Processes in Flexible Production Systems Based on OEE Metrics, with the Use of Computer Modelling and Simulation of AGVs. *Processes* **2020**, *8*, 1648. [CrossRef]
15. Lozano Santamaria, F.; Macchietto, S. Stability of Optimal Closed-Loop Cleaning Scheduling and Control with Application to Heat Exchanger Networks under Fouling. *Processes* **2020**, *8*, 1623. [CrossRef]
16. Alkahtani, M.; Omair, M.; Khalid, Q.S.; Hussain, G.; Sarkar, B. An Agricultural Products Supply Chain Management to Optimize Resources and Carbon Emission Considering Variable Production Rate: Case of Nonperishable Corps. *Processes* **2020**, *8*, 1505. [CrossRef]
17. Gradišar, D.; Glavan, M. Material Requirements Planning Using Variable-Sized Bin-Packing Problem Formulation with Due Date and Grouping Constraints. *Processes* **2020**, *8*, 1246. [CrossRef]
18. Paszkiewicz, A.; Bolanowski, M.; Budzik, G.; Przeszłowski, Ł.; Oleksy, M. Process of Creating an Integrated Design and Manufacturing Environment as Part of the Structure of Industry 4.0. *Processes* **2020**, *8*, 1019. [CrossRef]
19. Chen, Y.; Yang, O.; Sampat, C.; Bhalode, P.; Ramachandran, R.; Ierapetritou, M. Digital Twins in Pharmaceutical and Biopharmaceutical Manufacturing: A Literature Review. *Processes* **2020**, *8*, 1088. [CrossRef]
20. Sarkis, M.; Bernardi, A.; Shah, N.; Papathanasiou, M.M. Emerging Challenges and Opportunities in Pharmaceutical Manufacturing and Distribution. *Processes* **2021**, *9*, 457. [CrossRef]
21. Kumar, A.; Udugama, I.A.; Gargalo, C.L.; Gernaey, K.V. Why Is Batch Processing Still Dominating the Biologics Landscape? Towards an Integrated Continuous Bioprocessing Alternative. *Processes* **2020**, *8*, 1641. [CrossRef]

processes MDPI

Article

Investigating the Trade-Off between Design and Operational Flexibility in Continuous Manufacturing of Pharmaceutical Tablets: A Case Study of the Fluid Bed Dryer

Sheng-Long Jiang [1,2], Lazaros G. Papageorgiou [1], Ian David L. Bogle [1] and Vassilis M. Charitopoulos [1,*]

1 Centre for Process Systems Engineering, Department of Chemical Engineering, University College London, London WC1E 7JE, UK; sheng-long.jiang@ucl.ac.uk (S.-L.J.); l.papageorgiou@ucl.ac.uk (L.G.P.); d.bogle@ucl.ac.uk (I.D.L.B.)
2 College of Materials Science and Engineering, Chongqing University, Chongqing 400044, China
* Correspondence: v.charitopoulos@ucl.ac.uk

Abstract: Market globalisation, shortened patent lifetimes and the ongoing shift towards personalised medicines exert unprecedented pressure on the pharmaceutical industry. In the push for continuous pharmaceutical manufacturing, processes need to be shown to be agile and robust enough to handle variations with respect to product demands and operating conditions. In this paper we examine the use of operational envelopes to study the trade-off between the design and operational flexibility of the fluid bed dryer at the heart of a tablet manufacturing process. The operating flexibility of this unit is key to the flexibility of the full process and its supply chain. The methodology shows that for the fluid bed dryer case study there is significant effect on flexibility of the process at different drying times with the optimal obtained at 700 s. The flexibility is not affected by the change in volumetric flowrate, but only by the change in temperature. Here the method used a black box model to show how it could be done without access to the full model equation set, as this often needs to be the case in commercial settings.

Keywords: pharmaceutical manufacture; uncertainty; operational flexibility; operational envelopes; modeling

Citation: Jiang, S.-L.; Papageorgiou, L.G.; Bogle, I.D.L.; Charitopoulos, V.M. Investigating the Trade-Off between Design and Operational Flexibility in Continuous Manufacturing of Pharmaceutical Tablets: A Case Study of the Fluid Bed Dryer. *Processes* **2022**, *10*, 454. https://doi.org/10.3390/pr10030454

Academic Editors: Luis Puigjaner, Antonio Espuña Camarasa, Edrisi Muñoz Mata and Elisabet Capón García

Received: 7 January 2022
Accepted: 21 February 2022
Published: 24 February 2022

1. Introduction

The power of big data, emanating from the process and from customers, is having a number of effects on manufacturing. With coordinated access to reliable data, a manufacturer can respond more rapidly and efficiently to supply chain demands. However, with data comes the capability and often the demands from internal and external stakeholders (customers, shareholders, regulators, neighbours, etc.) for greater transparency of operations. Industry is going through something of a revolution to realise these aims. It is known as Smart Manufacturing, Industry 4.0 or Digitalisation because of the capabilities enabled by greater computing power, smarter algorithms, better measurement, and wider connectivity. The smart manufacturing revolution is said to have three phases [1,2]:

1. Factory and enterprise integration and plant-wide optimisation,
2. Exploiting manufacturing intelligence,
3. Creating disruptive business models.

For the process industries, all three phases are likely to drive significant change [1–6]. To a considerable extent, the first phase has been well underway for a decade or more, particularly plant wide optimisation. The exploitation of big data from enhanced process measurement, as well as using data for demand, supply and the operating environment, is enabling the second phase which is also to some extent underway. Key enablers are methods to manage flexibility and uncertainty, responsiveness and agility, robustness and security, the prediction of mixture properties and function, and new modelling and mathematics paradigms [2]. The third phase is less clear, but the drivers for personalised medicine may affect the pharmaceutical industry more rapidly. Over the last decade there has been an increasing industrial and research interest in the concept of continuous pharmaceutical manufacturing (CPM). CPM offers the benefits of better resource utilisation, reducing energy costs and the potential for operating at processing conditions that would otherwise be prohibitive within the conventional batch setting [7,8]. A key issue related to CPM is the systematic identification of the attainable regions, typically referred to as the design space, in order to employ optimisation for the design and operation of such processes [9].

Pharmaceutical processes involve a number of features which challenge current modelling and control paradigms. They involve multiple phases: solids, liquids and gases often with multiple liquid phases; they are typically combinations of batch and continuous units; and there are tighter regulatory frameworks for their operation than for chemical processes. Litster and Bogle [10] have highlighted the potential for Smart Manufacturing in processes for formulated products which is the form of many pharmaceuticals. Formulated products are structured, multiphase products (i.e., granules, tablets, emulsions, and suspensions) whose performance characteristics—critical quality attributes (CQAs)—are just as dependent on the product structure as they are on the chemical composition (see for example [11,12]). To this end, a variety of process systems engineering tools have been investigated for materialising Quality by Design (QbD) initiatives (see for example [13]). Diab and Gerogiorgis [14] surveyed recent development for the design space identification and visualisation for CPM while the same authors have proposed the use of flowsheeting for technoeconomic assessment for the synthesis and crystallisation of rufinamide [14] and nevirapine [15]. Recognising the inherent difficulty in accurately deriving first-principles mechanistic models for CPM units, Boukouvala et al. [8,9] proposed the use of Kriging data-driven models for the dynamic modelling of unit operations. In their work, dynamic Kriging models showed the ability to efficiently adapt across transition regimes and outperformed the accuracy of neural network modelling. Recently, Nagy et al. [16] presented a dynamic, integrated flowsheet model for the continuous manufacturing of acetylsalicylic acid which entailed a two-step flow synthesis and crystallisation.

Litster and Bogle [9] outlined the potential challenges and opportunities for Smart Manufacturing for formulated products. Pressures on healthcare providers is requiring greater efficiency and less inventory within a more changeable regulatory environment. Personalised medicine will require much more responsive manufacturing for specific patient groups. The industry is expected to bring products faster to market, as the recent pandemic has demonstrated for vaccines. This all requires greater agility and flexibility within the context of greater uncertainty of demand and of raw materials. This will need greater use of mature model-based tools—for design, control and supply chain optimization—to enable the managing of complexity and uncertainty. Many tools are available but there is a lack of experience and often concern about the fidelity of the models and their ability to predict with sufficient accuracy. This is exacerbated by the tendency of optimisers to push operations to the limits of well understood operation. Recently, Chen et al. [17] surveyed a variety of contributions from the process systems engineering community and outlined challenges and opportunities for the deployment of digital twins in pharmaceutical and biopharmaceutical manufacturing.

Uncertainty is caused by a wide range of factors: variability in quality and supply of raw materials, in customer demand, and in environmental and utility conditions, and in batch processes the effects of manual operations which is required. The potential impact of uncertainty on the quality of pharmaceutical products in the context of continuous pharmaceutical manufacturing has been widely recognized by the FDA [18,19]. Most plants are over-designed to cope with such uncertainty. When data are available through extensive experimentation, multivariate statistical methods such as PLS (partial least squares regression) and PCA (principal component analysis) [20,21] as well as Bayesian tools have been proposed [22]. Nonetheless, investigating the design space of a process through experimentation comes at very high costs, due to the associated raw material and energy utilisation, and is time consuming. To overcome this issue, model-based probabilistic frameworks have been examined. Laky et al. [23] presented two algorithms for the refinement of the flexibility test and index formulations, originally proposed by Swaney and Grossmann [24]. Kusumo et al. [25] examined the use of a nested sampling strategy to reduce the computational time required related to Bayesian approaches for the probabilistic characterisation of design space characterisation. In order to ensure operation within defined ranges it is important to define these regions for complex integrated batch processing schemes. Samsatli et al. [26] developed a multi-scenario optimisation method for determining operational envelopes for batch processes. Since formulated products have a range of critical quality attributes, it is necessary that these envelopes reflect a number of quality conditions. There has been work to include a more systematic approach to handling uncertainty: through stochastic methods which use knowledge of the likelihood of uncertain events or through defining more explicit operational windows where safety and quality can be guaranteed [27,28]. More recently, in the context of CPM work has been published on methods of global sensitivity analysis [29], flexibility analysis [23] and clustering techniques [30]. Finally, the importance of Quality by Control (QbC) has been highlighted by a number of research groups [31–34]

In this paper we examine the use of the concept of operational envelopes for a part of the tableting process for continuous pharmaceutical manufacturing, the fluidised bed dryer which helps control the quality of the tableting process shown in Figure 1. These envelopes can then be used within a schema for rapidly devising new optimal operating schedules for changes in the uncertain conditions which affect the ability to achieve a product of suitable quality. The remainder of the article is organised as follows: in Section 2 the main methodology is outlined, in Section 3 we apply the method of operating envelopes on a segmented fluidised bed dryer and finally in Section 4 conclusions are drawn.

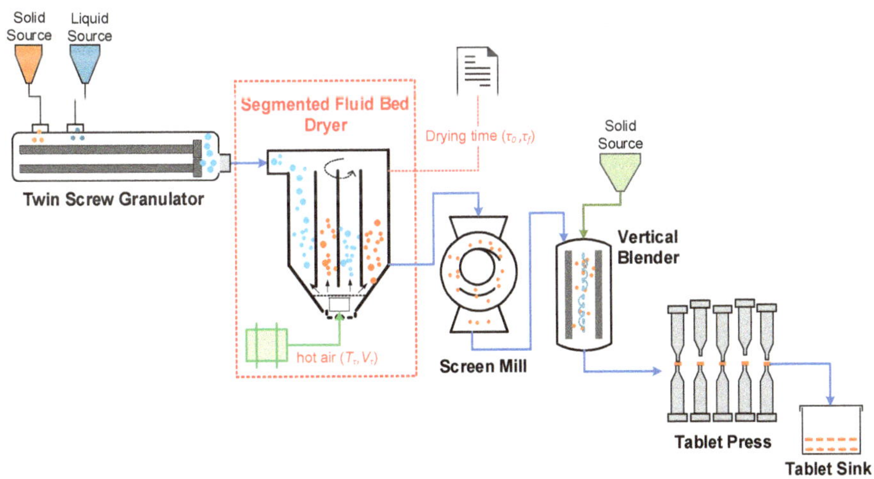

Figure 1. Flowsheet of continuous pharmaceutical process of tableting process (DiPP pilot plant).

2. Methodology

2.1. Description of the Mathematical Model

The dynamic model of the segmented fluidised bed dryer being explored here is implemented in the gPROMS modelling suite as part of the gPROMS FormulatedProducts® library [31]. The underlying mathematical formulation is based on the mechanistic model presented by Burgschweiger et al. [35,36] and model parameters have been validated using the Diamond Pilot Plant (DiPP) at the University of Sheffield. For the sake of brevity, we omit the presentation of the full mathematical model and the interested reader is referred to Burgschweiger and Tsotsas [36]. Regarding the underlying assumptions of this model, we summarise them as follows: (i) plug flow in the bubble phase; (ii) the particle-free bubble phase and the suspension phase within the bed are modelled separately, (iii) mass and heat transfer between drying gas and bubbles is significant and included in the model; (iv) heat transfer between the bed wall, particles, suspension gas, environment and bubble gas is also included.

2.2. Deriving the Operational Envelopes

As described in Samsatli et al. [26] the aim of deriving the operational envelopes of a process or unit operation is to find the maximum range of uncertain operating policies over which the design can be guaranteed to meet specific targets. The union of the maximum range of the uncertainty operating policies is referred to as the "operational envelope". This is particularly important for continuous pharmaceutical manufacturing as a multistage process, since through the use of such decoupled envelopes for each unit operation it can be ensured that the product specifications can be met if we restrict ourselves within the operating limits denoted through these envelopes.

The geometry of these envelopes can be arbitrary. However, in this work we employ hyperrectangular geometry for the sake of computational simplicity. Mathematically, if we denote by $b \in \left[b^{min},\ b^{max}\right]$ the vector of uncertain parameters and their respective limits, which can be inferred either by expert knowledge or based on past observations, we seek to maximise the following objective function:

$$z = \prod_{i=1}^{N_b} b_i^{max} - b_i^{min} \tag{1}$$

where the index $i = 1, \ldots,\ N_b$ is the index of the parameters under investigation. Instead of this objective function, which is non-convex, Samsatli et al. [26] proposed the use of a linear counterpart by introducing the difference in the magnitude of the ranges, i.e., $\Delta b_i = b_i^{max} - b_i^{min}\ \forall i$. Following this step, Equation (1) is replaced by the linear Equation (2) which reflects the scaled perimeter of the envelope.

$$f = \frac{1}{N_b} \sum_{i=1}^{N_b} \frac{\Delta b_i - \Delta b_i^{min}}{\Delta b_i^{max} - \Delta b_i^{min}} \tag{2}$$

Intuitively, since Equation (2) reflects a scaled perimeter the objective function range is [0,1] with an value of 0 reflecting the minimal envelope possible, i.e., $\Delta_{b_i} = \Delta b_i^{min}\ \forall i$, and the maximal envelope feasible is obtained at the value of 1 where $\Delta_{b_i} = \Delta b_i^{max}\ \forall i$. With this modification the overall problem that maximises f is given by model **(M1)**.

$$\begin{gathered}
\max_{\boldsymbol{a},\boldsymbol{b}^{\min},\boldsymbol{b}^{\max}} f = \frac{1}{N_b} \sum_{i=1}^{N_b} \frac{\Delta b_i - \Delta b_i^{min}}{\Delta b_i^{max} - \Delta b_i^{min}} \\
\text{Subject to} \\
\boldsymbol{\Phi}_0\left[\dot{\boldsymbol{x}}_0, \boldsymbol{x}_0, \boldsymbol{y}_0, \boldsymbol{a}_0, \boldsymbol{b}_0\right] = 0\ \ \forall \boldsymbol{b} \in \left[\boldsymbol{b}^{min}, \boldsymbol{b}^{max}\right] \\
\boldsymbol{h}(\dot{\boldsymbol{x}}, \boldsymbol{x}, \boldsymbol{y}, \boldsymbol{a}, \boldsymbol{b}) = 0\ \ \forall \boldsymbol{b} \in \left[\boldsymbol{b}^{min}, \boldsymbol{b}^{max}\right],\ t \in (0, \tau],\ \tau \in \boldsymbol{b} \\
\boldsymbol{g}(\dot{\boldsymbol{x}}, \boldsymbol{x}, \boldsymbol{y}, \boldsymbol{a}, \boldsymbol{b}) \geq 0\ \ \forall \boldsymbol{b} \in \left[\boldsymbol{b}^{min}, \boldsymbol{b}^{max}\right],\ t \in (0, \tau],\ \tau \in \boldsymbol{b} \\
\Delta \boldsymbol{b} = \boldsymbol{b}^{max} - \boldsymbol{b}^{min} \\
\Delta \boldsymbol{b}^{min} \leq \Delta \boldsymbol{b} \leq \Delta \boldsymbol{b}^{max}
\end{gathered} \tag{M1}$$

In model **(M1)**, $\boldsymbol{\Phi}_0$ represents the set of initial conditions for the system under study; $\boldsymbol{h}(\cdot)$ represents the vector of equality constraints which are part of the model, e.g., mass/energy balances; $\boldsymbol{g}(\cdot)$ represents the vector of inequality constraints, e.g., product specifications/resource limitations; $\boldsymbol{x}$ corresponds to differential state variables; $\dot{\boldsymbol{x}}$ their derivatives with respect to time (t); $\boldsymbol{y}$ represents algebraic state variables; while $\boldsymbol{a},\ \boldsymbol{b}$ represent time variant and time invariant controls, respectively. Notice that in **(M1)** the upper bound of the time horizon is also allowed to be an "envelope" variable in case one wanted to investigate suitable bounds, for example for drying times.

Model **(M1)** is a semi-infinite programming problem since it needs to be solved for all the possible values of the $\boldsymbol{b}$ vector of variables. To overcome this issue, a two-step multiscenario optimisation problem is solved in which the envelope variables are discretised as described in Samsatli et al. [26].

3. Case Study: Segmented Fluidised Bed Dryer

In this section we demonstrate the methodology using the digital model of the continuous pharmaceutical process of the Diamond Pilot Plant (DiPP) at the University of Sheffield, shown in Figure 1. The process is a tableting pilot plant at the heart of which is a fluidised bed dryer (FBD) which is critical to the production of consistent quality product. The fluidised bed dryer (FBD) fluidises the feed granules to reduce their moisture content. In the process high-pressure hot air is introduced through a perforated bed of moist solid granules. The wet solids are lifted from the bottom and when fluidised are suspended in a stream of air. Heat transfer is accomplished by direct contact between the wet solid and hot gases. The vaporised liquid is carried away by the gas stream. The temperature and rate of input gas can be adjusted to save energy by, for example, aiming to shorten the drying time and manipulate the desired product (pharmaceutical granules) quality subject to a required range for the moisture content. The FBD is typically divided into a number of vertical segments.

As the FBD is connected with continuous twin screw granulation, the segmented FBD will ensure the wet granules in one cell are dried whilst the incoming wet granules flow into the neighbouring cell. Once the drying process in one cell is finished, the respective cell is emptied pneumatically and then conveyed to the downstream unit, in this case a mill. More segments contribute to reducing moisture but consume more time. In this study we set the FBD equipment to have two segments. Each segment size is 0.035 m^3, with initial charge of 0.1 kg wet air and 0.1 kg granulates (lactose), with a particle density of 750 kg/m^2. With these equipment specifications and initial conditions, the drying time is fixed by setting the volume and mass of the FBD, while temperature and flowrate of input streams are time-varied operating variables for achieving the moisture content objective. We implemented a single-factor experiment using gPROMS to investigate the effect of drying times and the two operational parameters, temperature and flowrate of input gas, on the envelope size. Using these studies enables us to find a suitable design that consumes less time and energy but has a bigger operational envelope.

Within a time interval $\left[\tau_0, \tau_f\right]$, solid particles flow through cells of the FBD, and air with a temperature of $T(\tau)$ and a rate of $V(\tau)$ is continuously fed to the bottom of the FBD. Through fluidisation of the particles and consequent drying of the particles, the moisture content $\Gamma(\tau)$ of feed granules is reduced to the goal of a moisture content Γ (which could be a point or an interval). V is the volumetric flowrate and T is the temperature.

Employing the approach for traditional optimal control, we used the FBD model developed within gPROMS as a black box model [31], adding end point and path constraints. We used a black box model in order to show how it could be done without access to the full model equation set since this often needs to be the case in commercial settings.

The mathematical formulation is as follows:

$$\min_{x,y,T,V\varnothing} f = \Gamma_\varnothing$$

Subject to :

$$\Gamma(t) = \Phi(x(t), y, T(t), V(t), \tau), 0 \le t \le \tau$$

with **(M2)**

$$\text{End point constraints : } \Gamma^{\min} \le \Gamma_\tau \le \Gamma^{\max}\Gamma_\tau \le \Gamma^{\max}$$

$$\text{Path constraints : } \quad T^{\min} \le T(t) \le T^{\max}, \forall t \in [0,\tau]$$

$$V^{\min} \le V(t) \le V^{\max}, \ \forall t \in [0,\tau],$$

where min and max refer to the upper and lower bounds, respectively, for each operational variable that is controllable. x and y refer to other model parameters that are uncontrollable. The drying time τ_f is a design variable and is fixed.

For each fixed value of the drying time, we applied the methodology shown in Section 2 to find an optimal operating envelope. We were then able to explore the design sensitivity by varying the value of the drying time to find a suitable design that consumes less time and energy but has a bigger operational envelope. The selected design would be the one that consumes less energy and has more flexibility.

Using the methodology shown in Section 2, to obtain an optimal balance between design and operational variables, we let $b = \left[\left(T^{\min}, T^{\max}\right), \left(V^{\min}, V^{\max}\right)\right]$, and formulate the following problem to determine the optimal operating envelope:

$$\max_{y, b^{\min}, b^{\max}} f \equiv \frac{1}{N_b}\sum_{i=1}^{N_b} \frac{\Delta b_i - \Delta b_i^{\min}}{\Delta b_i^{\max} - \Delta b_i^{\min}}$$

Subject to :

$$\Gamma'(\tau) = f(x(\tau), y, b_i, \tau), \qquad \tau_0 \le \tau \le \tau_f$$

$$\Gamma^{min} \le \Gamma_{\tau_f} \le \Gamma^{max} \text{ or } \Gamma_{\tau_f} \le \Gamma^{max} \qquad \textbf{(M3)}$$

$$y^{\min} \le b_i \le y^{\max}$$

$$\Delta b_i = b_i^{\max} - b_i^{\min}$$

$$\Delta b_i^{\min} \le \Delta b_i \le \Delta b_i^{\max}$$

The process modeling tool gPROMS [29] was used to implement and solve the model to determine the optimal operating envelopes. The gPROMS modeling platform allows existing models of processes to be converted to the envelope form and optimise their dynamic operation. The solution steps are briefly illustrated as follows:

Step 1: fix the value of design variable τ, the upper and lower bounds $\Delta T, \Delta V$ and Γ, specify the interested range $\left(\left(\overline{T}^{\min}, \overline{T}^{\max}\right)\left(\overline{V}^{\min}, \overline{V}^{\max}\right)\right)$ of the bounded variables, and let

$$\overline{T}^{\min} \le T^{\min} \le T \le T^{\max} \le \overline{T}^{\max}$$

$$\overline{V}^{\min} \le V^{\min} \le V \le V^{\max} \le \overline{V}^{\max}$$

Step 2: generate N_S scenarios, each with a different set of operational variables (T, V). For scenario $k = 1, \cdots, N_s$, the values are given by:

$$T^{[i]} = T^{\min} + p^{[k]}\left(T^{\max} - T^{\min}\right)$$

$$V^{[i]} = V^{\min} + p^{[k]}\left(V^{\max} - V^{\min}\right)$$

where $p^{[k]}$ are normalized positions. For example, an optimization using two scenarios ($N_S = 2$), one corresponding to the bottom left and another to the top right of the feasible region, we specify:

$$p^{[1]} = (0, 0, \ldots, 0),\ p^{[2]} = (1, 1, \ldots, 1)$$

Step 3: Then we define the objective function, variables and constraints from the FBD model within gPROMS, and solve the optimization problem to obtain the best values of $\left(T^{\min}, T^{\max}\right)$ and $\left(V^{\min}, V^{\max}\right)$.

The algorithms were run on a personal computer with four 3.50 GHz processors and 16.0GB RAM using the Windows 10 operating system. The model and the approach can be used to optimise the steady-state and/or the dynamic behaviour of a continuous or batch process; in this case the fluid bed dryer is continuous.

The sampling technique employed in this work was a grid-based quasi-Monte Carlo sampling by using Sobol' low discrepancy sequences [37]. They have been shown to provide good distribution coverage even for fairly small sampling points. The design space was partitioned into a number of square grids and then within each grid sampling points were generated to evaluate feasibility The interested reader is referred to Kucherenko et al. [38] for an in-depth discussion on the subject. In brief, for a response variable $Y(X_1, X_2, \ldots, X_k)$ which is a function of a set of input variables $X_1, X_2, \ldots X_k$ a unit hypercube can be defined over the k-dimensions. Combining unit hypercubes over a grid-partitioned design space with quasi-random sequences is the most uniform possible solution to secure coverage. This is due to the fact that quasi-random points are selected from a sequence whilst knowing the position of the previous points and thus filling gaps between them [38]

We constructed an independent FBD model **(M2)**, to minimise drying time and moisture content, respectively, subject to it being in the interval $[10\%, 40\%]$. Next, we took the following steps:

Step 1: Specify the range of the operating variables:

$$\left(\left[T^{\min}, T^{\max}\right] = [20\ ^\circ\text{C}, 80\ ^\circ\text{C}], \left[V^{\min}, V^{\max}\right] = \left[240\ \text{m}^3/\text{h}, 480\ \text{m}^3/\text{h}\right]\right)$$

Step 2: Determine the feasible operating range with a drying time of 900 s which specifies a range of outputs of interest and hence a range of inputs. We uniformly sampled 13 temperatures in the range $[20, 80]$ °C and 25 flow rates in the range $[240, 480]$ m^3/h. Next, we simulated the FBD model to detect the feasible region (i.e., $13 \times 25 = 325$ points) that satisfies end point and path constraints. Finally, we found all feasible solutions where the moisture falls in the range $[10\%, 40\%]$. This is shown in Figure 2.

Step 3: Run the optimisation model **(M3)** with a drying time of 900 s to obtain the operating envelope for T and V.

(a) When ΔT and ΔV are allowed to vary freely we obtain the optimal operational envelope as shown in Figure 3 which maximises the area of the rectangle within the feasible boundary.
(b) When we constrain the variation that T and V can have to the following range $5 \leq \Delta T \leq 20$ °C and $10 \leq \Delta V \leq 60\text{m}^3/\text{h}$, solving **(M3)** gives the optimal operational envelope as shown in Figure 4. This maximises the envelope size while also maintaining the maximal distance to the feasible boundary using model **(M3)**.

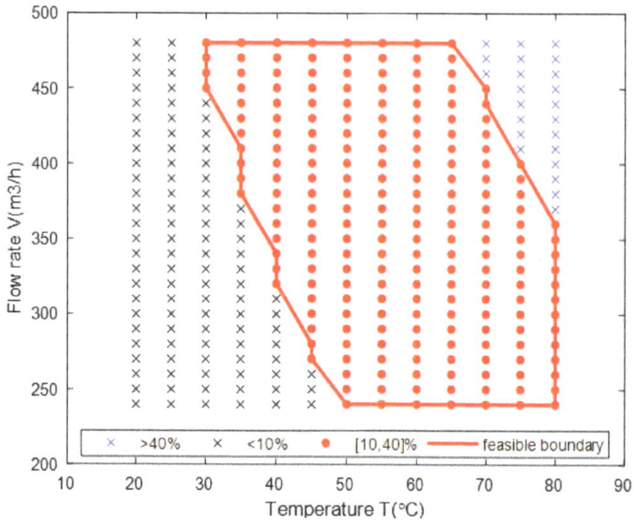

Figure 2. Feasible design range for T and V at a drying time of 900 s.

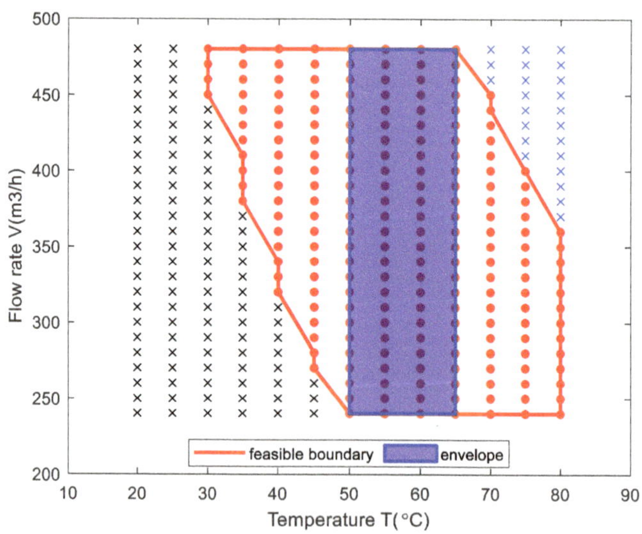

Figure 3. Operational envelope for a drying time of 900 s: $f = 0.77$.

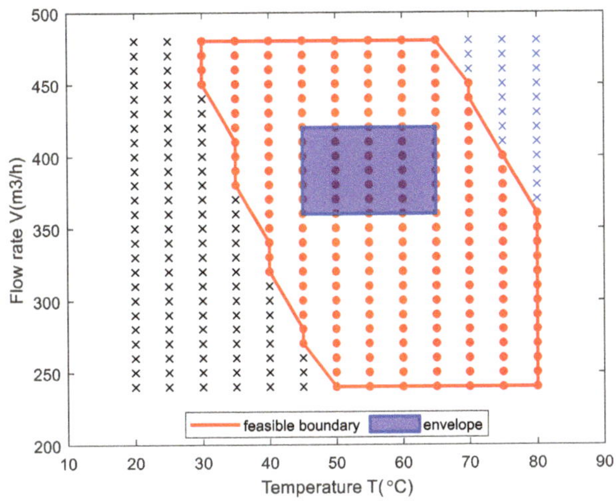

Figure 4. Operational envelope for a drying time of 900 s while maintaining the maximal distance to the feasible boundary: $f = 0.325$.

The final stage is to explore the trade-off between design and operational flexibility as measured by the envelope size. The FBD model indicates that the feasible design space varies with the drying time. Hence, we can select a best drying time by exploring the envelope size. To do this we used a scenario-based algorithm with 10 candidate drying times (600–1500 s) and allowed ΔT and ΔV to vary.

From the results shown in Figure 5, we found that the FBD process can obtain the maximal envelope size with 700 s (as shown Figure 6 where a larger number of sampling points, i.e., 1000, was used to increase the resolution of the results), which means that this design has the best flexibility using the chosen operating variables. Figure 5 shows that there is significant effect on the flexibility of the process at different drying times with the optimal obtained at 700 s. Interestingly, in this case, the flexibility is not affected by the change in ΔV but only by the change in temperature, for the specified ranges of uncertainty. Nonetheless, we should point out that in this work the related nonlinear programming models were solved with a local and not a global optimisation solver which could explain some of the irregularities shown in Figure 5 for design options and envelope sizes.

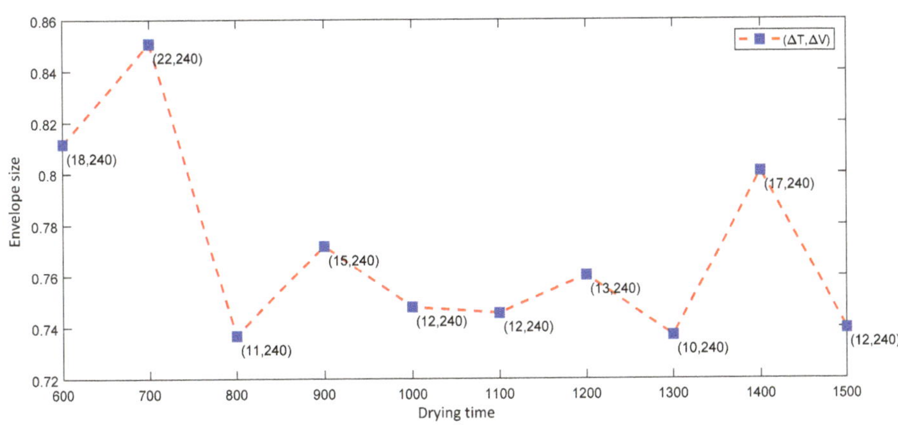

Figure 5. Result of a design selection by trade-off between envelope size and drying time.

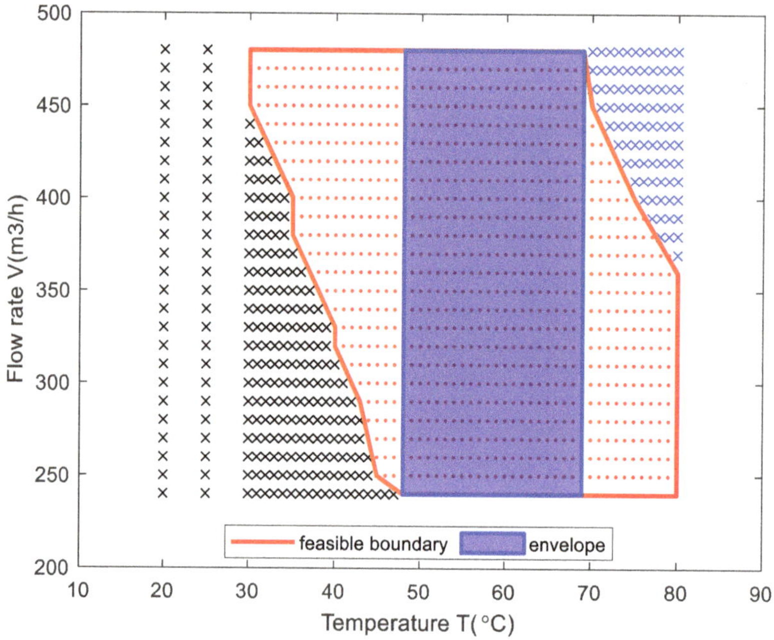

Figure 6. Operational envelope for a drying time of 700 s.

4. Conclusions

We have presented results for exploring the operational flexibility for a fluid bed drying unit that is at the heart of formulation processes for tablet manufacture. The methodology obtains a feasible operating envelope which is then reduced to one that allows constrained flexibility in two key parameters (T and V) but maintains an optimal distance from the feasible boundary. Finally, when using this optimal set of conditions, it is possible to explore the trade-off between the envelope size and a key parameter, the drying time. We have demonstrated the value of this approach to a process which is known to have considerable uncertainty and which is key to operational excellence. We aim to broaden the analysis to embrace all elements of the formulation process to explore operational flexibility and demonstrate the value of using a model-based optimisation approach to managing uncertainty in the pharmaceutical industry. It can add to the toolkit of the Quality by Design approach being brought in to pharmaceutical process development and operations. The approach seeks to support systematic development processes: in this case to systematically identify operating flexibility with robustness guarantees subject to model accuracy. Further work in tandem with experimental pilot plant work is needed to fully validate the approach within the tight regulatory regime of pharmaceutical manufacture.

Author Contributions: Conceptualization, V.M.C., I.D.L.B. and L.G.P.; methodology, V.M.C., L.G.P. and S.-L.J.; software, S.-L.J. and V.M.C.; validation, S.-L.J.; formal analysis, S.-L.J. and V.M.C.; writing—original draft preparation, S.-L.J. and I.D.L.B.; writing—review and editing, all; supervision, I.D.L.B.; project administration, I.D.L.B.; funding acquisition, S.-L.J. All authors have read and agreed to the published version of the manuscript.

Funding: National Natural Science Foundation of China Grant No. 61873042 funded S.J.-L. visit to UCL.

Acknowledgments: The authors acknowledge LiGe Wang of PSEnterprise Ltd. for his help with using gPROMS and the model and to PSEnterprise Ltd. for use of gPROMS software.

Conflicts of Interest: The authors declare no conflict of interest.

Nomenclature

Symbols	Definition
t	time variables.
x	differential state variables.
$\dot{x}$	derivatives of x with respect to time t.
y	algebraic state variables.
a	time-varying control and not bounded variables, which present the design decision variable in process.
b	time-varying control and bounded variables, which present the operational variable in processes.
Δb	sizes of the bound variables
N_b	number of bounded variables
τ	processing time.

References

1. Davis, J.; Edgar, T.; Porter, J.; Bernaden, J.; Sarli, M. Smart manufacturing, manufacturing intelligence and demand-dynamic performance. *Comput. Chem. Eng.* **2012**, *47*, 145–156. [CrossRef]
2. Bogle, I.D.L. A Perspective on Smart Process Manufacturing Research. Challenges for Process Systems Engineers. *Engineering* **2017**, *3*, 161–165. [CrossRef]
3. Li, D. Perspective for smart factory in petrochemical industry. *Comput. Chem. Eng.* **2016**, *91*, 136–148. [CrossRef]
4. Gamer, T.; Hoernicke, M.; Kloepper, B.; Bauer, R.; Isaksson, A.J. The autonomous industrial plant—Future of process engineering, operations and maintenance. *J. Process Control* **2020**, *88*, 101–110. [CrossRef]
5. Kamble, S.S.; Gunasekaran, A.; Gawankar, S.A. Sustainable Industry 4.0 framework: A systematic literature review identifying the current trends and future perspectives. *Process Saf. Environ. Prot.* **2018**, *117*, 408–425. [CrossRef]
6. Rafiei, M.; Ricardez-Sandoval, L.A. New frontiers, challenges, and opportunities in integration of design and control for enterprise-wide sustainability. *Comput. Chem. Eng.* **2020**, *132*, 106610. [CrossRef]
7. Wang, Z.; Escotet-Espinoza, M.S.; Ierapetritou, M. Process analysis and optimization of continuous pharmaceutical manufacturing using flowsheet models. *Comput. Chem. Eng.* **2017**, *107*, 77–91. [CrossRef]
8. Boukouvala, F.; Muzzio, F.J.; Ierapetritou, M.G. Design space of pharmaceutical processes using data-driven-based methods. *J. Pharm. Innov.* **2010**, *5*, 119–137. [CrossRef]
9. Boukouvala, F.; Muzzio, F.J.; Ierapetritou, M.G. Dynamic data-driven modeling of pharmaceutical processes. *Ind. Eng. Chem. Res.* **2011**, *50*, 6743–6754. [CrossRef]
10. Litster, J.; Bogle, I.D.L. Smart Process Manufacturing for Formulated Products. *Engineering* **2019**, *5*, 1003–1009. [CrossRef]
11. Lakio, S.; Ervasti, T.; Tajarobi, P.; Wikström, H.; Fransson, M.; Karttunen, A.; Ketolainen, J.; Folestad, S.; Abrahmsén-Alami, S.; Korhonen, O. Provoking an end-to-end continuous direct compression line with raw materials prone to segregation. *Eur. J. Pharm. Sci.* **2017**, *109*, 514–524. [CrossRef] [PubMed]
12. Singh, R.; Muzzio, F.J.; Ierapetritou, M.; Ramachandran, R. A Combined feed-forward/feed-back control system for a QbD-based continuous tablet manufacturing process. *Processes* **2015**, *3*, 339–356. [CrossRef]
13. Oka, S.; Sahay, A.; Meng, W.; Muzzio, F. Diminished segregation in continuous powder mixing. *Powder Technol.* **2017**, *309*, 79–88. [CrossRef]
14. Diab, S.; Gerogiorgis, D.I. Process modelling, simulation and technoeconomic evaluation of crystallisation antisolvents for the continuous pharmaceutical manufacturing of rufinamide. *Comput. Chem. Eng.* **2018**, *111*, 102–114. [CrossRef]
15. Diab, S.; McQuade, D.T.; Gupton, B.F.; Gerogiorgis, D.I. Process design and optimization for the continuous manufacturing of nevirapine, an active pharmaceutical ingredient for HIV treatment. *Org. Process. Res. Dev.* **2019**, *23*, 320–333. [CrossRef]
16. Nagy, B.; Szilágyi, B.; Domokos, A.; Vészi, B.; Tacsi, K.; Rapi, Z.; Pataki, H.; Marosi, G.; Nagy, Z.K.; Nagy, Z.K. Dynamic flowsheet model development and digital design of continuous pharmaceutical manufacturing with dissolution modeling of the final product. *Chem. Eng. J.* **2021**, *419*, 129947. [CrossRef]
17. Chen, Y.; Yang, O.; Sampat, C.; Bhalode, P.; Ramachandran, R.; Ierapetritou, M. Digital twins in pharmaceutical and biopharmaceutical manufacturing: A literature review. *Processes* **2020**, *8*, 1088. [CrossRef]
18. Food and Drug Administration. *Pharmaceutical cGMPs for the 21st Century—A Risk-Based Approach*; Technical Report; U.S. Department of Health and Human Services, Food and Drug Administration, Center for Drug Evaluation and Research (CDER): Rockville, MD, USA, 2004.
19. Food and Drug Administration. *Guidance for Industry Q8 Pharmaceutical Development*; Technical Report August; U.S. Department of Health and Human Services, Food and Drug Administration, Center for Drug Evaluation and Research (CDER): Rockville, MD, USA, 2009.

20. Huang, J.; Kaul, G.; Cai, C.; Chatlapalli, R.; Hernandez-Abad, P.; Ghosh, K.; Nagi, A. Quality by design case study: An integrated multivariate approach to drug product and process development. *Int. J. Pharmaceut.* **2009**, *382*, 23–32. [CrossRef]
21. Garcia-Munoz, S.; Luciani, C.V.; Vaidyaraman, S.; Seibert, K.D. Definition of design spaces using mechanistic models and geometric projections of probability maps. *Org. Process. Res. Dev.* **2015**, *19*, 1012–1023. [CrossRef]
22. Peterson, J.J. A Bayesian approach to the ICH Q8 definition of design space. *J. Biopharm. Stat.* **2008**, *18*, 959–975. [CrossRef]
23. Laky, D.; Xu, S.; Rodriguez, J.S.; Vaidyaraman, S.; García Muñoz, S.; Laird, C. An optimization-based framework to define the probabilistic design space of pharmaceutical processes with model uncertainty. *Processes* **2019**, *7*, 96. [CrossRef]
24. Swaney, R.E.; Grossmann, I.E. An index for operational flexibility in chemical process design. Part I: Formulation and theory. *AIChE J.* **1985**, *31*, 621–630. [CrossRef]
25. Kusumo, K.P.; Gomoescu, L.; Paulen, R.; García Muñoz, S.; Pantelides, C.C.; Shah, N.; Chachuat, B. Bayesian approach to probabilistic design space characterization: A nested sampling strategy. *Ind. Eng. Chem. Res.* **2019**, *59*, 2396–2408. [CrossRef]
26. Samsatli, N.J.; Sharif, M.; Shah, N.; Papageorgiou, L.G. Operational envelopes for batch processes. *AIChE J.* **2001**, *47*, 2277–2288. [CrossRef]
27. Sharifian, S.; Sotudeh-Gharebagh, R.; Zarghami, R.; Tanguy, P.; Mostoufi, N. Uncertainty in chemical process systems engineering: A critical review. *Rev. Chem. Eng.* **2019**, *37*, 687–714. [CrossRef]
28. Charitopoulos, V.M.; Aguirre, A.M.; Papageorgiou, L.G.; Dua, V. Uncertainty aware integration of planning, scheduling and multi-parametric control. *Comput. Aid. Chem. Eng.* **2018**, *44*, 1171–1176.
29. Wang, Z.; Ierapetritou, M. Global sensitivity, feasibility, and flexibility analysis of continuous pharmaceutical manufacturing processes. *Comput. Aid. Chem. Eng.* **2018**, *41*, 189–213.
30. von Stosch, M.; Schenkendorf, R.; Geldhof, G.; Varsakelis, C.; Mariti, M.; Dessoy, S.; Vandercammen, A.; Pysik, A.; Sanders, M. Working within the Design Space: Do Our Static Process Characterization Methods Suffice? *Pharmaceutics* **2020**, *12*, 562. [CrossRef]
31. *gPROMS*; v.7.0.1; PSEnterprise: London, UK, 2021. Available online: https://www.psenterprise.com/products/gproms (accessed on 6 September 2021).
32. Djuris, J.; Djuric, Z. Modeling in the quality by design environment: Regulatory requirements and recommendations for design space and control strategy appointment. *Int. J. Pharm.* **2017**, *533*, 346–356. [CrossRef]
33. Su, Q.; Ganesh, S.; Moreno, M.; Bommireddy, Y.; Gonzalez, M.; Reklaitis, G.V.; Nagy, Z.K. A perspective on Quality-by-Control (QbC) in pharmaceutical continuous manufacturing. *Comput. Chem. Eng.* **2019**, *125*, 216–231. [CrossRef]
34. Szilágyi, B.; Borsos, Á.; Pal, K.; Nagy, Z.K. Experimental implementation of a Quality-by-Control (QbC) framework using a mechanistic PBM-based non-linear model predictive control involving chord length distribution measurement for the batch cooling crystallization of l-ascorbic acid. *Chem. Eng. Sci.* **2019**, *195*, 335–346. [CrossRef]
35. Burgschweiger, J.; Groenewold, H.; Hirschmann, C.; Tsotsas, E. From hygroscopic single particle to batch fluidized bed drying kinetics. *Can. J. Chem. Eng.* **1999**, *77*, 333–341. [CrossRef]
36. Burgschweiger, J.; Tsotsas, E. Experimental investigation and modelling of continuous fluidized bed drying under steady-state and dynamic conditions. *Chem. Eng. Sci.* **2002**, *57*, 5021–5038. [CrossRef]
37. Kucherenko, S.; Albrecht, D.; Saltelli, A. Exploring multi-dimensional spaces: A comparison of Latin hypercube and quasi Monte Carlo sampling techniques. *arXiv* **2015**, arXiv:1505.02350.
38. Saltelli, A.; Annoni, P.; Azzini, I.; Campolongo, F.; Ratto, M.; Tarantola, S. Variance based sensitivity analysis of model output. Design and estimator for the total sensitivity index. *Comput. Phys. Com.* **2010**, *181*, 259–270. [CrossRef]

processes

MDPI

Article

Evaluation of a Combined MHE-NMPC Approach to Handle Plant-Model Mismatch in a Rotary Tablet Press

Yan-Shu Huang [1], M. Ziyan Sheriff [1,*], Sunidhi Bachawala [2], Marcial Gonzalez [2,3], Zoltan K. Nagy [1] and Gintaras V. Reklaitis [1]

[1] Davidson School of Chemical Engineering, Purdue University, West Lafayette, IN 47907, USA; huan1289@purdue.edu (Y.-S.H.); znagy@purdue.edu (Z.K.N.); reklaiti@ecn.purdue.edu (G.V.R.)
[2] School of Mechanical Engineering, Purdue University, West Lafayette, IN 47907, USA; sbachawa@purdue.edu (S.B.); marcial-gonzalez@purdue.edu (M.G.)
[3] Ray W. Herrick Laboratories, Purdue University, West Lafayette, IN 47907, USA
* Correspondence: sherifm@purdue.edu

Abstract: The transition from batch to continuous processes in the pharmaceutical industry has been driven by the potential improvement in process controllability, product quality homogeneity, and reduction of material inventory. A quality-by-control (QbC) approach has been implemented in a variety of pharmaceutical product manufacturing modalities to increase product quality through a three-level hierarchical control structure. In the implementation of the QbC approach it is common practice to simplify control algorithms by utilizing linearized models with constant model parameters. Nonlinear model predictive control (NMPC) can effectively deliver control functionality for highly sensitive variations and nonlinear multiple-input-multiple-output (MIMO) systems, which is essential for the highly regulated pharmaceutical manufacturing industry. This work focuses on developing and implementing NMPC in continuous manufacturing of solid dosage forms. To mitigate control degradation caused by plant-model mismatch, careful monitoring and continuous improvement strategies are studied. When moving horizon estimation (MHE) is integrated with NMPC, historical data in the past time window together with real-time data from the sensor network enable state estimation and accurate tracking of the highly sensitive model parameters. The adaptive model used in the NMPC strategy can compensate for process uncertainties, further reducing plant-model mismatch effects. The nonlinear mechanistic model used in both MHE and NMPC can predict the essential but complex powder properties and provide physical interpretation of abnormal events. The adaptive NMPC implementation and its real-time control performance analysis and practical applicability are demonstrated through a series of illustrative examples that highlight the effectiveness of the proposed approach for different scenarios of plant-model mismatch, while also incorporating glidant effects.

Keywords: continuous pharmaceutical manufacturing; model predictive control; state estimation; quality-by-control (QbC); glidant effects; plant-model mismatch

Citation: Huang, Y.-S.; Sheriff, M.Z.; Bachawala, S.; Gonzalez, M.; Nagy, Z.K.; Reklaitis, G.V. Evaluation of a Combined MHE-NMPC Approach to Handle Plant-Model Mismatch in a Rotary Tablet Press. *Processes* **2021**, *9*, 1612. https://doi.org/10.3390/pr9091612

Academic Editor: Luis Puigjaner

Received: 16 August 2021
Accepted: 3 September 2021
Published: 8 September 2021

1. Introduction

Pharmaceutical manufacturing processes have traditionally employed the batch operation mode, in which fixed amounts of raw materials are run through different unit operations to obtain the final drug product. Quality attributes of the final drug product were originally tested at the end of each batch processing step, where quality control essentially followed a quality-by-testing approach (QbT) [1], e.g., mixing uniformity is tested at the conclusion of the blending process. Over the last few years several factors have driven a shift from batch towards continuous pharmaceutical manufacturing. These factors include a reduction in the development cost for new medicines, making it both desirable and feasible to produce smaller annual volumes of targeted solutions for smaller patient populations, as well as improving product quality, decreasing cycle time, and better controlled processing, to name a few popular drivers [2]. An economic analysis provided by Schaber and co-workers [3] highlights that continuous operation is able to provide estimated overall savings that can range from 9 to 40%, depending on the drug loading and process chosen, when compared to traditional batch operation. Additionally, encouraged by the regulatory agencies to modernize pharmaceutical manufacturing processes, academia and industry have invested significant time and resources to study different aspects required to successfully shift from batch to continuous operation mode. These efforts were made possible through various collaborations and consortiums [4–6].

In 2012, Gernaey and co-workers [7] identified the design and implementation of continuous pharmaceutical processes as one of the many issues that remain unresolved. Advanced process understanding is critical to the implementation of continuous pharmaceutical manufacturing applications [8]. To address this requirement, a quality-by-design (QbD) approach was pursued over the last decade [9]. QbD is a multi-step procedure that involves: (i) definition of quality target product profiles (QTPPs) and critical quality attributes (CQAs), (ii) identification of critical material attributes (CMAs) and critical process parameters (CPPs), (iii) linking of the CMAs and CPPs with the CQAs, (iv) examination of the design space and required control strategies, (v) validation, scale-up, and production [10]. While QbT primarily focused on end-stage testing, QbD revolved around advanced product and process understanding for systematic design of the operating space using mechanistic models and design of experiments (DoE). However, more recently there has been a shift towards quality-by-control (QbC), wherein quantitative and predictive understanding is leveraged for active process control with robust process design and operation, enabling smart manufacturing [11].

A goal of the QbC approach is real-time process monitoring and management, wherein advanced process control strategies are utilized to handle disturbances and exceptional events [11]. Process analytical technology (PAT) methods play a crucial role in monitoring a variety of CQAs in order to accomplish this [8,9]. Monitoring and control of CQAs such as tensile strength and tablet porosity are critical as they are linked to dissolution profiles of the manufactured tablets, which are ultimately linked to patient safety and treatment efficacy [12–16]. Tablet tensile strength and dissolution profile are affected by various factors such as particle size, API concentration, and addition of lubricants and glidant [17,18]. Glidants are added to improve the flowability of the blend. However, glidants and lubricants are also known to impact other product parameters, such as bulk density, compactibility, and compressibility. An objective of this work is to incorporate the impact of the use and control of glidants while assuring that the critical properties, such as tensile strength of the manufactured tablets, are maintained at desirable levels. In the context of continuous manufacturing, when a glidant feeder is used, it is important to use calibrated mechanistic models to handle the variations of glidant concentration. Therefore, it is essential to explore a variety of different control strategies to address the changes in CQAs of a tablet that may arise when glidants are used. It is worth noting that even though these challenges and control strategies are also relevant to lubricants, the development of mechanistic models and relevant control strategies associated with lubricants is beyond the scope of this work.

The identification and handling of plant-model mismatch (PMM) is an important component of any real-time process monitoring and control approach, and it has been an area of interest for decades. PMM can arise in the continuous tablet manufacturing process for a variety of reasons, e.g., the feeder refill step is known to introduce disturbances that affect CMAs such as the bulk density [19–21]. Since this may result in a deviation in the CQAs, PMM needs to be monitored and algorithms to mitigate it need to be developed and implemented. In order to monitor PMM, Harris initially presented a minimum variance-based assessment criterion to assess the condition of the working control loop [22]. This approach has gained popularity, but it is limited to single-input-single-output (SISO) systems [23]. More recently, data-driven methods that examine autocovariance and solve an optimization problem formulated to address the mismatch estimates in MIMO systems have been developed. These methods minimize the discrepancy between the autocovariance of the output and the actual autocovariance of the mean-centered output variable [24–26]. Partial correlation coefficient (PCC)-based methods to identify PMM have also received attention in the literature, where PCC is well-suited as it is able to handle cases with high correlation in the manipulated variables [27,28]. As model re-identification is a critical, and often a time-consuming step once PMM has been identified, hybrid machine learning approaches have been proposed in order to aid model selection [28]. While there is great depth in the literature associated with the identification of PMM, there is limited discussion on practical approaches that would be applicable to the continuous pharmaceutical manufacturing industry in terms of management of the PMM [29,30]. CQAs and CMAs need to be tracked online during plant operation but they may be unmeasurable in practice through existing PAT sensing methods (e.g., the bulk density within a unit operation); therefore, alternative solutions are required. This work proposes novel state estimation methods to accurately track states and model parameters online and, hence, guide operating decisions.

Additionally, most work in the continuous tablet manufacturing domain utilizes linear model predictive control strategies, often resulting from the linearization of a nonlinear system, which may not be adequate for some strongly nonlinear processes [31–33]. A literature review of traditional MPC application for different unit operations in the continuous pharmaceutical manufacturing industry can be found in [34], including end-to-end in silico and implementation studies. Since there is limited implementation of nonlinear model predictive control strategies for the continuous pharmaceutical manufacturing industry, a main objective of this work is to develop and present a moving horizon estimation-based nonlinear model predictive control (MHE-NMPC) framework to serve the dual requirement of accurate estimation and effective control. Model predictive control strategies are also advantageous over traditional proportional-integral-derivative (PID)-based control strategies, as they are able to effectively handle constraints, loop interactions and non-square control systems that may be encountered in manufacturing of pharmaceutical solid dosage forms [35–38]. A main practical concern for any developed framework is the need to ensure that the optimization problem can be solved in real-time, particularly for relatively quick processes such as those in the continuous pharmaceutical manufacturing industry. Therefore, an additional objective is to examine and discuss the real-time feasibility of the developed framework in controlling a rotary tablet press.

It is important to note that once non-conforming quality attributes have been identified, a long-term goal is the integration of control frameworks similar to the MHE-NMPC structure with residence time distribution (RTD)-based modeling frameworks that are currently being developed to guide tablet product diversion in the continuous pharmaceutical manufacturing industry and truly enhance and enable smart manufacturing operations [39,40].

To summarize, the primary objective of this work is to develop and present a moving horizon estimation-based nonlinear model predictive control (MHE-NMPC) framework to serve the dual requirement of accurate estimation and effective control, and to demonstrate its practical applicability by discussing its implementation feasibility in controlling a rotary tablet press. A secondary objective of this work is to examine different control strategies that are required when incorporating glidant feeders to further control tablet properties.

The rest of this work is organized as follows. In Section 2, mathematical modeling and optimization approaches for state estimation and control will be briefly discussed, along with the proposed monitoring algorithm and its advantages, i.e., the MHE-NMPC framework to monitor CMAs and CQAs and determine control actions will be presented. Section 3 will illustrate the robustness of the proposed MHE-NMPC framework with two examples of application. Specifically, the studies will showcase monitoring and control of a rotary tablet press in the presence of (i) plant-model mismatch and (ii) uncertainty in the glidant concentration. Section 4 will provide concluding remarks and directions for future work.

2. Material and Methods

2.1. State Estimation

State and parameter estimation methods have been utilized to enhance process monitoring capabilities in a number of industrial applications, ranging from bioreactors to robotics to continuous pharmaceutical manufacturing [30,41,42]. State and parameter estimation is a powerful tool in scenarios where process states or model parameters cannot be directly measured with sensors.

A nonlinear state-space model is defined as follows:

$$\dot{x} = g(x, u, \theta, w) \tag{1}$$

$$y = l(x, u, \theta, v) \tag{2}$$

where x, u, θ, and y are the state variable vector, input variable vector, model parameter vector, and measurement vector, respectively [43]. The process and measurement noise are denoted by w and v, respectively. A schematic illustration of conventional state estimation algorithms is presented in Figure 1, where the nonlinear model is initialized based on the state values at the previous time step ($k-1$) in order to obtain a prediction of the states and model parameters at the current time (k). State measurements are obtained from available sensors and are utilized to obtain a more accurate estimate of the states and parameter values by correcting the predictions from the model.

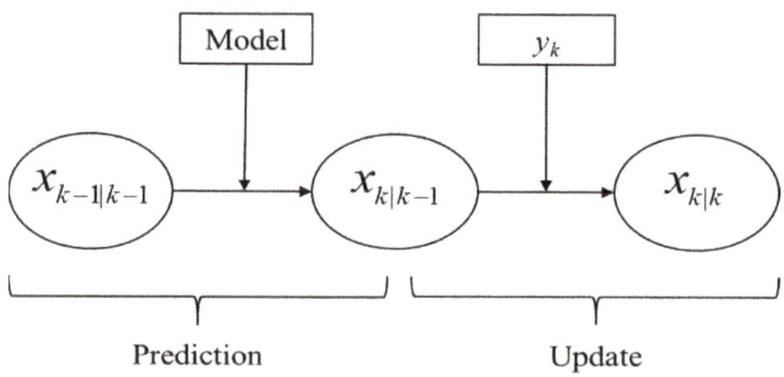

Figure 1. Conventional state estimation algorithm.

A number of alternative algorithms to carry out state estimation have been developed, e.g., the Kalman filter (KF), extended Kalman filter (EKF), unscented Kalman filter (UKF), particle filter (PF), and moving horizon estimation (MHE) are among the more popular approaches [44–47]. While the KF and EKF algorithms are suitable for linear or approximately linear applications, the UKF, PF, and MHE algorithms are able to handle processes that are more nonlinear in nature. The KF, EKF, and UKF algorithms also assume that error distributions are Gaussian in nature, while this assumption does not have to be satisfied for the PF and MHE algorithms [46]. Unlike the conventional approach illustrated in Figure 1, MHE utilizes a window, or moving horizon, of previous measurements in order to estimate the current states and model parameters, often providing improved performance when compared to the other algorithms. MHE is also capable of handling measurements collected from sensors at different sampling intervals or frequencies, which is advantageous for industries that utilize a variety of sensors to track physical attributes, as is the case in continuous pharmaceutical manufacturing. Therefore, MHE will be the choice of state estimation algorithm utilized in this work, as it is able to effectively track plant-model mismatch caused by deviations in model parameters, such as variations in the bulk density due to uncertainty in upstream unit operations (e.g., refilling of feeders).

The importance of monitoring powder feeder dosing in continuous pharmaceutical manufacturing is investigated in great detail by Destro and co-workers [29], wherein an MHE-based state estimation approach is implemented to reconcile mass measurements that are available from loss-in-weight (LIW) feeders with downstream measurements that are available from a PAT instrument and, thereby, to obtain practically continuous measurements as opposed to sampled measurements provided from the PAT instrument. The authors also demonstrate that the MHE approach is superior to one that utilizes statistical filters instead of the state estimator. Similarly, robust estimators were incorporated within the MHE skeleton of a feeding-blending system to handle dynamic systems with gross errors [30]. While the results presented in both case studies are promising, neither discusses or elaborates on the importance of their integration with efficient control strategies. To the knowledge of the authors, there has been no work in the continuous pharmaceutical manufacturing domain that has examined the integration of state estimation strategies with efficient control strategies, and therefore, this will be the central focus of this work, as discussed in the following sections.

2.2. Model Predictive Control (MPC)—Linear and Nonlinear

Model predictive control methods have been employed by various industries over the past few decades [48–52]. MPC relies on the dynamic model of the process. This model can either be linear or linearized models obtained through system identification, as in the case of the linear implementation of MPC, or be nonlinear and derived using first principles or semi-empirically (using a hybrid model), as in the case of NMPC [53]. Both MPC algorithms utilize a finite time horizon to optimize the control input at the current time iteration, while keeping future time iterations in mind. This ability makes MPC predictive in nature due to its ability to anticipate future events and take control actions accordingly, which is not possible using traditional PID controllers [54].

While nonlinear model predictive control (NMPC) methods have been utilized by some industries, to the knowledge of the authors their implementation has not been explored extensively for continuous pharmaceutical manufacturing of solid dosage forms, cf. [31–33], where linear or hybrid implementations of MPC are utilized. Since processes in the continuous pharmaceutical manufacturing industry are known to be nonlinear in nature, it is therefore desirable to develop and implement an NMPC approach.

It should also be noted that these predictive control strategies are particularly advantageous for cases of non-square systems, i.e., where the number of manipulated variables exceeds the number of the controlled variables, since these methods are able to effectively manage nonlinear relationships [35–37]. These cases cannot be straightforwardly handled using traditional PID control strategies [38].

The following section will present the developed MHE-NMPC framework that seeks to accomplish the dual requirement of accurate estimation and efficient control. Since real-time implementation feasibility is an area of interest, a discussion on the practical applicability of the framework developed will also be presented.

2.3. Moving Horizon Estimation-Based Nonlinear Model Predictive Control (MHE-NMPC) Framework

The algorithm proposed in this work seeks to combine the effective estimation capabilities of MHE with the control abilities of NMPC, through the MHE-NMPC framework illustrated in Figure 2. Real-time measurements of output variables (y) and input variables (u) are first collected to monitor the process. Since disturbances, either known or unknown, can always exist in a real plant, mismatches may arise between the sensor measurements and model values. As elaborated previously, the goal of state estimation is to obtain a 'true state' value by utilizing the information from both measurements and process models. The 'true state' can be either measurable, e.g., API concentration at the blender exit using NIR sensors [55–57], or unmeasured, e.g., powder holdup in the blender. Through the updating of uncertain model parameters, which have changes due to upstream disturbances, MHE enables the handling of plant-model mismatch, thus allowing the controller to receive estimated output variables $(\hat{y})$ with less uncertainty. The NMPC control algorithm then minimizes the error between setpoints (y_{sp}) and estimated output variables $(\hat{y})$ by deciding the optimal control move (u) for the process to reach both setpoint tracking and disturbance rejection, i.e., the control objectives, while updating the model parameter $(\hat{\theta}_k)$ and median of the error distribution in the past time window (ζ).

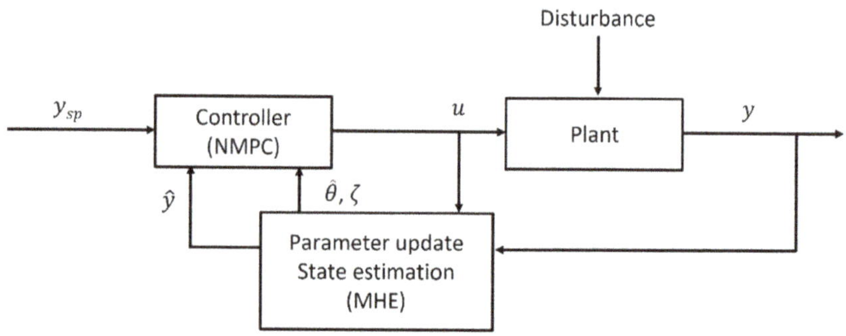

Figure 2. Adaptive control framework of MHE-NMPC.

A schematic illustration of the MHE-NMPC framework at time $t = k$ is shown in Figure 3, with N_{past} measurements available in the past window and N_p estimations in the prediction window. The MHE is then formulated as follows:

$$\min_{\hat{\theta}_k} J = \sum_{t=k-N_{past}}^{k} (\epsilon_t)^T W_E \, \epsilon_t + (\hat{\theta}_k - \hat{\theta}_{k-1})^T W_\theta (\hat{\theta}_k - \hat{\theta}_{k-1}) \tag{3a}$$

subject to

$$\hat{x}_{k-N_{past}+j+1} = f(\hat{x}_{k-N_{past}+j}, u_{k-N_{past}+j}, \hat{\theta}_k) \tag{3b}$$

$$\hat{y}_{k-N_{past}+j} = h(\hat{x}_{k-N_{past}+j}) \tag{3c}$$

$$\epsilon_{k-N_{past}+j} = y_{k-N_{past}+j} - \hat{y}_{k-N_{past}+j} \tag{3d}$$

$$\hat{x}_{k-N_{past}+j+1} \in \mathbb{X}, \quad \epsilon_{k-N_{past}+j} \in \Omega_\epsilon, \quad \hat{\theta}_k \in \Omega_\theta \tag{3e}$$

$$j = 0, 1, \ldots, N_{past} \tag{3f}$$

where $\hat{\theta}_k$ are estimated uncertain parameters, which are bounded in the compact set Ω_θ. In the above formulation, y_t and u_t are measurements of output variables and input variables at time t, respectively; $\hat{y}_t$ and $\hat{x}_t$ are estimated output and state values, respectively; ε_t are output disturbances, which are bounded in the compact set Ω_ϵ; and W_E and W_θ are the weighting matrices. After the MHE optimization problem is solved at time $t = k$, the estimated state $\hat{x}_{k-N_{past}+1|t=k}$ is chosen as the initial state value of next time step $t = k + 1$, i.e., $\hat{x}_{k-N_{past}+1|t=k+1} = \hat{x}_{k-N_{past}+1|t=k}$ [58].

While an error distribution of output variables $y_t - \hat{y}_t$ in the past time window can be obtained from Equation (3d), a single point estimate of the output $\hat{y}_t$ is of most interest in many applications, instead of the whole error distribution [59]. When no probabilistic process models are used, it is easier to use a single point estimate of the output $\hat{y}_t$ to visualize and control process dynamics. In this study, the median of the error distribution in the past time window is used to represent output disturbances ζ_k at time $t = k$, i.e.,

$$\zeta_k = median\left\{\epsilon_{k-N_{past}+j}\right\}, \text{ for } j = 0, 1, \ldots, N_{past} \tag{4}$$

Therefore, with estimated states $\hat{x}_k$, output disturbances ζ_k, and updated uncertain optimal parameters $\hat{\theta}_k$, the NMPC framework at time $t = k$ is given by:

$$\min_{\Delta u_t} J = \sum_{t=k}^{k+N_p} (\hat{y}_t - y_{sp})^T W_y (\hat{y}_t - y_{sp}) + \sum_{t=k}^{k+N_c-1} (\Delta u_t^T W_{\Delta u} \Delta u_t) \quad (5a)$$

subject to

$$\hat{x}_{k+j+1} = f(\hat{x}_{k+j}, \hat{u}_{k+j}, \hat{\theta}_k) \quad (5b)$$

$$\hat{y}_{k+j} = h(\hat{x}_{k+j}) + \zeta_k \quad (5c)$$

$$\Delta u_{k+j} = \hat{u}_{k+j+1} - \hat{u}_{k+j} \quad (5d)$$

$$\hat{x}_{k+j} \in \mathbb{X}, \quad \hat{u}_{k+j} \in \mathbb{U}, \quad \Delta u_{k+j} \in \Omega_{\Delta u} \quad (5e)$$

$$j = 0, 1, \ldots, N_p - 1 \quad (5f)$$

where N_c is the length of the control time window, and y_{sp} are the setpoints of the output variables. W_y and $W_{\Delta u}$ are the weighting matrices. Control movements Δu are constrained in the compact set $\Omega_{\Delta u}$. The control window N_c is usually smaller than the prediction window N_p and has to be chosen considering a compromise between computational burden and stability requirements. Control movements Δu_{k+j} in control window N_c vary according to results of optimization, but those beyond the control window are zero, i.e., $\Delta u_{k+N_c} = \Delta u_{k+N_c+1} = \cdots = \Delta u_{k+N_p-1} = 0$, which implies that $\hat{u}_{k+N_c} = \hat{u}_{k+N_c+1} = \cdots = \hat{u}_{k+N_p}$. In other words, while the predicted $\hat{y}_{k+j}$ can still be calculated using Δu_{k+j} and $\hat{u}_{k+j}$ in the prediction window N_p, only Δu_{k+j} in control window N_c is considered in the objective function. It should be noted that models of estimated output variables $\hat{y}_t$ are different in Equations (3c) and (5c). In the future time window, output disturbances ζ_k are added to the model of the process, allowing zero steady-state offset in controlled output variables y [59,60]. A schematic illustration of MHE-NMPC is provided in Figure 3, where at each iteration, the MHE is utilized to obtain a more accurate representation of the true state of the process and plant-model mismatch, and the NMPC is utilized to find the optimal first move for each input variable u. This framework thus allows for both accurate estimation and efficient control.

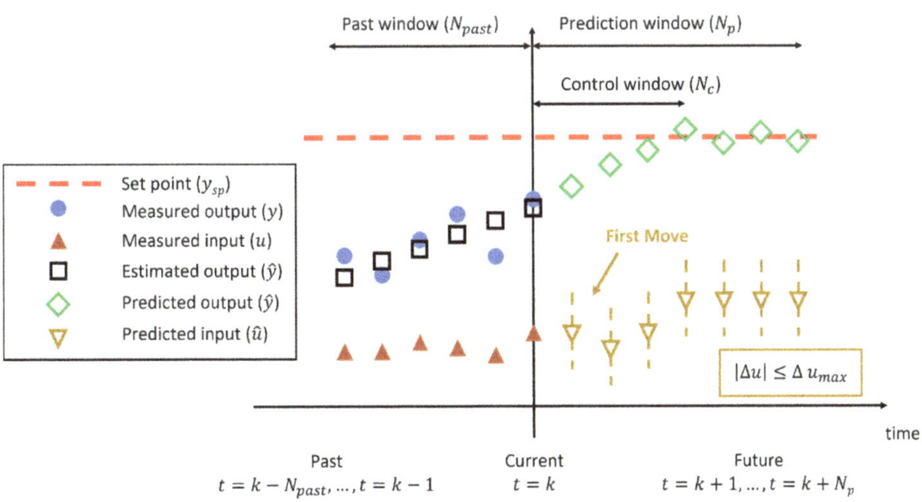

Figure 3. Illustration of MHE-NMPC coupling at each time interval.

2.4. Implementation of a Real-Time Feasible MHE-NMPC Framework

The MHE-NMPC framework is implemented in MATLAB (MathWorks R2018a) and the MATLAB built-in *fmincon* function is used to solve the optimization problem in each iteration. The computation is performed on a 64-bit ASUS VivoBook with Intel® Core™ i7-8550U @1.80 GHz processor and 8GB of total memory. In all simulated results, the time unit for each step is 1 s, the past time window (N_{past}) used in MHE is 30 time units, and the NMPC is tuned with prediction time window (N_p) chosen to be 60 time units and control time window (N_c) to be 10 time units. Sensor measurements are also assumed to be available at 1 s intervals. It should be noted that the average computation time for each iteration is 0.7 s, indicating that the optimization problem can be solved and implemented in real time. These results demonstrate the feasibility of the proposed framework, and its ability to achieve real-time process control.

The following section will explore the applicability of the developed MHE-NMPC framework to track plant-model mismatch and to efficiently control a key process unit operation in the continuous manufacturing line of solid dosage forms, i.e., the rotary tablet press.

3. Examples of Application to Continuous Direct Compression

The applicability of the developed MHE-NMPC framework will be demonstrated through two case studies. The first case study will highlight the importance of monitoring model parameters in real time and how this is enabled via state estimation, as opposed to the use of fixed model parameters. Different degrees of plant-model mismatch will be used. The second case study will present the applicability of the framework in the practical scenario of having uncertainty in the glidant feeding conditions.

Both case studies will focus primarily on the tablet press unit operation of the direct compression line. A hierarchical implementation of the three-level quality-by-control (QbC) framework of control systems for the continuous direct compression line is illustrated in Figure 4, whose unit operations are comprised of feeders, blenders, and the tablet press. For this line, Level 0 control is generally implemented through programmable logic control (PLC) systems built into the unit operation equipment in order to control CPPs. Level 1 control relies on PAT tools to monitor and control CQAs and can encompass multiple unit operations designed to reduce the impact of disturbances that may propagate downstream. Level 1 control supervises the Level 0 control, typically accomplished through SISO control loops which aim to maintain desired setpoints for the CQAs. A distributed control system (DCS) is employed to integrate process equipment such as the feeders and tablet press and any instrumentation that measures material properties. More advanced approaches are applied at Level 2 and use mathematical models such as MHE for validating process measurements, with the ability to predict the effects of disturbances and changes in the CPPs on the CQAs. Additional functionalities provided at Level 2 can include NMPC, a quality management system (QMS), and real-time optimization (RTO).

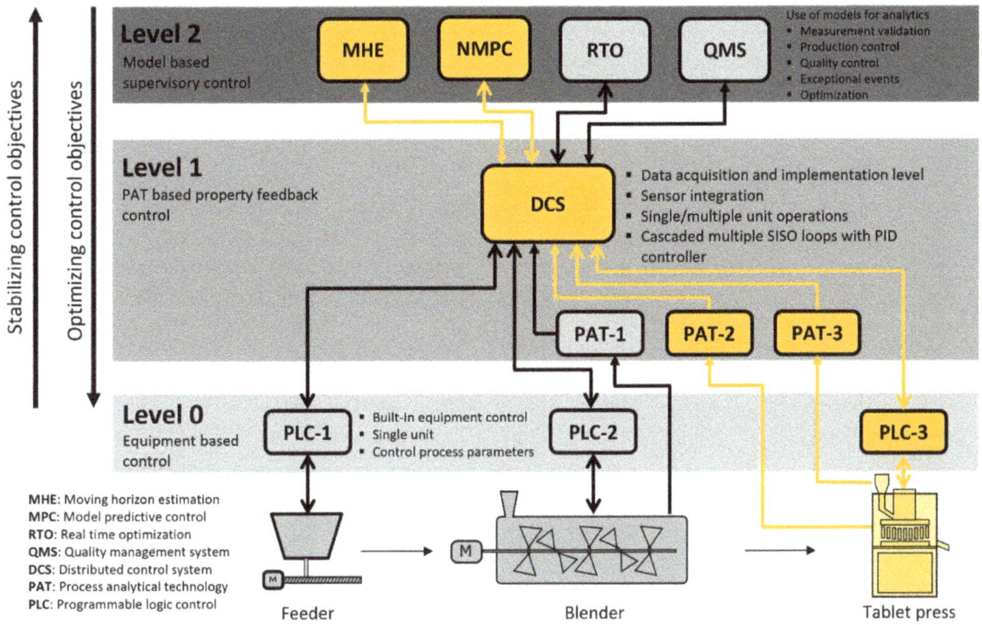

Figure 4. A 3-level hierarchical implementation of control systems for the continuous direct compression process (modified from source: [11]).

3.1. Tablet Press Model

The rotary tablet press and the lubricant/glidant feeder are key unit operations, where the latter is used to reduce frictional loses and facilitate powder flow during die filling and formation of solid tablets via mechanical compression. Therefore, models for glidant effects in die filling and compression processes will be used to monitor and control the porosity and tensile strength of tablets. Specifically, these mechanistic models capture the effects of glidant concentration and mixing conditions [61,62].

The weight of a convex tablet, W, formed using Natoli D-type tooling with shallow cup depth can be computed as follows:

$$W = \rho_b V_{fill}\left(1 - \xi_1 \frac{n_T}{n_F} + \xi_2 \frac{H_{fill}}{D}\right) \tag{6}$$

where D, V_{fill}, H_{fill}, ρ_b, n_T, and n_F, refer to the diameter of the die, volume of the die cavity, dosing position, bulk density of the powder, turret speed, and feed frame speed, respectively [62]. In Equation (6), ξ_1 and ξ_2 are model parameters to be estimated from experimental data. The bulk density depends on glidant concentration and mixing conditions, but its dependency is beyond the scope of this work. For the D-type tooling, the volume of the die cavity is given by:

$$V_{fill} = \frac{\pi D^2 H_{fill}}{4} + \frac{\pi h\left(\frac{3D^2}{4} + h^2\right)}{6} \tag{7}$$

where h is the cup depth. The tablet production rate, $\dot{m}_{tablet}$, is given by:

$$\dot{m}_{tablet} = W n_T N_{station} \tag{8}$$

where $N_{station}$ is the number of turret stations available. For a blend composed of MCC (Avicel PH200), APAP (acetaminophen) and silica experimental evidence suggests that both pre-compression and main compression forces do not show a dependency on glidant conditions [62]. The pre-compression force (PCF) is then computed as follows [63]:

$$F_{pc} = \frac{\pi D^2}{4b}\left[\frac{\rho^{pc} - \rho_c}{\rho^{pc}(a-1) + \rho_c}\right] \tag{9}$$

where parameters a and b are Kawakita constants [63], which represent the maximum degree of compression and the reciprocal of the pressure applied to attain this degree of compression, respectively. In Equation (9), ρ_c and ρ^{pc} refer to the critical density and the pre-compression relative density, respectively. The pre-compression relative density is computed as follows:

$$\rho^{pc} = \frac{W}{V^{pc}\rho_t} \tag{10}$$

and

$$V^{pc} = \frac{\pi D^2 H^{pc}}{4} + \frac{\pi h\left(\frac{3D^2}{4} + h^2\right)}{3} \tag{11}$$

where ρ_t and H^{pc} refer to the true density of the powder and the pre-compression thickness, respectively. Similarly, the main compression force (F_{punch}) is computed as follows:

$$F_{punch} = \frac{\pi D^2}{4b}\left[\frac{\rho^{in-die} - \rho_c}{\rho^{in-die}(a-1) + \rho_c}\right] \tag{12}$$

with the in-die relative density ρ^{in-die} given by:

$$\rho^{in-die} = \frac{W}{V^{in-die}\rho_t} \tag{13}$$

and

$$V^{in-die} = \frac{\pi D^2 H^{in-die}}{4} + \frac{\pi h\left(\frac{3D^2}{4} + h^2\right)}{3} \tag{14}$$

where H^{in-die} refers to the main compression thickness. The tablet density, or out-of-die relative density of the tablet, ρ^{tablet}, is then obtained utilizing the elastic recovery, ε_ρ, of the tablet as follows:

$$\rho^{tablet} = (1 - \varepsilon_\rho)\rho^{in-die} \tag{15}$$

The elastic recovery model is not sensitive to the glidant mixing conditions [61,62], and it is governed by:

$$\varepsilon_\rho = \varepsilon_0 \frac{\rho^{in-die} - \rho_{c,\varepsilon}}{1 - \rho_{c,\varepsilon}} \tag{16}$$

where ε_0 and $\rho_{c,\varepsilon}$ are the in-die elastic recovery at full compaction and the relative density at which tablets do not exhibit elastic recovery, respectively [64]. The tensile strength σ_t exhibits dependency on glidant conditions and it is computed as follows

$$\sigma_t = \sigma_0\left[1 - \left(\frac{1 - \rho^{tablet}}{1 - \rho_{c,\sigma_t}}\right)e^{(\rho^{tablet} - \rho_{c,\sigma_t})}\right] \tag{17}$$

where σ_0 is the tensile strength at zero porosity and ρ_{c,σ_t} is the critical relative density at which tablets do not exhibit any the tensile strength, i.e., the relative density at which a tablet starts forming [17]. It bears emphasis that these parameters are functions of glidant conditions, specifically:

$$\rho_{c,\sigma_t} = \frac{\rho_{c,0} - \rho_{c,\infty}}{1 + C_\sigma} + \rho_{c,\infty} \tag{18}$$

$$\sigma_0 = \frac{\sigma_{0,\phi}}{1 + C_\sigma} \tag{19}$$

$$C_\sigma = \frac{c_l^{b_1} \gamma^{b_2}}{b_3} \tag{20}$$

where $\rho_{c,0}$, $\rho_{c,\infty}$, $\sigma_{0,\phi}$, b_1, b_2, and b_3 are model parameters estimated from experimental data. In Equation (20), c_l and γ are the glidant concentration and total shear imparted to the blend, respectively. For simplicity, the total shear strain is represented by an equivalent mixing time, which, in turn, is estimated as follows

$$\gamma = \gamma_0 + \frac{m_{f,h}}{\dot{m}_{tablet}} \tag{21}$$

where γ_0 and $m_{f,h}$ are a total shear strain base line, expressed in term of mixing time, and the mass hold up in the feed-frame and hopper, respectively. Specifically, for the purpose of these case studies, the model parameter γ_0 is the glidant mixing time used in the rotary Tote blender when the blend was prepared. A 5L rotary Tote blender was employed. The mean residence time in the feed-frame, i.e., $\frac{m_{f,h}}{\dot{m}_{tablet}}$, is used to estimate the additional shear, or mixing time, imparted inside the tablet press.

The dependency of the bulk density on the glidant concentration, c_l, can be incorporated through the following equation [62]:

$$\rho_b = \rho_{b,\infty} - \frac{\rho_{b,\infty} - \rho_{b,0}}{1 + C_\rho} \tag{22}$$

where $\rho_{b,\infty}$ and $\rho_{b,0}$ represent the bulk densities when the shear strain imparted is infinite and zero, respectively. C_ρ is a lumped parameter that defines the glidant mixing conditions computed as follows [62]:

$$C_\rho = \frac{c_l^{r_1} (\gamma + \gamma_0)^{r_2}}{r_3} \tag{23}$$

where γ and γ_0 are the shear imparted to the powder during mixing and the initial shear strain imparted prior to mixing, respectively. r_1, r_2, and r_3 are fitting parameters.

A Natoli NP-400 tablet press and SOTAX AT4 tablet tester were used in this work to fabricate tablets and gather experimental data under steady-state conditions. The experimental data were then used to carry out parameter fitting using the *fmincon* function in MATLAB to obtain realistic model parameters values that could be used for the simulations presented in case studies 1 and 2, which are summarized in Table 1.

Table 1. Summary of model parameters for case studies 1 and 2.

	Case Study 1			Case Study 2
Purpose	**Assess Control Performance in the Presence of Different Levels of PMM**			**Assess Control Performance When Uncertainty in Glidant Concentration Is Present**
Assumption	**Glidant Concentration Can Be Manipulated**			**Glidant Concentration Needs to Be Estimated**
Model Parameters	**No PMM**	**Mild PMM**	**High PMM**	**Nominal Operation**
ξ_1	0.036	0.036	0.036	0.036
ξ_2	0.030	0.030	0.050	0.030
ρ_b (g/cm^3)	0.365	0.390	0.410	0.365
ρ_c	0.265	0.290	0.230	0.265
Kawakita: *a*	0.80	0.77	0.84	0.80

Table 1. *Cont.*

	Case Study 1			Case Study 2
Purpose	Assess Control Performance in the Presence of Different Levels of PMM			Assess Control Performance When Uncertainty in Glidant Concentration Is Present
Assumption	Glidant Concentration Can Be Manipulated			Glidant Concentration Needs to Be Estimated
Model Parameters	No PMM	Mild PMM	High PMM	Nominal Operation
Kawakita: 1/b (MPa)	10.26	10.26	8.55	10.26
ρ_t (g/cm^3)	1.53	1.53	1.51	1.53
ε_0	0.08	0.08	0.08	0.08
$\rho_{c,\varepsilon}$	0.57	0.57	0.57	0.57
σ_0 (MPa)	11.67	11.67	11.67	11.67
ρ_0	0.57	0.57	0.57	0.57
ρ_∞	0.61	0.61	0.61	0.61
b_1	0.31	0.31	0.31	0.31
b_2	0.38	0.38	0.38	0.38
b_3	8.40	8.40	8.40	8.40
$\rho_{b,\infty}$ (g/cm^3)	N/A			0.450
$\rho_{b,0}$ (g/cm^3)				0.330
r_1				0.361
r_2				1.394
r_3				23.326

3.2. *Case Study 1: Monitoring and Control of the Rotary Tablet Press in the Presence of Plant-Model Mismatch*

Monitoring powder bulk density in the tablet press is of critical importance, as it affects the tablet properties [12]. Sources of variability can be introduced during any of the unit operations upstream, e.g., in the feeder unit operations during refill, as the feeder switches from gravimetric mode to volumetric mode, leading to either increases in bulk density due to compression or decreases in bulk density due to aeration [19–21].

For this case study, a four-by-five system was employed as it would enable the incorporation of an extra manipulated input for added control benefits, i.e., glidant concentration. It is assumed that the direct compression line has the ability to utilize the glidant concentration as a manipulated variable through changes in the glidant flowrate. In practice this would be implemented in the hierarchical three-level QbC framework, by using a level-one PID control, that would use the glidant concentration measurement and adjust the glidant flowrate to follow the concentration setpoint set by the level-two NMPC. In this case the four-by-five non-square level-e control system is comprised of controlled variables consisting of the tablet weight, pre-compression force, production rate, and tensile strength and manipulated variables consisting of the dosing position, pre-compression thickness, main compression thickness, turret speed, and glidant concentration. It is assumed that measurements for the tablet weight, pre-compression force, main compression force, and production rate are all available every second [61]. In this simulation, it should be noted that the main compression force was not a directly controlled variable with specified set points. This is because the tensile strength could not be maintained while simultaneously fixing the main compression force. Given the objective to maintain the tensile strength at desired levels due to its link to patient safety, the tensile strength was chosen over the main compression force as a controlled variable. Since measurements of the main compression force are available, they were utilized in the MHE framework only for the purpose of parameter estimation. As maintaining the CQAs is important, and since the tensile strength measurements are not available in real time, a soft sensor based on Equation (17) is utilized in order to track this particular state in real time. In practice, the SOTAX AT4 tablet tester can be utilized in order to obtain measurements of the tensile strength. However, since the diametrical compression test is destructive, tensile strength measurements are available at a lower frequency than one of the PAT sensors, which in turn also drives the need for a soft sensor.

A summary of the controlled variables, manipulated variables, measured variables, and uncertain model parameter is provided in Table 2. In order to examine the performance of the MHE-NMPC framework under PMM, three different scenarios will be examined, namely: nominal operation (no PMM), operation with mild PMM, and operation with high PMM. A summary of model parameters for each scenario is provided in Table 1, where the MHE-NMPC tuning parameters are those described in detail in Section 2.4. Mild PMM is simulated by introducing mismatch to three model parameters: ρ_b, ρ_c and Kawakita parameter a. High PMM is simulated by introducing mismatch to six model parameters: ξ_2, ρ_b, ρ_c, Kawakita parameters a and b, and ρ_t. In this simulation, the 'model' and 'plant' share the same equations detailed in Section 3.1. Different parameter values were assigned to the 'model' and 'plant' in order to simulate mismatch. Additionally, sensor measurement noise in the plant was simulated by adding normally distributed error with zero mean and variance analogous to the variability of a real sensor. This variability was obtained from historical plant data.

Table 2. Summary of variables and uncertain model parameters for case study 1.

Controlled variables	Tablet weight, pre-compression force, production rate, tensile strength
Manipulated variables	Dosing position, pre-compression thickness, main compression thickness, turret speed, silica concentration
Measured variables	Tablet weight, pre-compression force, main compression force, production rate
Uncertain model parameters	Bulk density, critical density, a: maximum degree of compression

For all three scenarios, three model parameters are tracked, i.e., ρ_b, ρ_c and Kawakita parameter a. The bulk density was monitored due to its influence on a number of other model parameters and states, while the relative critical density and Kawakita parameter a are both known to influence the compression forces, making them critical parameters that also need to be tracked in real time.

Simulation results of the process outputs for all three scenarios are presented in Figure 5a–c, respectively. The MHE-NMPC framework is utilized for all simulations and includes open-loop control from 0–100 s (indicated by red shading in all plots), state estimation using MHE and open loop control from 100–200 s (indicated by yellow shading in all plots), and implementation of the MHE-NMPC framework from 200 s until the end of the simulation (indicated by gray shading in all plots). Setpoint changes are introduced for the tablet weight from 210 mg to 240 mg at 400 s, for the pre-compression force from 0.3 kN to 0.6 kN at 600 s, for the production rate from 6.9 kg/h to 8 kg/h at 800 s, and for the tensile strength from 4.2 MPa to 6 MPa at 600 s, respectively.

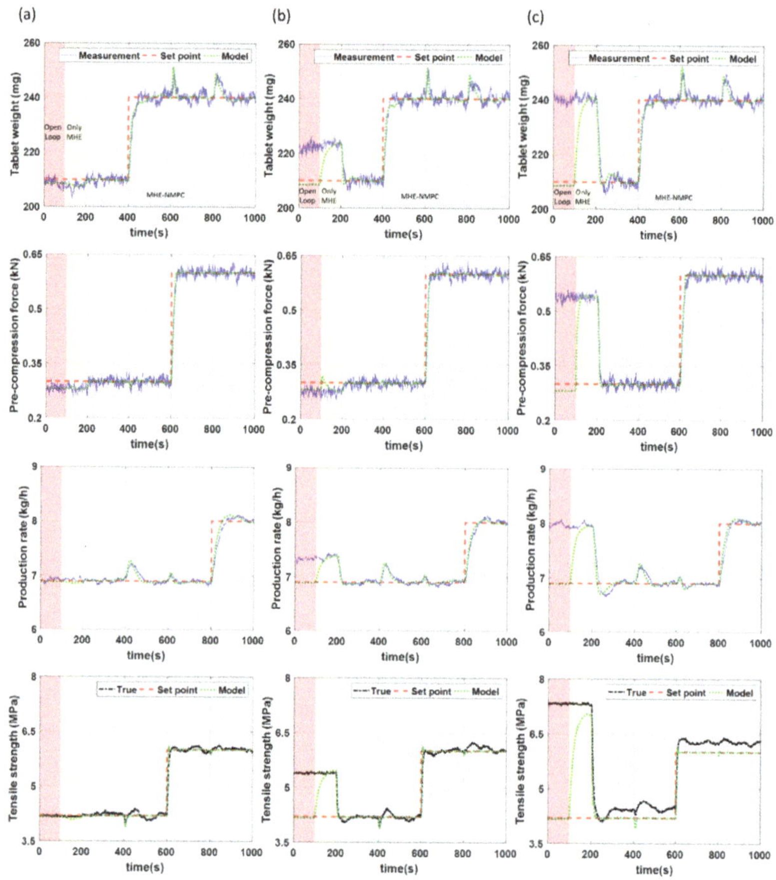

Figure 5. Controlled variables for case study 1 under scenarios with (**a**) no PMM, (**b**) mild PMM, and (**c**) high PMM.

For the scenario where there is no PMM, accurate setpoint tracking can be achieved for all states (see Figure 5a). This is also true for the case where there is mild PMM (see Figure 5b). This is primarily enabled due to the ability of the MHE-NMPC framework to accurately track variations in the uncertain model parameters in real time as illustrated in Figure 6b, allowing the impact of PMM to be effectively managed. In the case where there is high PMM (see Figure 5c), fairly accurate setpoint tracking can still be achieved despite there being mismatch in more parameters than those being tracked. This is also observed from the parameter estimation results in Figure 6c, although there is a slight offset in the model parameters being tracked to compensate for the variation in the additional model parameters that are not being tracked. The corresponding plots of the manipulated variables for all three scenarios are presented in Figure 7.

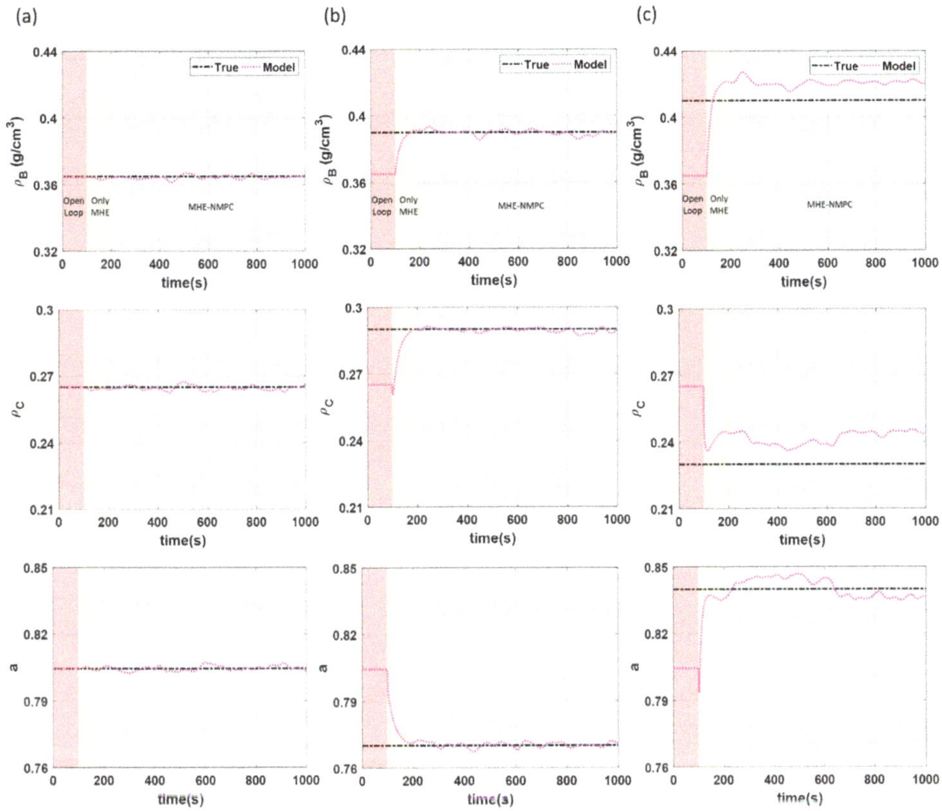

Figure 6. Parameter estimation of model parameters for case study 1 under scenario with (**a**) no PMM, (**b**) mild PMM, and (**c**) high PMM.

Control performance metrics are required in order to assess the performance of the development framework and accurately assess the impact of PMM for each scenario. Beyond the typical controller performance metric, i.e., integral of absolute error (IAE), some additional metrics are used to quantify the control performance in each scenario, which include the duration-to-reject (D2R) and magnitude to product (M2P) [65]. These metrics are able to quantify control performance in a manner that is more relevant for the continuous pharmaceutical industry. D2R is the duration of time that the process requires to smooth out the process disturbance or to reach a new set point for the CQA/CPP. M2P describes the maximum deviation in the CQA/CPP from the target setpoint. Larger values of all these performance metrics indicate worse or degraded control performance. The IAE values are calculated from t = 300 s to t = 1000 s. A summary of the control performance metrics is provided in Table 3.

When mild PMM exists, the control performance of the MHE-NMPC framework is comparable to the scenario without PMM, implying that the framework is able to sufficiently handle the PMM. However, when there is excessive PMM in multiple parameters as in the scenario with high PMM, significantly higher values of IAE and M2P are obtained, particularly in the tensile strength. This can be attributed to the fact that there was mismatch in more model parameters than those being tracked for this particular simulation, as can be noted from Table 1, resulting in an offset in the estimates of the model parameters to compensate for the added uncertainty (see Figure 6c). It should also be noted that the setpoints for the tablet weight and production rate track reasonably well, even in the presence of high PMM as demonstrated by the comparable IAE values for both states. However, since the tensile strength is an important CQA that is linked to the dissolution profile of the tablets, once high PMM begins to cause an offset in the tensile strength, it can serve as an indicator for the requirement to carry out model re-identification. This case study was able to demonstrate the strength of the MHE-NMPC framework in its ability to handle PMM in multiple model parameters.

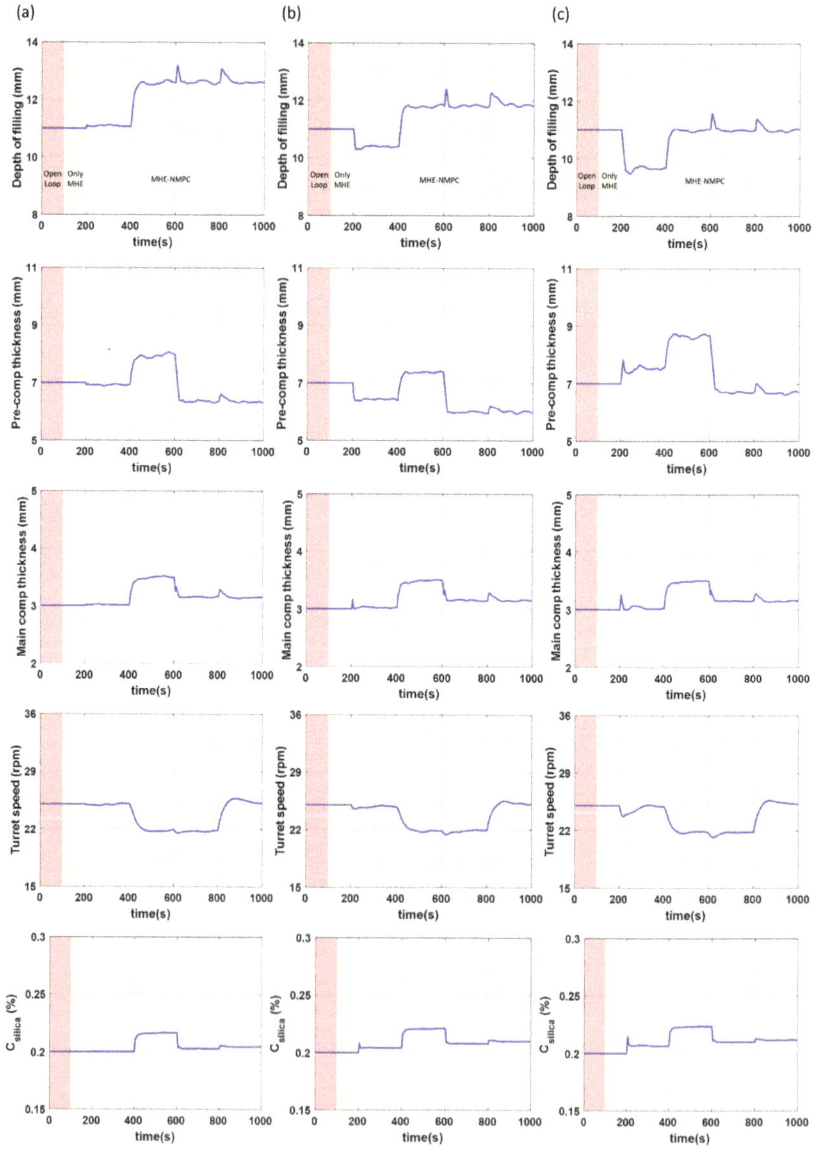

Figure 7. Manipulated variables for case study 1 under scenarios with (**a**) no PMM, (**b**) mild PMM, and (**c**) high PMM.

Table 3. Control performance of the MHE-NMPC framework for case study 1 under different levels of PMM.

Controlled Variables	Performance Metrics	No PMM	Mild PMM	High PMM
Tablet Weight	IAE	6.83	7.00	7.05
	M2P (%)	3.31	3.19	3.61
	D2R (s)	76	78	74
Tensile Strength	IAE	9.95	10.18	39.07
	M2P (%)	5.25	5.23	10.46
	D2R (s)	82	81	90
Production Rate	IAE	8.84	8.26	8.41

3.3. Case Study 2: Monitoring and Control of the Rotary Tablet Press in the Presence of Uncertainty in the Glidant Concentration

This scenario aims to explore a more practical concern with regards to the incorporation of the glidant feeder in the control scheme. In practice in some applications, it might not be possible to accurately control the concentration of the glidant in the direct compression process, due to the low concentrations used and, thus, the small feeding rates needed. Uncertainty in the glidant concentration is important, as it leads to variations in the bulk density upstream of the rotary tablet press. Therefore, accurate monitoring and control of these variations is required.

Since the glidant concentration can no longer be treated as a manipulated input for this scenario, the original system is revised to form a four-by-four MIMO system with the controlled variables consisting of the tablet weight, pre-compression force, production rate, and tensile strength. The manipulated variables consist of the dosing position, pre-compression thickness, main compression thickness, and turret speed. Once again it is assumed that only measurements for the tablet weight, pre-compression force, main compression force, and production rate are available every second. As tensile strength measurements are not available in real time, a soft sensor based on Equation (17) is once again utilized for this particular state. The concentration of silica is then assumed to be an uncertain parameter. A summary of the controlled variables, manipulated variables, measured variables, and uncertain model parameters is provided in Table 4. A summary of the model parameters was provided in Table 1, where the MHE-NMPC tuning parameters are those detailed in Section 2.4.

Table 4. Summary of variables and uncertain model parameters for case study 2.

Controlled variables	Tablet weight, pre-compression force, production rate, tensile strength
Manipulated variables	Dosing position, pre-compression thickness, main compression thickness, turret speed
Measured variables	Tablet weight, pre-compression force, main compression force, production rate
Uncertain model parameters	Silica concentration

For this particular case study, mismatch is introduced through positive and negative step changes in the silica concentration from its nominal value of 0.2% to 0.35% between 300 and 700 s and from 0.2% to 0.05% between 1100 and 1500 s, respectively. Step changes in either direction are introduced in order to examine if and how the direction of the mismatch affects the control performance. Simulation results of the process outputs for open loop control, closed loop control using only NMPC control, and closed loop estimation and control using the proposed MHE-NMPC framework are presented in Figure 8a–c, respectively. The simulation for the NMPC framework includes open loop control from 0 to 200 s, with a closed loop NMPC strategy implemented from 200 s until the end of the simulation. The simulation for the MHE-NMPC framework includes open loop control from 0 to 100 s, state estimation using MHE and open loop control from 100 to 200 s, and implementation of the MHE-NMPC framework from 200 s until the end of the simulation.

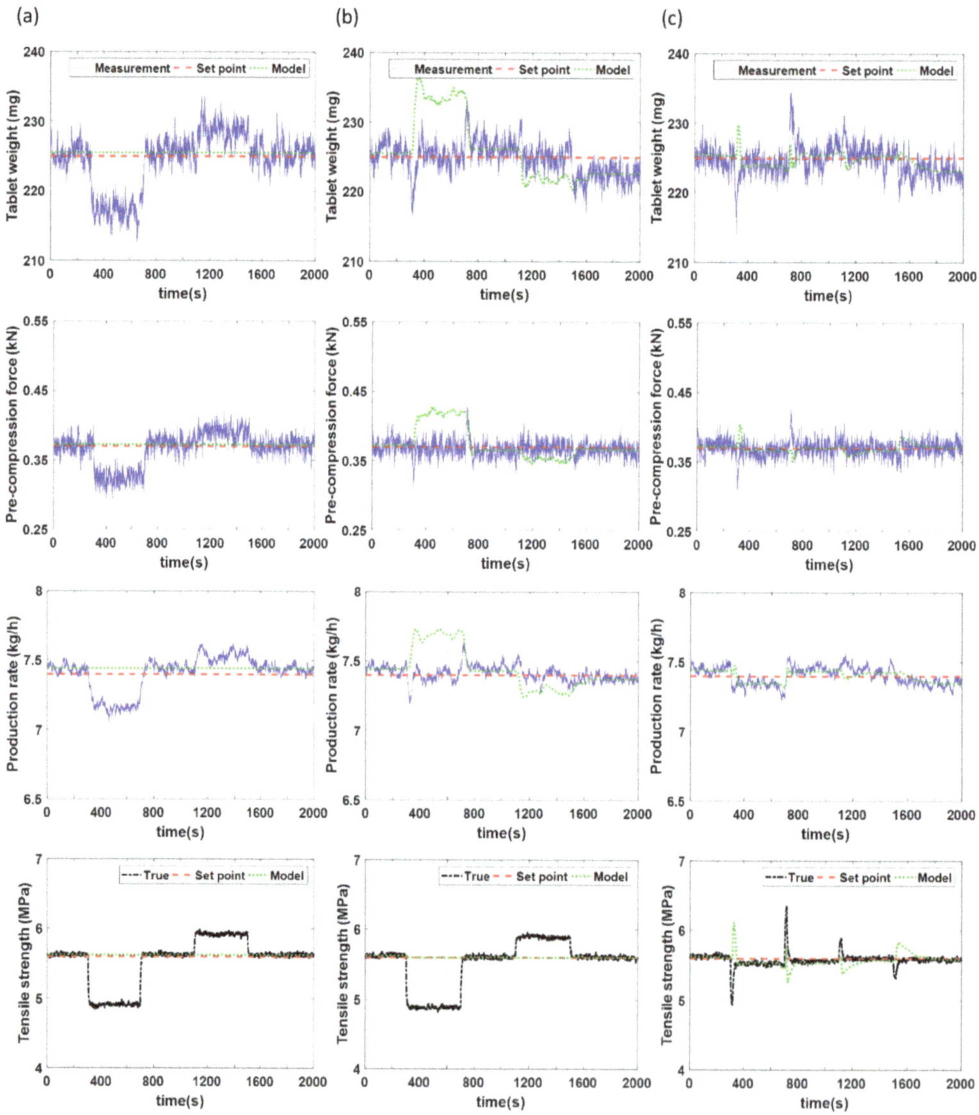

Figure 8. Controlled variables for case study 2 under scenarios using (**a**) open loop control, (**b**) only NMPC, and (**c**) MHE-NMPC.

Figure 8a demonstrates failure of the open loop control to maintain the controlled variables at their setpoints and, thus, the need to implement effective control strategies. Due to the disturbance terms utilized in the NMPC framework, offset free control is achieved for three of the four controlled variables when only NMPC is employed, as illustrated in Figure 8b. However, since real-time measurements are unavailable for the tensile strength, a soft sensor is employed. Since the soft sensor does not incorporate a disturbance term, an offset can be observed between the soft sensor values and the set point due to the mismatch caused by the assumption of fixed model parameters in the NMPC framework. In contrast, as illustrated in Figure 8c the MHE-NMPC framework is able to achieve offset free control for all four states, including the tensile strength. This is attributed to the ability of the MHE-NMPC framework to provide (i) real-time and accurate estimation of uncertain model parameters, enabled by MHE, and (ii) efficient control, enabled by NMPC.

Figure 9 shows results for the uncertain model parameter estimation, i.e., the estimation of silica concentration. These results demonstrate the ability of the MHE-NMPC framework to accurately track variations in silica concentration. It should be noted that the sluggish behavior at higher concentrations of silica is due to the nonlinear effect silica has on the process. Since in practice the true value of the concentration of silica might be unknown, this case study demonstrates the advantage of utilizing the MHE-NMPC framework to achieve accurate estimation of both measurable and unmeasurable states and parameters, such as the concentration of silica, while executing realistic and effective control strategies.

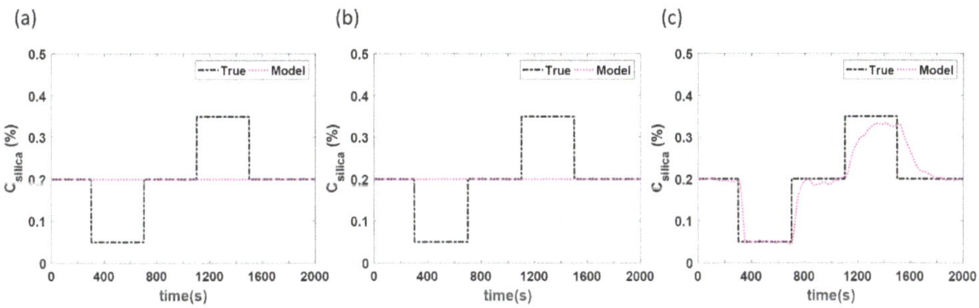

Figure 9. Parameter estimation for case study 2 under scenarios using (**a**) open loop control, (**b**) only NMPC, and (**c**) MHE-NMPC.

Figure 10 shows the manipulated variables for the different control strategies studied in this section. All changes in the manipulated variables presented in these results are realistic in nature and can be achieved during normal operation of the tablet press. It should be noted that the variations in the manipulated variables are larger in the case where only NMPC is utilized (see Figure 10b) compared to the case where the MHE-NMPC framework is utilized (see Figure 10c). This may be attributed to a less effective linear output disturbance model implemented in the NMPC framework, when compared to the MHE-NMPC framework that also carries out parameter updating incorporating more directly nonlinear effects of the PMM in the scheme.

This case study was able to demonstrate the ability of the MHE-NMPC framework to track and manage fluctuations in the glidant concentration, often caused by upstream disturbances, thereby providing an efficient solution to a common process upset faced when operating the rotary tablet press.

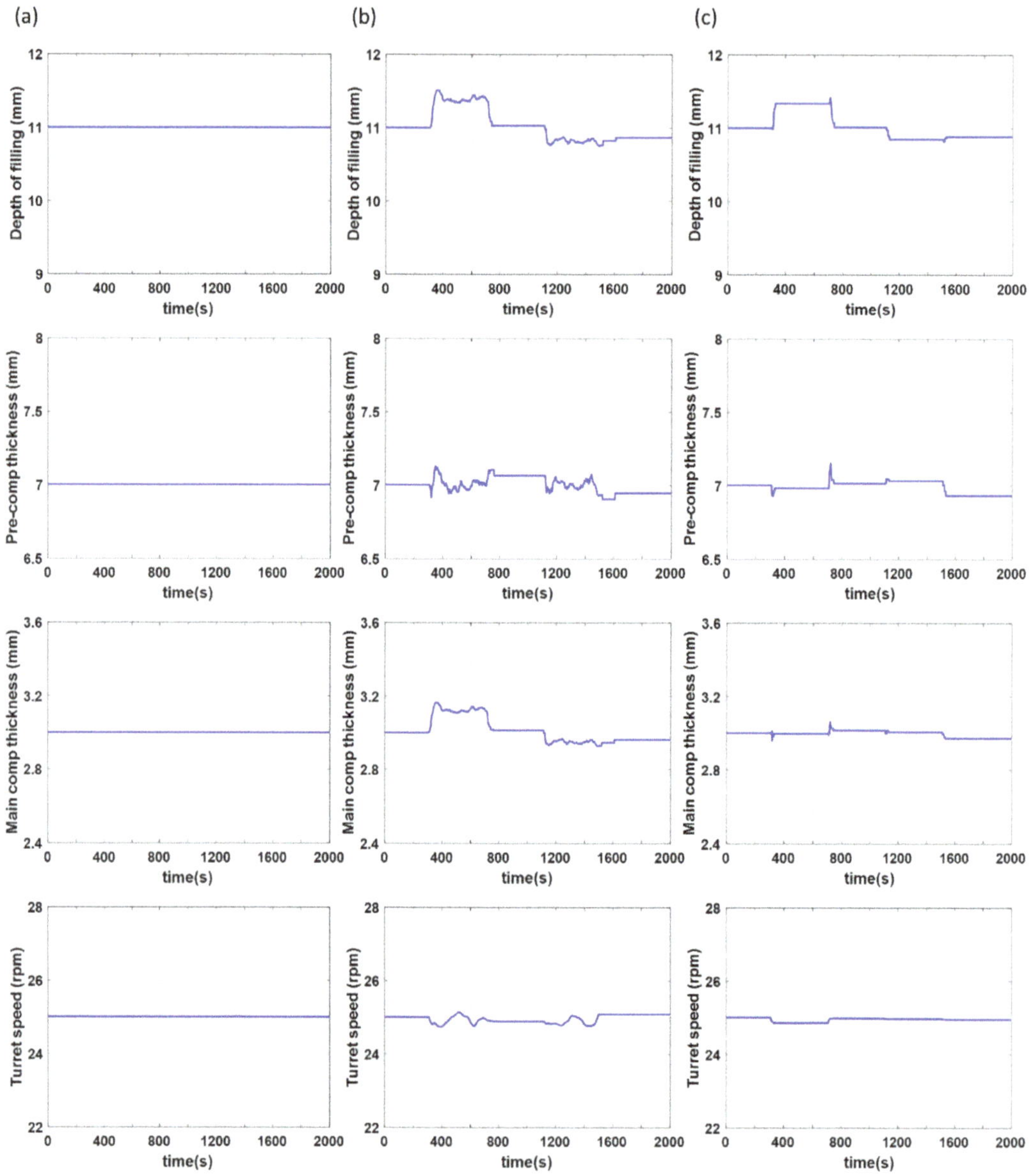

Figure 10. Manipulated variables for case study 2 under scenarios using (**a**) open loop control, (**b**) only NMPC, and (**c**) MHE-NMPC.

4. Conclusions

The continuous pharmaceutical manufacturing industry is in need of improved real-time process monitoring and management strategies that, specifically, are able to effectively identify and handle plant-model mismatch (PMM). In order to enable the quality-by-control (QbC) paradigm to move forward, this work developed and presented a moving horizon estimation-based nonlinear model predictive control (MHE-NMPC) framework to accomplish the dual requirement of accurate estimation and efficient control. Real-time implementation feasibility of the developed framework was also discussed, and the ability of the proposed framework to solve the optimization problem at each time step in a manner that enabled real-time implementation was highlighted. The practical applicability of the developed framework was corroborated through two realistic case studies that incorporated the effects of glidant to better control CQAs such as the tensile strength. Both examples demonstrated the ability of the framework to achieve reasonable control performance despite the presence of varying sources and degrees of PMM.

Future work includes further demonstration of the practical applicability of the proposed MHE-NMPC framework utilizing the rotary tablet press at Purdue University, including the application of the framework to the entire direct compression line. While a soft sensor was utilized in this work to track the tensile strength, in practice, due to low-frequency measurement availability from the SOTAX AT4 tablet tester, sensor fusion methods might be required to integrate and efficiently utilize all available plant data. This improved strategy would also require additional studies to determine how frequently to collect measurement data from the SOTAX AT4 tablet tester, due to the destructive nature of the testing method.

Author Contributions: Conceptualization, Y.-S.H. and M.Z.S.; methodology, Y.-S.H. and M.Z.S.; investigation: Y.-S.H. and S.B.; writing—original draft preparation, Y.-S.H., M.Z.S. and S.B.; writing—review and editing, M.G., Z.K.N., G.V.R.; funding acquisition, G.V.R., Z.K.N., M.G. All authors have read and agreed to the published version of the manuscript.

Funding: This research was funded by United States Food and Drug Administration grant number 1U01FD006487-01.

Institutional Review Board Statement: Not applicable.

Informed Consent Statement: Not applicable.

Data Availability Statement: Not applicable.

Conflicts of Interest: The authors declare no conflict of interest.

References

1. Hubert, C.; Lebrun, P.; Houari, S.; Ziemons, E.; Rozet, E.; Hubert, P. Improvement of a stability-indicating method by quality-by-design versus quality-by-testing: A case of a learning process. *J. Pharm. Biomed. Anal.* **2014**, *88*, 401–409. [CrossRef]
2. Burcham, C.L.; Florence, A.J.; Johnson, M.D. Continuous manufacturing in pharmaceutical process development and manufacturing. *Annu. Rev. Chem. Biomol. Eng.* **2018**, *9*, 253–281. [CrossRef] [PubMed]
3. Schaber, S.D.; Gerogiorgis, D.I.; Ramachandran, R.; Evans, J.M.B.; Barton, P.I.; Trout, B.L. Economic analysis of integrated continuous and batch pharmaceutical manufacturing: A case study. *Ind. Eng. Chem. Res.* **2011**, *50*, 10083–10092. [CrossRef]
4. Piñeiro, D.P.; Nikolakopoulou, A.; Jäschke, J.; Braatz, R.D. Self-optimizing control of a continuous-flow pharmaceutical manufacturing plant. *IFAC-PapersOnLine* **2020**, *53*, 11601–11606. [CrossRef]
5. Lee, S.L.; O'Connor, T.F.; Yang, X.; Cruz, C.N.; Chatterjee, S.; Madurawe, R.D.; Woodcock, J. Modernizing pharmaceutical manufacturing: From batch to continuous production. *J. Pharm. Innov.* **2015**, *10*, 191–199. [CrossRef]
6. Karttunen, A.P.; Poms, J.; Sacher, S.; Sparén, A.; Ruiz Samblás, C.; Fransson, M.; Martin De Juan, L.; Remmelgas, J.; Wikström, H.; Hsiao, W.K.; et al. Robustness of a continuous direct compression line against disturbances in feeding. *Int. J. Pharm.* **2020**, *574*, 118882. [CrossRef]
7. Gernaey, K.V.; Cervera-Padrell, A.E.; Woodley, J.M. Development of continuous pharmaceutical production processes supported by process systems engineering methods and tools. *Future Med. Chem.* **2012**, *4*, 1371–1374. [CrossRef] [PubMed]
8. Ierapetritou, M.; Muzzio, F.; Reklaitis, G. Perspectives on the continuous manufacturing of powder-based pharmaceutical processes. *AIChE J.* **2016**, *62*, 1846–1862. [CrossRef]
9. Kovács, B.; Péterfi, O.; Kovács-Deák, B.; Székely-Szentmiklósi, I.; Fülöp, I.; Bába, L.-I.; Boda, F. Quality-by-design in pharmaceutical development: From current perspectives to practical applications. *Acta Pharm.* **2021**, *71*, 497–526. [CrossRef]
10. Beg, S.; Hasnain, M.S.; Rahman, M.; Swain, S. Introduction to Quality by Design (QbD): Fundamentals, principles, and applications. In *Pharmaceutical Quality by Design*; Elsevier: Amsterdam, The Netherlands, 2019; pp. 1–17.
11. Su, Q.; Ganesh, S.; Moreno, M.; Bommireddy, Y.; Gonzalez, M.; Reklaitis, G.V.; Nagy, Z.K. A perspective on Quality-by-Control (QbC) in pharmaceutical continuous manufacturing. *Comput. Chem. Eng.* **2019**, *125*, 216–231. [CrossRef]
12. Adolfsson, Å.; Nyström, C. Tablet strength, porosity, elasticity and solid state structure of tablets compressed at high loads. *Int. J. Pharm.* **1996**, *132*, 95–106. [CrossRef]

13. Liu, L.; Zhang, K.; Sun, Z.; Dong, Q.; Li, L.; Zang, H. A new perspective in understanding the dissolution behavior of nifedipine controlled release tablets by NIR spectroscopy with aquaphotomics. *J. Mol. Struct.* **2021**, *1230*, 129872. [CrossRef]
14. de Meira, R.Z.C.; Maciel, A.B.; Murakami, F.S.; de Oliveira, P.R.; Bernardi, L.S. In vitro dissolution profile of dapagliflozin: Development, method validation, and analysis of commercial tablets. *Int. J. Anal. Chem.* **2017**, *2017*, 1–6. [CrossRef]
15. Tsunematsu, H.; Hifumi, H.; Kitamura, R.; Hirai, D.; Takeuchi, M.; Ohara, M.; Itai, S.; Iwao, Y. Analysis of available surface area can predict the long-term dissolution profile of tablets using short-term stability studies. *Int. J. Pharm.* **2020**, *586*, 119504. [CrossRef] [PubMed]
16. Galata, D.L.; Farkas, A.; Könyves, Z.; Mészáros, L.A.; Szabó, E.; Csontos, I.; Pálos, A.; Marosi, G.; Nagy, Z.K.; Nagy, B. Fast, spectroscopy-based prediction of in vitro dissolution profile of extended release tablets using artificial neural networks. *Pharmacy* **2019**, *11*, 400. [CrossRef] [PubMed]
17. Razavi, S.M.; Gonzalez, M.; Cuitiño, A.M. Quantification of lubrication and particle size distribution effects on tensile strength and stiffness of tablets. *Powder Technol.* **2018**, *336*, 360–374. [CrossRef]
18. Olowosulu, A.K.; Apeji, Y.E. Quantifying the effect of glidant on the compaction and tableting properties of paracetamol granules. *J. Res. Pharm.* **2020**, *24*, 1–12. [CrossRef]
19. Blackshields, C.A.; Crean, A.M. Continuous powder feeding for pharmaceutical solid dosage form manufacture: A short review. *Pharm. Dev. Technol.* **2018**, *23*, 554–560. [CrossRef]
20. Engisch, W.E.; Muzzio, F.J. Feedrate deviations caused by hopper refill of loss-in-weight feeders. *Powder Technol.* **2015**, *283*, 389–400. [CrossRef]
21. Hanson, J. Control of a system of loss-in-weight feeders for drug product continuous manufacturing. *Powder Technol.* **2018**, *331*, 236–243. [CrossRef]
22. Harris, T.J. Assessment of control loop performance. *Can. J. Chem. Eng.* **1989**, *67*, 856–861. [CrossRef]
23. Chen, M.; Xie, L.; Su, H. Impact of model-plant mismatch to minimum variance benchmark in control performance assessment. In Proceedings of the 2020 39th Chinese Control Conference (CCC)—IEEE, Shenyang, China, 27–30 July 2020; pp. 2252–2257.
24. Wang, S.; Simkoff, J.M.; Baldea, M.; Chiang, L.H.; Castillo, I.; Bindlish, R.; Stanley, D.B. Autocovariance-based plant-model mismatch estimation for linear model predictive control. *Syst. Control Lett.* **2017**, *104*, 5–14. [CrossRef]
25. Wang, S.; Simkoff, J.M.; Baldea, M.; Chiang, L.H.; Castillo, I.; Bindlish, R.; Stanley, D.B. Autocovariance-based MPC model mismatch estimation for systems with measurable disturbances. *J. Process Control* **2017**, *55*, 42–54. [CrossRef]
26. Xu, X.; Simkoff, J.M.; Baldea, M.; Chiang, L.H.; Castillo, I.; Bindlish, R.; Ashcraft, B. Data-driven plant-model mismatch quantification for MIMO MPC systems with feedforward control path. In Proceedings of the 2020 American Control Conference (ACC), Denver, CO, USA, 1–3 July 2020; pp. 2760–2765.
27. Badwe, A.S.; Gudi, R.D.; Patwardhan, R.S.; Shah, S.L.; Patwardhan, S.C. Detection of model-plant mismatch in MPC applications. *J. Process. Control.* **2009**, *19*, 1305–1313. [CrossRef]
28. Chen, Y.; Ierapetritou, M. A framework of hybrid model development with identification of plant-model mismatch. *AIChE J.* **2020**, *66*. [CrossRef]
29. Destro, F.; García Muñoz, S.; Bezzo, F.; Barolo, M. Powder composition monitoring in continuous pharmaceutical solid-dosage form manufacturing using state estimation—proof of concept. *Int. J. Pharm.* **2021**, *605*, 120808. [CrossRef] [PubMed]
30. Liu, J.; Su, Q.; Moreno, M.; Laird, C.; Nagy, Z.; Reklaitis, G. Robust state estimation of Feeding—blending systems in continuous pharmaceutical manufacturing. *Chem. Eng. Res. Des.* **2018**, *134*, 140–153. [CrossRef]
31. Singh, R.; Ierapetritou, M.; Ramachandran, R. System-wide hybrid MPC–PID control of a continuous pharmaceutical tablet manufacturing process via direct compaction. *Eur. J. Pharm. Biopharm.* **2013**, *85*, 1164–1182. [CrossRef]
32. Mesbah, A.; Paulson, J.A.; Lakerveld, R.; Braatz, R.D. Model predictive control of an integrated continuous pharmaceutical manufacturing pilot plant. *Org. Process Res. Dev.* **2017**, *21*, 844–854. [CrossRef]
33. Singh, R.; Sahay, A.; Karry, K.M.; Muzzio, F.; Ierapetritou, M.; Ramachandran, R. Implementation of an advanced hybrid MPC–PID control system using PAT tools into a direct compaction continuous pharmaceutical tablet manufacturing pilot plant. *Int. J. Pharm.* **2014**, *473*, 38–54. [CrossRef] [PubMed]
34. Jelsch, M.; Roggo, Y.; Kleinebudde, P.; Krumme, M. Model predictive control in pharmaceutical continuous manufacturing: A review from a user's perspective. *Eur. J. Pharm. Biopharm.* **2021**, *159*, 137–142. [CrossRef]
35. Jacob, N.C.; Dhib, R. Unscented kalman filter based nonlinear model predictive control of a LDPE autoclave reactor. *J. Process. Control* **2011**, *21*, 1332–1344. [CrossRef]
36. Magni, L.; Opizzi, S.; Scattolini, R. Tracking of non-square nonlinear systems via model predictive control. In Proceedings of the 2001 European Control Conference (ECC), Porto, Portugal, 4–7 September 2001; pp. 951–956.
37. Magni, L.; Scattolini, R. Tracking of non-Square nonlinear continuous time systems with piecewise constant model predictive control. *J. Process Control* **2007**, *17*, 631–640. [CrossRef]
38. Zhang, L.; Hao, Y.; Han, H.; Tan, T. PID Control of Non-Square Systems and Its Application in the Fuel Cell Voltage. In Proceedings of the 2012 24th Chinese Control and Decision Conference (CCDC), Taiyuan, China, 23–25 May 2012; pp. 4091–4094.
39. Billups, M.; Singh, R. Systematic framework for implementation of material traceability into continuous pharmaceutical tablet manufacturing process. *J. Pharm. Innov.* **2020**, *15*, 51–65. [CrossRef]
40. Bhaskar, A.; Singh, R. Residence Time Distribution (RTD)-based control system for continuous pharmaceutical manufacturing process. *J. Pharm. Innov.* **2019**, *14*, 316–331. [CrossRef]

41. De Assis, A.J.; Maciel Filho, R. Soft sensors development for on-line bioreactor state estimation. *Comput. Chem. Eng.* **2000**, *24*, 1099–1103. [CrossRef]
42. Papachristos, C.; Dang, T.; Khattak, S.; Mascarich, F.; Khedekar, N.; Alexis, K. Modeling, control, state estimation and path planning methods for autonomous multirotor aerial robots. *Found. Trends Robot.* **2018**, *7*, 180–250. [CrossRef]
43. Mansouri, M.M.; Nounou, H.N.; Nounou, M.N. Nonlinear control and estimation in induction machine using state estimation techniques. *Syst. Sci. Control Eng.* **2014**, *2*, 642–654. [CrossRef]
44. Shenoy, A.V.; Prakash, J.; Prasad, V.; Shah, S.L.; McAuley, K.B. Practical issues in state estimation using particle filters: Case studies with polymer reactors. *J. Process Control* **2013**, *23*, 120–131. [CrossRef]
45. Fang, H.; Tian, N.; Wang, Y.; Zhou, M.; Haile, M.A. Nonlinear bayesian estimation: From kalman filtering to a broader horizon. *IEEE/CAA J. Autom. Sin.* **2018**, *5*, 401–417. [CrossRef]
46. Rawlings, J.B.; Bakshi, B.R. Particle filtering and moving horizon estimation. *Comput. Chem. Eng.* **2006**, *30*, 1529–1541. [CrossRef]
47. Mohd Ali, J.; Ha Hoang, N.; Hussain, M.A.; Dochain, D. Review and classification of recent observers applied in chemical process systems. *Comput. Chem. Eng.* **2015**, *76*, 27–41. [CrossRef]
48. Xia, C.; Pan, Z.; Zhang, S.; Polden, J.; Wang, L.; Li, H.; Xu, Y.; Chen, S. Model predictive control of layer width in wire arc additive manufacturing. *J. Manuf. Process.* **2020**, *58*, 179–186. [CrossRef]
49. Cavanini, L.; Ippoliti, G.; Camacho, E.F. Model predictive control for a linear parameter varying model of an UAV. *J. Intell. Robot. Syst.* **2021**, *101*, 1–18. [CrossRef]
50. Hewing, L.; Wabersich, K.P.; Menner, M.; Zeilinger, M.N. Learning-based model predictive control: Toward safe learning in control. *Annu. Rev. Control Robot. Auton. Syst.* **2020**, *3*, 269–296. [CrossRef]
51. Drgoňa, J.; Arroyo, J.; Cupeiro Figueroa, I.; Blum, D.; Arendt, K.; Kim, D.; Ollé, E.P.; Oravec, J.; Wetter, M.; Vrabie, D.L.; et al. All you need to know about model predictive control for buildings. *Annu. Rev. Control* **2020**, *50*, 190–232. [CrossRef]
52. Yang, S.; Wan, M.P.; Ng, B.F.; Dubey, S.; Henze, G.P.; Chen, W.; Baskaran, K. Experimental study of model predictive control for an air-conditioning system with dedicated outdoor air system. *Appl. Energy* **2020**, *257*, 113920. [CrossRef]
53. Zavala, V.M.; Biegler, L.T. The advanced-step NMPC controller: Optimality, stability and robustness. *Automatica* **2009**, *45*, 86–93. [CrossRef]
54. Salem, F.; Mosaad, M.I. A comparison between MPC and optimal PID controllers: Case studies. In Proceedings of the Michael Faraday IET International Summit 2015, Kolkata, India, 12–14 September 2015.
55. Martínez, L.; Peinado, A.; Liesum, L.; Betz, G. Use of near-infrared spectroscopy to quantify drug content on a continuous blending process: Influence of mass flow and rotation speed variations. *Eur. J. Pharm. Biopharm.* **2013**, *84*, 606–615. [CrossRef] [PubMed]
56. Vanarase, A.U.; Alcalà, M.; Jerez Rozo, J.I.; Muzzio, F.J.; Romañach, R.J. Real-time monitoring of drug concentration in a continuous powder mixing process using NIR spectroscopy. *Chem. Eng. Sci.* **2010**, *65*, 5728–5733. [CrossRef]
57. Kirchengast, M.; Celikovic, S.; Rehrl, J.; Sacher, S.; Kruisz, J.; Khinast, J.; Horn, M. Ensuring tablet quality via model-based control of a continuous direct compaction process. *Int. J. Pharm.* **2019**, *567*, 118457. [CrossRef]
58. López-Negrete, R.; Biegler, L.T. A moving horizon estimator for processes with multi-rate measurements: A nonlinear programming sensitivity approach. *J. Process. Control.* **2012**, *22*, 677–688. [CrossRef]
59. Rao, C.V.; Rawlings, J.B.; Mayne, D.Q. Constrained state estimation for nonlinear discrete-time systems: Stability and moving horizon approximations. *IEEE Trans. Autom. Control* **2003**, *48*, 246–258. [CrossRef]
60. Huang, R.; Biegler, L.T.; Patwardhan, S.C. Fast offset-free nonlinear model predictive control based on moving horizon estimation. *Ind. Eng. Chem. Res.* **2010**, *49*, 7882–7890. [CrossRef]
61. Su, Q.; Bommireddy, Y.; Shah, Y.; Ganesh, S.; Moreno, M.; Liu, J.; Gonzalez, M.; Yazdanpanah, N.; O'Connor, T.; Reklaitis, G.V.; et al. Data reconciliation in the Quality-by-Design (QbD) implementation of pharmaceutical continuous tablet manufacturing. *Int. J. Pharm.* **2019**, *563*, 259–272. [CrossRef]
62. Medina-González, S.; Huang, Y.-S.; Bachawala, S.; Bommireddy, Y.; Gonzalez, M.; Reklaitis, G.V.; Nagy, Z.K. A NMPC strategy applied to a continuous direct compaction tablet manufacturing. In Proceedings of the AIChE 2020 Annual Meeting, Virtual, 16–20 November 2020; p. 716b.
63. Kawakita, K.; Lüdde, K.-H. Some considerations on powder compression equations. *Powder Technol.* **1971**, *4*, 61–68. [CrossRef]
64. Gonzalez, M. Generalized loading-unloading contact laws for elasto-plastic spheres with bonding strength. *J. Mech. Phys. Solids* **2019**, *122*, 633–656. [CrossRef]
65. Su, Q.; Moreno, M.; Giridhar, A.; Reklaitis, G.V.; Nagy, Z.K. A systematic framework for process control design and risk analysis in continuous pharmaceutical solid-dosage manufacturing. *J. Pharm. Innov.* **2017**, *12*, 327–346. [CrossRef]

processes

MDPI

Article

Research on Measurement and Application of China's Regional Logistics Development Level under Low Carbon Environment

Zixue Guo [1,2,*], Yu Tian [2,*], Xinmei Guo [1] and Zefang He [3]

1 Research Center for Capital Commercial Industry, Beijing Technology and Business University, Beijing 100048, China; guoxinmei@126.com
2 School of Management, Hebei University, Baoding 071002, China
3 School of Information, Beijing Wuzi University, Beijing 101149, China; hezefang2013@163.com
* Correspondence: guozx12345@163.com (Z.G.); tian_yu0216@163.com (Y.T.); Tel.: +86-134-6342-1681 (Z.G.); +86-172-6292-0305 (Y.T.)

Abstract: To solve the problem of fuzziness and randomness in regional logistics decarbonization evaluation and accurately assess regional logistics decarbonization development, an evaluation model of regional logistics decarbonization development is established. First, the evaluation index of regional logistics decarbonization development is constructed from three dimensions: low-carbon logistics environment support, low-carbon logistics strength and low-carbon logistics potential. Second, the evaluation indexes are used as cloud model variables, and the cloud numerical characteristic values and cloud affiliation degrees are determined according to the cloud model theory. The entropy weight method is used to determine the index weights, and the comprehensive determination degree of the research object affiliated to the logistics decarbonization level is calculated comprehensively. Finally, Beijing-Tianjin-Hebei region is used as an example for empirical evidence, analyzing the development logistics decarbonization and its and temporal variability in Beijing, Tianjin and Hebei provinces and cities. The results of the study show that the development logistics decarbonization in Beijing, Tianjin and Hebei Province has been improved to different degrees during 2013–2019, but the development is uneven. Developing to 2019, the three provinces and cities of Beijing, Tianjin and Hebei still have significant differences in terms of economic environment, logistics industry scale, logistics industry inputs and outputs, and technical support.

Keywords: regional logistics; low-carbon economy; cloud model; comprehensive evaluation; Beijing-Tianjin-Hebei region

Citation: Guo, Z.; Tian, Y.; Guo, X.; He, Z. Research on Measurement and Application of China's Regional Logistics Development Level under Low Carbon Environment. *Processes* **2021**, *9*, 2273. https://doi.org/10.3390/pr9122273

Academic Editors: Luis Puigjaner, Antonio Espuña Camarasa, Edrisi Muñoz Mata and Elisabet Capón García

Received: 17 November 2021
Accepted: 10 December 2021
Published: 17 December 2021

1. Introduction

China has entered a new stage of high-quality development; the people's demand for ecological environment is getting higher and higher, and the importance and urgency of promoting green development has become more and more prominent. In 2020, General Secretary Xi Jinping solemnly declared to the world at the United Nations General Assembly that China' s carbon dioxide emissions will peak by 2030 and strive to achieve carbon neutrality by 2060. As a high-end service industry, logistics has the characteristics of high energy consumption and high emission. The development path of logistics must follow low-carbon development, focusing on green logistics, low-carbon logistics and intelligent informatization. With the rise of the low-carbon revolution and the official advocation of green environment at the Copenhagen environment conference, low-carbon logistics has become the focus of academic research at home and abroad. The research of low-carbon logistics focuses on four aspects: carbon emission accounting, carbon emission driver identification, low-carbon logistics capability measurement and low-carbon logistics development strategy. In carbon emission accounting of logistics process, Butner K, Dada A, Piecyk M I adopt the method of carbon emission measurement based on whole life cycle and design the analytical of carbon emission measurement including structural factors and commercial factors [1–3]. Wang LP and Liu Y calculated the carbon emission from energy data of Chinese provinces from 1997–2004 and 2007–2013 [4,5]. Concerning identifying drivers of carbon emissions in the logistics industry, Timilsina and others studied on the growth of carbon emissions in the transport in selected Asian countries from 1980 to 2005 [6]. Lei Yang takes Shenzhen port as an example and measures the carbon emission in the port comprehensive logistics system [7]. Yang YW, Li FG, Men D et.al explore the driving causes of carbon emission growth by using LMDI model decomposition analysis [8–10]. In the low carbon logistics capability, Jessica Wehner takes an interactive approach to capacity utilization to contribute to sustainable freight transport and logistics [11]. The Chinese scholars mainly focus on the fuzzy comprehensive evaluation method, entropy weight TOPSIS model, DEA evaluation model, and Malmquist model static measurement methods to evaluate [12–15]. In the development strategy of low-carbon logistics, relevant scholars analyze the current situation and problems of low-carbon logistics development from different perspectives and put forward suggestions to promote the development of low-carbon logistics [16–20]. At present, scholars have conducted fewer studies related to the low-carbon development of regional logistics. Ma YY used data envelopment analysis to study the total factor productivity of China's logistics industry under low-carbon constraints [21]. Xie F and Gao FF analyzed the low carbonization of China's logistics industry and related industries by constructing an index system for the coordinated development of logistics industry and low carbon economy and using a coordinated development model [22,23]. Yu Q analyzed the logistics efficiency and its influencing factors in 30 provinces and cities, as well as the eastern, central and western regions of China based on the DEA-Tobit two-stage method [24]. Song Lina used a combined model of principal component analysis and data envelopment analysis to evaluate the regional low-carbon logistics performance of provinces along the Silk Road Economic Belt in China [25]. Wang X, taking Anhui Province as an example, explored the mechanism of the low-carbon development of regional logistics using the theoretical analysis framework of "development dynamics-measurement criteria-acting subject" [26].

To solve the fuzzy and stochastic problems in the process of low carbonization evaluation of regional logistics, the fuzzy and stochastic properties were converted into a definite value by the cloud generator, which broke the limitation of qualitative and quantitative research and made the evaluation more hierarchical [27].

2. Theoretical Basis

2.1. Entropy Weighting Method

Entropy is a measure of the disorder degree of a system. According to defined entropy, we can use the size of entropy to judge the discreteness degree of an index. The smaller the entropy value is, the greater the influence of the index on the comprehensive evaluation (i.e., the weight). Therefore, information entropy is a tool that can be used to objectively empower multiple signs to provide the basis for a comprehensive evaluation:

1. Standardized processing of data: assume that m evaluation objects, n evaluation signals, get the original evaluation, $X = (X_{ij})_{m\times n}$, make

$$U_{ij} = \begin{cases} \frac{X_{ij}-\min\limits_{i}X_{ij}}{\max\limits_{i}X_{ij}-\min\limits_{i}X_{ij}}, Negative\ indicators \\ \frac{\max\limits_{i}X_{ij}-X_{ij}}{\max\limits_{i}X_{ij}-\min\limits_{i}X_{ij}}, Positive\ Indicators \end{cases} \quad (1)$$

   where X_{ij} denotes the j indicator of the i evaluator in a given locality, U_{ij} it is the standardized data.

2. Calculation of weights for each indicator:

$$P_{ij} = \frac{U_{ij}}{\sum\limits_{i=1}^{m} U_{ij}} \quad (2)$$

3. Calculation of entropy for each indicator:

$$e_j = -\frac{1}{\ln m}\sum_{i=1}^{m} P_{ij} \ln P_{ij} \quad (3)$$

4. Determination of weights for each indicator:

$$w_j = \frac{1 - e_j}{n - \sum_{j=1}^{n} e_j} \quad (4)$$

2.2. Cloud Models

2.2.1. The Cloud Models

Li DY and others are the basis of cloud computing, reasoning, and control, and it is a model for the transformation of uncertainty between qualitative concepts and quantitative descriptions [27–29]. It is widely used in risk assessment, data mining, and performance evaluation and so on [30–33]. Let O be a quantitative set represented by a numerical value. I is a qualitative concept in O space. If the quantitative value $x \in O$ and x is a stochastic implementation in the qualitative concept I, the determinacy of x to I: $\mu(x) \in [0,1]$, It is a stochastic number with a tendency to stability:

$$\mu : O \to [0,1], \forall x \in O, x \to \mu(x)$$

Then the distribution of x in the set is called the cloud model, with each x being a cloud drop.

2.2.2. Numerical Characteristics of Clouds

Cloud models represent the primitive-language values in natural language, and the three numeric features of cloud models—*Ex* (expectation), *En* (entropy), and *He* (supers entropy)—represent the numerical characteristics of language values, thus achieving the goal of integrating the fuzziness and randomness of objects studied. Among them, *Ex* is the expectation of cloud droplet distribution in the domain. It is the central value of cloud droplet in a given set space distribution. *En* indicates the uncertainty measure of qualitative concept, which reflects the dispersion degree of cloud droplet, which is determined by the ambiguity and randomness of qualitative concept. *He* is a measure of the fuzziness of entropy, the size of which indirectly reflects the thickness of cloud droplets and the fuzziness and randomness of entropy [34–36].

2.2.3. Cloud Generator

The mutual transformation between qualitative concept and quantitative data in cloud model needs to be realized by cloud generator. Typically, a cloud generator includes a forward cloud generator, a reverse cloud generator and a conditional cloud generator.

Forward Cloud Generator: A mapping from a qualitative concept to a quantitative value, a process by which cloud droplets are generated from the numerical eigenvalues of a cloud model, as shown in Figure 1.

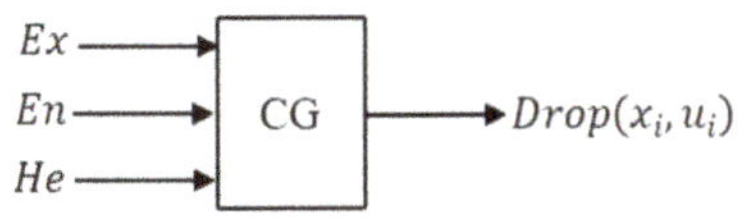

Figure 1. Positive cloud generator.

In Figure 1, *CG* means the forward cloud generator, x_i is the cloud droplet, and μ_i is its affiliation degree.

Reverse Cloud Generator: Mapping from quantitative values to directed ideas, that is, converting exact data into the suitable qualitative language (Ex, En, He), as shown in Figure 2.

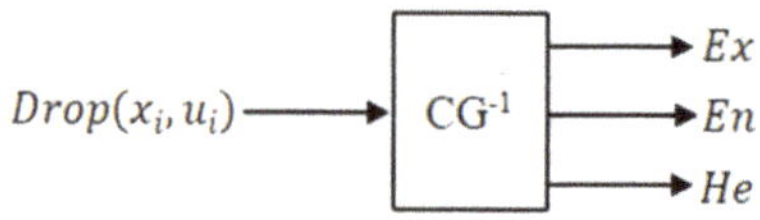

Figure 2. Inverse cloud generator.

In Figure 2, CG^{-1} notes a reverse cloud generator, x_i is the cloud droplet, and μ_i is its affiliation degree.

X Conditional Cloud Generator: In the numerical domain space of a given set, the three digital eigenvalues of the known cloud, Ex,En,He, and contain a specified condition $x = x_0$, this is called Conditional Cloud Generator. As shown in Figure 3.

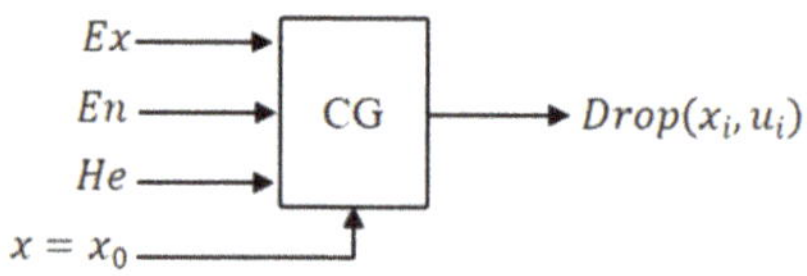

Figure 3. Conditional cloud generator.

2.3. Carbon Emission Measurement

At present, there is no uniform standard for carbon emission measurement in the world. This paper adopts the more extensive estimation method of IPCC, also known as the IPCC inventory coefficient method. This method is based on the final energy consumption, and considering the waste gas emitted during the logistics process includes not only carbon dioxide, but also carbon monoxide, hydrocarbons, etc. In this paper, the carbon emission of the logistics industry is estimated by energy consumption. This is done by multiplying the various energy consumption of the logistics industry by their respective standard coal coefficient and then by their respective carbon emission factors to arrive at the total carbon emissions for a given year in the region:

$$C = \sum_i C_i = \sum_i \delta_i \theta_i E_i \tag{5}$$

Of which: C_i means carbon emissions from type i energy sources, E_i denotes consumption of type i energy sources, θ_i marks coefficient of fractional standard coal for type i energy sources, δ_i stands for carbon emission factors for type i energy sources, and $\theta_i E_i$ denotes amount of fractional standard coal for type i energy sources.

3. Regional Logistics Decarbonization Development Evaluation Model Construction

3.1. Evaluation Index System for Low-Carbon Development of Regional Logistics

Low-carbonization of regional logistics means building a regional logistics system which is based on low-carbon economy and green logistics and supports the concept of "sustainable development" and"carbon emissions". It meets the regional economic and political development and has a supporting system of logistics information and organization and operation, while possessing the characteristics of green, balanced and efficient. Related scholars have different focuses and starting points for the research on the level of regional logistics decarbonization, such as Lai, Ma Shihua et al. from the logistics system [37,38], and Daugherty and Wang Ming from the level of enterprises [39,40] to define the low carbon logistics capacity. This paper argues that the level of logistics at the regional level is essentially a kind of competitiveness, which should not only focus on the current existing strengths, but also on the potential for future development, and should pay attention to both its own capacity building and the influence of the growth environment. According to the *China Logistics Development Report 2019–2020* and the Low Carbon Logistics Development Guidelines, the low-carbon logistics development focuses on the following subjects: railway freight transport, low-carbon automobile transport, logistics rationalization, common distribution, recycling of waste facilities, green packaging, industrial waste disposal and information e-commerce. According to the quantitative nature of the action guide and the availability of data, following the principles of systematism, scientificity and application of the selection of indicators, this paper summarizes three first-level indicators to evaluate the level of regional logistics decarbonization. Low-carbon logistics environmental support is an external factor that affects the level of regional low-carbon logistics capacity, which is influenced by the economic and policy environment. Low-carbon logistics environmental support is to evaluate the existing competitiveness of regional logistics low-carbon development, mainly in terms of infrastructure construction, logistics industry scale and logistics industry efficiency. The potential of low-carbon logistics is the sustainable driving force for the decarbonization of regional logistics, which includes the potential of regional logistics in terms of input, output and demand, and is mainly measured by the growth rate indicator. The specific indicators are shown in Table 1.

Table 1. Regional Logistics Decarbonization Evaluation Index System.

Target Layer	First Level Indicator Layer	Secondary Indicator Layer	Three-Level Indicator Layer
Evaluation of regional logistics decarbonization development X	Low carbon logistics environment support force X_1	Economic environment	Gross regional product per capita $X_{1,1}$ Fiscal revenue per capita $X_{1,2}$ Total retail sales of social goods per capita$X_{1,3}$
		Policy environment	The part of local financial expenses on environmental protection to total expenses $X_{1,4}$ Logistics industry as a proportion of fixed investment $X_{1,5}$
	Low carbon logistics strength X_2	Logistics infrastructure	Road Density $X_{2,1}$ Rail Density $X_{2,2}$
		Logistics industry scale	Logistics operations per head $X_{2,3}$ E-commerce sales per capita $X_{2,4}$ Increase in logistics per capita$X_{2,5}$ The proportion of logistics employees in the workforce $X_{2,6}$ Cargo turnover per capita $X_{2,7}$
		Logistics industry efficiency	Contribution of logistics industry to GDP $X_{2,8}$ Value added of logistics industry per logistics employee $X_{2,9}$ Carbon emissions per unit of added value in the logistics industry $X_{2,10}$
	Low carbon logistics potential X_3	Logistics industry input	Growth rate of new fixed asset investment in logistics industry $X_{3,1}$ Logistics workforce growth rate $X_{3,2}$
		Logistics output	Value added growth rate of logistics industry $X_{3,3}$
		Logistics industry demand	GDP per capita growth rate$X_{3,4}$
		Technical support	Technology Market Turnover Growth Rate$X_{3,5}$ R&D expenditure growth rate $X_{3,6}$

3.2. Construction of Evaluation Model

3.2.1. Defining the Object and Domain of Cloud Model Evaluation

The evaluation object is established as the regional logistics decarbonization evaluation, showed by X. According to the regional logistics index evaluation index system constructed in Table 1, the factor domain of the criterion layer is determined as $X = \{X_1, X_2, X_3\}$, and the index layer domains are $X_1 = \{X_{1,1}, X_{1,2}, \cdots, X_{1,5}\}$, $X_2 = \{X_{2,1}, X_{2,2}, \cdots, X_{2,10}\}$ and $X_3 = \{X_{3,1}, X_{3,2}, \cdots, X_{3,7}\}$.

3.2.2. Settle the Evaluation Level of Each Indicator

For each index evaluation level domain A, to more clearly represent the average level of the research object and the degree of distinction, the general number of levels p is an odd number not greater than 7. Therefore, this paper divides each evaluation index into 5 levels according to the relevant literature and index characteristics: $T = \{low,\ lower,\ general, higher, high\}$.

3.2.3. Decide the Cloud Numerical Eigenvalues of Each Evaluation Index and Cloud Model Map

Factor evaluation is carried out between the various hierarchical domains corresponding to each evaluation indicator, and the fuzzy relationship matrix is obtained by generating cloud numerical eigenvalues through a forward cloud generator. Let the upper and lower critical values of the rank $T_k(k = 1, 2, \cdots, p)$ corresponding to the evaluation indicator $j(j = 1, 2, \cdots, 21)$ be $[G_{\min}, G_{\max}]$. The normal cloud model for the rank k corresponding to the evaluation indicator j is

$$Ex = (G_{\min} + G_{\max})/2 \tag{6}$$

The critical value is the transition value of two adjacent levels, which belong to the two corresponding levels at the same time, so the affiliation of the two levels is equal:

$$\exp\left[-\frac{(G_{\max} - G_{\min})^2}{8(E_n)^2}\right] = 0.5 \tag{7}$$

$$En = \frac{G_{\max} - G_{\min}}{2.355} \tag{8}$$

The super entropy He reflects the thickness of the cloud layer, which is a measure of the uncertainty of entropy, and the final value is determined by repeated trials according to the magnitude of entropy. According to the obtained fuzzy relationship matrix, MATLAB programming is applied to obtain the cloud model map corresponding to each metric.

3.2.4. Determine the Affiliation of Each Evaluation Index

Using X conditional cloud generator, we calculate the affiliation degree of each index corresponding to different levels, form the corresponding cloud model affiliation matrix. Select the largest affiliation degree as the evaluation level of the index. The corresponding cloud affiliation degree is

$$v_{jk} = \exp\left\{-\frac{(x_0 - Ex)^2}{2(En')^2}\right\} \tag{9}$$

where En' is a normal random number with En as the expected value and He^2 as the variance, i.e., $En' \sim N(En, He^2)$. The affiliation matrix is denoted as $V = (V_{jk})_{n\times p}$ and V_{jk} denotes the affiliation of the k th rank of the j th evaluation index, and in order to optimize the evaluation accuracy, the average of different affiliations under the repeated N times conditional cloud generator is used, i.e.,

$$V_{jk} = \frac{1}{N}\sum_{q=1}^{N} v_{jk}^{q} \tag{10}$$

3.2.5. Entropy Weighting Method to Assign the Index Weights

According to the calculation steps of the entropy weighting method mentioned in 2.1 above, the weighting values of each indicator are determined in conjunction with the regional logistics low carbon development evaluation index system.

3.2.6. Determine the Comprehensive Evaluation Level of Regional Logistics Decarbonization Development

In this paper, the comprehensive determination degree of regional logistics decarbonization development level is obtained according to the following formula.

$$C_{ik} = w_j V_{jk}^{i} (i = 1, 2, \cdots, m) \tag{11}$$

where V_{jk}^{i} is the affiliation degree of an region in year i and w_j is the weight of the index. According to the principle of maximum degree of certainty, the level where the maximum degree of certainty is selected is the final comprehensive evaluation level of regional logistics decarbonization development.

4. Empirical Analysis and Pathway Study

As a pioneer area of green and low-carbon development in China, the development of low-carbon logistics in Beijing, Tianjin and Hebei can play a typical demonstration and promotion role in the country. Beijing, Tianjin and Hebei have significantly different logistics capabilities due to their regional characteristics and differences in economic and political levels, and it is urgent to establish a mechanism for the collaborative development of low-carbon logistics in the region. Therefore, this paper takes Beijing, Tianjin and Hebei as an example to evaluate the development of low-carbon logistics in each region, find out the differences between them, and then discover the main factors affecting the development of low-carbon logistics in each city, so as to provide theoretical support for the development of low-carbon logistics.

4.1. Data Sources and Carbon Emission Measurement in Beijing, Tianjin and Hebei

4.1.1. Data Sources

This paper analyzes the data of Beijing-Tianjin-Hebei region from 2013 to 2019 as samples, and the relevant raw data are obtained from the annual data of National Bureau of Statistics by province, *China Economic Statistical Yearbook* and *China Energy Statistical Yearbook*.

4.1.2. Carbon Emission Measurement in Beijing, Tianjin and Hebei

According to the relevant data of China Energy Statistical Yearbook, the logistics industry in Beijing, Tianjin and Hebei mainly consumes 11 types of energy, including raw coal, gasoline, kerosene, diesel, fuel oil, liquefied petroleum gas, natural gas, liquefied natural gas, heat, electricity and other energy sources; among them, the carbon emission coefficients of liquefied natural gas, heat and other energy sources have not been found for the time being, and the consumption of these three types of energy sources accounts for a small part. The carbon emission coefficients of LNG, heat and other energy sources are not available, and these three types of energy sources account for little consumption, so their carbon emissions are not counted. Due to limited space, the raw data are shown in Table 2, taking the Beijing area as an example.

Table 2. Energy consumption by region in Beijing.

Energy Name	2013	2014	2015	2016	2017	2018	2019
Raw Coal (million tons)	15.86	16.03	12.30	7.97	3.22	0.94	0.41
Gasoline (million tons)	45.40	46.45	44.65	41.62	42.41	42.57	49.74
Kerosene (million tons)	476.51	507.07	543.78	593.66	643.31	690.47	697.17
Diesel (million tons)	124.28	126.56	118.00	109.92	106.98	110.11	99.81
Fuel Oil (million tons)	1.59	1.88	1.79	1.49	1.50	0.08	0.27
Liquefied Petroleum Gas (million tons)	0.35	0.32	0.38	0.28	1.17	1.55	17.41
Natural Gas (billion kilowatt hours)	2.35	3.17	2.11	1.99	1.80	3.72	3.42
Power (billion kilowatt hours)	44.64	45.02	47.31	50.61	53.29	582.03	57.98

According to the *2006 IPCC Guidelines for National Greenhouse Gas Inventories*, the reference coefficients for the conversion of standard coal and carbon emission coefficients for various energy sources are shown in Table 3.

Table 3. Reference factors for the conversion of standard coal and carbon emission factors for various energy sources.

Energy Name	Discount Factor for Standard Coal	Unit	Carbon Emission Factor	Unit
Raw Coal	0.7143	million tons of standard coal/million tons	0.7559	Tonnes of carbon/tonne of standard coal
Gasoline	1.4714	million tons of standard coal/million tons	0.5538	Tonnes of carbon/tonne of standard coal
Kerosene	1.4714	million tons of standard coal/million tons	0.5714	Tonnes of carbon/tonne of standard coal
Diesel	1.4571	million tons of standard coal/million tons	0.5821	Tonnes of carbon/tonne of standard coal
Fuel Oil	1.4286	million tons of standard coal/million tons	0.6185	Tonnes of carbon/tonne of standard coal
Liquefied Petroleum Gas	1.7143	million tons of standard coal/million tons	0.5042	Tonnes of carbon/tonne of standard coal
Natural Gas	13.3	million tons of standard coal/billion cubic meters	0.4483	Tonnes of carbon/tonne of standard coal
Power	1.229	million tons of standard coal/billion kilowatt hours	2.2132	Tonnes of carbon/tonne of standard coal

Substitute the data into Equation (5) for calculation to get the carbon emissions of each region from 2013–2019, which are divided by the unit value added of logistics industry as the raw data of carbon emissions per unit value added of indicator logistics industry and show the calculation results in Table 4.

Table 4. Carbon emissions from Beijing, Tianjin and Hebei regions.

Region	2013	2014	2015	2016	2017	2018	2019
Beijing	688.7401	723.4680	743.4722	781.6607	825.9406	2315.8326	904.9456
Tianjin	283.6145	302.9090	326.6324	339.6521	350.2898	362.5607	377.9557
Hebei	724.9221	699.0801	519.9094	791.0719	762.9008	809.7153	1003.4259

4.2. Evaluation of the Low Carbon Development of Logistics in Beijing, Tianjin and Hebei

4.2.1. Selection of Indicator Samples

According to the index system constructed in Table 1, select the relevant statistics of Beijing, Tianjin and Hebei to analysis, and the following table takes the raw data of Beijing as an example, as shown in Table 5.

Table 5. Raw data of each indicator in Beijing.

Indicators	2013	2014	2015	2016	2017	2018	2019
$X_{1,1}$	9.9927	10.6533	11.4137	12.4442	13.7646	15.3695	16.4555
$X_{1,2}$	1.7310	1.8714	2.1759	2.3384	2.5015	2.6861	2.7006
$X_{1,3}$	4.1948	4.4786	4.7619	5.0645	5.3318	6.6956	6.9934
$X_{1,4}$	0.0331	0.0472	0.0529	0.0567	0.0672	0.0535	0.0417
$X_{1,5}$	0.0966	0.1117	0.0960	0.0965	0.1359	0.1601	0.1389
$X_{2,1}$	1.3207	1.3314	1.3336	1.3422	1.3544	1.3562	1.3629
$X_{2,2}$	0.0778	0.0783	0.0783	0.0709	0.0770	0.0770	0.0833
$X_{2,3}$	0.4023	0.4821	0.6281	0.5686	0.7341	1.1534	1.6162
$X_{2,4}$	3.6093	4.2612	4.8934	5.5397	8.4610	8.4114	10.7873
$X_{2,5}$	0.3171	0.3368	0.3408	0.3639	0.4150	0.4716	0.4693
$X_{2,6}$	0.0798	0.0796	0.0772	0.0735	0.0710	0.0735	0.0746
$X_{2,7}$	0.4970	0.4817	0.4152	0.3799	0.4415	0.4801	0.5058
$X_{2,8}$	0.0317	0.0316	0.0299	0.0292	0.0302	0.0307	0.0285
$X_{2,9}$	11.3277	12.0399	12.3300	13.5876	15.6153	16.8754	17.1322
$X_{2,10}$	−1.0271	−0.9982	−1.0050	−0.9884	−0.9167	−2.2796	−0.8953
$X_{3,1}$	−0.0568	0.1692	−0.0691	0.0653	0.4825	0.1130	−0.1540
$X_{3,2}$	0.0242	0.0169	−0.0033	−0.0300	−0.0086	0.0433	−0.0199
$X_{3,3}$	0.0552	0.0808	0.0207	0.0689	0.1394	0.1275	−0.0050
$X_{3,4}$	0.0854	0.0638	0.0669	0.0859	0.1050	0.1126	0.0749
$X_{3,5}$	0.1599	0.1001	0.1010	0.1410	0.1385	0.1050	0.1487
$X_{3,6}$	0.0796	0.0959	0.0453	0.0441	0.0559	0.0183	0.0408

Note: To ease the subsequent ranking, the data related to the negative item is added with a negative sign to make it a positive indicator.

4.2.2. Determine the Level of Each Evaluation Index

In this study, the domain was divided into five evaluation levels, and the maximum and minimum values of the indicator data of 31 provinces were taken as the range of evaluation factors, and then the range was reasonably divided into five levels to determine the upper and lower critical values $[G_{min}, G_{max}]$ of each level, and the results are shown in Table 6.

Table 6. Classification of the evaluation level of each indicator.

Grade	Low	Lower	General	Higher	High
$X_{1,1}$	(0, 3)	(3, 6)	(6, 9)	(9, 13)	(13, 17)
$X_{1,2}$	(0, 0.6)	(0.6, 1.2)	(1.2, 1.8)	(1.8, 2.4)	(2.4, 3)
$X_{1,3}$	(0, 1)	(1, 2)	(2, 4)	(4, 6)	(6, 8)
$X_{1,4}$	(0, 0.02)	(0.02, 0.03)	(0.03, 0.05)	(0.05, 0.06)	(0.06, 0.07)
$X_{1,5}$	(0, 0.02)	(0.02, 0.05)	(0.05, 0.1)	(0.1, 0.15)	(0.15, 0.2)
$X_{2,1}$	(0, 0.4)	(0.4, 0.8)	(0.8, 1.2)	(1.2, 1.6)	(1.6, 2)
$X_{2,2}$	(0, 0.02)	(0.02, 0.04)	(0.04, 0.06)	(0.06, 0.08)	(0.08, 0.1)
$X_{2,3}$	(0, 0.3)	(0.3, 0.6)	(0.6, 0.9)	(0.9, 1.3)	(1.3, 1.7)
$X_{2,4}$	(0, 1)	(1, 3)	(3, 5)	(5, 8)	(8, 11)
$X_{2,5}$	(0, 0.12)	(0.12, 0.24)	(0.24, 0.36)	(0.36, 0.48)	(0.48, 0.6)
$X_{2,6}$	(0, 0.02)	(0.02, 0.04)	(0.04, 0.06)	(0.06, 0.08)	(0.08, 0.1)
$X_{2,7}$	(0, 1)	(1, 3)	(3, 5)	(5, 7)	(7, 13)
$X_{2,8}$	(0, 0.02)	(0.02, 0.04)	(0.04, 0.06)	(0.06, 0.08)	(0.08, 0.1)
$X_{2,9}$	(0, 25)	(25, 50)	(50, 75)	(75, 100)	(100, 110)
$X_{2,10}$	(−2.5, −1.5)	(−1.5, −1)	(−1, −0.6)	(−0.6, −0.3)	(−0.3, 0)
$X_{3,1}$	(−1, 0)	(0, 0.1)	(0.1, 0.3)	(0.3, 0.4)	(0.5, 0.6)
$X_{3,2}$	(−1, 0)	(0, 0.05)	(0.05, 0.1)	(0.1, 0.15)	(0.15, 0.2)
$X_{3,3}$	(−1, 0)	(0, 0.05)	(0.05, 0.1)	(0.1, 0.15)	(0.15, 0.2)
$X_{3,4}$	(−1, 0)	(0, 0.05)	(0.05, 0.1)	(0.1, 0.15)	(0.15, 0.2)
$X_{3,5}$	(−1, 0)	(0, 0.5)	(0.5, 1)	(1, 2)	(2, 3)
$X_{3,6}$	(−1, 0)	(0, 0.05)	(0.05, 0.1)	(0.1, 0.15)	(0.15, 0.2)

4.2.3. Determine the Cloud Digital Characteristic Value of Each Evaluation Index and Cloud Model Map

According to the ranking of each indicator in Table 6, the upper and lower critical values $[G_{min}, G_{max}]$ were substituted into Equations (6)–(8) to obtain the numerical characteristic values of the cloud model for each indicator, as shown in Table 7. The number of cloud drops per cloud was set to 3000, and the cloud model plots for each evaluation metric were derived by plotting the normal cloud model with MATLAB software. The cloud model diagrams for each of the five evaluation indicators included under the level of low carbon logistics environmental support are shown in Figures 4–8, for example. The horizontal coordinates represent the range of values of the evaluation factors, the vertical coordinates represent the corresponding affiliation degrees, and the curves from left to right represent the clouds represented by the evaluation levels of "low", "low", "average", "high", and "high".

Table 7. Numerical feature values of each indicator cloud model.

Grade	Low	Lower	General	Higher	High
$X_{1,1}$	(1.5, 1.2739, 0.2)	(4.5, 1.2739, 0.2)	(7.5, 1.2739, 0.2)	(11, 1.6985, 0.3)	(15, 1.6985, 0.3)
$X_{1,2}$	(0.3, 0.2548, 0.05)	(0.9, 0.2548, 0.05)	(1.5, 0.2548, 0.05)	(2.1, 0.2548, 0.05)	(2.7, 0.2548, 0.05)
$X_{1,3}$	(0.5, 0.4246, 0.1)	(1.5, 0.4246, 0.1)	(3, 0.8493, 0.15)	(5, 0.8493, 0.15)	(7, 0.8493, 0.15)
$X_{1,4}$	(0.01, 0.0085, 0.0015)	(0.025, 0.0042, 0.001)	(0.04, 0.0085, 0.0015)	(0.055, 0.0042, 0.001)	(0.065, 0.0042, 0.001)
$X_{1,5}$	(0.01, 0.0085, 0.0015)	(0.035, 0.0127, 0.002)	(0.075, 0.0212, 0.003)	(0.125, 0.0212, 0.003)	(0.175, 0.0212, 0.003)
$X_{2,1}$	(0.2, 0.1699, 0.03)	(0.6, 0.1699, 0.03)	(1, 0.1699, 0.03)	(1.4, 0.1699, 0.03)	(1.8, 0.1699, 0.03)
$X_{2,2}$	(0.01, 0.0085, 0.0015)	(0.03, 0.0085, 0.0015)	(0.05, 0.0085, 0.0015)	(0.07, 0.0085, 0.0015)	(0.09, 0.0085, 0.0015)
$X_{2,3}$	(0.15, 0.1274, 0.02)	(0.45, 0.1274, 0.02)	(0.75, 0.1274, 0.02)	(1.1, 0.1699, 0.03)	(1.5, 0.1699, 0.03)
$X_{2,4}$	(0.5, 0.4246, 0.1)	(2, 0.8493, 0.15)	(4, 0.8493, 0.15)	(6.5, 1.2739, 0.2)	(9.5, 1.2739, 0.2)
$X_{2,5}$	(0.06, 0.0510, 0.01)	(0.18, 0.0510, 0.01)	(0.3, 0.0510, 0.01)	(0.42, 0.0510, 0.01)	(0.54, 0.0510, 0.01)
$X_{2,6}$	(0.01, 0.0085, 0.0015)	(0.03, 0.0085, 0.0015)	(0.05, 0.0085, 0.0015)	(0.07, 0.0085, 0.0015)	(0.09, 0.0085, 0.0015)
$X_{2,7}$	(0.5, 0.4246, 0.1)	(2, 0.8493, 0.15)	(4, 0.8493, 0.15)	(6, 0.8493, 0.15)	(10, 2.5478, 0.4)
$X_{2,8}$	(0.01, 0.0085, 0.0015)	(0.03, 0085, 0.0015)	(0.05, 0.0085, 0.0015)	(0.07, 0.0085, 0.0015)	(0.09, 0.0085, 0.0015)
$X_{2,9}$	(12.5, 10.6157, 1.8)	(37.5, 10.6157, 1.8)	(62.5, 10.6157, 1.8)	(87.5, 10.6157, 1.8)	(105, 4.2463, 0.7)
$X_{2,10}$	(−2, 0.4246, 0.1)	(−1.25, 0.2123, 0.03)	(−0.8, 0.1699, 0.03)	(−0.45, 0.1274, 0.02)	(−0.15, 0.1274, 0.02)
$X_{3,1}$	(−0.5, 0.4246, 0.1)	(0.05, 0.0425, 0.01)	(0.2, 0.0850, 0.015)	(0.35, 0.0425, 0.01)	(0.55, 0.0425, 0.01)
$X_{3,2}$	(−0.5, 0.4246, 0.1)	(0.025, 0.0212, 0.003)	(0.075, 0.0212, 0.003)	(0.125, 0.0212, 0.003)	(0.175, 0.0212, 0.003)
$X_{3,3}$	(−0.5, 0.4246, 0.1)	(0.025, 0.0212, 0.003)	(0.075, 0.0212, 0.003)	(0.125, 0.0212, 0.003)	(0.175, 0.0212, 0.003)
$X_{3,4}$	(−0.5, 0.4246, 0.1)	(0.025, 0.0212, 0.003)	(0.075, 0.0212, 0.003)	(0.125, 0.0212, 0.003)	(0.175, 0.0212, 0.003)
$X_{3,5}$	(−0.5, 0.4246, 0.1)	(0.25, 0.2123, 0.03)	(0.75, 0.2123, 0.03)	(1.5, 0.4246, 0.1)	(2.5, 0.4246, 0.1)
$X_{3,6}$	(−0.5, 0.4246, 0.1)	(0.025, 0.0212, 0.003)	(0.075, 0.0212, 0.003)	(0.125, 0.0212, 0.003)	(0.175, 0.0212, 0.003)

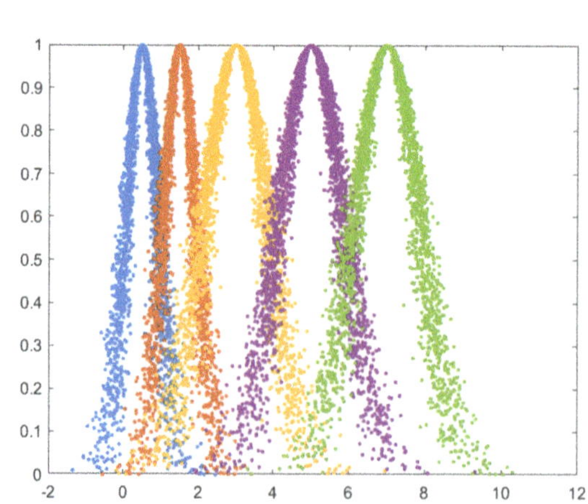

Figure 4. Per capita gross regional product cloud model.

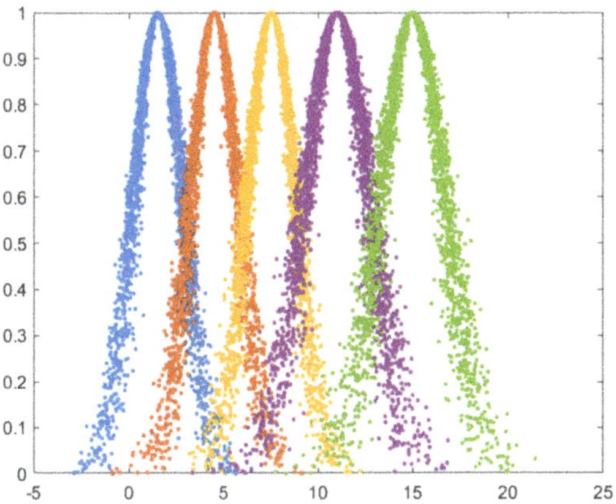

Figure 5. Per capita fiscal revenue cloud model.

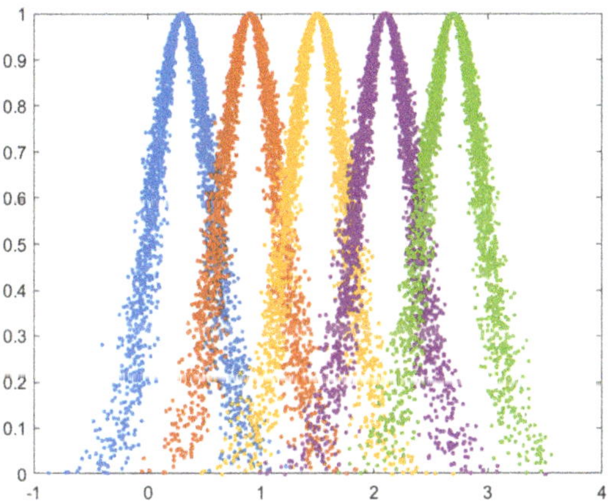

Figure 6. Retail sales of social goods per capita.

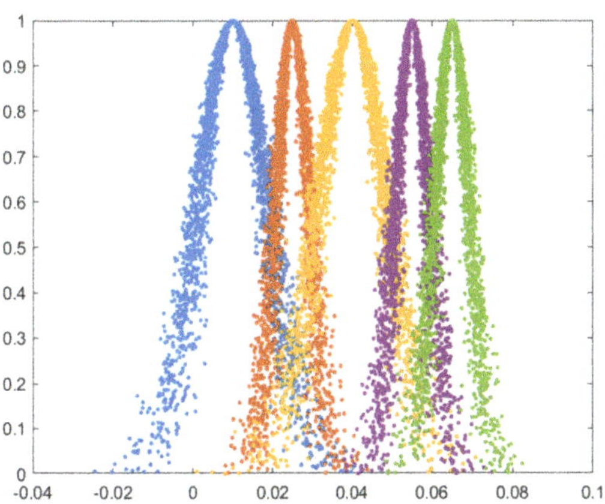

Figure 7. Cloud model of total expenditure ratio of per capita financial environmental protection expenditure.

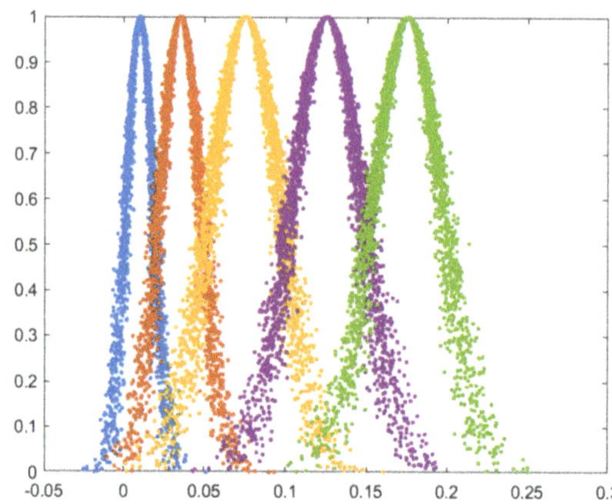

Figure 8. Logistics industry as a percentage of fixed asset investment cloud model.

4.2.4. Calculate the Affiliation Degree of Each Index

After getting the cloud model of each evaluation index in the regional logistics low carbonization evaluation index system, we use the X conditional cloud generator of the cloud model by MATLAB programming and take N = 3000 to get the affiliation degree of different levels corresponding to each evaluation index of the province. According to the principle of maximum affiliation degree, select the level corresponding to the maximum of the affiliation degree as the index level, taking Beijing in 2013 as an example, show the results in Table 8.

Table 8. Indicator affiliation with Beijing 2013 as an example.

Grade	Low	Lower	General	Higher	High	Grade
$X_{1,1}$	0.0000	0.0006	0.1504	0.8250	0.0222	Higher
$X_{1,2}$	0.0000	0.0122	0.6359	0.3419	0.0034	General
$X_{1,3}$	0.0000	0.0000	0.3626	0.6251	0.0095	Higher
$X_{1,4}$	0.0360	0.1714	0.7032	0.0002	0.0000	General
$X_{1,5}$	0.0000	0.0001	0.5851	0.4010	0.0027	General
$X_{2,1}$	0.0000	0.0009	0.1732	0.8858	0.0276	Higher
$X_{2,2}$	0.0000	0.0000	0.0104	0.6376	0.3473	Higher
$X_{2,3}$	0.1492	0.9276	0.0324	0.0013	0.0000	Lower
$X_{2,4}$	0.0000	0.1711	0.8907	0.0850	0.0002	General
$X_{2,5}$	0.0001	0.0411	0.9385	0.1421	0.0009	General
$X_{2,6}$	0.0000	0.0000	0.0060	0.4999	0.4717	Higher
$X_{2,7}$	1.0000	0.2158	0.0013	0.0000	0.0028	low
$X_{2,8}$	0.0501	0.9780	0.1101	0.0005	0.0000	Higher
$X_{2,9}$	0.9933	0.0598	0.0002	0.0000	0.0000	low
$X_{2,10}$	0.0888	0.5643	0.3939	0.0003	0.0000	Higher
$X_{3,1}$	0.5553	0.0633	0.0191	0.0000	0.0000	low
$X_{3,2}$	0.4427	0.9992	0.0651	0.0001	0.0000	Higher
$X_{3,3}$	0.4102	0.3580	0.6335	0.0079	0.0000	General
$X_{3,4}$	0.3705	0.0239	0.8801	0.1793	0.0006	General
$X_{3,5}$	0.2948	0.9096	0.0285	0.0179	0.0001	Higher
$X_{3,6}$	0.3812	0.0439	0.9752	0.1069	0.0003	General
C	0.2233	0.2623	0.3704	0.2311	0.0322	General

4.2.5. Entropy Weighting Method to Determine the Weights

Based on the entropy weighting method to calculate the weight of each index in the system, substitute the data of each index into the Equations (1)–(4) by MATLAB programming to calculate the weight of each index, and the results are shown in Table 9.

Table 9. Standardization of raw data for each indicator in Beijing.

Indicators	2013	2014	2015	2016	2017	2018	2019	Weights
$X_{1,1}$	0.0000	0.1022	0.2199	0.3793	0.5836	0.8320	1.0000	0.0513
$X_{1,2}$	0.0000	0.1448	0.4588	0.6264	0.7947	0.9850	1.0000	0.0394
$X_{1,3}$	0.0000	0.1014	0.2026	0.3108	0.4063	0.8936	1.0000	0.0572
$X_{1,4}$	0.0000	0.4135	0.5806	0.6921	1.0000	0.5982	0.2522	0.0333
$X_{1,5}$	0.0094	0.2449	0.0000	0.0078	0.6225	1.0000	0.6693	0.0886
$X_{2,1}$	0.0000	0.2536	0.3057	0.5095	0.7986	0.8412	1.0000	0.0376
$X_{2,2}$	0.5565	0.5968	0.5968	0.0000	0.4919	0.4919	1.0000	0.0270
$X_{2,3}$	0.0000	0.0657	0.1860	0.1370	0.2733	0.6187	1.0000	0.0714
$X_{2,4}$	0.0000	0.0908	0.1789	0.2689	0.6759	0.6690	1.0000	0.0566
$X_{2,5}$	0.0000	0.1275	0.1534	0.3029	0.6337	1.0000	0.9851	0.0574
$X_{2,6}$	1.0000	0.9773	0.7045	0.2841	0.0000	0.2841	0.4091	0.0399
$X_{2,7}$	0.9301	0.8086	0.2804	0.0000	0.4893	0.7959	1.0000	0.0320
$X_{2,8}$	1.0000	0.9688	0.4375	0.2188	0.5313	0.6875	0.0000	0.0366
$X_{2,9}$	0.0000	0.1227	0.1727	0.3893	0.7387	0.9558	1.0000	0.0535
$X_{2,10}$	0.9048	0.9257	0.9208	0.9327	0.9845	0.0000	1.0000	0.0221
$X_{3,1}$	0.1527	0.5078	0.1334	0.3445	1.0000	0.4195	0.0000	0.0526
$X_{3,2}$	0.7394	0.6398	0.3643	0.0000	0.2920	1.0000	0.1378	0.0450
$X_{3,3}$	0.4169	0.5942	0.1780	0.5118	1.0000	0.9176	0.0000	0.0388
$X_{3,4}$	0.4426	0.0000	0.0635	0.4529	0.8443	1.0000	0.2275	0.0547
$X_{3,5}$	1.0000	0.0000	0.0151	0.6839	0.6421	0.0819	0.8127	0.0668
$X_{3,6}$	0.7899	1.0000	0.3479	0.3325	0.4845	0.0000	0.2899	0.0380

4.2.6. Determine the Comprehensive Evaluation Level of Regional Logistics Index

First, the weights and affiliation degrees of each evaluation index are substituted into Equation (11) to obtain the comprehensive determination degree and evaluation grade, for example, the determination degree of each evaluation grade in Beijing in 2013 is $C = (0.2204, 0.2689, 0.3709, 0.2260, 0.0315)$. Second, according to the principle of maximum determination degree, select the evaluation grade with the maximum determination degree as the final comprehensive evaluation result, as shown in Tables 10–12. Finally, the Beijing-Tianjin-Hebei regional logistics decarbonization development grade from 2013 to 2019 do the comparison, as shown in Figure 9.

Table 10. Evaluation Results of Logistics Decarbonization Development in Beijing from 2013–2019.

Grade	Low	Lower	General	Higher	High	Evaluation Results
2013	0.2204	0.2689	0.3709	0.2260	0.0315	Genera
2014	0.2011	0.2365	0.3887	0.2838	0.0347	Genera
2015	0.2199	0.2262	0.2756	0.3306	0.0398	Higher
2016	0.2052	0.2422	0.2594	0.3657	0.0530	Higher
2017	0.1812	0.1463	0.2054	0.3581	0.1863	Higher
2018	0.2093	0.1874	0.1329	0.3762	0.2357	Higher
2019	0.2202	0.1678	0.1777	0.1642	0.2708	High

Table 11. Evaluation Results of Logistics Decarbonization Development in Tianjin 2013–2019.

Grade	Low	Lower	General	Higher	High	Evaluation Results
2013	0.2440	0.3131	0.2515	0.2546	0.0808	Lower
2014	0.2166	0.3202	0.2887	0.2628	0.0486	Lower
2015	0.1852	0.3268	0.2735	0.2599	0.0594	Lower
2016	0.2378	0.3301	0.2310	0.2435	0.0656	Lower
2017	0.1437	0.3075	0.3556	0.2153	0.0661	General
2018	0.1284	0.3849	0.3815	0.1642	0.0623	Lower
2019	0.0741	0.2191	0.3531	0.2450	0.1517	General

Table 12. Evaluation Results of Logistics Decarbonization Development in Hebei Province from 2013 to 2019.

Grade	Low	Lower	General	Higher	High	Evaluation Results
2013	0.2859	0.2655	0.2323	0.1925	0.1081	Low
2014	0.3137	0.3102	0.2573	0.1368	0.0664	Low
2015	0.2849	0.3590	0.2781	0.1271	0.0834	Lower
2016	0.2826	0.2778	0.3544	0.1294	0.0526	General
2017	0.2692	0.2677	0.3071	0.1574	0.1307	General
2018	0.1639	0.3536	0.3213	0.1601	0.1493	Lower
2019	0.1619	0.2760	0.3480	0.2277	0.1552	General

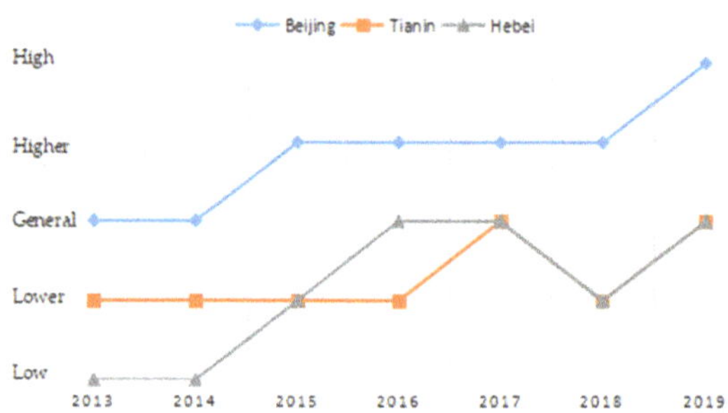

Figure 9. 2013–2019 Beijing-Tianjin-Hebei Regional Logistics Decarbonization Development Grade Comparison.

From the time dimension of the comprehensive evaluation results, the overall development of logistics decarbonization in the Beijing-Tianjin-Hebei region from 2013 to 2019 is on an upward trend, with all three regions showing different degrees of improvement. Among them, the development of logistics decarbonization in Beijing develops from average level to high-level, that in Tianjin develops from lower level to average level, and that in Hebei develops from low-level to average level; relatively speaking, the development in Tianjin is slow, which is not in line with its economic level. From the spatial dimension of the comprehensive evaluation results, the development of logistics low-carbon within the Beijing-Tianjin-Hebei region is not balanced, the specific performance is Beijing $\succ$ Tianjin $\succ$ Hebei. There has been a level difference in the development of logistics decarbonization within the Beijing-Tianjin-Hebei region between 2013 and 2019, and the development to 2019, Beijing is at a high-level of development nationwide, while Tianjin and Hebei Province are still at an average level of development, which is two levels away from Beijing in the same region.

4.3. Determination of Influencing Factors and Suggestions for Countermeasures

4.3.1. Determination of Influencing Factors

According to the development trend of each index and the horizontal comparison with the three provinces and cities in Beijing, Tianjin and Hebei, the shortcomings of each region in the development of logistics low carbonization are identified, and the main factors affecting the development of logistics low carbonization in the city are found, so as to provide theoretical support for the development of logistics low carbonization. Due to space limitations, the evaluation grade of each indicator is displayed in Beijing region as an example, as shown in Table 13; meanwhile, the evaluation grade of each indicator in Beijing, Tianjin and Hebei in 2019 is compared, as shown in Figure 10.

Table 13. Evaluation level of each indicator in Beijing region for example.

Indicators	2013	2014	2015	2016	2017	2018	2019
$X_{1,1}$	Higher	Higher	Higher	Higher	High	High	High
$X_{1,2}$	General	Higher	Higher	Higher	High	High	High
$X_{1,3}$	Higher	Higher	Higher	Higher	Higher	High	High
$X_{1,4}$	General	General	Higher	Higher	High	Higher	General
$X_{1,5}$	General	General	General	General	General	General	General
$X_{2,1}$	Higher	Higher	Higher	Higher	Higher	Higher	Higher
$X_{2,2}$	Higher	Higher	Higher	Higher	Higher	Higher	High
$X_{2,3}$	Lower	Lower	General	Lower	General	Higher	High
$X_{2,4}$	General	General	General	Higher	High	High	High
$X_{2,5}$	General	General	General	Higher	Higher	Higher	Higher
$X_{2,6}$	Higher	Higher	Higher	Higher	Higher	Higher	Higher
$X_{2,7}$	Low	Low	Low	Low	Low	Low	Low
$X_{2,8}$	Lower	Lower	Lower	Lower	Lower	Lower	Lower
$X_{2,9}$	Low	Low	Low	Low	Low	Low	Low
$X_{2,10}$	Lower	General	Lower	General	General	Low	General
$X_{3,1}$	Low	General	Low	Lower	High	General	Low
$X_{3,2}$	Lower	Lower	Low	Low	Low	Lower	Low
$X_{3,3}$	General	General	Lower	General	Higher	Higher	Low
$X_{3,4}$	General	General	General	General	Higher	Higher	General
$X_{3,5}$	Lower	Lower	Lower	Lower	Lower	Lower	Lower
$X_{3,6}$	General	General	Lower	Lower	General	Lower	Lower
C	General	General	Higher	Higher	Higher	Higher	High

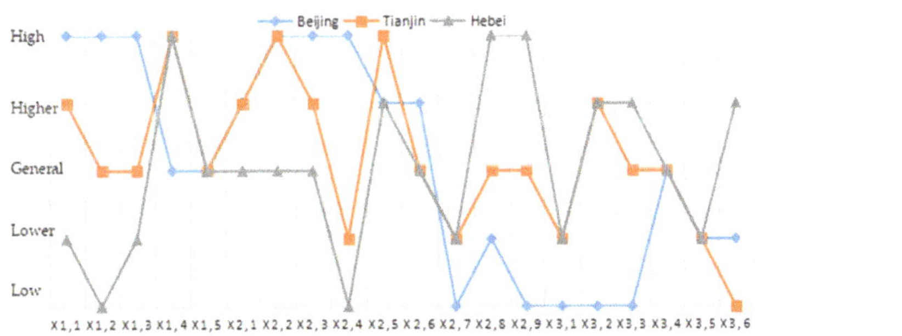

Figure 10. 2019 Beijing-Tianjin-Hebei comparison of evaluation ratings for each indicator.

From the evaluation grade of each indicator in Beijing, Tianjin and Hebei provinces and cities, the five indicators of per capita cargo turnover, the contribution rate of logistics industry to GDP, the part of logistics personnel in the workforce, the growth rate of logistics personnel and the growth rate of technical market turnover in Beijing are below the national average level all year, and the development is slow. By 2019, the efficiency of logistics industry, logistics industry input, logistics industry output and technical support are the indicators under the four secondary indicators are still below the national average level, and the shortcomings are more obvious. Tianjin region has been at a low or lower level nationally in the five indicators of per capita e-commerce sales, per capita cargo turnover, growth rate of new fixed asset investment in logistics industry, growth rate of technology market turnover, and growth rate of R&D funding during 2013–2019, and the development has been neglected, a large gap between the levels of the indicators under the low-carbon logistics environment support power and Beijing. Hebei Province has the most obvious gap in low-carbon logistics environment support power relative to neighboring Beijing and Tianjin, mainly in the form of per capita fiscal revenue per year at a low national level, per capita gross regional product and per capita total retail sales of social goods per year at a low-level. In addition, the three indicators of per capita e-commerce sales, per capita turnover of goods and growth rate of technology market turnover in 2019 are still at a low or lower level nationwide.

As can be seen from Figure 10, the development of each indicator in Beijing, Tianjin and Hebei provinces and cities is still in an unbalanced state by 2019, with the biggest difference between Beijing and the other two provinces and cities, as shown because the evaluation levels of each indicator under the two secondary indicators of economic environment and logistics infrastructure are higher than those of Tianjin and Hebei, while the levels of each indicator in logistics industry efficiency and logistics industry input and output are significantly lower than those of the other two provinces and cities; Tianjin is generally higher than Hebei in the four secondary indicators of economic environment, policy environment, logistics infrastructure, and logistics industry scale, but not higher than Hebei in the indicators of low carbon logistics potential. Thus, it seems that although the three provinces and cities in Beijing, Tianjin and Hebei have made breakthroughs in cooperation, they still lack synergy in the development of low-carbon logistics due to the large differences in administrative division, consciousness and economic development level.

4.3.2. Suggestions for Countermeasures to the Low-Carbon Development of Logistics in Beijing-Tianjin-Hebei Region

(1) From the shortcomings of the development of low-carbon logistics in Beijing, Tianjin and Hebei in recent years, Beijing needs to strengthen two aspects of low-carbon logistics strength and low-carbon logistics potential, especially the three modules of logistics industry efficiency, logistics industry input and demand, and technical support. Tianjin should start with a balanced approach to logistics industry efficiency, input, output, demand and technical support in order to improve the overall low-carbon development of logistics. Hebei Province should strengthen the development of logistics economy, improve the practice base of logistics enterprises, promote industrial clusters and create a logistics ecological chain while improving economic strength, so as to enhance the level of logistics low carbonization in all aspects.
(2) Strengthen the division of labor and cooperation between Beijing, Tianjin and Hebei in logistics. In the 13th Five-Year Plan, Beijing, Tianjin and Hebei are planned as a whole region, and the respective positions of the three provinces and cities have been clarified. In this context, the logistics industry synergy among the three provinces and cities should optimize the logistics network and divide the work according to the characteristics of each region. Beijing gives full play to the advantages of science and technology and innovates the development of logistics industry while improving the consumer-oriented end logistics system. Tianjin focuses on building a port logistics base in the context of the linkage of three ports. Compare with Beijing and Tianjin, Hebei Province is rich in resources, so it should undertake the transfer of Beijing-Tianjin trade logistics and build Hebei into an important base for modern trade logistics in the country.
(3) The government increases the policy support for developing low-carbon logistics. The development of low-carbon logistics in Beijing, Tianjin and Hebei needs the cooperation and joint planning of the three regions, and government departments should give support and guidance in policies, such as encouraging the development of ecological logistics industry chain, providing relevant enterprises with corresponding technical support or improving the reasonableness of taxation and financing policy preferences, etc. In addition, while developing regional logistics and economy at high-speed, we should actively promote the idea of green logistics, change the traditional concept of consumers, advocate low-carbon consumption and raise the low-carbon awareness of the logistics industry.
(4) Improve the level of informatization of low-carbon logistics in Beijing, Tianjin and Hebei. Informatization is an important feature of modern logistics and an effective way to achieve low carbon regional logistics. In the process of integrated development and communication, Beijing, Tianjin and Hebei provinces and cities should break the information silos, establish and improve the logistics information exchange platform,

and share and freely exchange logistics information so as to connect the information of each node of the supply chain and give full play to the advantages of regional informatization, to reduce logistics costs and improve logistics efficiency.

5. Conclusions

(1) Twenty-one indicators are selected from the three dimensions of low-carbon logistics environment support, low-carbon logistics strength and low-carbon logistics potential to establish the regional logistics low-carbonization development evaluation index system. Combined with the cloud model and entropy weight method to build the index evaluation model, which solves the problem of fuzziness and randomness in the process of regional logistics low-carbonization development evaluation.
(2) The evaluation model of regional logistics decarbonization development can show the development changes of each region in spatial and temporal dimensions and also solve the problem of horizontal comparison between different regions, giving quantitative results of different regions and different times. Then, according to the quantitative results, can discover the shortcomings of regional logistics decarbonization development and provide theoretical support for the further development of regional logistics.
(3) The entropy weight-cloud model method uses the characteristic indicators that can reflect the complex relationship between multiple factors to derive the corresponding evaluation level. It makes the evaluation results more intuitive and accurate through the cloud diagram and calculation of evaluation level. At the same time, it provides reference for the shortcomings of regional logistics decarbonization development, which is of positive significance to enhance the development of regional logistics decarbonization.
(4) The development of regional logistics decarbonization is a complex and continuously changing process, and future research can further improve the evaluation index system, optimize the evaluation model, and enhance the accuracy and applicability of the evaluation model.

Author Contributions: Conceptualization, Z.G. and Y.T.; methodology, Z.G. and Y.T.; software, Y.T.; validation, Z.G., Y.T.; formal analysis, Z.G.; investigation, Y.T.; resources, Z.G., X.G. and Z.H.; data curation, Y.T.; writing—original draft preparation, Y.T.; writing—review and editing, Z.G.; project administration, X.G. and Z.H.; funding acquisition, Z.G. All authors have read and agreed to the published version of the manuscript.

Funding: This paper is supported by the National Social Science Foundation of China (NO. 20BTJ012), the Social Science Foundation of Hebei province of China (No. HB18GL008), Beijing Intelligent Logistics System Collaborative Innovation Center (BILSCIC-2019KF-15), the Opening Project of Research Center of Capital Commercial Industry of China (JD-KFKT-2020-003), and Philosophy and Social Science key cultivation project of Hebei University (2019HPY035).

Institutional Review Board Statement: Not applicable.

Informed Consent Statement: Not applicable.

Data Availability Statement: The datasets used and/or analyzed during the current study are available from the corresponding author on reasonable request.

Conflicts of Interest: It is declared by the authors that this article is free of conflict of interest.

References

1. Butner, K.; Geuder, D.; Hittner, J. *Mastering Carbon Management: Balancing Trade-Offs to Optimize Supply Chain Efficiencies*; IBM Institute for Business Value: New York, NY, USA, 2008; pp. 2–12.
2. Dada, A.; Staake, T. Carbon footprints from enterprises to product instances: The potential of the EPC network. In *Beherrschbare Systeme-Dank Informatik*; Jahrestagung der Gesellschaft für Informatik: München, Germany, 2008; pp. 873–878.
3. Piecyk, M.I.; Mc Kinnon, A.C. Forecasting the carbon footprint of road freight transport in 2020. *Int. J. Prod. Econ.* **2010**, *128*, 31–42. [CrossRef]

4. Wang, L.P.; Liu, M.H. Research on carbon emission measurement and influencing factors of China's logistics industry based on input-output method. *Resour. Sci.* **2018**, *40*, 195–206.
5. Liu, Y.; Li, L. Study on decoupling and influencing factors of carbon emissions from logistics industry in China. *Environ. Sci. Technol.* **2018**, *41*, 177–181.
6. Timilsina, G.R.; Shrestha, A. Transport sector CO_2 emissions growth in Asia: Underlying factors and policy options. *Energy Policy* **2009**, *37*, 4523–4539. [CrossRef]
7. Yang, L.; Cai, Y.; Zhong, X.; Shi, Y.; Zhang, Z. A Carbon Emission Evaluation for an Integrated Logistics System—A Case Study of the Port of Shenzhen. *Sustainability* **2017**, *9*, 462. [CrossRef]
8. Yang, Y.W.; Wu, A.L.; Zhu, Y.Y. Decomposition and dynamic simulation of carbon emission drivers: Taking Inner Mongolia autonomous region as an example. *Stat. Decis. Mak.* **2020**, *36*, 76–80.
9. Li, F.G.; Wu, L.J. Decomposition of carbon emission drivers based on LMDI method. *Stat. Decis. Mak.* **2019**, *35*, 101–104.
10. Men, D.; Huang, X. Study on influencing factors of carbon emission in Jiangxi Province—Based on LMDI decomposition method. *Ecol. Econ.* **2019**, *35*, 31–35.
11. Wehner, J. Energy Effificiency in Logistics: An Interactive Approach to Capacity Utilisation. *Sustainability* **2018**, *10*, 1727. [CrossRef]
12. Peng, L. Evaluation of low carbon logistics capability of logistics enterprises in China. *Bus. Econ. Res.* **2017**, *10*, 87–89.
13. Xiang, C.; Gong, B.G.; Wu, B.L. Low carbon behavior capability evaluation model of logistics service supply chain. *Stat. Decis. Mak.* **2017**, *5*, 67–71.
14. Jian, L.; Jian, T. Research on the impact of China's industrial structure adjustment on low carbon logistics efficiency—An empirical analysis based on super efficiency DEA low carbon logistics efficiency evaluation model. *Price Theory Pract.* **2017**, *12*, 130–133.
15. Yang, C.M. Efficiency measurement of Jiangsu logistics industry under low carbon constraints. *East China Econ. Manag.* **2018**, *32*, 27–32.
16. Chaabane, A.; Ramudhin, A.; Paquet, M. Design of sustainable supply chains under the emission trading scheme. *Int. J. Prod. Econ.* **2012**, *135*, 37–49. [CrossRef]
17. Li, C. Research on the development status and countermeasures of low carbon logistics in China. *Logist. Eng. Manag.* **2018**, *40*, 1–3.
18. Dong, Y.X. Research on the development status and trend of China's low carbon logistics from an international perspective. *Price Mon.* **2018**, *12*, 46–49.
19. Zhang, H. Analysis of low carbon logistics development countermeasures from the perspective of Beijing Tianjin Hebei synergy. *Reform Strategy* **2016**, *32*, 118–120.
20. Zhao, S.L.; Yang, X.Y.; Song, W. General strategy of transportation and logistics integration in Beijing, Tianjin and Hebei from the perspective of low-carbon economy. *China Stat.* **2017**, *12*, 6–8.
21. Ma, Y. *Research on Total Factor Productivity of China's Logistics Industry under Low Carbon Constraints*; Northeast University of Finance and Economics: Dalian, China, 2014.
22. Fei, X. *Research on the Coordinated Development of Logistics Industry and Economic Decarbonization*; Nanchang University: Nanchang, China, 2015.
23. Gao, F. *Research on the Coordinated Development of China's Logistics Industry and Low-Carbon Economy*; Tianjin University of Technology: Tianjin, China, 2016.
24. Qin, Y. *Analysis of Regional Logistics Efficiency and Its Influencing Factors under Low Carbon Economy*; East China Jiaotong University: Nanchang, China, 2015.
25. Lina, S. *Research on the Difference of Low-Carbon Logistics Performance of Regions along the Silk Road Economic Belt in China*; Zhengzhou University: Zhenzhou, China, 2018.
26. Wang, X. *Research on the Mechanism and Path of Low-Carbon Development of Regional Logistics*; Tianjin University: Tianjin, China, 2017.
27. Li, D.Y. *Uncertain Artificial Intelligence*; National Defense Industry Press: Beijing, China, 2005.
28. Wang, L.; Zhao, H.; Liu, X.; Zhang, Z.L.; Xu-Hui Steve, E. Optimal Remanufacturing Service Resource Allocation for Generalized Growth of Retired Mechanical Products: Maximizing Matching Efficiency. *IEEE Access* **2021**, *9*, 89655–89674.
29. Jia, L.; Yu, Y.; Li, Z.; Qin, S.; Guo, J.; Zhang, Y.; Jin, Y. Study on the Hg0 removal characteristics and synergistic mechanism of iron-based modified biochar doped with multiple metals. *Bioresour. Technol.* **2021**, *332*, 125086. [CrossRef] [PubMed]
30. Xu, Q.W.; Xu, K.L. Evaluation of ambient air quality based on synthetic cloud model. *Fresenius Environ. Bull.* **2018**, *27*, 141–146.
31. Yan, F.; Li, Z.J.; Dong, L.J.; Huang, R.; Cao, R.H.; Ge, J.; Xu, K.L. Cloud model-clustering analysis based evaluation for ventilation system of underground metal mine in alpine region. *J. Cent. South Univ.* **2021**, *28*, 796–815. [CrossRef]
32. Lin, C.J.; Zhang, M.; Li, L.P.; Zhou, Z.Q.; Liu, S.; Liu, C.; Li, T. Risk Assessment of Tunnel Construction Based on Improved Cloud Model. *J. Performence Constr. Facil.* **2020**, *34*, 04020028. [CrossRef]
33. Cong, X.H.; Ma, L. Performance Evaluation of Public-Private Partnership Projects from the Perspective of Efficiency, Economic, Effectiveness, and Equity: A Study of Residential Renovation Projects in China. *Sustainability* **2018**, *10*, 1951. [CrossRef]
34. Li, G.; Zhang, Y.; Wang, H.C. Civil aviation master data recognition method based on cloud model and rough set. *Comput. Eng. Des.* **2020**, *41*, 2338–2344.

35. Sun, Y.B.; Zhang, Y. Research on green development evaluation of coal listed companies based on cloud model. *China Min.* **2020**, *29*, 79–85.
36. Liu, F.; Zhu, X.; Hu, Y.; Ren, L.; Johnson, H. A Cloud Theory-Based Trust Computing Model in Social Networks. *Entropy* **2017**, *19*, 11. [CrossRef]
37. Lai, K. Service capability and performance of logistics service providers. *Transp. Res. Part E Logist. Transp. Rev.* **2004**, *40*, 385–399. [CrossRef]
38. Ma, S.; Meng, Q. Research status and development trend of supply chain logistics capability. *Comput. Integr. Manuf. Syst.* **2005**, *11*, 301–307.
39. Ming, W.; Hao, F. *Research on the Development Policy of China's Logistics Industry*; China Planning Press: Beijing, China, 2002; p. 17.
40. Jiang, J.; Liu, Z. Identification of three important capabilities of logistics enterprises. *Logist. Technol.* **2005**, *7*, 18–21.

processes

MDPI

Article

Multi Set-Point Explicit Model Predictive Control for Nonlinear Process Systems

Vassilis M. Charitopoulos, Lazaros G. Papageorgiou and Vivek Dua *

Centre for Process Systems Engineering, Department of Chemical Engineering, University College London, Torrington Place, London WC1E 7JE, UK; v.charitopoulos@ucl.ac.uk (V.M.C.); l.papageorgiou@ucl.ac.uk (L.G.P.)
* Correspondence: v.dua@ucl.ac.uk

Abstract: In this article, we introduce a novel framework for the design of multi set-point nonlinear explicit controllers for process systems engineering problems where the set-points are treated as uncertain parameters simultaneously with the initial state of the dynamical system at each sampling instance. To this end, an algorithm for a special class of multi-parametric nonlinear programming problems with uncertain parameters on the right-hand side of the constraints and the cost coefficients of the objective function is presented. The algorithm is based on computed algebra methods for symbolic manipulation that enable an analytical solution of the optimality conditions of the underlying multi-parametric nonlinear program. A notable property of the presented algorithm is the computation of exact, in general nonconvex, critical regions that results in potentially great computational savings through a reduction in the number of convex approximate critical regions.

Keywords: multi-parametric programming; explicit MPC; enterprise-wide optimisation; set-point tracking; algebraic geometry

Citation: Charitopoulos, V.M.; Papageorgiou, L.G.; Dua, V. Multi Set-Point Explicit Model Predictive Control for Nonlinear Process Systems. *Processes* **2021**, *9*, 1156. https://doi.org/10.3390/pr9071156

Academic Editor: Luis Puigjaner

Received: 9 March 2021
Accepted: 13 May 2021
Published: 2 July 2021

1. Introduction

High-fidelity and computationally efficient optimisation models are key for profitable decision making in process industries and have been the focus of extensive research over the years [1]. In recent years, the need for exploiting and explicitly considering interdependencies throughout the different layers of decision making has been underpinned by the enterprise-wide optimisation (EWO) concept [2]. Stemming from the progressively volatile and competitive market conditions, it is imperative for process industries to operate with agility in order to maximise their profitability [3]. EWO is aiming at increased profitability and resilience in process operations through the integration and simultaneous optimisation of existing information streams. Nonetheless, it comes at a considerable cost. Because of the multiple scales considered, EWO leads to computational challenges, thus preventing practitioners from harnessing the potential benefits such wide integration has to offer. Particularly, incorporating control considerations in an EWO fashion results in (mixed integer) nonconvex problems which are hard to solve.

By the same token, control considerations are ubiquitous in EWO problems. Figure 1 showcases how real-time optimisation and production scheduling exchange information with the layer of APC because of their interdependent decisions.

Real-time optimisation is concerned with the manipulation of systems' dynamics in order to achieve optimised profitability and operations. On the other hand, production scheduling determines the optimal allocation of resources for the completion of competing tasks. As indicated by Figure 1, both RTO and process scheduling exchange information with the layer of APC so as to achieve optimal dynamic operations. To this end, the research community has proposed different methods for their integration.

A common shortfall when focusing on integrating RTO and APC is that two different models are employed for the optimisation of the same system. Typically, a locally linear model of the initial nonlinear dynamics is used at the APC because of the need for fast solution rates while RTO considers the original nonlinear model. This leads subsequently to issues related to suboptimal trajectories and non-reachable states [4].

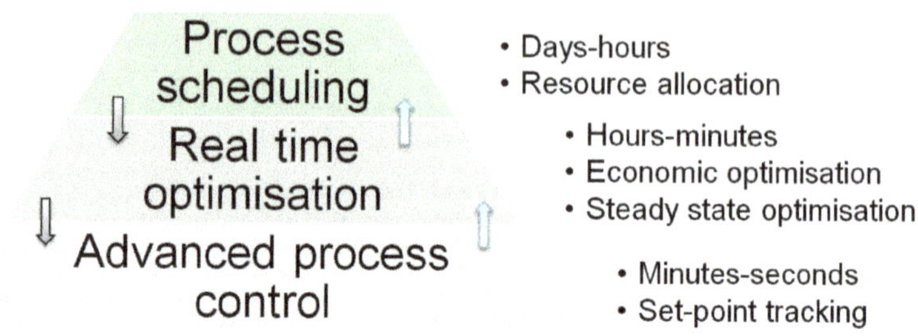

Figure 1. Interaction of APC with different layers of decision making in process industries.

Darby et al. [5], through their literature review regarding the integration of RTO and MPC, suggested that for a successful integration, common issues such as model mismatch among the layers of RTO and APC should be eliminated. Nonetheless, in real industrial processes, model degradation and other factors can result in model mismatch, so the consideration of parameter estimation and data reconciliation functionalities is needed to integrate RTO and MPC, as indicated by Figure 2.

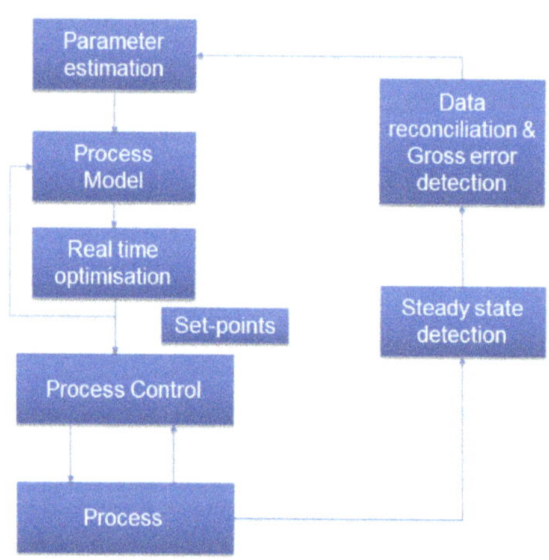

Figure 2. Interaction between advanced process control and real-time optimisation.

The interaction between real-time optimisation and model predictive control can be categorised broadly into three classes: (i) dynamic RTO (d-RTO), (ii) static RTO (s-RTO) and (iii) economic model predictive control (e-MPC). Both s-RTO and d-RTO are two-layer schemes where reference trajectories are passed to the layer of APC in the form of set-points [6]. While under the static real-time optimisation paradigm, the optimisation problem is solved at specific instances whenever new data become available or when steady state is achieved, in the d-RTO paradigm, the system's transient behaviour is explicitly considered, thus resulting in dynamic optimisation problems. e-MPC [7] refers to single-layer strategies which are incorporated into the control structure economic considerations. In that spirit, De Souza et al. [8] proposed the inclusion of the gradient of the economic objective function into the MPC objective as a single-layer strategy. Considering uncertain systems, Chachuat et al. [9] examined alternative model adaptation strategies.

This article is motivated by the abovementioned issues and aims at introducing a method for designing multi set-point explicit controllers for nonlinear systems through recent advances in multi-parametric programming. Multi-parametric programming (mp-P) has received considerable attention from the process systems engineering community because of its unique ability to aid in the design of explicit model predictive controllers and thus shift the computational burden associated with offline control [10]. We examine a case of multi-parametric nonlinear programs (mp-NLPs) that involve both endogenous uncertainty, in the form of left-hand side parameters (LHS), as well as exogenous uncertainty in the cost coefficient of the objective function (OFC), and, on the right-hand side of the constraints (RHS), uncertain parameters on the right-hand side (RHS). In engineering problems, LHS uncertainty arises from variations in model coefficients, due to parameter estimation errors or model mismatch; OFC uncertainty arises due to fluctuation in market prices or control penalties while RHS uncertainty can be due to varying system exogenous factors. The contribution of the present work is a novel framework for the design of multi set-point explicit controllers for nonlinear process systems.

The remainder of the article is organised as follows: in Section 2, an overview of the field of multi-parametric programming and explicit MPC is given, and then, in Section 3, the proposed algorithm is detailed and a framework for multi set-point explicit controllers is introduced. In Section 4, two case studies are examined so as to illustrate the main computational steps of the proposed methodology and, lastly, in Section 5, concluding remarks are made.

2. Background

2.1. Multi-Parametric Programming

Overall, multi-parametric programming problems are concerned with the effect of parametric perturbations on the optimal solution of an optimisation problem. Consider the following optimisation problem:

$$\begin{aligned} z(\boldsymbol{\theta}) = \quad & \min_{\mathbf{x} \in \mathbb{R}^{n_x}} \quad f(\mathbf{x}, \boldsymbol{\theta}) \\ & \text{subject to}: \quad \mathbf{g}(\mathbf{x}, \boldsymbol{\theta}) \leq 0 \\ & \qquad \boldsymbol{\theta} \in \mathbb{R}^{n_\theta} \end{aligned} \tag{1}$$

where $\boldsymbol{\theta}$ stands for the vector of uncertain parameters, which is n_θ-dimensional, $\mathbf{x}$ is the n_x-dimensional vector of continuous decision variables, $\mathbf{g}(\mathbf{x}, \boldsymbol{\theta})$ is the vector of inequality constraints and f is the objective function as a mapping $\mathbb{R}^{n_x \times n_\theta} \to \mathbb{R}$, both of which are assumed to be C^2 (twice continuously differentiable). Problem (1) is a multi-parametric program and its solution results in the partition of the parametric $\mathbb{R}^{n_\theta}$-space into a number of regions, also know as critical regions (CRs). Within each CR, the optimal solution and the objective value are given as functions of the uncertain parameters, i.e., $\mathbf{x}(\boldsymbol{\theta})$ and $\mathbf{z}(\boldsymbol{\theta})$, respectively. Even though mp-P has been studied quite actively, the class of multi-parametric nonlinear programming problems remains one of the most challenging ones [11,12]. Depending on the convexity of the nonlinear functions that form Problem (1), different solution techniques have been proposed in the literature to date.

Advances in the algorithms and theory of parametric nonlinear programs (p-NLPs) date back to the early works of Fiacco [13] and Bank et al. [14]. More specifically, in the books of Bank et al. [14] and Fiacco [13], a collection of the early research works for parametric NLPs can be found and invaluable theoretical foundations for some classes of convex p-NLPs with perturbations in the OFC and the right-hand side of the constraints are provided. Even though the term "parametric nonlinear optimisation/programming" was widely established from the aforementioned works, the early works on numerical stability analysis of NLPs by [15,16] and the work of Robinson [17,18] on generalised equations provided a significant way of studying the effect of parametric variations on the optimal solution of p-NLPs. Kyparisis [19] studied the uniqueness and differentiability of solutions of parametric nonlinear complementarity problems while in Ralph and Dempe [20], the directional derivatives of parametric nonlinear programs were used to characterise their explicit solution. However, the first algorithm for the multi-parametric case of convex NLPs was due to Dua and Pistikopoulos [21]. The authors, based on the findings about the convexity properties of the parametric value function ($z(\boldsymbol{\theta})$), devised an iterative procedure in which the integer variables were fixed by the solution of a primal mixed integer nonlinear program (MINLP) and the resulting mp-NLP was then transformed into an mp-LP following the outer approximation idea. Because of the value function's convexity property, the maximum error of the approximation occurs at the vertices of the CRs and if the error is greater than the prespecified tolerance, the CR is partitioned again; otherwise, integer and parametric cuts are implemented and then the algorithm iterates until the primal MINLP is infeasible. The same algorithm was revisited by Acevedo and Salgueiro [22], where the authors proposed heuristics to improve its computational efficiency while quadratic approximations were studied by Johansen [23] and Domínguez and Pistikopoulos [24]. An approximate algorithm for the solution of convex mp-NLPs was proposed by Johansen [25], who proposed the consecutive subdivision of the parametric space in hyper-rectangles and the interpolation of the parametric solution through the solution of 2^{n_θ} NLPs at each step. Further approaches involve the geometric vertex search by Narciso [26] and sub-gradient methods by Leverenz et al. [27]. For the nonconvex cases, Dua et al. [28] developed suitable parametric under/overestimators which were then incorporated into a spatial branch and bound routine for the global optimisation of the nonconvex problem within $\epsilon-$tolerance. For a more thorough discussion on the algorithms that have been proposed for the solution of mp-NLPs, the interested reader is directed to the review of Domínguez et al. [29] while Hale [30], in her doctoral thesis, also offers a thorough discussion on several classes of parametric optimisation. Fotiou et al. [31,32] initially studied the polynomial multi-parametric programming problem with application to control, however, their approach did not include the definition of final nonconvex CRs, while the mixed integer polynomial case was studied by Charitopoulos and Dua [33] and a procedure for the computation of exact nonconvex CRs was presented. Despite the aforementioned research effort, mp-NLP problems remain one of the most difficult to tackle and, as illustrated in Table 1, all the aforementioned algorithms can handle uncertain parameters only on the right-hand side (RHS) of the constraints. Recently, Pappas et al. [34], by generalising the basic sensitivity theorem of Fiacco [13], devised an algorithm for the exact solution of multi-parametric quadratically constrained quadratic programs.

Table 1. Summary of multi-parametric nonlinear programmming algorithms.

mp-NLP Solution Techniques	RHS	LHS	OFC	Comments
Dua and Pistikopoulos [21]	✓	-	-	Convex
Johansen [23]	✓	-	-	Convex
Acevedo and Salgueiro [22]	✓	-	-	Convex
Johansen [25]	✓	-	-	Convex
Dua et al. [28]	✓	-	-	Nonconvex
Fotiou et al. [31]	✓	-	-	Polynomial
Narciso [26]	✓	-	-	Convex
Domínguez and Pistikopoulos [24]	✓	-	-	Convex
Charitopoulos and Dua [33]	✓	-	-	Polynomial
Pappas et al. [34]	✓	-	-	QCQPs

Among the wide range of appplications that multi-parametric programming has been applied to, the invention of explicit model predictive control (mp-MPC) is undoubtedly the most dominant area where mp-P has had the biggest impact [10,12,35]. The main concept of mp-MPC is that instead of solving the optimisation problems related to standard MPC at each sampling instance, the state of the system is treated as an uncertain parameter and an mp-P can be solved offline to derive the explicit control solution once and for all [10,36].

The general formulation of mp-MPC for discrete time systems is shown by (2):

$$(mp - MPC)\begin{cases} \Phi(x(t_k)) = \min_{\mathbf{u}} \sum_{i=0}^{P_H-1} \mathcal{L}(x_t, u_t) + E(x_N) \\ \text{subject to:} \quad x_{t|t=0} = x(t_k) \\ \quad x_{t+1} = f(x_t, u_t) \quad t = 0, 1, \ldots, P_H - 1 \\ \quad y_{t+1} = h(x_t, u_t) \quad t = 0, 1, \ldots, P_H - 1 \\ \quad \mathbf{A}x_t \le \boldsymbol{\alpha} \quad t = 0, 1, \ldots, P_H \\ \quad \mathbf{B}y_t \le \boldsymbol{\beta} \quad t = 0, 1, \ldots, P_H \\ \quad \mathbf{C}u_t \le \gamma \quad t = 0, 1, \ldots, P_H \end{cases} \tag{2}$$

where $\mathbf{x}_t$, $\mathbf{u}_t$, $\mathbf{z}_t$ are the state, control input and system output vectors, respectively, at every sampling point, t, and are n_x, n_u, n_y-dimensional. **A**, **B**, **C** are matrices of appropriate dimensions and $\boldsymbol{\alpha}$, $\boldsymbol{\beta}$, γ vectors of pertinent dimensions which represent inequality constraints for the state, output and control inputs while $\mathcal{L}$: $\mathbb{R}^{n_x+n_u} \to \mathbb{R}$ is a stage cost and $E : \mathbb{R}^{n_x} \to \mathbb{R}$ is a terminal cost function. The repetitive solution of Problem (2) provides the optimal cost $\Phi(\mathbf{x}(t_k))$ and the optimisation vector, i.e., the sequence of optimal control inputs $\mathbf{u}^* = \left[\mathbf{u}_1^*, \mathbf{u}_2^*, \ldots, \mathbf{u}_{P_H-1}^*\right]$ over the finite prediction horizon P_H. Compared to the conventional model predictive control fashion, in which an optimisation problem is solved at each sampling point, through the mp-MPC notion, the explicit control law is calculated offline once and for all. The solution of the resulting mp-P problem results in the optimal control inputs as explicit functions of the (uncertain) parameters, i.e., the state of the system at each sampling instance, along with the corresponding CRs, as shown by Equation (3).

$$\mathbf{u}^* = \begin{cases} v_1(\mathbf{x}(\mathbf{t_k})) & if\ \mathbf{x}(\mathbf{t_k}) \in CR_1 \\ v_2(\mathbf{x}(\mathbf{t_k})) & if\ \mathbf{x}(\mathbf{t_k}) \in CR_2 \\ \vdots & \vdots \\ v_\omega(\mathbf{x}(\mathbf{t_k})) & if\ \mathbf{x}(\mathbf{t_k}) \in CR_\omega \end{cases} \tag{3}$$

For the case of MPC for linear systems, instead of solving a quadratic program at each sampling instance, the explicit MPC requires the offline solution of an mp-QP while online, so only simple function evaluations are required [10,37,38]. This concept is also known as online via offline optimisation and it is shown in Figure 3.

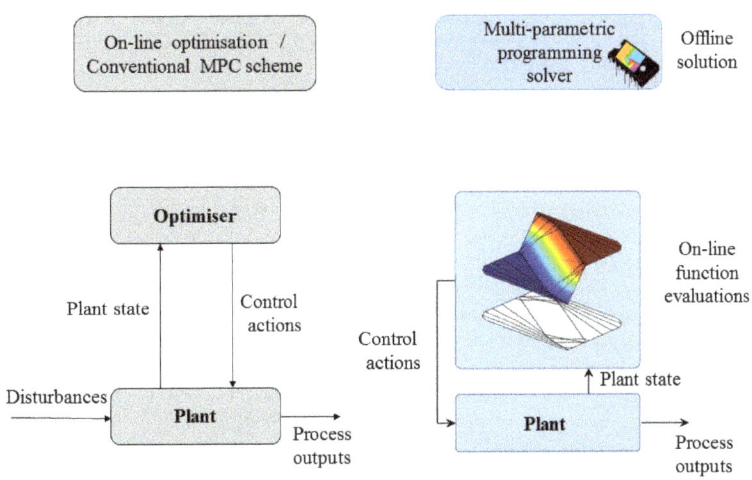

Figure 3. Online via offline optimisation framework [10,39].

Even though mp-MPC is the niche area of mp-P, designing mp-MPCs of nonlinear systems for set-point tracking is still a computationally strenuous task as one has to design an mp-MPC for each set-point based on the algorithms that exist in the literature to date [40]. Next, we review two mathematical techniques that will enable the development of novel multi set-point explicit controllers though the algorithm we propose in the present work.

2.2. Computer Algebra

2.2.1. Gröbner Bases Theory

The idea of the present work is to devise an algorithm for the solution of mp-NLPs from an analytical and not numerical perspective, and the reason is two-fold. Firstly, for the case that we are interested in, i.e., mp-NLPs with combined uncertainty on the RHS and OFC, no theoretical foundations exist for the computation of the explicit solution like the basic sensitivity theorem of Fiacco [13] which serves as the basis of the numerical mp-P approaches. Secondly, because of the nonconvex nature of the parametric problem, the numerical solution would require global optimisation techniques similar to the ones presented in Dua et al. [28] and would lead to an explosion in the number of convex approximate CRs.

Gröbner bases theory was introduced by the doctoral research of Bruno Buchberger [41] as a method of analytically solving systems of multivariate polynomial equations. In brief, Gröbner bases and the Buchberger algorithm can be considered as the polynomial counterpart of Gaussian elimination for the case of nonlinear systems. Before formally defining what a Gröbner basis is, it would be useful to define some preliminary concepts.

Definition 1. *Power products*

Let $\mathbb{R}$ be any field and let $\mathbb{R}[x_1, \ldots, x_n]$ be the ring of polynomials in n-indeterminates. Any polynomial can be described as a sum of terms of the form: $\alpha x_1^{\beta_1} \cdots x_n^{\beta_n}$ with $\alpha \in \mathbb{R}$ and $\beta_i \in \mathbb{N}$, $i = 1, \ldots, n$ and the term $x_1^{\beta_1} \cdots x_n^{\beta_n}$ is called a power product.

Definition 2. *Term order*

A term order is defined with regard to a set of power products (T_n) and imposes a total order $<$ on the set in compliance with the conditions below:

1. $1 < x^\beta$ *for all* $x^\beta \in T_n$
2. *If* $x^\alpha < x^\beta \rightarrow x^\alpha x^\gamma < x^\beta x^\gamma$, *for all* $x^\gamma \in T_n$

A number of alternative power product orderings exist but the most commonly employed is the the lexicographic one due to its computational efficiency [42]. Lastly, the notion of *ideals* is crucial within the Gröbner bases theory.

Definition 3. *Ideals*

Let $\mathbb{R}$ be a field and $\mathbb{R}[x_1, x_2, \ldots, x_n]$ be a ring over the field of n-variate polynomials. Let a finite subset of the field, $G = \{g_1, g_2, \ldots, g_t\}$, then an ideal I can be generated by G as follows:

$$I = \{\sum_{i=1}^{n} u_i g_i | \; u_i \in \mathbb{R}[x_1, x_2, \ldots, x_n],\; g_i \in \mathbb{R}, \forall i$$

For problems that accept analytical solutions, their existence is guaranteed by the Hilbert basis theorem [43], which also guarantees that algorithms used to compute Gröbner bases can terminate in a finite number of steps.

Definition 4. *Gröbner basis [41]*

A set of non-zero polynomials $G = \{g_1, \ldots, g_t\}$, contained in an ideal I, is called a Gröbner basis for I if and only if for all $g \in I$, such that $g \neq 0$, there exists $i \in \{1, \ldots, t\}$ such that $lp(g_i)$ divides $lp(g)$, where $lp(\cdot)$ stands for the leading power product of a polynomial function.

For the calculation of Gröbner bases, apart from Buchberger's algorithm, Faugère has presented algorithms F4 [44] and F5 [45] as two variants of another algorithm. They exploit concepts from linear algebra and represent polynomials using matrix forms, thus enabling successive truncated Gröbner bases to be created. Lastly, software implementations of different algorithms that compute Gröbner bases computations can be found in freely available CAS such as Singular, SymPy and SageMath as well as commercial tools like Maple and Mathematica.

2.2.2. Cylindrical Algebraic Decomposition (CAD)

The notion of cylindrical algebraic decomposition was presented by Collins in 1975 [46] in an effort to solve the problem of quantifier elimination over real closed fields. In this article, as will be shown later on, CADs are used for computing nonconvex regions in the space of parameters. Thus, we provide the following definitions for ease of exposition in the algorithmic steps that we detail later on in the manuscript.

Definition 5. *Semi-Algebraic Sets [43].*

Let $\mathbb{R}[x_1, x_2, \ldots, x_n]$ indicate the ring of polynomials in n-indeterminates with real coefficients. If, for example, a subset $\mathcal{S}$ of $\mathbb{R}^n$ can be developed by a finite number of applications of the complementation, union and intersection operations, it is called semi-algebraic and can have the following form:
$\{\mathbf{x} \in \mathbb{R}^n | \, \mathbf{g}(\mathbf{x}) \leq 0\}$, where $\mathbf{g} \in \mathbb{R}[X]$

Definition 6. *Standard atomic formula*
A formula that includes a functional relation over a polynomial ring in either of the ways shown below is called a standard atomic formula:

$$g(\mathbf{x}) = 0,\ g(\mathbf{x}) \neq 0,\ g(\mathbf{x}) < 0,\ g(\mathbf{x}) > 0,\ g(\mathbf{x}) \leq 0,\ g(\mathbf{x}) \geq 0$$

Proposition 1 ([43]). *Semi-algebraic sets of $\mathbb{R}^n$ can be written as a finite union of semi algebraic sets of the form:*

$$\{\mathbf{x} \in \mathbb{R}^{nx} | \, g_1(\mathbf{x}) = \ldots = g_\omega(\mathbf{x}) = 0,\ g_{\omega+1}(\mathbf{x}) > 0, \ldots, g_t(\mathbf{x}) > 0\}$$

where $g_1 \ldots, g_\omega, g_{\omega+1}, \ldots, g_t$ are in $g \in \mathbb{R}[X]$.

The proof of the proposition can be found in the book of Bochnak et al. [43].

Using the definitions and propositions given above, in summary, one can use CAD routines to compute the solution to polynomial inequalities. In the process of computations, one partitions the related space over finite cells and qualifies whether or not standard atomic formulas hold. A comprehensive exposition on the solution of polynomial inequalities using CAD is given at the book of Jirstrand [47].

3. Algorithms and MPC

3.1. Multi Set-Point Explicit Controller via Multi-Parametric Programming

In the context of multi-parametric model predictive control, the state of the system at each sampling point is treated as an uncertain parameter and as a result an mp-P with RHS uncertainty arises [10,37,38,40]. Its solution results in the explicit control law, i.e., the control decisions as explicit functions of the system's initial conditions at a sampling instance along with the related CRs.

In many applications, however, particularly those related to continuous manufacturing, there is a great need for fast calculations in order to communicate decisions between the different layers of decision making in an effective manner. For instance, set-point tracking goals for APC are, most of the time, passed down from either the functionality of process scheduling or RTO [39,48]. In these cases, it becomes obvious that explicit MPC can provide a significant advantage in computational time by treating the set-points or estimated model inputs as uncertain parameters. One way to design such explicit controllers, assuming the existence of the n_u set-point, is to solve n_u mp-P problems and thus design n_u mp-MPCs. Another way, which has not been investigated in the literature, is to consider the set-points and/or model inputs as uncertain parameters and thus derive a multi set-point explicit MPC. Conceptually, by doing so, we would design a layered controller as given by Figure 4.

Conventionally, when mp-MPC is employed for set-point tracking of nonlinear systems, one would have to compute a different mp-MPC for each of those set-points as well as account for any time delay in the offline solution of the related mp-P should a new set-point arise. Stringent market regulations and an increasingly volatile market environment lead process industries to constantly optimise their operations and give rise to new set-points from a control perspective which in turn hinder the deployment of mp-MPC. As illustrated by Figure 4, in this article, by considering the set-points as uncertain parameters which lie within prespecified bounds, we overcome the abovementioned drawback of explicit MPC since, by solving one mp-NLP, we can design a "multi set-point" mp-MPC for nonlinear systems.

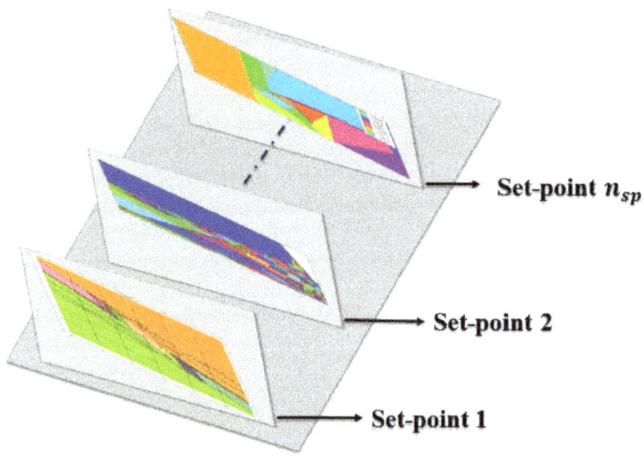

Figure 4. Concept of a multi set-point mp-MPC setting where instead of separate explicit controllers one designs a universal explicit MPC for all the set-points.

The design problem of a multi set-point mp-MPC can be formulated as in (4).

$$\begin{cases} Y(x(t_k),\ x_{sp}) = \min\limits_{\mathbf{u}} \sum\limits_{t=0}^{P_H-1} \mathcal{L}(x_t, u_t, x_{sp}) + E(x_N, x_{sp}) \\ \text{subject to:} \quad x_{t|t=0} = x(t_k) \\ \qquad x_{t+1} = f(x_t, u_t) \qquad t = 0, 1, \ldots, P_H - 1 \\ \qquad y_{t+1} = h(x_t, u_t) \qquad t = 0, 1, \ldots, P_H - 1 \\ \qquad \mathbf{A}x_t \leq \boldsymbol{\alpha} \qquad t = 0, 1, \ldots, P_H \\ \qquad \mathbf{B}y_t \leq \boldsymbol{\beta} \qquad t = 0, 1, \ldots, P_H \\ \qquad \mathbf{C}u_t \leq \boldsymbol{\gamma} \qquad t = 0, 1, \ldots, P_H \\ \qquad x_{sp} \in \mathbb{R}^{n_{sp}} \end{cases} \tag{4}$$

The notation adopted is the same as in the previous section, aside from the following: we treat the set-points together with the initial state of the system as uncertain parameters, thus resulting in a problem with simultaneous variations on the RHS and OFC. As indicated by (4), we consider both the initial states ($x(t_k)$) and the various set-points (x_{sp}) as uncertain parameters. Notice that within an EWO framework, the deployment of such universal controllers is of great importance since they allow for rapid communication between the layer of control with RTO and scheduling. For instance, when integrating scheduling with control, the changeover times as well as the production rates are immediate results of the dynamic decisions made through the control system. In the next section, an algorithm for the design of such "multi set-point" explicit MPC is presented.

3.2. Solution Algorithm for Analytical mp-NLPs with Global Uncertainty

Here, we present an algorithm that can solve multi-parametric nonlinear programs with non transcendental nonlinear terms, i.e., nonlinear terms that have closed-form solutions. The proposed method can be seen as a generalisation of the our previous work on the solution of multi-parametric mixed integer polynomial programs [33] as well as the algorithm of Fotiou et al. [32] for multi-parametric polynomial programs. However, both of these methods were employed only for instances that the uncertainty is present on the right-hand side of the constraints and the latter does not compute the critical regions associated with each explicit solution.

The main idea of the algorithm proposed herein can be explained as follows: given a multi-parametric nonlinear program with analytical terms, or terms that can be expressed in a nontranscendental fashion, derive the first order KKT conditions and compute its solution using Gröbner bases by treating the uncertain parameters as symbols. The output of this step is a collection of candidate solutions which are explicit in $\boldsymbol{\theta}$ and encompass: global and local optima as well as infeasible solutions. For these solutions, examine their dual and primal feasibility along with a constraint qualification so as to remove infeasible candidate solutions. Lastly, in order to report only the globally optimal solutions, perform a comparison procedure [33].

Problem (1) details a general formulation of multi-parametric programs. The case that f and/or $\mathbf{g}$ are analytically nonlinear and the uncertain parameters are in the OFCs along with the RHS and LHS of the constraints is used.

Deriving the 1st order KKT conditions of Problem (1) returns a system of equations that is square and is given by Equations (5) and (6).

$$\nabla_{\mathbf{x}} L(\mathbf{x}, \boldsymbol{\theta}) = \mathbf{0} \tag{5}$$

$$\boldsymbol{\lambda}^T \mathbf{g}(\mathbf{x}, \boldsymbol{\theta}) = \mathbf{0} \tag{6}$$

$L(\mathbf{x}, \boldsymbol{\theta})$ is the Lagrangian function of Problem (1), $\nabla_{\mathbf{x}}$ is the nabla operator with respect to the decision variables and $\boldsymbol{\lambda}$ are the Lagrange multipliers corresponding to the constraints. Because of the assumption that the nonlinearities have an analytical solution, Gröbner bases can be employed for the solution of the square system of equations because of its elimination property. Even though a tailored implementation of one of the already existing algorithms for computing Gröbner bases may be advantageous from a computational standpoint, it is beyond the scope of the present work and thus Mathematica 10 was employed as the computer algebra software in which the calculations were performed.

By solving the system of Equations (5) and (6), a number of candidate solution sets are returned. Note that although the original optimisation problem involves n_x variables, in the current step, the variables for which we compute the explicit solution are $n_x + n_g$. The candidate solutions include the Lagrange multipliers together with the optimisation variables as explicit functions of the uncertain parameters, i.e., $\boldsymbol{\lambda}(\boldsymbol{\theta})$ and $\mathbf{x}(\boldsymbol{\theta})$, respectively.

Definition 7. *Candidate solutions [49]*

A solution of the Problem (1) is said to be candidate if it satisfies the system of Equations (5) and (6) along with a constraint qualification, e.g., linear independence constraint qualification [15].

In this part of the algorithm, due to strict complementary slackness, the collection of candidate solutions indicates the active and inactive constraints for each solution. Until this step, the set of solutions computed may be infeasible, local or global optima. By evaluating the primal and dual feasibility of the candidate solutions, the infeasible solutions can be rejected, i.e., Equations (7) and (8).

$$\mathbf{g}(\boldsymbol{\theta}) \leq \mathbf{0} \Longrightarrow \textit{feasibility conditions} \tag{7}$$

$$\boldsymbol{\lambda}(\boldsymbol{\theta}) \geq \mathbf{0} \Longrightarrow \textit{optimality conditions} \tag{8}$$

Conditions (7) and (8) are evaluated by substituting the explicit expressions of the optimisation variables and they form a collection of parametric constraints. If for a candidate solution there exists a subset of the initial parametric space such that the aforementioned inequalities are satisfied, then this region is called the CR of the candidate feasible solution; otherwise, the candidate solution is infeasible and thus removed from further consideration. Note that the evaluation performed at this step, from a computer algebra perspective, is equivalent to computation of the corresponding CAD.

Definition 8. *Critical region [39,49]*
A critical region (CR) is a partition of the parametric space where Conditions (7) and (8) are satisfied for a specific candidate solution. A critical region is characterised by a set of inactive/active constraints and can be discontinuous or nonconvex.

In Algorithm 1, the pseudo-code of the presented method is given. The comparison procedure is outlined in [33].

3.3. Illustrative Example

The proposed methodology will be motivated through the following modified example by Domínguez et al. [29].

$$\min_{x_1,x_2} x + 2x_1^2 - 5x_1 + x_2^2 - 3\theta_1 x_2 - 6$$

Subject to:

$$\begin{gathered} 2x_1 + x_2 \leq 2.4 - \theta_2 \\ 0.5\theta_3 x_1 + x_2 \leq 1.5 \\ x_1 \geq 0,\ x_2 \geq 0 \\ 0 \leq \theta_1 \leq 6,\ 0 \leq \theta_2 \leq 4,\ 0 \leq \theta_3 \leq 2 \end{gathered} \tag{9}$$

In the beginning, the first order KKT conditions of (9) are formulated and we derive the square system of Equations (10) and (14)

$$\nabla_{\mathbf{x}} L(\mathbf{x}, \boldsymbol{\lambda}, \boldsymbol{\theta}) = \mathbf{0} \tag{10}$$

$$\lambda_1(2x_1 + x_2 - 2.5 + \theta_2) = 0 \tag{11}$$

$$\lambda_2(0.5\theta_3 x_1 + x_2 - 1.5) = 0 \tag{12}$$

$$\lambda_3(-x_1) = 0 \tag{13}$$

$$\lambda_4(-x_2) = 0 \tag{14}$$

where $L(\mathbf{x}, \boldsymbol{\lambda}, \boldsymbol{\theta})$ is the Langangian of Problem (9). Equations (10) and (14) can be analytically solved through symbolic computations which return the dual and primal variables as functions of the uncertain parameters, i.e., $\boldsymbol{\lambda}(\boldsymbol{\theta})$ and $\boldsymbol{x}(\boldsymbol{\theta})$. Systems (10)–(14) are solved in 0.006 s and, as shown in Table A1, fifteen candidate solutions are computed.

The primal and dual feasibility of the candidate solutions is examined by computing the CAD of the related disjunctions. If the result of this step is "False" the solution violates feasibility (in either the primal or dual sense), otherwise, a collection of explicit inequalities is returned which characterises the candidate solution's CR. The CAD computations of this example take 5.33 s and nine of them are nonempty. Despite the fact that nine candidate solutions are primal and dual feasible, their global optimality is not guaranteed given the nonconvex nature of the problem. To this effect, for those regions, a new set of CAD computations is performed so as to identify overlaps.

Algorithm 1: mp-NLP under global uncertainty

Input: $f, \mathbf{g}, \mathbf{x}, \boldsymbol{\theta}$
Output: $\mathbf{x}(\boldsymbol{\theta}), CRs$

Formulate 1st order KKT conditions of mp-NLP
Solve the 1st order KKT conditions of mp-NLP using Gröbner Bases
$TEMP \leftarrow solutions$, i.e., $\mathbf{x}(\boldsymbol{\theta}),\ \boldsymbol{\lambda}(\boldsymbol{\theta}),\ \mathbf{z}(\boldsymbol{\theta})$
if $TEMP = \emptyset$ **then**
  mp-NLP is infeasible.
else
  for ($i \in range(1, \ldots, Length[TEMP])$ **) {**
    Evaluate with a first order constraint qualification, e.g. Linear Independence Constraint Qualification (LICQ)
    Evaluate with primal and dual feasibility conditions (**Cylindrical Algebraic Decomposition computation**):
    $CR = \{\exists\, \boldsymbol{\theta} \text{ such that } [\boldsymbol{\lambda}_i(\boldsymbol{\theta}) \geq 0] \wedge [\mathbf{g}_i(\boldsymbol{\theta}) \leq 0]\}$
    if $CR = \emptyset$ **then** Candidate solution i is infeasible and discard from TEMP.
    else
      Candidate solution i is feasible and append CR_i to TEMP.
    i+=1
  for ($(k, j) \in range(1, \ldots, Length[TEMP]) \wedge k \neq j$ **) {**
    Identify if any overlapping CRs exist (**Cylindrical Algebraic Decomposition computation**):
    $CR_{int} = \{\boldsymbol{\theta} | CR_k \wedge CR_j\}$
    if $CR_{int} = \emptyset$ **then** The two CRs are not overlapping
    else
      Follow the **comparison procedure** from Charitopoulos et al. (2016) so as to remove the overlaps
Collect the final non-overlapping CRs and the corresponding explicit solutions, i.e., $x(\theta)$

It was found that CR_{10} and CR_{11} were overlapping, as shown by Figure 5, where the overlap (CR_{int}) is shown as the dark partition in between the two CRs.

For the elimination of the resulting overlap, the comparison procedure is invoked and the logic disjunction, as illustrated below, is used for the CAD, as shown by Equations (15) and (16).

$$\exists \boldsymbol{\theta} |\ \{\theta_1, \theta_2, \theta_3\} \in CR_{int}\ \wedge\ z_{CR10}(\boldsymbol{\theta}) \leq z_{CR11}(\boldsymbol{\theta}) \tag{15}$$

or

$$\exists \boldsymbol{\theta} |\ \{\theta_1, \theta_2, \theta_3\} \in CR_{int}\ \wedge\ z_{CR10}(\boldsymbol{\theta}) \geq z_{CR11}(\boldsymbol{\theta}) \tag{16}$$

where $z_{CRi}(\boldsymbol{\theta})$ denotes the optimal explicit value within CR_i. The result of this step is partitioning of the parametric space where each explicit solution is the globally optimal. In this case, CR_{10} was shown to be dominant within the common parametric space and thus the overlap was subtracted from CR_{11}. The algorithm terminates once no more overlapping critical regions are identified. In Table 2, an overview of the explicit solutions is presented, while in Figure 6, the final critical regions are shown. Practically, one would consult the CR column of Table 2 to identify based on the uncertain parameter values where the uncertainty is realised, i.e., CR1 or CR2, and then the optimal cost can be computed by evaluating the corresponding expression from the "Explicit solution" column.

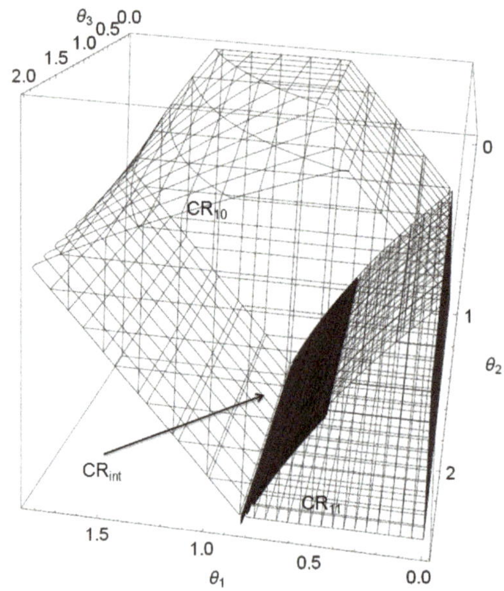

Figure 5. Instance of overlap between CRs.

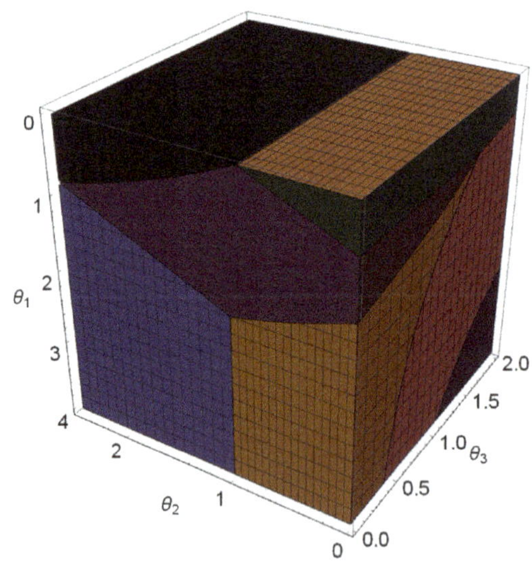

Figure 6. Visualition of the critical regions for motivating example.

Table 2. Optimal explicit solutions of motivating example.

CRs	Explicit Solution
$CR_1 = \begin{cases} 2.667 \leq \theta_1 \leq 4. \\ 0 \leq \theta_2 \leq 1 \\ \frac{10}{3\theta_1 - 3} \leq \theta_3 \leq 2 \end{cases}$	$z_1(\boldsymbol{\theta}) = -3.75 - 4.5\theta_1$
$CR_2 = \begin{cases} \begin{cases} 0.666667\sqrt{19 - 18\theta_1} + \theta_2 \geq 3.8333 \\ 0 \leq \theta_1 \leq 2.005302497 \cdot 10^{-16} \\ 1.5\theta_1 + \theta_2 \leq 0.927401 \\ 0 \leq \theta_3 \leq 2 \end{cases} \\ \\ \begin{cases} 3.2118 \cdot 10^{-8}\sqrt{8.186 \times 10^{15} - 7.7552 \times 10^{15}\theta_1} + \theta_2 \geq 3.8333 \\ 0 \leq \theta_1 \leq 0.8333 \\ \theta_2 \leq 2.5 \\ 0 \leq \theta_3 \leq 2 \end{cases} \end{cases}$	$z_2(\boldsymbol{\theta}) = 0.015625(5 - 2\theta_2)^3 + 0.125(5 - 2\theta_2)^2 - 1.25(5 - 2\theta_2) - 6$

4. Case Studies

Here, we examine two case studies from process systems and their corresponding explicit controllers are designed. The computatonal experiments were performed on a workstation with 24 GB RAM, a 3.80 GHz processor and a Windows 10 64-bit operating system. For the symbolic calculation, the computer algebra system that was employed was Mathematica 10.2.

4.1. Multiple-Input Multiple-Output Non-Isothermal CSTR

We examine a non-isothermal MIMO multi-product CSTR where the decomposition reaction A $\rightarrow$R happens under the kinetic law: $-\mathscr{R}_b = k_r C_b$. Additional details on the design and kinetics can be found in Camacho and Alba [50] and the data used for this case study can be found in Table 3. The system has two control inputs: the liquid (F_l) and coolant (F_c) flow rates, whereas the system's states are the temperature of the liquid (T_l) and the concentration of the decomposition product (C_b). Using the mass and energy balances, the dynamic model of the system is derived as given by Equations (17) and (18).

$$\frac{d(V_l C_b)}{dt} = V_l k_r (C_{a0} - C_b) - F_l C_b \tag{17}$$

$$\frac{d(V_l \rho_l C_{pl} T_l)}{dt} = F_l \rho_l C_{pl} T_{l0} - F_l \rho_l C_{pl} T_l + F_c \rho_c C_{pc} (T_{c0} - T_c) + V_l k_r (C_{a0} - C_b) H \tag{18}$$

Table 3. Data of the multiple-input multiple-output CSTR case study.

k_r reaction constant	$26\ \frac{1}{h}$
V_l tank volume	$24\ L$
ρ_l liquid density	$800\ \frac{kg}{m^3}$
ρ_c coolant density	$1000\ \frac{kg}{m^3}$
C_{pl} specific heat of liquid	$3\ \frac{kJ}{kg \cdot K}$
C_{pc} specific heat of coolant	$4.19\ \frac{kJ}{kg \cdot K}$
T_{l0} entering liquid temperature	283 K
T_{c0} inlet coolant temperature	273 K
T_c outlet coolant temperature	303 K
C_{a0} initial concentration of the reactant	$4\ \frac{mol}{L}$

Firstly, the systems (17) and (18) are transformed into an algebraic one in order to design the explicit controllers. Using the forward Euler method, the MPC problem of the discretised system is shown by Equations (19) and (20).

$$\min_{\mathbf{u}} J(\boldsymbol{\theta}) = \sum_{t=0}^{P_H} \left\| \boldsymbol{x}(t) - \boldsymbol{x}_{ref} \right\|_2 \tag{19}$$

Subject to:

$$\begin{cases} C_{b_{t+1}} = C_{b_t} + \frac{h_e(k_r(C_{a0}-C_{b_t})-F_{l_t}C_{b_t})}{V_l}, \ 0 \le t \le P_H - 1 \\ T_{l_{t+1}} = T_{l_t} + h_e \frac{F_{l_t}\rho_l C_{pl} T_{l0} - F_{l_t}\rho_l C_{pl} T_{l_t} + F_{c_t}\rho_c C_{pc}(T_{c0}-T_c) + V_l k_r (C_{a0}-C_{b_t})}{V_l \rho_l C_{pl}} \\ 0.8 \le C_{b_t} \le 3.5, \ 0 \le t \le P_H \\ 280 \le T_{l_t} \le 400, \ 0 \le t \le P_H \\ 0 \le F_{c_t} \le 1000, \ 0 \le t \le P_H \\ 0 \le F_{l_t} \le 2000, \ 0 \le t \le P_H \\ C_{b_{t|t=0}} = \theta_1, T_{l_{t|t=0}} = \theta_2 \\ C_b^{ref} = \theta_3, T_l^{ref} = \theta_4, \end{cases} \tag{20}$$

By employing the presented solution algorithm for mp-NLPs, the related KKT system is solved analytically using Gröbner bases. It takes 0.76 s to compute 29 candidate solutions explicit in θ_1, θ_2 and a collection is shown in Table 4.

Table 4. Collection of candidate solutions of MIMO mp-MPC.

Candidate Solution	F_c	F_l
1	0	0
2	2000	$-0.2011\theta_1 - 38.178\theta_2 - 0.00796\theta_4 + 10807.5$
3	$\frac{23376(\theta_1 - 0.714579)}{\theta_1}$	$-\frac{0.2011(\theta_1^2 + 2219.4\theta_1\theta_2 - 628088.\theta_1 - 1585.96\theta_2 + 448827)}{\theta_1}$
4	$\frac{23376.\theta1 - 24000.\theta3 + 2496.}{\theta_1}$	$\frac{-0.201\theta_1{}^2 - 446.32\theta_1\theta_2 + 126307\theta_1 + 458.23\theta_2\theta_3 - 47.66\theta_2 - 129680\theta_3 + 13486.7}{\theta_1}$

As mentioned in the previous section, the proposed algorithm can facilitate both continuous as well as discrete set-points for the solution of the resulting problem. For the MIMO case study, we consider 8 different set-points for which the explicit control law is derived. In Table A2, we outline the explicit solutions for the different set-points while the optimal partition of the uncertainty space is shown in Figure 7.

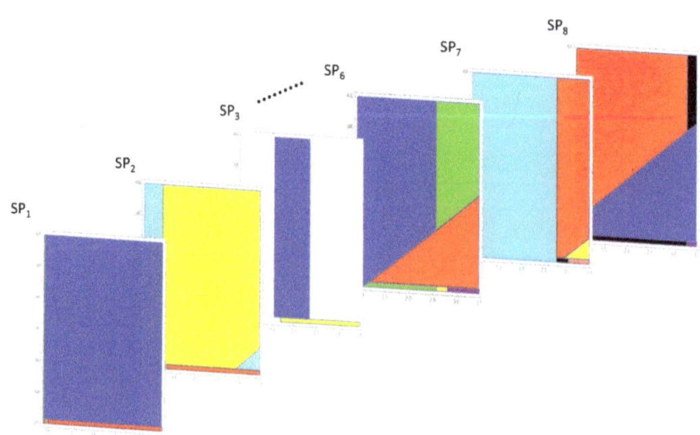

Figure 7. Critical regions for the mp-MPC controller for the MIMO CSTR.

We validate the performance of the explicit multi set-point controller by examining the transition between two steady states. We benchmark the controller's predictions against the globally optimal solution as computed using the BARON 14.4 solver. In Figure 8, the control and state evolution can be seen.

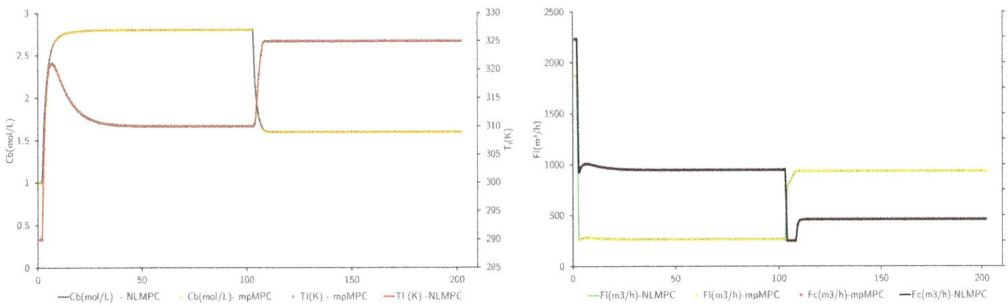

(**a**) $[C_b, T_l] = f(t)$ (**b**) $[F_c, F_l] = f(t)$

Figure 8. Comparative graphs of the control performance of the mp-MPC vs. NLMPC for case study 1.

4.2. *Isothermal Polymerisation CSTR*

Next, we examine the design of a multi set-point controller for the grade transition problem with polymerisation CSTRs. The model nonlinearities involve bilinear terms and square roots. The free radical polymerisation reaction happens in a CSTR that operates isothermally at 335 K, where methyl methacrylate (MMA) is produced [51,52]. The mathematical model is given by Equations (21) and (25). The system has 4 state variables, i.e., the concentrations of the monomer (C_m) and the initiator (C_I), the dead chains' molar concentration (D_0) and the dead chains' mass concentration (D_I). The control input is the flow rate of the initiator (F_I) and one output, i.e., the molecular weight of the polymer produced (y). In the multiple steady states, different polymeric grades can be produced corresponding to alternative molecular weights. We provide the notation for this system in Table 5 and model parameter values can be found in Table 6.

$$\frac{dC_m}{dt} = -(k_p + k_{fm})\sqrt{\frac{2f^*k_IC_I}{k_{Td}+k_{Tc}}}C_m + \frac{F(C_{m_{in}} - C_m)}{V} \tag{21}$$

$$\frac{dC_I}{dt} = \frac{F_IC_{I_{in}} - FC_I}{V} - k_IC_I \tag{22}$$

$$\frac{dD_0}{dt} = (0.5k_{Tc} + k_{Td})\frac{2f^*k_IC_I}{k_{Td}+k_{Tc}}Cm + k_{fm}\sqrt{\frac{2f^*k_IC_I}{k_{Td}+k_{Tc}}}C_m - \frac{FD_0}{V} \tag{23}$$

$$\frac{dD_I}{dt} = M_m(k_p + k_{fm})\sqrt{\frac{2f^*k_IC_I}{k_{Td}+k_{Tc}}}C_m - \frac{FD_I}{V} \tag{24}$$

$$y = \frac{D_I}{D_0} \tag{25}$$

Table 5. MMA CSTR notation.

C_m (kmol/m^3)	state: monomer concentration
C_I (kmol/m^3)	state: initiator concentration
D_0 (kmol/m^3)	state: molar concentration of dead chains
D_I (kg/m^3)	state: mass concentration of dead chains
F_I (m^3/h)	control: initiator flow rate
$y = D_I/D_0$	output: molecular weight

Table 6. Model parameters for the MMA polymerisation reactor.

$F = 10.0\ \text{m}^3/\text{h}$	monomer flow rate
$V = 10.0\ \text{m}^3$	reactor volume
$f^* = 0.58$	initiator efficiency
$k_p = 2.50 \times 10^6\ \frac{\text{m}^3}{\text{kmol}\cdot\text{h}}$	propagation rate constant
$k_{Td} = 1.09 \times 10^{11}\ \frac{\text{m}^3}{\text{kmol}\cdot\text{h}}$	termination by disproportionation rate constant
$k_{Tc} = 1.33 \times 10^{10}\ \frac{\text{m}^3}{\text{kmol}\cdot\text{h}}$	termination by coupling rate constant
$C_{I_{in}} = 8.00\ \text{kmol}/\text{m}^3$	inlet initiator concentration
$C_{m_{in}} = 6.00\ \text{kmol}/\text{m}^3$	inlet monomer concentration
$k_{fm} = 2.45 \times 10^3\ \frac{\text{m}^3}{\text{kmol}\cdot\text{h}}$	chain transfer to monomer rate constant
$k_I = 1.02 \times 10^{-1}\text{h}^{-1}$	initiation rate constant
$M_m = 100.12\ \text{kg}/\text{kmol}$	molecular weight of monomer

The model nonlinearities are not transcedental and, thus, the presented method for the design of the multi set-point mp-MPC can be used. This polymerisation system has been examined intensively by the research community and it has been noted that online computation of its optimal control law can be challenging due to numerical instabilities arising because of scaling issues [52]. As a trade-off between computational complexity and stability of the integration scheme, we employ the forward Euler method with a step size of $h = 36$ s.

Following the proposed method, the globally optimal solutions are computed, whereas employing off-the-self global optimisation solvers for online implementation leads to extensive computational times. In Table 7, the results using the BARON 14.4 solver in GAMS with different prediction horizons are given.

Table 7. Computational effort for varying prediction horizons using BARON 14.4 [53].

Prediction Horizon	CPU (s)
1	603
2	1906
5	3600 †
10	3600 †
20	3600 †

† Reached time limit.

In this case study, eight uncertain parameters were considered, one for each state and set-point. In Table 8, we provide a mapping of the uncertain parameters of the control problem. Whilst having the set-point as continuous, uncertain parameters increase the computational complexity of the mp-P and one could argue that in the context of systems integration where the APC receives data by the real-time optimisation functionality, the same set-points may not always be realised. In such cases, following conventional mp-MPC frameworks, the explicit laws would have to be recomputed from the beginning (mp-P solution and implementation of the explicit solutions, possibly in a microchip), whereas, following the proposed framework, if the bounds of the set-points remain within the prespecified ranges, then the same multilayer controller can be readily used.

Table 8. MMA CSTR uncertain parameters.

Parameter	Range	Notation
θ_1	$\in [0, 5]$	$C_{m\|t=0}$
θ_2	$\in [0, 0.5]$	$C_{I\|t=0}$
θ_3	$\in [0, 0.05]$	$D_{0\|t=0}$
θ_4	$\in [0, 300]$	$D_{I\|t=0}$
θ_5	$\in [0, 5]$	Set-points for C_m
θ_6	$\in [0, 0.5]$	Set-points for C_I
θ_7	$\in [0, 0.05]$	Set-points for D_0
θ_8	$\in [0, 300]$	Set-points for D_I

The resulting mp-NLP involves one optimisation variable, eight uncertain parameters and ten constraints when a prediction horizon of unity is considered. Overall, we seek analytical solutions to eleven variables, i.e., the Lagrange multipliers and the optimisation variables. The solution of the mp-NLP returns five candidate solutions, as shown in Table A3.

Substituting the explicit expressions into the constraints, the feasible region is projected into the uncertainty space. Candidate solutions that satisfy primal/dual feasibility are considered for the next step of the algorithm; otherwise, they are discarded as infeasible. For instance, the 6th candidate solution violates dual feasibility as any value of θ_6 would result in negative λ_8.

Subsequently, the explicit inequalities for the remaining solutions are examined. The intersection of the feasible regions defined by the parametric inequalities defines the critical regions of the candidate solution. Because of the nonconvex nature of the problem, it is likely that explicit solutions may be valid in the same uncertainty space, thus overlapping. In order to compute only the global explicit solutions, we employ the comparison procedure. Three overlapping solutions were identified. An example of the inequalities defining the overlap between CR_1, CR_2 is shown by Equation (26).

$$CR_{int} := CR_1 \cap CR_2 = \begin{cases} 0 \leq \theta_3 \leq 0.05 \\ 0 \leq \theta_4 \leq 300 \\ 0 \leq \theta_5 \leq 5 \\ 0 \leq \theta_7 \leq 0.05 \quad \begin{cases} 4.9899 \leq \theta_1 \leq 5 \\ 1617.74 + \frac{40280.2}{\theta_1^2} \leq \frac{16144.7}{\theta_1} + \theta_2 \\ \theta_2 \leq \frac{1.48197}{\theta_1^2} \end{cases} \\ 0 \leq \theta_4 \leq 300 \\ \theta_2 + 0.00242674 = 1.01114\theta_6 \\ 0 \leq \theta_2 \leq 0.199802 \\ 0 \leq \theta_1 \leq 2.72345 \end{cases} \tag{26}$$

For illustration purposes, the mathematical definition of CR_2 is given by Equation (A1) in the Appendix A. Due to the extensive set of inequalities defining the rest of the CRs, we do not detail them in the manuscript for the sake of space.

After the algorithm's convergence and with the optimal explicit solutions reported, the performance of the multi set-point mp-MPC's explicit control law is compared to that of conventional MPC. The solution of the online MPC is found by implementing the related NLP in GAMS and solving it to global optimality using BARON 14.4. As can be seen in Figure 9, the state and control evolution of the system are in perfect agreement when the two schemes are compared, thus highlighting the accuracy and correctness of the proposed framework while in Figure 10 the stability of the resulting control policy can be envisaged.

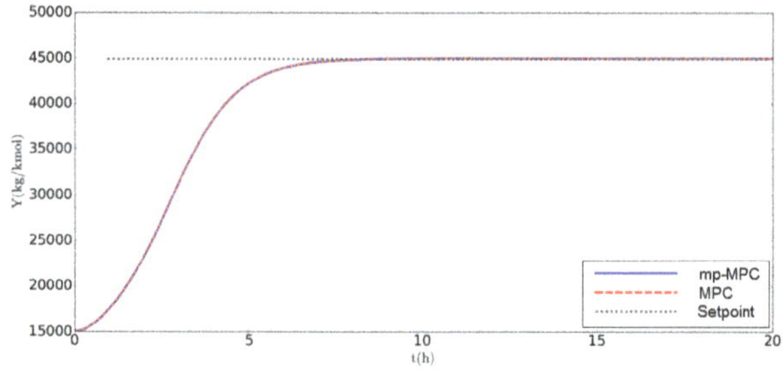

(**a**) $y = f(t)$

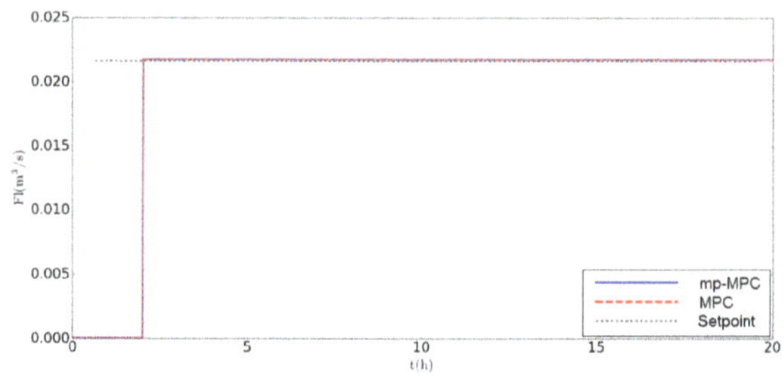

(**b**) $F_I = f(t)$

Figure 9. Plots comparing the control solutions computed by the proposed method (mp-MPC) and online NLMPC for the polymerisation CSTR.

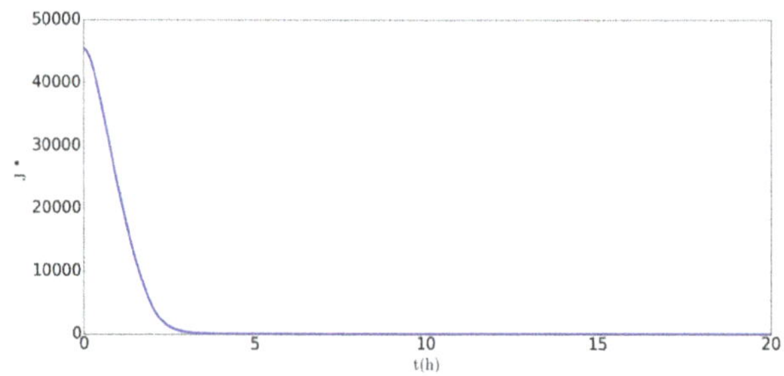

Figure 10. Graph of the control cost function vs. time for the polymerisation CSTR.

We assess the performance of the controller for set-point tracking between two set-points. At the beginning, we assume the CSTR to be operated at a steady state of $y = 15{,}000\frac{kg}{kmol}$ and then controlled towards $y = 45{,}000\frac{kg}{kmol}$ where it is regulated for nine hours to produce a specific polymer grade. Next, the controller steers the system to the next set-point ($y = 19{,}250\frac{kg}{kmol}$), in which steady state another polymer is produced. The performance of the set-point tracking can be seen in Figure 11.

Finally, with respect to the scalability of the proposed method, a number of systems and prediction horizon settings were examined and, as shown in Table 9, for the current state of the art in computer algebra software, only small- to medium-scale systems can be efficiently facilitated.

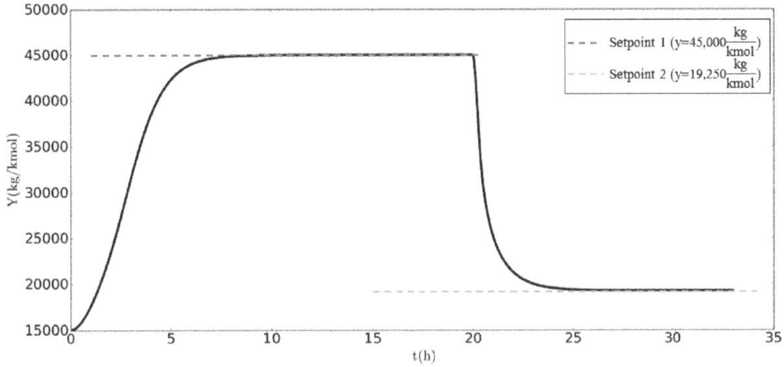

Figure 11. Set-point tracking performance of the polymerisation reactor case study.

Table 9. Computational statistics of the proposed method for different case studies.

Case Study	Problem Statistics			Uncertainty	CPU (s)
	n_x	n_g	n_θ		
SISO CSTR ($P_H = 1$)	1	2	2	OFC, RHS	1.32
SISO CSTR ($P_H = 2$)	2	4	2	OFC, RHS	65.8
SISO CSTR ($P_H = 2$)	3	6	2	OFC, RHS	4652
MMA CSTR ($P_H = 1$)	1	10	8	OFC, RHS	1.86
MMA CSTR ($P_H = 1$)	1	10	9	OFC, RHS, LHS	4.86
MMA CSTR ($P_H = 2$)	2	20	8	OFC, RHS	Memory limit
MIMO CSTR ($P_H = 1$)	2	8	4	OFC, RHS	2.76
MIMO CSTR ($P_H = 2$)	4	16	4	OFC, RHS	192.75
MIMO CSTR ($P_H = 3$)	6	24	4	OFC, RHS	5929

5. Conclusions

We have presented a computer algebra-based algorithm for the analytical solution of mp-NLPs that involve uncertain parameters on the RHS and OFC as well as the LHS of the constraints. In the first step, Gröbner bases are used for symbolically expressing the optimisation variables and the Lagrange multipliers as functions of the uncertain parameters. Next, by computing cylindrical algebraic decompositions, the globally optimal CRs are defined. Building upon the proposed algorithm, we introduce a framework for the design of multi set-point explicit MPC for nonlinear systems. The proposed technique expands the scope of mp-MPCs, as we illustrate that it is feasible to design a single "multi-layer" controller for capturing set-point tracking problems and potentially new model parameter estimations. Ongoing research focuses on the latter and how current progress in algebraic geometry can alleviate the related computational burden and allow for solutions of large-scale studies. Specifically, the application of machine learning techniques for faster evaluations of standard atomic formulas and thus reductions in the computational expense of CAD calculations is a promising direction.

Author Contributions: Conceptualization: V.M.C. and V.D.; Data curation: V.M.C.; Formal Analysis: V.M.C.; Funding acquisition: L.G.P. and V.D.; Investigation: V.M.C.; Methodology V.M.C., L.G.P. and V.D.; Visualization: V.M.C.; Writing—original draft: V.M.C.; Writing—review & editing: V.M.C. and V.D. All authors have read and agreed to the published version of the manuscript.

Funding: The authors gratefully acknowledge financial support from EPSRC grants EP/M027856/1 and EP/M028240/1.

Acknowledgments: The authors gratefully acknowledge financial support from EPSRC grants EP/M027856/1 and EP/M028240/1.

Conflicts of Interest: The authors declare no conflict of interest.

Nomenclature

APC	Advanced Process Control
CAD	Cylindrical Algebraic Decomposition
CAS	Computer Algebra Software
CR	Critical Region
CSTR	Continuous Stirred Tank Reactor
e-MPC	Economic Model Predictive Control
EWO	Enterprise Wide Optimisation
KKT	Karush–Kuhn–Tucker
LHS	Left-hand Side
MIMO	Multiple Input Multiple Output
MINLP	Mixed Integer Nonlinear Program
MMA	Methyl Methacrylate
mp-(NL)P	Multi-parametric (Nonlinear) Program
NLMPC	Nonlinear Model Predictive Control
OFC	Objective Function Coefficient
RHS	Right-hand Side
(s/d)-RTO	(Static/Dynamic) Real-time Optimisation

Appendix A

Table A1. Candidate solutions of motivating example.

	x_1	x_2	λ_1
1	-2.12	0	0
2	-2.12	$1.5\theta_1$	0
3	0	1.5	0
4	0	$0.5(5-2\theta_2)$	$3\theta_1+2\theta_2-5$
5	0	0	0
6	0	$1.5\theta_1$	0
7	0.786	0	0
8	0.786	$1.5\theta_1$	0
9	$-0.577\sqrt{-6\theta_1-4\theta_2+27}-2$	$0.5(2.309\sqrt{-6\theta_1-4\theta_2+27}-2\theta_2+13)$	$0.333(-6.9282\sqrt{-6\theta_1-4\theta_2+27}+9\theta_1+6\theta_2-39)$
10	$0.333(1.732\sqrt{-6\theta_1-4\theta_2+27}-6)$	$0.166(-6.928\sqrt{-6\theta_1-4\theta_2+27}-6\theta_2+39)$	$2.309\sqrt{-6\theta_1-4\theta_2+27}+3\theta_1+2\theta_2-13$
⋮	⋮	⋮	⋮
14	$0.0833\left(-\sqrt{\theta_3^4-72\theta_1\theta_3+16\theta_3^2+72\theta_3+304}-\theta_3^2-8\right)$	$0.0416\left(\theta_3\sqrt{\theta_3^4-72\theta_1\theta_3+16\theta_3^2+72\theta_3+304}+\theta_3+8\theta_3+36\right)$	0
15	$0.0833\left(\sqrt{\theta_3^4-72\theta_1\theta_3+16\theta_3^2+72\theta_3+304}-\theta_3^2-8\right)$	$0.0416\left(-\theta_3\sqrt{\theta_3^4-72\theta_1\theta_3+16\theta_3^2+72\theta_3+304}+\theta_3+8\theta_3+36\right)$	0

	λ_2	λ_3	λ_4
1	0	0	$-3\theta_1$
2	0	0	0
3	$3(\theta_1-1)$	$0.5(3\theta_1\theta_3-3\theta_3-10)$	0
4	0	$6\theta_1+4\theta_2-15$	0
5	0	-5	$-3\theta_1$
6	0	-5	0
7	0	0	$-3\theta_1$
8	0	0	0
9	0	0	0
10	0	0	0
⋮	⋮	⋮	⋮
14	$0.0833(-\theta_3\sqrt{\theta_3^4-72\theta_1\theta_3+16\theta_3^2+72\theta_3+304}+36\theta_1-\theta_3^3-8\theta_3-36)$	0	0
15	$0.0833(\theta_3\sqrt{\theta_3^4-72\theta_1\theta_3+16\theta_3^2+72\theta_3+304}+36\theta_1-\theta_3^3-8\theta_3-36)$	0	0

Table A2. Final CRs and explicit solutions for $P_H = 1$ of the MIMO CSTR for two of the set-points.

Set-Point	Explicit Solution
$(\theta_3=1, \theta_4=290)$	$\wedge\begin{cases} (\theta_1,\theta_2)\in CR_1^{sp1} & \begin{cases} F_{c(t=0)} = 0 \\ F_{l(t=0)} = 0.00796\theta_2\text{-}0.2011\theta_1 - 0.00796\theta_4 + 0.8044 \end{cases} \\ (\theta_1,\theta_2)\in CR_2^{sp1} & \begin{cases} F_{c(t=0)} = 2000 \\ F_{l(t=0)} = 10807.5 - 0.2011\theta_1 - 38.178\theta_1 - 0.00796\theta_4 \end{cases} \\ (\theta_1,\theta_2)\in CR_3^{sp1} & \begin{cases} F_{c(t=0)} = \frac{23376(\theta_1-3.487)}{\theta_1} \\ F_{l(t=0)} = 0 \end{cases} \\ (\theta_1,\theta_2)\in CR_4^{sp1} & \begin{cases} F_{c(t=0)} = \frac{23376\theta_1^2+\theta_1(-3.495\cdot 10^{10}\theta_2-24000\theta_3+9.889\cdot 10^{12})+\theta_2\times 10^9(1.38\theta_2-1.38\theta_4-251)+3.913\times 10^{11}\theta_4-3.95\times 10^{13}}{\theta_1^2+3.318\times 10^9\theta_2^2-1.878\times 10^{12}\theta_2+2.658\times 10^{14}} \\ F_{l(t=0)} = 0 \end{cases} \\ (\theta_1,\theta_2)\in CR_5^{sp1} & \begin{cases} F_{c(t=0)} = \frac{23376\theta_1-24000\theta_3+2496}{\theta 1} \\ F_{l(t=0)} = \frac{-0.2011\theta_1^2+\theta_1(-446.31\theta_2-0.00796\theta_4+126309)+\theta_2(458.24\theta_3-47.66)-129680\theta_3+13486.7}{\theta_1} \end{cases} \end{cases}$
$(\theta_3=1.6, \theta_4=325)$	$\wedge\begin{cases} (\theta_1,\theta_2)\in CR_1^{sp2} & \begin{cases} F_{c(t=0)} = 2000 \\ F_{l(t=0)} = 10807.5 - 0.2011\theta_1 - 38.178\theta_1 - 0.00796\theta_4 \end{cases} \\ (\theta_1,\theta_2)\in CR_2^{sp2} & \begin{cases} F_{c(t=0)} = \frac{23376(\theta_1-3.487)}{\theta_1} \\ F_{l(t=0)} = 0 \end{cases} \end{cases}$

Table A3. Candidate solutions for MMA CSTR problem.

	F_l	λ_1	λ_2	λ_3	λ_4	λ_5	λ_6	λ_7	λ_8	λ_9	λ_{10}
1	$125\theta_6 - 124$	0	0	0	0	0	0	0	0	0	0
2	0.4	$-0.0158237\theta_2 + 0.016\theta_6 - 0.0000512$	0	0	0	0	0	0	0	0	0
3	0	0	$0.0158237\theta_2 - 0.016\theta_6$	0	0	0	0	0	0	0	0
4	$-124\theta_2$	0	0	0	$-2\theta_6$	0	0	0	0	0	0
5	$62.5 - 124\theta_2$	0	0	0	0	0	0	0	$2\theta_6 - 1$	0	0

$$CR_2 = \begin{cases} 0 \le \theta_3 \\ 0 \le \theta_4 \\ 0 \le \theta_5 \le 5 \\ 0.0032 + 0.988\theta_2 \le \theta_6 \\ \theta_6 \le 0.5 \\ 0 \le \theta_7 \le 0.05 \\ 0 \le \theta_8 \le 300 \end{cases} \begin{cases} \theta_1 = 1.87695 \begin{cases} \begin{cases} 0.42 \le \theta_2 \le 0.5 \\ 0.00113\theta_2 + 0.0000457\sqrt{\theta_2} + \theta_3 \le 0.05 \\ 4.6722\sqrt{\theta_2} + \theta_4 \le 303.03 \end{cases} \\ \begin{cases} 0 \le \theta_2 \le 0.42 \\ \theta_3 \le 0.05 \\ \theta_4 \le 300 \end{cases} \end{cases} \\ 0.0000243\sqrt{\theta_1{}^2\theta_2} + 0.00113\theta_2 + \theta_3 \le 0.050 \begin{cases} \begin{cases} \theta_4 + 2.489\sqrt{\theta_2}\theta_1 \le 303.03 \\ \theta_2 \le 0.5 \end{cases} \begin{cases} \begin{cases} \frac{1.48}{\theta_1^2} < \theta_2 \\ 1.72 < \theta_1 < 1.877 \end{cases} \\ \begin{cases} 1.87695 \le \theta_1 \le 5 \\ 0.0002319\theta_1^2 + 0.446 < 8.8528 \cdot 10^{-16}\sqrt{6.864 \times 10^{22}\theta_1^4 + 2.645 \times 10^{26}\theta_1^2} + \theta_2 \end{cases} \\ \begin{cases} 0 \le \theta_1 \le 1.72 \\ 0.0002319\theta_1^2 + 0.446 < 8.8528 \cdot 10^{-16}\sqrt{6.864 \times 10^{22}\theta_1^4 + 2.645 \times 10^{26}\theta_1^2} + \theta_2 \\ \theta_4 \le 300 \end{cases} \end{cases} \\ \begin{cases} 1.72 < \theta_1 \\ 0.0002319\theta_1^2 + 0.446 < 8.8528 \cdot 10^{-16}\sqrt{6.864 \times 10^{22}\theta_1^4 + 2.645 \times 10^{26}\theta_1^2} + \theta_2 \\ \theta_2 \le \frac{1.48}{\theta_1^2} \end{cases} \end{cases} \\ \theta_{3 \le 0.05} \begin{cases} 8.8528 \cdot 10^{-16}\sqrt{6.864 \times 10^{22}\theta_1^4 + 2.64509 \times 10^{26}\theta_1^2} + \theta_2 \le 0.0002319\theta_1^2 + 0.446 \begin{cases} \begin{cases} 1.877 < \theta_1 \le 5 \\ \theta_2 > \frac{1.48}{\theta_1^2} \\ \theta_4 + 2.489\sqrt{\theta_2}\theta_1 \le 303.03 \end{cases} \\ \begin{cases} 0 \le \theta_1 < 1.879 \\ \theta_4 \le 300 \end{cases} \end{cases} \\ \theta_2 \le \frac{1.48}{\theta_1^2} \begin{cases} \begin{cases} 1.87 < \theta_1 \le 4.989 \\ \theta_4 \le 300 \end{cases} \\ \begin{cases} \theta_1 > 4.989 \\ \frac{40280.2 - 16144.7\theta_1}{\theta_1^2} + 1617.74 \le \theta_2 \end{cases} \end{cases} \end{cases} \end{cases} \quad (A1)$$

References

1. Cutler, C.R.; Perry, R. Real time optimization with multivariable control is required to maximize profits. *Comput. Chem. Eng.* **1983**, *7*, 663–667. [CrossRef]
2. Grossmann, I. Enterprise-wide optimization: A new frontier in process systems engineering. *AIChE J.* **2005**, *51*, 1846–1857. [CrossRef]
3. Bauer, M.; Craig, I.K. Economic assessment of advanced process control—A survey and framework. *J. Process Control* **2008**, *18*, 2–18. [CrossRef]
4. Young, R.E. Petroleum refining process control and real-time optimization. *IEEE Control Syst.* **2006**, *26*, 73–83.
5. Darby, M.L.; Nikolaou, M.; Jones, J.; Nicholson, D. RTO: An overview and assessment of current practice. *J. Process Control* **2011**, *21*, 874–884. [CrossRef]
6. Pontes, K.V.; Wolf, I.J.; Embirucu, M.; Marquardt, W. Dynamic real-time optimization of industrial polymerization processes with fast dynamics. *Ind. Eng. Chem. Res.* **2015**, *54*, 11881–11893. [CrossRef]
7. Ellis, M.; Durand, H.; Christofides, P.D. A tutorial review of economic model predictive control methods. *J. Process Control* **2014**, *24*, 1156–1178. [CrossRef]
8. De Souza, G.; Odloak, D.; Zanin, A.C. Real time optimization (RTO) with model predictive control (MPC). *Comput. Chem. Eng.* **2010**, *34*, 1999–2006. [CrossRef]
9. Chachuat, B.; Srinivasan, B.; Bonvin, D. Adaptation strategies for real-time optimization. *Comput. Chem. Eng.* **2009**, *33*, 1557–1567. [CrossRef]
10. Pistikopoulos, E.N. From multi-parametric programming theory to MPC-on-a-chip multi-scale systems applications. *Comput. Chem. Eng.* **2012**, *47*, 57–66. [CrossRef]
11. Pistikopoulos, E. Perspectives in multiparametric programming and explicit model predictive control. *AIChE J.* **2009**, *55*, 1918–1925. [CrossRef]
12. Pistikopoulos, E.N.; Diangelakis, N.A.; Oberdieck, R.; Papathanasiou, M.M.; Nascu, I.; Sun, M. PAROC—An integrated framework and software platform for the optimisation and advanced model-based control of process systems. *Chem. Eng. Sci.* **2015**, *136*, 115–138. [CrossRef]
13. Fiacco, A.V. *Introduction to Sensitivity and Stability Analysis in Nonlinear Programming*; Academic Press: Cambridge, MA, USA, 1983.

14. Bank, B.; Guddart, J.; Klatte, D.; Kummer, B.; Tammer, K. *Non-Linear Parametric Optimization*; Springer Academie: Berlin/Heidelberg, Germany, 1983.
15. Kojima, M. *Strongly Stable Stationary Solutions in Nonlinear Programs*; Academic Press: Cambridge, MA, USA, 1980; Volume 43, pp. 93–138.
16. Levitin, E. Differentiability with respect to a parameter of the optimal value in parametric problems of mathematical programming. *Cybern. Syst. Anal.* **1976**, *12*, 46–64.
17. Robinson, S.M. Strongly regular generalized equations. *Math. Oper. Res.* **1980**, *5*, 43–62. [CrossRef]
18. Robinson, S.M. Generalized equations and their solutions, Part I: Basic theory. In *Point-to-Set Maps and Mathematical Programming*; Springer: Berlin/Heidelberg, Germany, 1979; pp. 128–141.
19. Kyparisis, J. Uniqueness and differentiability of solutions of parametric nonlinear complementarity problems. *Math. Prog.* **1986**, *36*, 105–113. [CrossRef]
20. Ralph, D.; Dempe, S. Directional derivatives of the solution of a parametric nonlinear program. *Math. Prog.* **1995**, *70*, 159–172. [CrossRef]
21. Dua, V.; Pistikopoulos, E.N. Algorithms for the solution of multiparametric mixed-integer nonlinear optimization problems. *Ind. Eng. Chem. Res.* **1999**, *38*, 3976–3987. [CrossRef]
22. Acevedo, J.; Salgueiro, M. An efficient algorithm for convex multiparametric nonlinear programming problems. *Ind. Eng. Chem. Res.* **2003**, *42*, 5883–5890. [CrossRef]
23. Johansen, T.A. On multi-parametric nonlinear programming and explicit nonlinear model predictive control. In Proceedings of the 41st IEEE Conference on Decision and Control, Las Vegas, NV, USA, 10–13 December 2002; Volume 3, pp. 2768–2773.
24. Domínguez, L.F.; Pistikopoulos, E.N. A quadratic approximation-based algorithm for the solution of multiparametric mixed-integer nonlinear programming problems. *AIChE J.* **2013**, *59*, 483–495. [CrossRef]
25. Johansen, T.A. Approximate explicit receding horizon control of constrained nonlinear systems. *Automatica* **2004**, *40*, 293–300. [CrossRef]
26. Narciso, D.A. Developments in Nonlinear Multiparametric Programming and Control. Ph.D. Thesis, Imperial College, London, UK, 2009.
27. Leverenz, J.; Xu, M.; Wiecek, M.M. Multiparametric optimization for multidisciplinary engineering design. *Struct. Multidiscipl. Optim.* **2016**, *54*, 795–810. [CrossRef]
28. Dua, V.; Papalexandri, K.P.; Pistikopoulos, E.N. Global optimization issues in multiparametric continuous and mixed-integer optimization problems. *J. Glob. Optim.* **2004**, *30*, 59–89. [CrossRef]
29. Domínguez, L.F.; Narciso, D.A.; Pistikopoulos, E.N. Recent advances in multiparametric nonlinear programming. *Comput. Chem. Eng.* **2010**, *34*, 707–716. [CrossRef]
30. Hale, E.T. Numerical Methods for d-Parametric Nonlinear Programming with Chemical Process Control and Optimization Applications. Ph.D. Thesis, The University of Texas at Austin, Austin, TX, USA, 2005.
31. Fotiou, I.A.; Parrilo, P.A.; Morari, M. Nonlinear parametric optimization using cylindrical algebraic decomposition. In Proceedings of the 44th IEEE Conference on Decision and Control and 2005 European Control Conference, Seville, Spain, 15 December 2005; pp. 3735–3740.
32. Fotiou, I.A.; Rostalski, P.; Parrilo, P.A.; Morari, M. Parametric optimization and optimal control using algebraic geometry methods. *Int. J. Control* **2006**, *79*, 1340–1358. [CrossRef]
33. Charitopoulos, V.M.; Dua, V. Explicit model predictive control of hybrid systems and multiparametric mixed integer polynomial programming. *AIChE J.* **2016**, *62*, 3441–3460. [CrossRef]
34. Pappas, I.; Diangelakis, N.A.; Pistikopoulos, E.N. The exact solution of multiparametric quadratically constrained quadratic programming problems. *J. Glob. Optim.* **2020**, 1–27. [CrossRef]
35. Charitopoulos, V.M.; Dua, V. A unified framework for model-based multi-objective linear process and energy optimisation under uncertainty. *Appl. Energy* **2017**, *186*, 539–548. [CrossRef]
36. Bretti, G.; Piccoli, B. A tracking algorithm for car paths on road networks. *SIAM J. Appl. Dyn. Syst.* **2008**, *7*, 510–531. [CrossRef]
37. Patrinos, P.; Sarimveis, H. A new algorithm for solving convex parametric quadratic programs based on graphical derivatives of solution mappings. *Automatica* **2010**, *46*, 1405–1418. [CrossRef]
38. Sun, M.; Chachuat, B.; Pistikopoulos, E.N. Design of multi-parametric NCO tracking controllers for linear dynamic systems. *Comput. Chem. Eng.* **2016**, *92*, 64–77. [CrossRef]
39. Charitopoulos, V.M. *Uncertainty-Aware Integration of Control with Process Operations and Multi-Parametric Programming under Global Uncertainty*; Springer Nature: Berlin/Heidelberg, Germany, 2020.
40. Pappas, I.; Kenefake, D.; Burnak, B.; Avraamidou, S.; Ganesh, H.S.; Katz, J.; Diangelakis, N.A.; Pistikopoulos, E.N. Multiparametric Programming in Process Systems Engineering: Recent Developments and Path Forward. *Front. Chem. Eng.* **2021**, 2. [CrossRef]
41. Buchberger, B. Bruno Buchberger's PhD thesis 1965: An algorithm for finding the basis elements of the residue class ring of a zero dimensional polynomial ideal. *J. Symb. Comput.* **2006**, *41*, 475–511. [CrossRef]
42. Lazard, D. Gröbner bases, Gaussian elimination and resolution of systems of algebraic equations. In *Computer Algebra: EUROCAL83, European Computer Algebra Conference London*; van Hulzen, J.A., Ed.; Springer: Berlin/Heidelberg, Germany, 1983; pp. 146–156.

43. Bochnak, J.; Coste, M.; Roy, M.F. *Real Algebraic Geometry*; Springer Science & Business Media: Berlin/Heidelberg, Germany, 2013; Volume 36, pp. 23–54.
44. Faugere, J.C. A new efficient algorithm for computing Gröbner bases (F4). *J. Pure Appl. Algebr.* **1999**, *139*, 61–88. [CrossRef]
45. Faugere, J.C. *Computing Gröbner Basis without Reduction to Zero (F5)*; Technical Report; LIP6: Paris, France, 1998.
46. Collins, G.E. Quantifier elimination for real closed fields by cylindrical algebraic decompostion. In *Automata Theory and Formal Languages 2nd GI Conference*; Springer: Berlin/Heidelberg, Germany, 1975; pp. 134–183.
47. Jirstrand, M. *Cylindrical Algebraic Decomposition—An Introduction*; Linköping University: Linköping, Sweden, 1995.
48. Charitopoulos, V.M.; Dua, V.; Papageorgiou, L.G. Traveling Salesman Problem-Based Integration of Planning, Scheduling, and Optimal Control for Continuous Processes. *Ind. Eng. Chem. Res.* **2017**, *56*, 11186–11205. [CrossRef]
49. Charitopoulos, V.M.; Papageorgiou, L.G.; Dua, V. Multi-parametric linear programming under global uncertainty. *AIChE J.* **2017**, *63*, 3871–3895. [CrossRef]
50. Camacho, E.F.; Alba, C.B. *Model Predictive Control*; Springer Science & Business Media: Berlin/Heidelberg, Germany, 2013.
51. Daoutidis, P.; Soroush, M.; Kravaris, C. Feedforward/feedback control of multivariable nonlinear processes. *AIChE J.* **1990**, *36*, 1471–1484. [CrossRef]
52. BenAmor, S.; Doyle, F.J.; McFarlane, R. Polymer grade transition control using advanced real-time optimization software. *J. Process Control* **2004**, *14*, 349–364. [CrossRef]
53. Kılınç, M.R.; Sahinidis, N.V. Exploiting integrality in the global optimization of mixed-integer nonlinear programming problems with BARON. *Optim. Meth. Soft.* **2018**, *33*, 540–562. [CrossRef]

processes

Article

Understanding the Evolution and Applications of Intelligent Systems via a Tri-X Intelligence (TI) Model

Min Zhao [1], Zhenbo Ning [1], Baicun Wang [2,*], Chen Peng [2], Xingyu Li [3] and Sihan Huang [4]

1 Towards-Intelligence Research Institute, Beijing 100085, China; mike_zhaomin@139.com (M.Z.); ningzhenbo@163.com (Z.N.)
2 State Key Lab of Fluid Power & Mechatronic Systems, Zhejiang University, Hangzhou 310027, China; pcme@zju.edu.cn
3 Department of Mechanical Engineering, University of Michigan, Ann Arbor, MI 48105, USA; lixingyu@umich.edu
4 School of Mechanical Engineering, Beijing Institute of Technology, Beijing 100081, China; huangsihan@bit.edu.cn
* Correspondence: baicunw@zju.edu.cn

Abstract: The evolution and application of intelligence have been discussed from perspectives of life, control theory and artificial intelligence. However, there has been no consensus on understanding the evolution of intelligence. In this study, we propose a Tri-X Intelligence (TI) model, aimed at providing a comprehensive perspective to understand complex intelligence and the implementation of intelligent systems. In this work, the essence and evolution of intelligent systems (or system intelligentization) are analyzed and discussed from multiple perspectives and at different stages (Type I, Type II and Type III), based on a Tri-X Intelligence model. Elemental intelligence based on scientific effects (e.g., conscious humans, cyber entities and physical objects) is at the primitive level of intelligence (Type I). Integrated intelligence formed by two-element integration (e.g., human-cyber systems and cyber-physical systems) is at the normal level of intelligence (Type II). Complex intelligence formed by ternary-interaction (e.g., a human-cyber-physical system) is at the dynamic level of intelligence (Type III). Representative cases are analyzed to deepen the understanding of intelligent systems and their future implementation, such as in intelligent manufacturing. This work provides a systematic scheme, and technical supports, to understand and develop intelligent systems.

Keywords: Tri-X Intelligence; cyber-physical systems; human-cyber systems; intelligent systems; intelligent manufacturing

Citation: Zhao, M.; Ning, Z.; Wang, B.; Peng, C.; Li, X.; Huang, S. Understanding the Evolution and Applications of Intelligent Systems via a Tri-X Intelligence (TI) Model. *Processes* **2021**, *9*, 1080. https://doi.org/10.3390/pr9061080

Academic Editors: Luis Puigjaner, Antonio Espuña Camarasa, Edrisi Muñoz Mata and Elisabet Capón García

Received: 2 May 2021
Accepted: 18 June 2021
Published: 21 June 2021

1. Introduction

In recent decades, intelligence has been a hot topic in various areas including human science, biology, computer and information science and social science [1]. Intelligence is well defined for its capabilities of perception and cognition, as well as its wide application of all living systems to natural laws [2–4]. This broad definition maximizes coverage of a variety of intelligent phenomena. The common definition of intelligence is to realize and maximize the value of the function of artificial systems according to human desires by natural laws, to be activated upon a human's request. Intelligence represents the system's responsiveness to environmental changes through an autonomous decision-making process, which enables the system to react using proper actions at the proper time in the proper way to achieve the objectives. Norbert Wiener published his book "Cybernetics: control and communication in the animal and the machine" in 1948 [5]. Wiener tried to analyze the difference between humans and machines. He stated that the special abilities of humans are in recognizing and adapting to the environment changes. In his opinion, artificial systems and living systems share a similar logic, in which the human is a control and communication system as is a machine. In his book, "cybernetics" is a concept with special meaning, including control, feedback, communication and interaction. It is a process followed by a series of procedures, including constant acquisition of condition changes, reaction, and continuous optimization. It is an autonomous process that an intelligent entity adapts to by control algorithms, unifying recognition, decision, and feedback to handle environmental uncertainties. The word "cyber" is closely related to cybernetics; automatic control systems in both machines and living things. Compared to human intelligence, the characteristics of machine intelligence can be interpreted as data circulation rather than human movement, machine computing rather than human brainpower, automated machining rather than manual operation. Driven by complex business processes, limited time windows and surge labor costs, the value of the above three characteristics is increased by an order of magnitude [6]. For example, the concept of intelligent manufacturing was proposed to liberate humans from tasks that can be done by machines. Much evidence indicates that machines can perform better in certain tasks compared to humans [4,7].

The level of system intelligence is measured by the ability for decision-making. For example, a higher level indicates more situations that a system can handle. Five basic features, including state recognition, real-time analysis, autonomous decision-making, accurate execution and promotion through learning, indicate the level of system intelligence [8]. As an extension of Wiener's idea, we designed five features to measure the intelligence of a physical entity, a consciousness of humans, and a cyber entity for determining their intelligence levels. According to the five features, Hu et.al [8] classified the intelligent systems into three levels including primitive level (Type I), normal level (Type II), and dynamic level (Type III), as shown in Figure 1. A system with state recognition, real-time analysis, and accurate execution is classified as a primitive-intelligent system. An advanced intelligent system has additional features regarding autonomous decision-making. A system with all five features is an open-intelligent system, also known as a system with a complete level of intelligence.

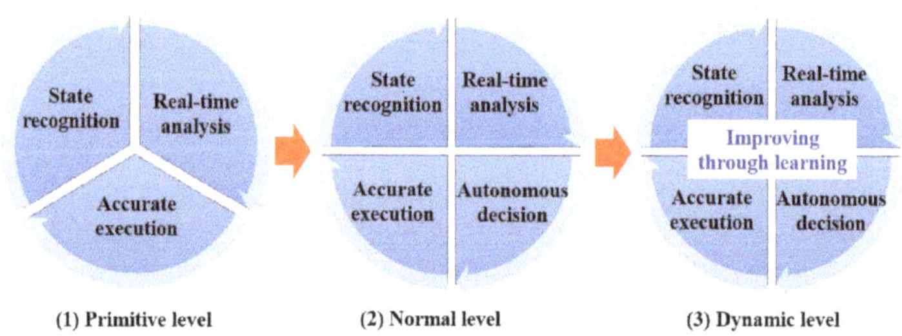

Figure 1. Three levels of intelligence in intelligent systems.

Intelligence has been discussed from the perspectives of life, control theory, artificial intelligence and industrial applications [1,5,9–11]. In dynamic systems, humans may not perform as well as robots in repeated tasks, but they are able to adapt to change, and can often invent out-of-the-box solutions. However, there is no consensus on the evolution of intelligence with the incorporation of human intelligence and its importance. Even though the human's role and full integration in these systems is often overlooked, the human is an indispensable component in the intelligent systems, especially for supervising and enforcing the intelligence of machines. To address this research gap, the Tri-X Intelligence (TI) model is proposed to systematically analyze the intelligence of humans, the physical world, the cyber world and their interactions. The proposed model consists of three intelligent elements: conscious humans, physical objects and cyber entities (Figure 2). In Figure 2, physical objects include natural substances and artificial systems based on physical materials. Conscious humans can be defined as biological systems with brainpower and awareness. A Cyber system is an advanced digital logic system in a computer with network facilities to drive the software and hardware.

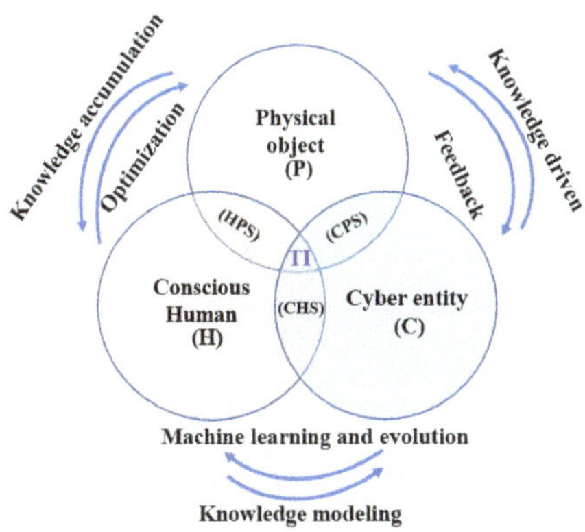

Figure 2. The Tri-X Intelligence (TI) model.

The goal and application area of this work focus on the industrial field including intelligent manufacturing, intelligent energy and intelligent transportation. The rest of the paper is organized as follows. In Sections 2–4, elemental intelligence, integrated intelligence and complex intelligence are discussed based on the hierarchy provided by an HCPS (human–cyber–physical systems) model. In Section 5, representative examples of HCPS are presented in detail. In Section 6, we conclude this work and summarize future research directions.

2. Elemental Intelligence Based on Scientific Effects

2.1. Physical Object

A physical object is one of the original intelligent systems or the zero-generation of intelligent systems. Taking the natural ecosystem as examples, a rock, tree, mountain, water, and even the planet, can recognize outside information, exchange materials/energy, and operate according to natural laws through scientific phenomenon or effect. Intelligence of a physical object can be shown in a scientific manner through geometry, physics, chemistry or biology. The interaction results from different materials following natural laws. The intelligence of a physical object is consistent with primitive intelligence, as shown in Figure 3. An old example of physical intelligence is the steam engine invented in the first industrial revolution [12].

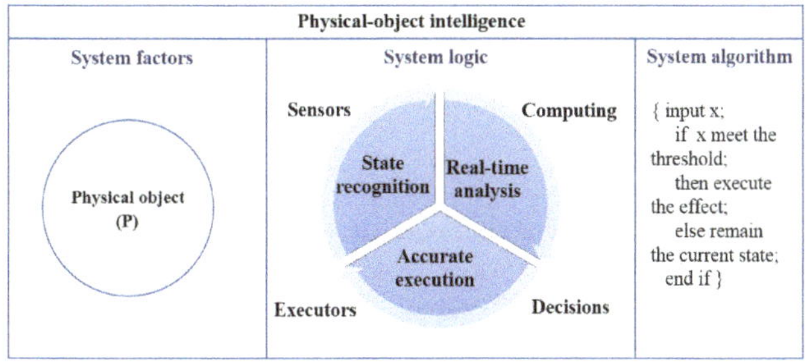

Figure 3. Physical object intelligence (Type I).

In recent years, the advancement of physical object intelligence in the form of intelligent/smart materials has drawn increasing attention. For example, intelligent fibers can recognize changes in the outside environment and inner states and respond to them in a certain manner [13]. Intelligent skin is made of super-thin (nanometer) film polyimide and monocrystalline silicon, which is equipped with tactile sensors to detect changes in temperature, humidity, pressure and transformation [14]. These intelligent materials are produced following interdisciplinary physical laws. Interactions among physical entities are very common in the industry. A variety of physical objects constitute many manufacturing facilities and products, which realize their functions via physical object intelligence. The intelligence of physical objects is often constant over time and relies on other intelligence for continuous improvement and dynamic innovation.

2.2. Conscious Humans

The living intelligence of human is attained from the continuous recognition of nature. It is a type of inherent intelligence developed during evolution. Conscious humans recognize outside information using sense organs and react to outside stimulation through subconscious actions, unconscious actions, or conscious actions that are recognized and controlled by the brain, as shown in Figure 4. For example, humans react immediately when touching extra-hot, frozen, or sharp objects. More importantly, humans learn how to make decisions based on past experiences [15]. Interactions among humans are common in society and determine the basic contents of human lives. Interactions and cooperation among humans create groups, domains and relationships. More importantly, emotional intelligence, also known as emotional quotient (EQ), is the ability of humans to recognize their own emotions and those of others, to discern between different feelings. and to label them appropriately. Emotional information helps to guide thinking and behavior and to manage emotions in order to adapt to various environments or achieve goals [16]. However, there are many known and well-documented human cognitive biases that plague human intelligence and the ability to reason consistently, to make decisions based on evidence, and to make accurate predictions of the future [16]. Other disadvantages of human labor include behavioral differences, forgetting information, mistakes and errors [17].

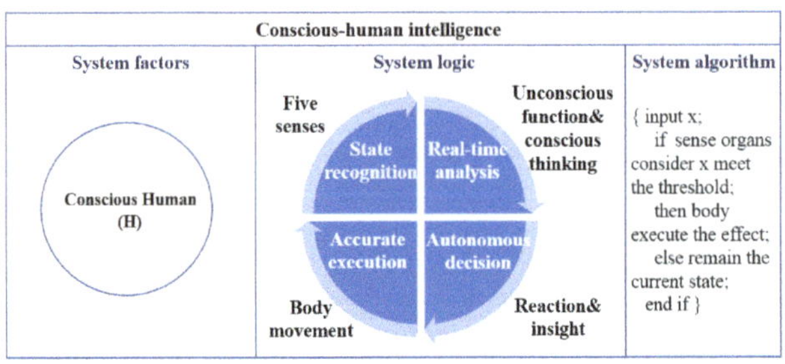

Figure 4. Human intelligence (Type I).

2.3. Cyber Entity

A cyber entity consists of software, hardware and a network that enables digital intelligence or computation intelligence on machines, as shown in Figure 5. For example, computers take inputs through the keyboard, mouse and camera. Autonomous decisions are enabled by the processor unit which is designed to analyze the signal, voice and image in real-time. Computers can execute commands following exact rules, including data storage, image capture and camera angle adaption. Initially, the computer was used for simple calculation and data storage. In the intelligent age, computers have become smarter with the capacity for communication, self-learning and super-computing. Moreover, knowledge systems can be obtained from collaborative learning from interactions among cyber entities [18–20]. However, cyber-entity intelligence (or called machine intelligence) has no setting for creativity, playfulness, fun or curiosity, which are the source of many inventions and breakthroughs [15].

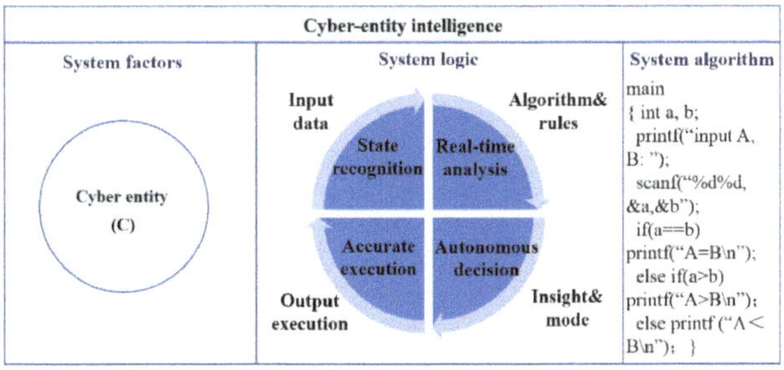

Figure 5. Cyber entity intelligence (Type I).

Today, physical object intelligence commonly exists in areas including new materials, super materials and intelligent materials. Cyber entity intelligence benefits from the development of algorithms and computation capacity. Artificial intelligence with learning ability is growing rapidly and is becoming comparable to human intelligence [21–26]. In summary, due to their own advantages and shortcomings, physical entity intelligence, conscious human intelligence and cyber entity intelligence should be integrated and synergetic in high-level intelligent systems. We seek to confirm that machine intelligence can interact and fuse with other types of intelligence, leading to a more advanced and complex intelligence.

3. Integrated Intelligence Formed by Two-Elements Integration

3.1. Human-Physical System (HPS) Intelligence

Humans can not only design physical objects through physical and mental work but can also generate knowledge in this process. Meanwhile, humans can use the acquired knowledge to create new physical products. In other words, development history is a process of recognizing, exploiting and changing physical objects, as shown in Figure 6. For example, colored pottery encompasses knowledge from hundreds of years ago. The knowledge in the brain and the product is implicit, which is different from explicit knowledge, such as an image or text. Benefiting from the development of explicit knowledge, the physical machine has become increasingly advanced to replace parts aspects of human labor. However, the development of implicit and explicit knowledge HPS is limited due to the restriction of knowledge carriers. The interaction mode with HPS is the typical mode of "human in the loop". Human and physical machines are the main system elements that keep improving HPS during evolution.

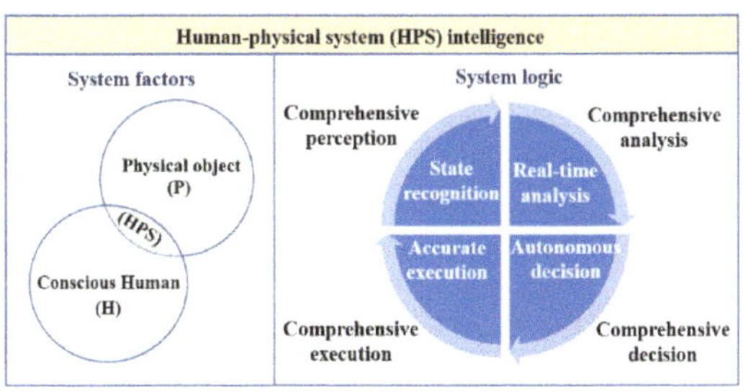

Figure 6. HPS intelligence (Type II).

3.2. Human-Cyber System (HCS) Intelligence

One goal of developing intelligent systems is to increase the interaction of efficiency between humans and cyber systems (e.g., computers) in the form of human-cyber systems (HCS). There are various interaction methods in HCS, such as programmable software [27], brain-computer interfaces [28], and inserted chips [29] between human and cyber systems. Software is a method to transform human intelligence into machine intelligence. Explicit knowledge is the main source of machine intelligence. The software intermediary interprets the humans' implicit knowledge into explicit knowledge to equip the cyber entity with reasoning ability. The brain-computer interface is a method that extracts brain awareness to control the physical entity via a cyber system. Related technologies have been investigated including communications from brain to machine, from machine to brain, and from brain to brain. An inserted chip is an intrusive connection method. In the future, with the development of super chips, it is possible to realize an interbrain network through inserted super chips. Action recognition is an indirect method to obtain human awareness through various sensors. The language, facial expression, gestures and other information of human awareness can be converted into digital information in cyber entity systems [30]. Taking WeChat as an example [31], recognition and software intermediary tools have been designed to convert screen touch actions into texts to be sent to people via cyber technologies. Interactions between humans and cyber entities to realize HCS intelligence are shown in Figure 7. Although many scientists have focused on brain science, the thinking mechanism of the mind is still unclear [32]. Interactions between human awareness and cyber entities still involve interpreting implicit knowledge to explicit knowledge in order to strengthen digital intelligence. This is a process to convert human intelligence to machine intelligence for more powerful knowledge-based tools.

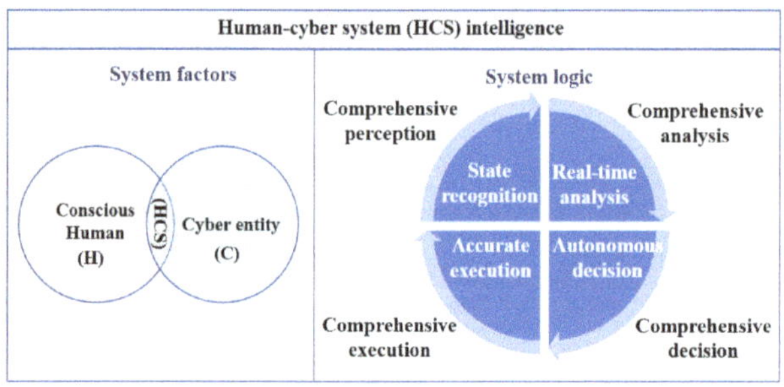

Figure 7. HCS intelligence (Type II).

3.3. Cyber-Physical System (CPS) Intelligence

Interactions between physical objects and cyber entities result in a cyber-physical system (CPS), which is a milestone to promote the development of intelligent systems. CPS was proposed by Helen Gill [33,34] and was introduced into industry by Germany to support Industry 4.0 initiatives [35]. CPS models not only the interaction between physical objects and cyber entities but also a scheme that converts human intelligence to machine intelligence in artificial systems. However, the influence of human intelligence will never disappear and keeps influencing the artificial systems via software and knowledge engines, as shown in Figure 8.

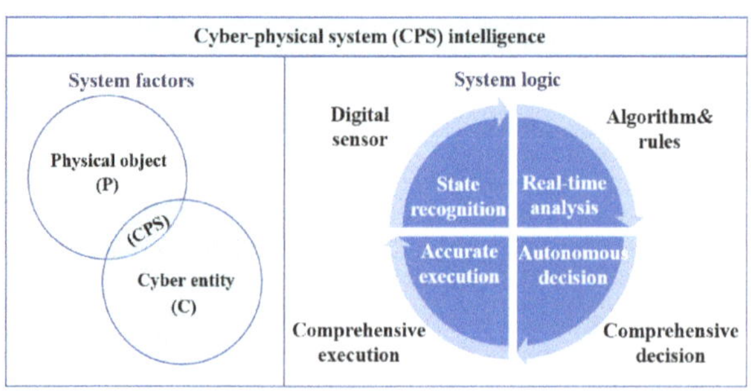

Figure 8. CPS intelligence (Type II).

For instance, CPS is the core technology of smart manufacturing (or intelligent manufacturing) [36]. The reference framework (RAMI 4.0) of CPS proposed by Germany Industry 4.0 consists of a physical layer, integration layer, communication layer, information layer and a function layer, in which the core is the digital technology and network technology [37]. RAMI 4.0 elaborates the concept of an administration shell that is an intermediate software platform including a communication layer, information layer and a function layer. The administration shell is a cyber system to support CPS, which can be applied to a physical object to constitute a CPS. Software is the crucial carrier of a cyber system, which defines new rules and stores knowledge within the restriction of hardware. Human intelligence and artificial intelligence define the majority of reasoning and judging rules in software. The information of physical entities flows into the digital space to create the cyber system. In turn, the cyber system participates in the activities of physical objects through software, which is called the digital twin [38,39]. In the future, more and more physical objects will fuse with digital entities, and more and more digital entities will be adopted to test and control physical objects.

4. Complex Intelligence Formed by Ternary-Interaction

Interactions within physical objects, conscious humans and cyber entities cocreate complex intelligent systems, called the intelligence of system-of-systems (SoS). The different focuses of the components create different applications, as shown in Figure 9. Most of the scenarios in Industry 3.0 and 4.0 have resulted from the fusion of physical objects, conscious humans and cyber entities, in design, production and service [40].

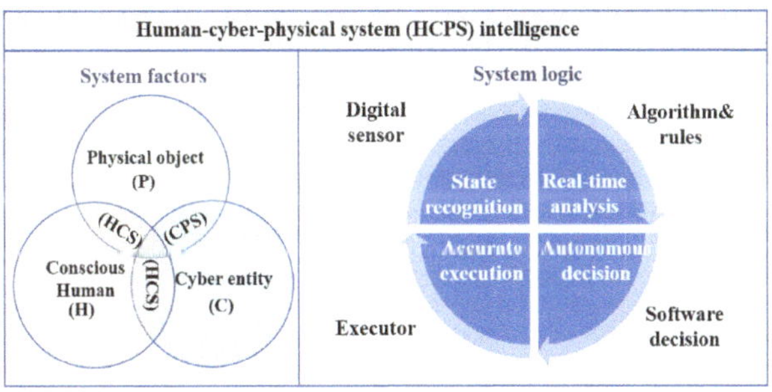

Figure 9. HCPS intelligence (Type III).

An advanced case of ternary-fusion HCPS intelligence is the self-driving automobile [41]. In practice, there are many self-driving automobiles that can handle most of the situations under supervision. Moreover, the 100% self-driving automobile has already been developed at the lab level. Here, AI takes over the driving position of the human operator and operates the self-driving system based on data analytics of the environment and human behavior. This type of intelligent system can not only practice the intelligent circle including recognition, analysis, decision, execution, but is also equipped with learning ability. Human-machine hybrid intelligence is an advanced form of human-machine intelligence. A typical case is Alpha AI software developed by Psibernetix, which can beat American pilots in simulation environments [42]. The chance of making mistakes will increase when a pilot is in control of a supersonic aircraft at 12,000 m and a speed of over 1200 km per hour. However, Alpha AI can increase error tolerance through tactical plan optimization in a dynamic environment. The responsiveness of Alpha AI is 250 times faster than that of a pilot. Alpha AI can be controlled by language commands. The most significant aspect of Alpha AI is that it can learn from other Alpha AI data installed in different places and in different versions to enhance its own performance. Another example is human-robot collaboration [43]. Human-robot collaboration can release human workers from heavy tasks if effective communication channels between humans and robots are established [44]. With the help of sensor technologies, gesture identification, gesture tracking and gesture classification, human-robot collaboration allows human workers and robots to work together in a shared manufacturing environment.

In summary, the single entity (conscious human, physical object or cyber entity) shows primitive intelligence (Type I) at the unit level. A two-entity integrated system may create normal-level intelligence (Type II) at a system level. Three-entity fusion can generate dynamic-level intelligence (Type III) at the SoS level. Therefore, when considering development from primitive intelligence, intelligentization has evolved over more than 200 years. The development of intelligence will be accelerated in the future resulting in hybrid intelligence and swarm intelligence.

5. Implementation and Applications of Intelligent Systems

Physical systems with primitive intelligence are the oldest intelligent systems; however, their control is limited [6] and the corresponding technologies are easy to e generalize. In the Wiener era [5], electricity was adopted for sensing information and driving motors and machinery, which broke through the obstacle between information and the physical entity to increase technology commonality. Due to technology limitations, only simple objects described by differential equations could be controlled in that era. With the development of computational technologies, digital/cyber intelligence has been applied to control more complex objects. In the following section, the implementation and applications of intelligent artificial systems are analyzed based on the evolution and development of system elements.

An artificial system is a set of elements with interaction and interconnection to realize specific functions in the forms of machine, product, workpiece and plants. Its components can be described as four basic subsystems, as shown in Figure 10: power unit, control unit, transmission unit and actuator unit. The executive device (actuator unit) is used for executing actions, the power device for producing and converting energy, the transmission device for transmitting energy and the control device for adjusting the operating parameters of subsystems to allow executive devices to react accurately. In the past decades, industrial evolution occurred with technological developments in artificial systems. The emergence of new machines, new tools and facilities created continuously improving productivity. The four basic components are also evolving constantly. For example, the executive device is updated by introducing new structures and new materials (e.g., intelligent fiber and super materials).

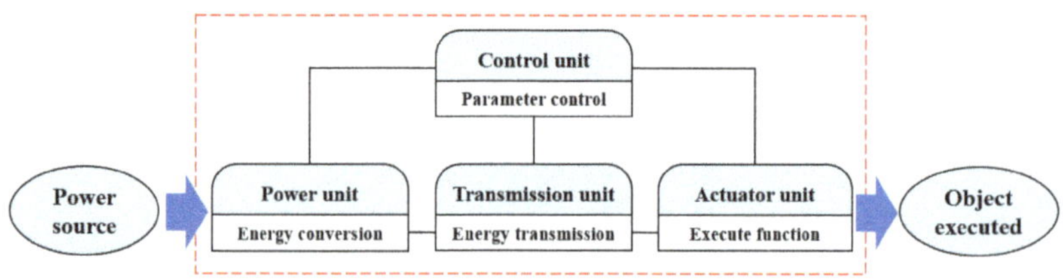

Figure 10. Four basic subsystems of artificial systems.

In recent decades, the control unit in artificial systems has evolved fastest compared to other components in artificial systems. The latest advance in control devices is related to cyber systems. The core technology of control devices has evolved through a mechanical → electromechanical → digital → software → cloud route. The continuous introduction of new technologies into control systems finally achieves CPS, as shown in Figure 11. The evolution of the control device is consistent with the fusion and integration of the administration shell and the physical facility in RAMI 4.0, which is how CPS is constructed.

There are many scenarios driving the development of intelligent systems (e.g., intelligent manufacturing) [45]. Nowadays, intelligent manufacturing is evolving into a new state based on next-generation artificial intelligence. This can be termed new-generation intelligent manufacturing (NGIM) [4]. Traditionally, artificial intelligence has been defined as a branch of computer science to simulate the thinking processes and intelligent actions of humans. However, new-generation artificial intelligence extends traditional digital intelligence to big-data intelligence, crowd intelligence and human-machine hybrid intelligence. These new-generation AI technologies have greater content and can be applied in more domains. For example, big-data intelligence originated from the operation information of cyber systems under the close collaboration among three entities, which cannot be processed by humans, to reveal the mode and inner laws [46]. Crowd intelligence is generated among different entities, and it is hard to determine which one is the controller, and which one is controlled [9].

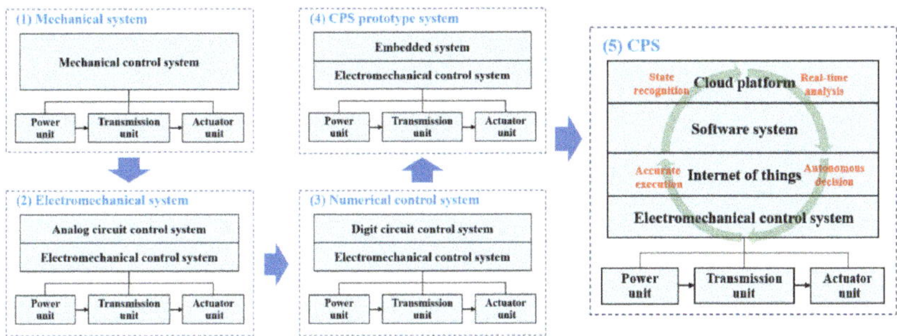

Figure 11. The evolution of control system to CPS.

6. Conclusions

As a demarcation of the past, present and future of intelligent systems, a Tri-X Intelligence (TI) model is proposed in this paper to state the mechanism, factors and connotation of three main entities (conscious humans, physical objects, and cyber entities) including single-X intelligence, two-X integrated intelligence and three-X complex intelligence. Every single entity shows primitive intelligence. Two-entity integration creates integrated intelligence. Three-entity fusion generates advanced intelligence. The intelligentization mechanism of artificial systems continuously converts human intelligence to machine intelligence via different channels and interfaces. With the increasing use of machine intelligence, humans will gradually play a less significant role in intelligent systems. However, human intelligence will keep influencing artificial systems in the form of software/algorithms to drive intelligent systems. Therefore, we cannot take humans out of the systems given the accelerating development of technology. The key to success is to adapt humans to new work environments, i.e., not to replace but to enhance. According to the Tri-X Intelligence (TI) model, humans need to think more about how to collaborate with cyber systems rather than training operators to work like computers.

The proposed Tri-X model (e.g., HCPS) will integrate the intelligence in a the complex system with a combination of human-cyber-physical and machine subsystems. In future research, modeling intelligence in experiments or simulations is critical. Different cognitive architectures, such as LIDA of Stan Franklin, ACT-R of CMU, SOAR from the University of Michigan, Subsumption Architecture of the MIT AI lab, or BDI (belief, desire and intention) provide structure to create intelligent actions. Different methodologies like neural networks, genetic algorithms, simulated annealing, the Monte Carlo method and swarm intelligence are approaches to create actions that could result in intelligent behavior. The ultimate goal of HCPS, or Tri-X modelling and implementation, is to achieve effective and efficient symbioses among humans, cyber systems and physical systems.

Author Contributions: Conceptualization, M.Z. and B.W.; methodology, Z.N.; investigation, B.W.; resources, Z.N.; writing—original draft preparation, C.P., and S.H.; writing—review and editing, B.W. and X.L.; supervision, B.W. and M.Z.; project administration, Z.N.; funding acquisition, B.W. All authors have read and agreed to the published version of the manuscript.

Funding: This research was funded by Zhejiang University Hundreds of Talents Program, grant number 0020886.

Institutional Review Board Statement: Not applicable.

Informed Consent Statement: Not applicable.

Data Availability Statement: Not applicable.

Conflicts of Interest: The authors declare no conflict of interest.

References

1. Wang, L. From Intelligence Science to Intelligent Manufacturing. *Engineering* **2019**, *5*, 615–618. [CrossRef]
2. Albus, J.S. *A Reference Model Architecture for Intelligent Systems Design*; The National Institute of Standards and Technology: Gaithersburg, MD, USA, 1993; pp. 27–56.
3. Albus, J.S. Outline for a theory of intelligence. *IEEE Trans. Syst. Man Cybern.* **1991**, *214*, 73–509. [CrossRef]
4. Zhou, J.; Li, P.; Zhou, Y.; Wang, B.; Zang, J.; Meng, L. Toward New-Generation Intelligent Manufacturing. *Engineering* **2018**, *4*, 11–20. [CrossRef]
5. Wiener, N. *Cybernetics or Control and Communication in the Animal and the Machine*; MIT Press: Boston, MA, USA, 1961.
6. Zhou, J.; Zhou, Y.; Wang, B.; Zang, J. Human–Cyber–Physical Systems (HCPSs) in the Context of New-Generation Intelligent Manufacturing. *Engineering* **2019**, *4*, 624–636. [CrossRef]
7. Wang, B.; Zang, J.; Qu, X.; Dong, J.; Zhou, Y. Research on New-Generation Intelligent Manufacturing based on Human-Cyber-Physical Systems. *Strateg. Study Chin. Acad. Eng.* **2018**, *20*, 29–34.
8. Hu, H.; Zhao, M.; Ning, Z. *Three-Body Intelligence Revolution*; China Machine Press: Beijing, China, 2016.
9. Li, W.; Wu, W.J.; Wang, H.M.; Cheng, X.Q.; Chen, H.J.; Zhou, Z.H.; Ding, R. Crowd intelligence in AI 2.0 era. *Front. Inform. Technol. Elect. Eng.* **2017**, *18*, 15–43. [CrossRef]
10. Wright, P.K.; Bourne, D.A. *Manufacturing Intelligence*; Addison-Wesley Longman Publishing Co., Inc.: Boston, MA, USA, 1988.
11. Wang, B.; Tao, F.; Fang, X.; Liu, C.; Liu, Y.; Freiheit, T. Smart Manufacturing and Intelligent Manufacturing: A Comparative Review. *Engineering* **2020**. in Press. [CrossRef]
12. Dickinson, H.W. *A Short History of the Steam Engine*; Cambridge University Press: Cambridge, UK, 2011.
13. Bedeloglu, A.; Demir, A.; Bozkurt, Y.; Sariciftci, N.S. A photovoltaic fiber design for smart textiles. *Text. Res. J.* **2010**, *80*, 1065–1074. [CrossRef]
14. Tan, T.Y.; Zhang, L.; Neoh, S.C.; Lim, C.P. Intelligent skin cancer detection using enhanced particle swarm optimization. *Knowl. Based Syst.* **2018**, *158*, 118–135. [CrossRef]
15. Zhou, Y.; Chuah, K. Human intelligence: The key factor for successful intelligent manufacturing. *Integr. Manuf. Syst.* **2000**, *11*, 30–41. [CrossRef]
16. Sanders, N.R.; Wood, J.D. *The Humachine: Humankind, Machines, and the Future of Enterprise*; Routledge: New York, NY, USA, 2019.
17. Elmaraghy, W.H.; Nada, O.A.; ElMaraghy, H.A. Quality prediction for reconfigurable manufacturing systems via human error modelling. *Int. J. Comput. Integr. Manuf.* **2008**, *21*, 584–598. [CrossRef]
18. Emmanouilidis, C.; Pistofidis, P.; Bertoncelj, L.; Katsouros, V.; Fournaris, A.; Koulamas, C.; Ruiz-Carcel, C. Enabling the human in the loop: Linked data and knowledge in industrial cyber-physical systems. *Annu. Rev. Control* **2019**, *47*, 249–265. [CrossRef]
19. Zhuang, Y.; Wu, F.; Chen, C.; Pan, Y. Challenges and opportunities: From big data to knowledge in AI 2.0. *Front. Inform. Technol. Elect. Eng.* **2017**, *18*, 3–14. [CrossRef]
20. Jardim-Goncalves, R.; Sarraipa, J.; Agostinho, C.; Panetto, H. Knowledge framework for intelligent manufacturing systems. *J. Intell. Manuf.* **2011**, *22*, 725–735. [CrossRef]
21. Pan, Y. Heading toward artificial intelligence 2.0. *Engineering* **2016**, *2*, 409–413. [CrossRef]
22. Li, B.H.; Hou, B.C.; Yu, W.T.; Lu, X.B.; Yang, C.W. Applications of artificial intelligence in intelligent manufacturing: A review. *Front. Inf. Technol. Electron. Eng.* **2017**, *18*, 86–96. [CrossRef]
23. Zhang, D.; Han, X.; Deng, C. Review on the research and practice of deep learning and reinforcement learning in smart grids. *CSEE J. Power Energy Syst.* **2018**, *4*, 362–370. [CrossRef]
24. Zheng, N.N.; Liu, Z.Y.; Ren, P.J.; Ma, Y.Q.; Chen, S.T.; Yu, S.Y.; Xue, J.R.; Chen, B.D.; Wang, F.Y. Hybrid-augmented intelligence: Collaboration and cognition. *Front. Inf. Technol. Electron. Eng.* **2017**, *18*, 153–179. [CrossRef]
25. Peng, Y.X.; Zhu, W.W.; Zhao, Y.; Xu, C.S.; Huang, Q.M.; Lu, H.Q.; Zheng, Q.H.; Huang, T.J.; Gao, W. Cross-media analysis and reasoning: Advances and directions. *Front. Inf. Technol. Electron. Eng.* **2017**, *18*, 44–57. [CrossRef]
26. Zhang, T.; Li, Q.; Zhang, C.S.; Liang, H.W.; Li, P.; Wang, T.M.; Li, S.; Zhu, Y.L.; Wu, C. Current trends in the development of intelligent unmanned autonomous systems. *Front. Inf. Technol. Electron. Eng.* **2017**, *18*, 68–85. [CrossRef]
27. Chantem, T.; Guan, N.; Liu, D. Sustainable embedded software and systems. *Sustain. Comput. Infor.* **2019**, *22*, 152–154. [CrossRef]
28. Feng, S.; Tang, M.; Quivira, F.; Dyson, T.; Cuckov, F.; Schirner, G. EEGu2: An Embedded Device for Brain/Body Signal Acquisition and Processing. In Proceedings of the 2016 27th International Symposium on Rapid System Prototyping, Pittsburg, PA, USA, 6–7 October 2016; IEEE: New York, NY, USA, 2016; pp. 19–25. [CrossRef]

29. Vassanelli, S.; Mahmud, M.; Girardi, S.; Maschietto, M. On the way to large-scale and high-resolution brain-chip interfacing. *Cogn. Comput.* **2012**, *4*, 71–81. [CrossRef]
30. Poppe, R. A survey on vision-based human action recognition. *Image Vis. Comput.* **2010**, *28*, 976–990. [CrossRef]
31. Nunes, D.S.; Zhang, P.; Silva, J.S. A Survey on Human-in-the-Loop Applications towards an Internet of All. *IEEE Commun. Surv. Tutor.* **2015**, *17*, 944–965. [CrossRef]
32. Lu, H.; Li, Y.; Chen, M.; Kim, H.; Serikawa, S. Brain intelligence: Go beyond artificial intelligence. *Mob. Netw. Appl.* **2018**, *23*, 368–375. [CrossRef]
33. Gill, H.; National Science Foundation, Alexandria, VA, USA. Cyber-Physical Systems: Beyond ES, SNs, SCADA. Personal Communication, 2010.
34. Gill, H.; National Science Foundation, Alexandria, VA, USA. NSF Perspective and Status on Cyber-Physical Systems. Personal Communication, 2006.
35. Xu, L.D.; Xu, E.L.; Li, L. Industry 4.0: State of the art and future trends. *Int. J. Prod. Res.* **2018**, *56*, 2941–2962. [CrossRef]
36. Yao, X.; Zhou, J.; Lin, Y.; Li, Y.; Yu, H.; Liu, Y. Smart manufacturing based on cyber-physical systems and beyond. *J. Intell. Manuf.* **2019**, *30*, 2805–2817. [CrossRef]
37. Hankel, M.; Rexroth, B. The reference architectural model industrie 4.0 (rami 4.0). *ZVEI* **2015**, *2*, 4.
38. Tao, F.; Zhang, H.; Liu, A.; Nee, A.Y.C. Digital twin in industry: State-of-the-art. *IEEE Trans. Ind. Inform.* **2018**, *15*, 2405–2415. [CrossRef]
39. Qi, Q.; Tao, F. Digital twin and big data towards smart manufacturing and industry 4.0: 360 degree comparison. *IEEE Access* **2018**, *6*, 3585–3593. [CrossRef]
40. Tao, F.; Cheng, J.; Qi, Q.; Zhang, M.; Zhang, H.; Sui, F. Digital twin-driven product design, manufacturing and service with big data. *Int. J. Adv. Manuf. Technol.* **2018**, *94*, 3563–3576. [CrossRef]
41. Litman, T. *Autonomous Vehicle Implementation Predictions*; Victoria Transport Policy Institute: Victoria, BC, Canada, 2017.
42. Fawkes, A.J. Developments in Artificial Intelligence: Opportunities and Challenges for Military Modeling and Simulation. In Proceedings of the 2017 NATO M&S Symposium, Lisbon, Portugal, 14–16 May 2017; pp. 11.11–11.14.
43. Liu, H.; Wang, L. Gesture recognition for human-robot collaboration: A review. *Int. J. Ind. Ergon.* **2018**, *68*, 355–367. [CrossRef]
44. Wang, X.V.; Wang, L. Augmented Reality Enabled Human–Robot Collaboration. In *Advanced Human-Robot Collaboration in Manufacturing*; Wang, L., Wang, X.V., Váncza, J., Kemény, Z., Eds.; Springer: Cham, Switzerland, 2021. [CrossRef]
45. Zhu, J.; Huang, T.; Chen, W.; Gao, W. The future of artificial intelligence in China. *Commun. ACM* **2018**, *61*, 44–45. [CrossRef]
46. Kusiak, A. Smart manufacturing must embrace big data. *Nature* **2017**, *544*, 23–25. [CrossRef]

processes

MDPI

Article

Predictive Process Adjustment by Detecting System Status of Vacuum Gripper in Real Time during Pick-Up Operations

Sujeong Baek * **and Dong Oh Kim**

Department of Industrial Management Engineering, Hanbat National University, 125 Dongseo-Daero, Yuseong-Gu, Daejeon 34158, Korea; 20151470@edu.hanbat.ac.kr
* Correspondence: sbaek@hanbat.ac.kr; Tel.: +82-42-821-1228

Abstract: In manufacturing systems, pick-up operations by vacuum grippers may fail owing to manufacturing errors in an object's surface that are within the allowable tolerance limits. In such situations, manual interference is required to resume system operation, which results in considerable loss of time as well as economic losses. Although vacuum grippers have many advantages and are widely used in the industry, it is highly difficult to directly monitor the current machine status and provide appropriate recovery feedback for stable operation. Therefore, this paper proposes a method to detect the success or failure of a suction operation in advance by analyzing the amount of outlet air pressure in the Venturi line. This was achieved by installing an air pressure sensor on the Venturi line to predict whether the current suction action will be successful. Through empirical experiments, it was found that downward movements in the z-axis of the vacuum gripper can easily rectify a faulty gripper suction operation. Real-time monitoring results verified that predictive process adjustment of the pick-up operation can be performed by modifying the z-position of the vacuum gripper.

Keywords: real-time monitoring; fault detection; predictive process adjustment; vacuum gripper; sensor data

Citation: Baek, S.; Kim, D.O. Predictive Process Adjustment by Detecting System Status of Vacuum Gripper in Real Time during Pick-Up Operations. *Processes* **2021**, *9*, 634. https://doi.org/10.3390/pr9040634

Academic Editor: Luis Puigjaner

Received: 9 March 2021
Accepted: 1 April 2021
Published: 5 April 2021

1. Introduction

1.1. Theoretical Background

In industrial production, as manufacturing systems have become more complicated, the concept of operations and maintenance (O&M) has become important in preventing unforeseen faults and errors and for smooth operations [1,2]. Lin et al. (2021) found that O&M investments are continuously increasing in the power plant industry owing to concerns regarding the safe and reliable operation of power plants [3]. Accordingly, the authors investigated the major influencing factors in the O&M of power plants by constructing a fishbone diagram and analyzing the gray correlation between the derived factors. In addition, O&M is relevant not only to physical products or systems but to cyber systems as well. In this regard, Furumoto et al. (2020) discussed how to assess and prevent cyber risks to achieve reliable ship operation [4]. In manufacturing processes, the prevention of unknown and sudden faults in real time is critical [5,6].

Grippers play an important role in an industrial manufacturing system [7–9]. They can be classified into two types: mechanical type and vacuum type. Mechanical grippers are also known as robotic grippers [10]. As their shape and operating mode mimic human hands and fingers, they are usually used for performing complex or delicate operations. For example, Vedhagiri et al. (2019) proposed a mechanical gripper with five fingers in order to handle objects with complex shapes, such as a coin, cosmetic cream, and fruit [11]. A mechanical gripper with three fingers has also been proposed to conduct micromanipulation [12]. The authors [12] applied a piezoceramic transducer for effectively handling objects as small as 10 to 800 μm, such as glass hollow microspheres and iron spheres.

On the other hand, vacuum grippers utilize suction cup(s), air pipe(s), and Venturi line(s) to handle objects, and they require an external air compressor to supply compressed air, as illustrated in Figure 1. In particular, a discrete manufacturing process is typically performed by several common operations using vacuum grippers, and the grippers are widely used for lifting, transporting, and inspections [13–15] as they are simple, inexpensive to install, easy to operate with fragile objects, and can handle various types of object shapes [16,17].

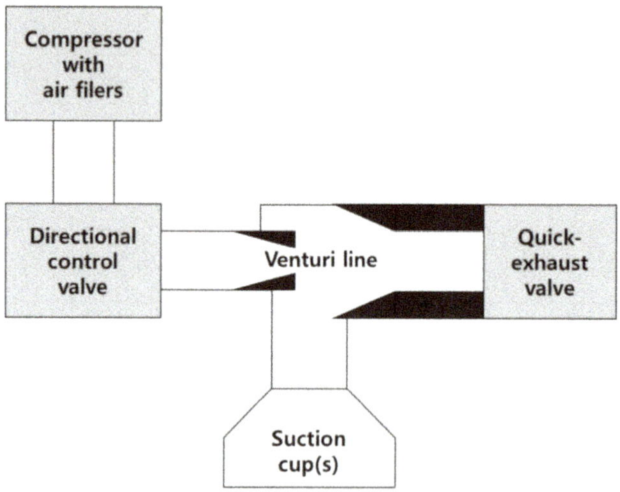

Figure 1. Illustration of a vacuum gripper system (composed with a suction cup and the connected Venturi line).

For example, a vacuum gripper with four suction cups placed in a two by two arrangement has been proposed for handling heavy objects with complex shapes [18]. Amend et al. (2012) developed a positive pressure gripper that uses a balloon instead of a suction cup and can handle objects of various shapes [10]. Since this universal gripper can utilize both positive and negative pressure, it showed good performance in gripping objects of different size, shape, weight, and fragility. Reinforcement materials such as magnetorheological elastomers have also been utilized in suction cups of vacuum gripper systems to increase their performance [19]. As can be deduced from the above, vacuum grippers with one or more suction cups are frequently used to handle different types of objects. For example, Nakamoto et al. (2018) validated that a vacuum gripper can handle objects of 40 different shapes, sizes, and materials without requiring any changes in the gripper system setup [20].

As above mentioned, several studies have focused on how to effectively use a vacuum gripper and how to prevent a faulty gripper operation in advance. Operational parameters that affect a manufacturing process are usually categorized into machine parameters and product parameters. Whereas product parameters pertain to the production of the handled object in the process (such as the three-dimensional lengths of the object), machine parameters are related to the motion control of the installed actuators (for example, the speed of actuator movement). In general, for the effective execution of manufacturing processes, appropriate machine parameters are usually analyzed, instead of product parameters, because product parameters are already optimized before being fed into the manufacturing system. For example, Wang et al. reported that the type of pipe material and operating temperature can determine the suction power during gripper operation [21]. The authors [21] also investigated the effects of cup size and length of suction time on the gripping operation. According to another study [22], the performance of the gripping operation can depend on the shape of suction cups (universal, flat, or deep). In addition, considerable research has been performed on the effects of machine parameters, including suction cup size [23,24], number of cups [25], and initial applied pressure, on the suction operation [26].

However, these investigated machine parameters are usually "design variables" that are determined in the design phase of the product life cycle; thus, they cannot be easily modified later during production. In other words, it is difficult to adjust these parameters during each gripping operation in real time. Therefore, many studies have alternatively focused on approaches to detect defects in objects, such as dents or ripped sections on the objects, before the gripping operation because the prior detection of defects enables an error-free operation. For box-lifting operations, ultrasonic sensor signals have been utilized to identify the relative curvature of box surfaces [27]. Similarly, convolutional neural network models based on sensor signals have been proposed to classify normal and defective box surfaces [27]. Moreover, surface cleanliness has also been found to be important in the effective use of grippers [28]. The authors determined that the suction performance when using vacuum grippers can be determined by the number and size of contaminants on target surfaces. Since the condition of objects being handled cannot always be changed during production operations, the only feasible solution is to remove the defective objects in advance. In other words, it is still difficult to adjust certain conditions or parameters for recovering a fault state.

Therefore, this paper proposes a real-time predictive process adjustment method for pick-up operations using vacuum grippers. To this end, a pick-up operations testbed equipped with a two-way data monitoring and process parameter control module was developed. To analyze sensor signals in terms of faulty operation detection, a sensing module was devised by installing an air pressure sensor inside the Venturi line of a vacuum gripper, and the significant features that can help distinguish between normal and faulty gripper operation before the completion of the current pick-up operation or start of the next operation were characterized. Finally, an experiment to determine whether it is possible to recover faulty states by adjusting relevant process parameter(s) was performed.

The remainder of this paper is organized as follows: In Section 2, the developed system configuration to acquire appropriate sensor signals and provide feedback commands to the controller is explained in terms of hardware and software. In addition, air pressure sensor signals to obtain meaningful features that can help detect faulty operations in advance are analyzed. The experimental results are summarized in Section 3. The concept of predictive process adjustment in the pick-up operation of a vacuum gripper by controlling a related control parameter is also demonstrated. Lastly, in Section 4, the study is summarized and the scope for future work discussed.

1.2. Literature Review on Fault Detection Methods

Fault detection and diagnosis (FDD) approaches based on traditional statistical process control charts as well as advanced data mining techniques have been employed in various manufacturing applications [5,29–32]. As summarized in Table 1, statistical chart-based FDD approaches provide a satisfactory detection performance when sensor signals are only collected during the normal operation of a system, and the corresponding measurements are concentrated in one or several clusters. Principal component analysis (PCA) is a popular example of a fault diagnosis method based on traditional statistical process control charts where multivariate analog sensor signals are simultaneously collected from a manufacturing process [33–35]. This method reduces the dimensionality of original historical data according to eigenvectors decomposed via singular value decomposition [33]. Kim et al. (2020) employed PCA to derive a residual control chart to monitor the current system status [36]. Since highly correlated and dimensional sensor signals were fed as input data, the authors were able to extract hidden but meaningful features by combining traditional PCA with functional PCA. Since traditional PCA deals with linear relationships in the given multivariate dataset and functional PCA contains non-linear eigenfunctions and can handle highly correlated multivariate sensor signals, the derived features were used as indicators of process control charts; this yielded a superior fault detection performance.

Table 1. Classification of fault detection and diagnosis (FDD) approaches.

	Statistical Chart	Data Mining
Usage of fault state data	Unsupervised learning	Supervised learning
Mathematical model	Yes	No
Gradient relationship between measurements and system operation states	Yes	Not necessary
Popular decision criteria	Distance from the normal states	Similarity to the known signal behaviors during fault states

By contrast, data-mining-based FDD approaches are usually utilized when the sensor signals are excessively scattered regardless of the system operation state (normal or fault). Due to this characteristic, it is not easy to develop a robust representative mathematical model for quantifying the state of the system. That is, advanced machine learning techniques have been frequently applied when the collected sensor signals do not have distinguishable features between the normal and faulty system states [37]. In order to detect unknown faulty statuses more accurately, pattern classification with discrete state vectors (DSVs) was improved based on a similarity analysis [38]. Since previous DSV-based fault detection methods showed good performance when every DSV is discernible based on training data, the authors used Naïve Bayes approximation and the Brier score to evaluate unknown DSVs in terms of fault detection power. Based on the proposed technique, the faulty operation of a vehicle engine could be detected more precisely.

As the above approaches explained, most O&M studies have focused on offline fault diagnosis [5,35,36]. Online fault detection is usually performed by comparing the current operation state and the offline fault detection model, and consequently, it gives the user alarm when the current state is considered as a fault state (or, a fault state will occur in near future. For effective manufacturing operations, corrective recovery action(s) should be manually performed by operators to prevent the detected faults. For example, for the effective operation of an automated storage and retrieval system (ASRS), Internet-of-things-based controllers and sensors were installed [31]. Depending on the current system status determined by analyzing real-time vibration signals, the controller changes the relevant process parameter (i.e., motor speed in transporters) and, consequently, the ASRS can automatically maintain a failure-free status. In this regard, O&M approaches have evolved from corrective maintenance, through preventive maintenance and condition-based monitoring, to predictive maintenance [6,39]; however, it is still not easy to combine real-time fault detection/prediction and automatic process adjustment.

2. Materials and Methods

2.1. Materials: System Configuration for Monitoring Pick-Up Operations in Real Time

To identify the key machine parameters and test the predictive process adjustment with the corresponding characteristics, a testbed that performs a pick-up operation using a vacuum gripper [27], as depicted in Figure 2 was constructed. Furthermore, the monitoring and control modules were upgraded, as shown in Figure 2b. The detailed information of the system configuration is summarized in Table 2.

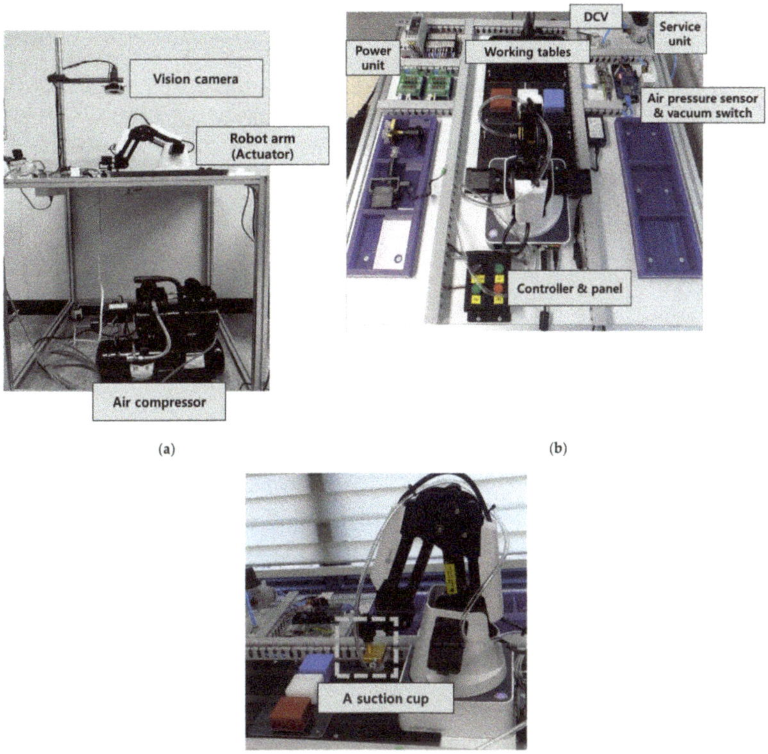

(a) (b) (c)

Figure 2. Testbed for performing the pick-up operation and monitoring the amount of applied air pressure during each operation cycle: (**a**) front view; (**b**) top view; (**c**) closed view for a suction cup.

Table 2. Summary of testbed configuration.

Unit	Function
Air compressor	Supply compressed air
Air pressure sensor	Measure air pressure during operation
Air pressure gauge	Measure an initially applied amount of air pressure
Vacuum switch	Detect success/failure of suction
Four-axis robot arm	Move vacuum gripper
Vacuum gripper with a suction cup and single solenoid directional control valve (DCV)	Conduct suction operation
Arduino-based controller	Collect signals and send them to the PC
PC-based controller	Control the robot arm and Arduino-based controller

The outlet air pressure was considered as an indicator to identify the current pick-up operation (i.e., suction operation) in real time, instead of the inlet air pressure, which is most commonly measured. Among various machine parameters, the outlet air pressure is the most appropriate variable because it changes depending on whether the suction is turned on or not. In a vacuum gripper, compressed air is supplied from an air compressor to the suction cup through a Venturi line. The Venturi line is designed to increase airflow speed by changing the size of the air pipe [40]. When the compressed air moves from the entrance to the exit of a Venturi line, a partial vacuum state occurs at the jet nozzle (an additional narrow vertical air pipe between the entrance and exit of the Venturi line). Regardless of whether a partial vacuum state was created, it was assumed that the magnitude of the outlet air pressure changed.

For predictive process adjustment according to real-time operation monitoring, the control modules and the corresponding control software were upgraded, as shown in Figure 3. Due to the upgrade, it was not only possible to acquire sensor signals in real time but also to simultaneously change machine parameters, such as the x-, y-, and z-positions corresponding to the suction start, during the operation. The workflow is as follows:

1. Make sure every module is turned on and connected to the PC-based controller via Ethernet communication.
2. Send a pick-up operation message to move the robot arm and start the suction operation.
3. When the suction is started by the PC-based controller, analog sensor signals (outlet air pressure) are simultaneously acquired in real time and can be monitored through the developed control software.
4. After performing the given pick-up operation (regardless of the operation success), the acquired sensor signals are exported as a spreadsheet per operation.

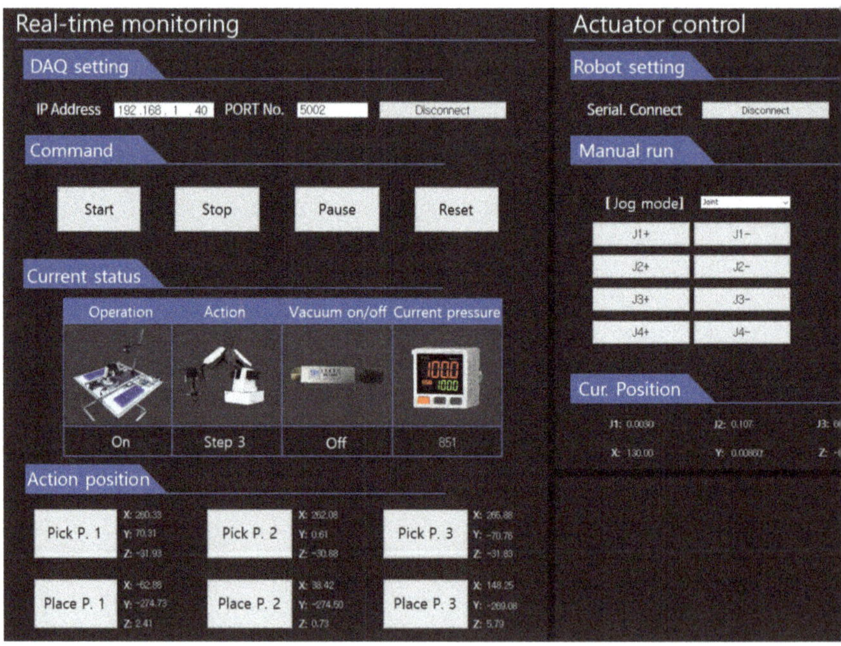

Figure 3. Software to control the testbed, collect sensor signals in real time, and send feedback control messages.

2.2. Methods

Predictive process adjustment is a representative approach in predictive maintenance. The concept of this approach is to automatically perform appropriate feedback actions, which are determined by detecting and diagnosing the current system status or predicting an upcoming fault status through real-time monitoring [31]. Therefore, in this paper, early fault detection and recovery actions are proposed in order to conduct predictive process adjustment in real time, as illustrated in Figure 4.

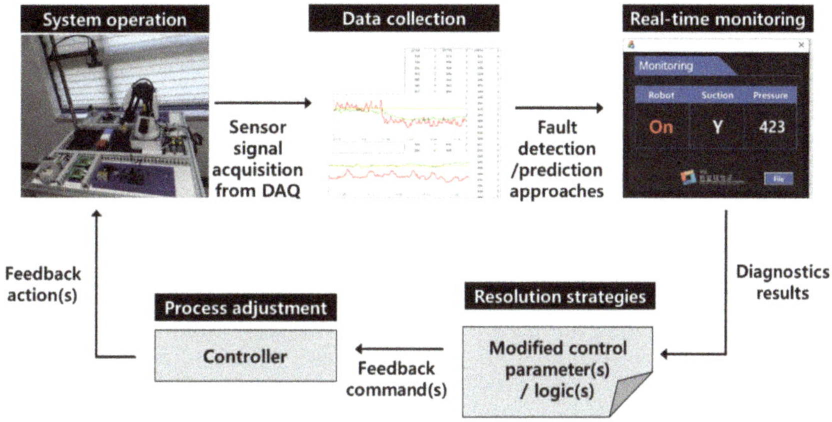

Figure 4. Framework of the predictive process adjustment method based on real-time data acquisition and feedback control.

First, the relevant measurements are acquired from the testbed when a vacuum gripper performs a pick-up operation according to the controller's command. During the training phase, the time-series sensor signals for every cycle are accumulated in a history dataset, which is employed to generate a fault detection model. To develop a fault detection model that can identify an unknown and sudden fault situation in advance, a detection threshold should be determined. Since sensor signals were already collected, it is possible to construct fault detection model using several fault detection/prediction approaches, from conventional statistical models to pattern extraction-based deep learning models. In this study, the simplest statistical-based fault detection model was employed using Equation (1) [41]:

$$Detection\ threshold = \overline{\overline{X}} + \frac{3}{d_2\sqrt{n}} \times (R), \tag{1}$$

where $\overline{\overline{X}}$ is (approximately) the mean of the sensor signals a short duration after the suction command is given to an actuator in normal operation; $\overline{R}$ is the mean of the differences of maximum and minimum measurements in the same dataset, which is used for calculating $\overline{\overline{X}}$; and d_2 is a constant value chosen by the number of sensor measurements (i.e., n).

For a statistical process control approach, the detection threshold is computed according to the $\overline{X}$-chart. The $\overline{X}$-chart is one of the control charts that is used for identifying whether measurements in the current batch are under the statistical control or not. It can be categorized as unsupervised learning, as history datasets collected during normal operations are used to develop a detection model, such as those based on $\overline{\overline{X}}$ and $\overline{R}$ computation. The $\overline{X}$-chart was utilized in this study because the time-series sensor signals collected during one pick-operation operation are considered as a dataset in one batch. In this regard, a series of sensor signals is collected for a short duration after the suction starts (during one operation cycle) in order to assume that the collected measurements are under identical conditions.

In common process control chart models [41], the upper and lower control limits are applied together to determine the statistical outliers. However, in this study, the direction of the failure status was clearly determined (e.g., a higher value indicates a failure of vacuum generation, and hence, an operation failure); therefore, only the upper control limit was used as the detection threshold. Using the derived detection threshold, the failure status of the pick-up operation was detected when the following condition is not satisfied:

$$x_i \leq \alpha \times Detection\ threshold, \tag{2}$$

where x_i is the current sensor value and α is the predefined confidence interval.

If a fault is detected during real-time monitoring based on the above detection model, the appropriate resolution strategy is consequently developed. In this study, the machine parameter was carefully controlled in order to overcome situations with faults.

3. Experimental Results

3.1. Result: Early Fault Detection for Pick-Up Operation

Using the testbed described in Section 2, a pick-up experiment was conducted to determine the characteristics of the collected sensor signals depending on the success or failure of the suction operation. The target operation was to lift rectangular boxes using one suction cup. To generate two different states (i.e., success and failure to lift), the following two different types of boxes were used, as shown in Figure 5:

- Box type I: Normal box with a surface that is sufficiently flat to be picked up with conventional vacuum grippers and judged to be acceptable based on product quality inspection.
- Box type II: Box that appears normal in the product quality inspection but is not flat enough to be picked up using a conventional vacuum gripper ("box type II" hereinafter) owing to a slight concave curvature in its contact surface, resulting in a faulty pick-up operation.

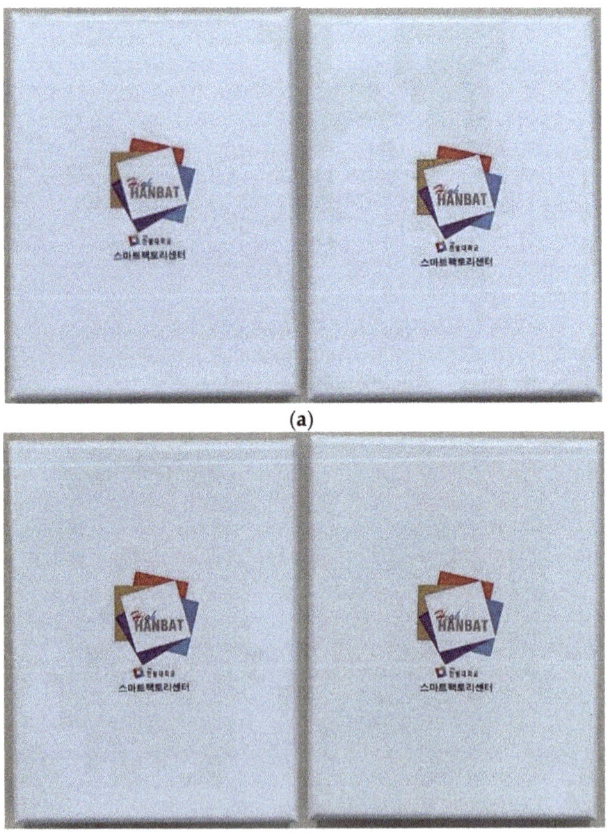

(b)

Figure 5. Target objects: (**a**) two samples of box type I; (**b**) two samples of box type II. (They do not show any significant difference in 2D images).

Both types of box cover, box type I and II, are considered as normal quality in product quality inspection phase. Although box type II should be used in the manufacturing process from the viewpoint of product quality, because of its very negligible concave on the contact surface, it results in a faulty pick-up operation. Importantly, even if the manufacturer attempts to install a system to remove box type II from the production line, it is highly difficult to distinguish between the two types using the naked eye or a 2D camera, as depicted in Figure 5.

The lifting operation was conducted 20 times for each box type, and the corresponding sensor signals were collected, as illustrated in Figure 6. The generated historical data were recorded with the corresponding timestamps. The duration was approximately 10 s, and the analog signals were recorded at a sampling rate of approximately 10 Hz. In addition, missing or noisy data in the collected raw signals were filtered using adjacent measurements.

Figure 6a describes the pressure signals of the airflow in the Venturi channel for every lifting iteration. Before starting the suction operation using the vacuum gripper, it was ensured that the initial pressure of the airflow in the Venturi line was almost identical to the original air pressure provided by the air service unit and air compressor. After starting the suction operation (before lifting the object), it can be observed in Figure 6a that the monitored air pressure decreased continuously (at approximately 5.6 to 6.0 s). When the suction was sufficiently succeeded, the decreased pressure was constantly maintained during the rest of the pick-up operation, until moving to the destination position, and the decreased level was identical in the 20 repetitions. Although certain fluctuations in the inlet pressure can be observed during the pick-up operation, the decrease in pressure was mostly constant until the suction operation was terminated. Signal collection automatically stopped after the robotic arm was moved with the vacuum gripper to the destination position and the vacuum switch was turned off to release the object.

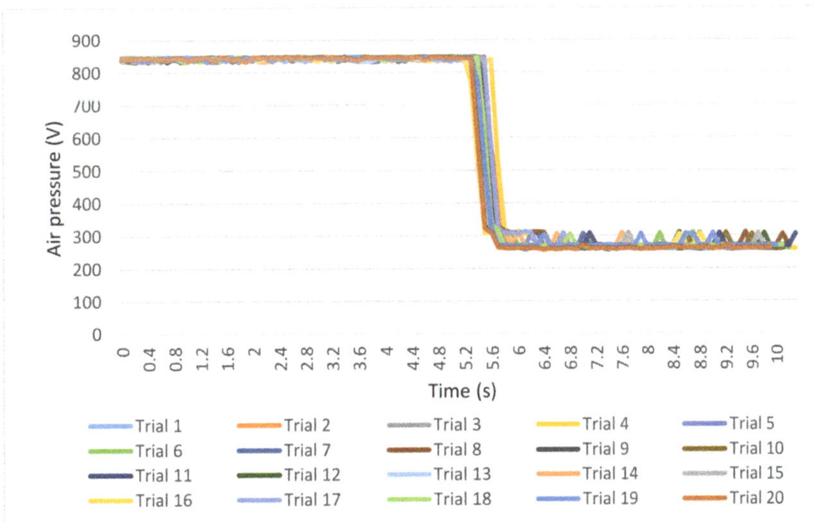

(a)

Figure 6. *Cont.*

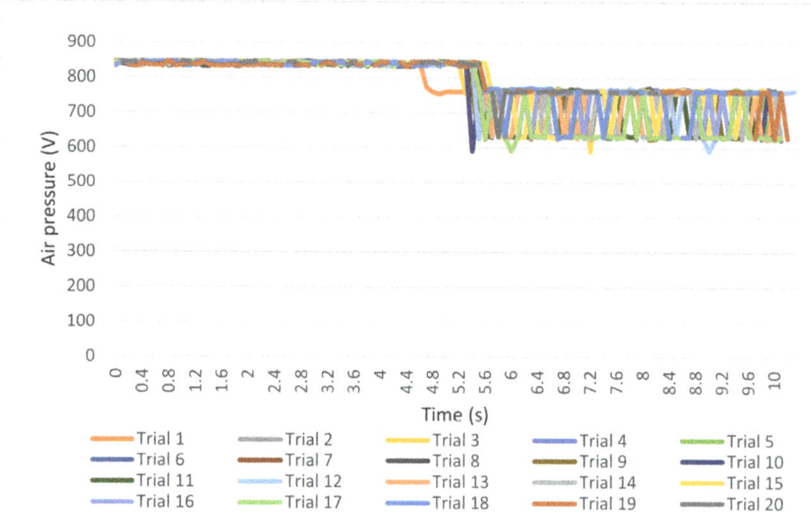

(b)

Figure 6. Measured air pressure over time: (**a**) successful lift for box type I; (**b**) failed lift for box type II.

For box type II, the gripper failed to pick-up the box in all repetitions, as listed in Table 3, even though the conditions were kept identical to those for box type I. As depicted in Figure 6b, the air pressure starts to decrease when the suction operation starts, but the level of decrease is different from that for box type I. As a result of the negligible concave curvature on the contact surface, it is not possible to generate sufficient suction, and the amount of outlet air pressure cannot decrease sufficiently for a successful gripper operation. In Figure 6b, the sensor signal for Trial 1 is slightly different from those for the other 19 trials; this is due to an inadvertent early manual control suction start command by the operator.

Table 3. Pick-up operation results.

Box Types	Operation Success Rate
Box type I	100%
Box type II	0%

In conventional operations with a vacuum gripper, a vacuum switch determines whether the suction operation was successful after the pick-up operation has been conducted. In other words, although the vacuum switch does not indicate that a sufficient vacuum has been successfully generated, the gripper (and the connected robotic arm) moves to the next position regardless of the completion of the current pick-up operation. However, upon using the information derived from this study, extraneous movements or operations without the target object can be prevented.

In summary, the process can be summarized as follows:

1. Start the suction operation and measure the amount of outlet air pressure in real time.
2. Check the decrease during approximately short time period (i.e., 5.6 to 6.0 s), which corresponds to suction start.

3.1. If there is a sufficient decrease in air pressure, then conduct the next step (usually, move the gripper to the next position).

3.2. If the air pressure does not decrease to the desired level, then conduct the predictive process adjustment until the object is lifted by the gripper.

3.2. Discussion: Conducting Appropriate Recovery Actions

After predicting the outcome of a pick-up operation (as already described in Section 3.1), it is necessary to identify which control parameter(s) can result in a failed pick-up operation. There are several control parameters whose value changes during the gripper operation, such as the initial applied pressure, length of suction time, and suction position [26,27]. In this study, the vertical position of the vacuum gripper at the suction start (hereafter called as the "z-position") was selected as the control parameter for the predictive process adjustment for the pick-up operation. The z-position is illustrated in Figure 7: it represents the bottom position of the suction cup attached on a vacuum gripper. In the case of more than one suction cup, the same z-position is recommended for all suction cups.

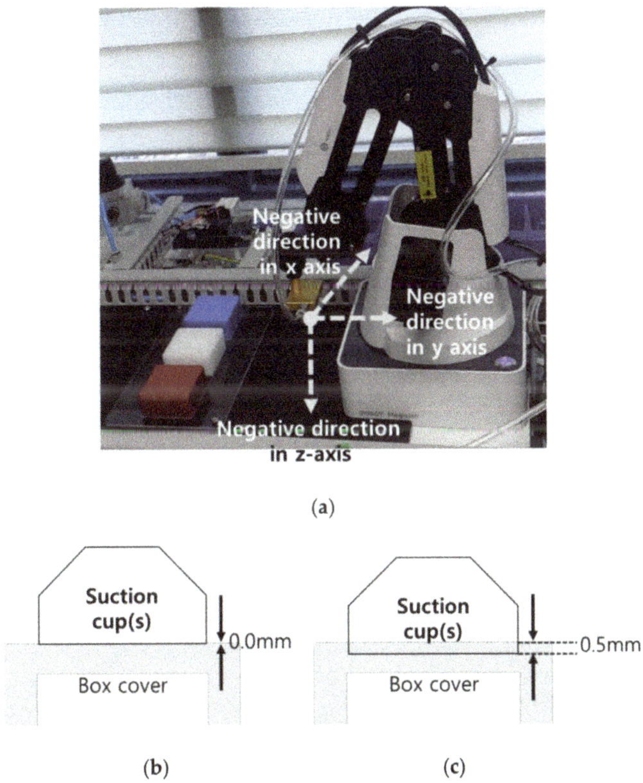

Figure 7. Suction cup positions for predictive process adjustment: (**a**) axis information for suction cup control; (**b**) diagram for the 0.0 mm z-position; (**c**) diagram for the −0.5 mm z-position.

Experiments were conducted to investigate the effect of the z-position on the pick-up operation, as summarized in Table 4. The pick-up operation consisted of the following consecutive steps: movement to the suction location, suction start, upward movement, movement to the destination, and suction finish.

Table 4. Experimental setup to determine effect of z-position on the pick-up operation.

Dependent variable	Amount of outlet air pressure at approximately 5.6 to 6.0 s
Independent variable	Z-position with four levels (0.0, −1.0, −1.5, and −0.0 mm)
Number of repetitions	10 times for each condition

Once the command was sent to the vacuum gripper, the sensor signals were collected in the form of a time series from step 2 to step 3. A total of 40 time-series datasets were recorded. Using mean values corresponding to approximately 5.6 to 6.0 s for each time-series dataset, a one-way analysis of variance (ANOVA) was conducted to determine the statistical effects of the dependent variable on the independent variable. ANOVA is a typical statistical test to determine the source of measurement differences, that is, whether the differences result from variance between groups (levels) or from variance within groups corresponding to independent variables [42]. It can be treated as the extended version of the t-test for more than three groups corresponding to an independent variable. This statistical test is prevalently employed to determine the effects of controllable parameters on the output [43–45].

The results are summarized in Table 5, where the z-position shows a significant difference in the pick-up operation (p-value < 0.05). As can be observed from Table 6, the success ratio of the operation increased from 0% to 100% as the vertical position at the suction start lowered. Specifically, the suction cup failed in nine trials except in the case of Trial 5 (–1.0 mm), whereas it failed in only three trials for box type II (i.e., Trials 1, 2, and 6) with the z-position at –1.5 mm. Finally, when the z-position was set as –2.0 mm, the gripper succeeded in picking the target up in every trial. Clearly, the outlet air pressure decreased by more (approximately 200 to 300 V) than the decrease in the cases of operation failures (approximately 600 to 700 V), and the outlet air pressure was continuously maintained until step 3, as depicted in Figure 8.

Table 5. ANOVA results of the effect of the z-position on the pick-up operation of the gripper.

Source	SS	df	MS	F	p-Value
Between groups	1,777,360	1	1,777,360	1284	<0.000
Within groups	52,616	38	1385		
Total	1,829,976	39			

* SS: sum of squares, df: degree of freedom, MS: mean squares.

Table 6. Experimental results of the predictive process adjustment by controlling the z-position of suction start with box type II.

Z-Position	Operation Success Rate
0.0 mm	0%
−1.0 mm	10%
−1.5 mm	70%
−2.0 mm	100%

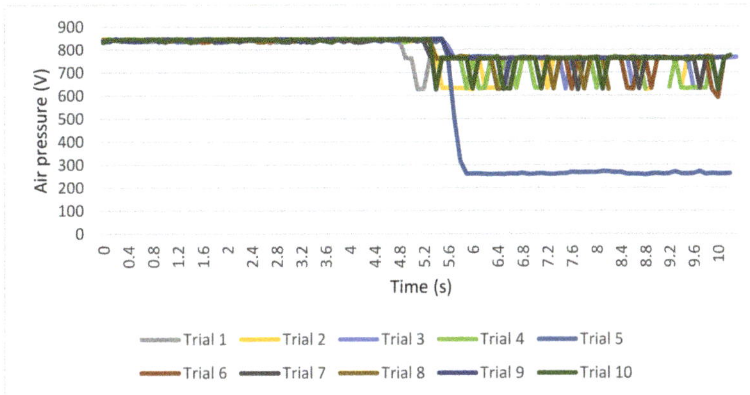

(a)

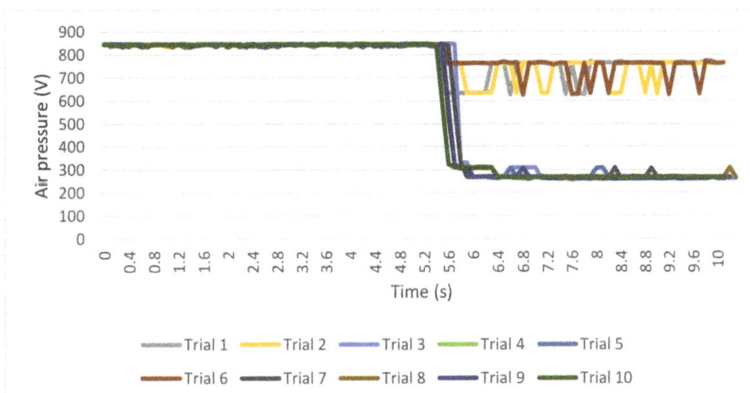

(b)

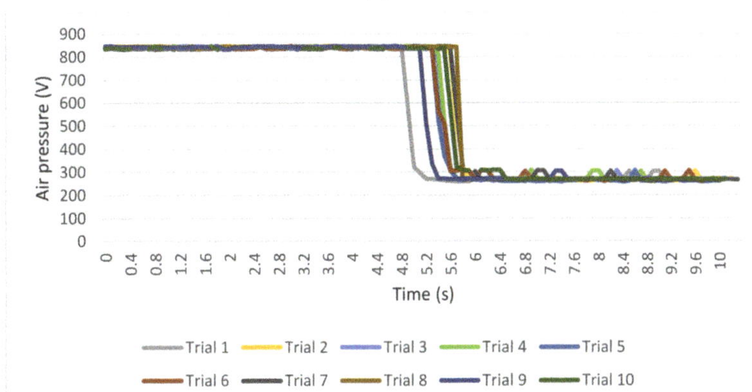

(c)

Figure 8. Collected air pressure signals depending on the z-position (during pick-up operation with box type II): z-position is set as (**a**) −1 mm; (**b**) −1.5 mm; and (**c**) −2.0 mm.

In summary, the predictive process adjustment with real-time operation monitoring can be conducted as follows:

1. Start the suction operation and measure the amount of outlet air pressure in real time.
2. Check the decrease during approximately a short time period (i.e., 5.6 to 6.0 s), which corresponds to suction start.

3.1. If the air pressure decreases sufficiently, then perform the next step (usually, move the gripper to the next position).
3.2. If a suction step is considered a failure (i.e., air pressure does not decrease to the predefined UCL), move the z-axis downward until the air pressure falls to the UCL.

However, as the z-position is further lowered, the probability of defects on the contact surface increases. A vacuum gripper that lowers excessively may result in a dent or a hole, and consequently, the object will be considered to be defective right before the pick-up operation. Therefore, it is essential to determine the appropriate z-position that does not cause any damage to the surface of the object while maximizing the success rate of the pick-up operation.

4. Conclusions

In this paper, a testbed for conducting a pick-up operation using a vacuum gripper with a single suction cup was prepared. Using the proposed method, the air pressure in the Venturi line was automatically monitored in real time. When a command for starting suction was provided to the gripper, a sharp decrease in the collected air pressure signals appeared at approximately 0.5 s. However, the same decrease was not observed in the signal for faulty box surfaces; consequently, the suction action and the corresponding gripper operation were not performed owing to insufficient contact between suction cup(s) and the contact surface of the object. Using the early detection results derived from the air pressure signal analysis, a prediction-based process adjustment method for the pick-up operation was proposed. Through pick-up experiments using the developed testbed, it was revealed that the z-position of the suction cup significantly affects whether an object is properly gripped by the vacuum gripper or not. Therefore, it is possible to determine a possible error situation in advance (before failure of the lifting operation) and provide appropriate feedback control commands so that the target operation is finished successfully without stopping machine operations.

However, for stable operation and generalization, it is necessary to conduct further research on the following: (i) identifying the appropriate depth for the z-position that does not generate any defects on a contact surface but maximizes the rate of success of the pick-up operation; (ii) generalizing the results for different materials, sizes, weights, or shapes of handled objects and various configurations of vacuum grippers; and (iii) combining the machine-status monitoring result and product defect detection result to improve the productivity and product quality of an industrial production system.

Author Contributions: Conceptualization, methodology, validation, formal analysis, investigation, resources, data curation, writing, visualization, supervision, project administration, funding acquisition, S.B.; Experiment, D.O.K. All authors have read and agreed to the published version of the manuscript.

Funding: This work was supported by the National Research Foundation of Korea (NRF) grant funded by the Korea government (MSIT) (No. NRF-2019R1G1A1097478) and was also supported by Korea Institute for Advancement of Technology (KIAT) grant funded by the Korea Government (MOTIE) (P0012744, The Competency Development Program for Industry Specialist). In addition, this research was supported by the research fund of Hanbat National University in 2020.

Data Availability Statement: Not applicable.

Acknowledgments: The staffs of GL Company (Republic of Korea) provided immense support in building the experimental system.

Conflicts of Interest: The authors declare no conflict of interest. The funders had no role in the design of the study; in the collection, analyses, or interpretation of data; in the writing of the manuscript; or in the decision to publish the results.

References

1. Bai, S.; Cheng, Z.; Guo, B. Maintenance optimization model with sequential inspection based on real-time reliability evaluation for long-term storage systems. *Processes* **2019**, *7*, 481. [CrossRef]
2. Rahman, M.; Zaccaria, V.; Zhao, X.; Kyprianidis, K. Diagnostics-oriented modelling of micro gas turbines for fleet monitoring and maintenance optimization. *Processes* **2018**, *6*, 216. [CrossRef]
3. Lin, M.; Ruigang, T.; Yan, R.; Lei, Z.; Xizhen, S. Research on identification and measurement methods of influencing factors of investment in operation and maintenance. In *E3S Web Conference 236, 3rd International Conference on Energy Resources and Sustainable Development (ICERSD 2020)*; 2021; p. 04010. Available online: https://www.e3s-conferences.org/articles/e3sconf/abs/2021/12/e3sconf_icersd2020_04010/e3sconf_icersd2020_04010.html (accessed on 9 February 2021).
4. Furumoto, K.; Kolehmainen, A.; Silverajan, B.; Takahashi, T.; Inoue, D.; Nakao, K. Toward automated smart ships: Designing effective cyber risk management. In Proceedings of the 2020 International Conferences on Internet of Things and IEEE Green Computing and Communications and IEEE Cyber, Physical and Social Computing and IEEE Smart Data and IEEE Congress on Cybermatics, Rhodes, Greece, 2–6 November 2020; pp. 100–105. [CrossRef]
5. Espadinha-Cruz, P.; Godina, R.; Rodrigues, E.M.G. A review of data mining applications in semiconductor manufacturing. *Processes* **2021**, *9*, 305. [CrossRef]
6. Kim, D.-Y.; Park, J.W.; Baek, S.; Park, K.-B.; Kim, H.-R.; Park, J.-I.; Kim, H.-S.; Kim, B.-B.; Oh, H.-Y.; Namgung, K.; et al. A modular factory testbed for the rapid reconfiguration of manufacturing systems. *J. Intell. Manuf.* **2020**, *31*, 661–680. [CrossRef]
7. Townsend, W. The BarrettHand Grasper—Programmably flexible part handling and assembly. *Ind. Robot Int. J.* **2000**, *27*, 181–188. [CrossRef]
8. Li, C.; Gu, X.; Ren, H. A cable-driven flexible robotic grasper with Lego-like modular and reconfigurable joints. *IEEE/ASME Trans. Mechatron* **2017**, *22*, 2757–2767. [CrossRef]
9. Shao, G.; Ware, H.O.T.; Huang, J.; Hai, R.; Li, L.; Sun, C. 3D printed magnetically-actuating micro-gripper operates in air and water. *Addit. Manuf.* **2021**, *38*, 101834.
10. Amend, R.; Brown, E.; Rodenberg, N.; Jaeger, H.M.; Lipson, H. A positive pressure universal gripper based on the jamming of granular material. *IEEE. Trans. Rob.* **2012**, *28*, 341–350. [CrossRef]
11. Vedhagiri, P.J.G.; Prituja, A.V.; Li, C.; Zhu, G.; Thakor, N.V.; Ren, H. Pinch grasp and suction for delicate object manipulations using modular anthropomorphic robotic gripper with soft layer enhancements. *Robotics* **2019**, *8*, 67. [CrossRef]
12. Chen, T.; Wang, Y.; Yang, Z.; Liu, H.; Liu, J.; Sun, L. A PZT actuated triple-finger gripper for multi-target micromanipulation. *Micromach* **2017**, *8*, 33. [CrossRef]
13. Matsuo, I.; Shimizu, T.; Nakai, Y.; Kakimoto, M.; Sawasaki, Y.; Mori, Y.; Sugano, T.; Ikemoto, S.; Miyamoto, T. Q-bot: Heavy object carriage robot for in-house logistics based on universal vacuum gripper. *Adv. Robot.* **2020**, *34*, 173–188. [CrossRef]
14. Jamaludin, A.S.; Razali, M.N.M.; Jasman, N.; Ghafar, A.N.A.; Hadi, M.A. Design of spline surface vacuum gripper for pick and place robotic arms. *J. Mod. Manuf. Syst. Technol.* **2020**, *4*, 48–55. [CrossRef]
15. Ashwini, K.; Rudraswamy, S.B. Automated inspection system for automobile bearing seals. *Mater. Today Proc.* **2020**. [CrossRef]
16. Seo, J.W.; Lee, J.K.; Lee, S.; Lee, K.C. Slip detection of robot gripper with flexible tactile sensor. *J. Korean Soc. Precis. Eng.* **2014**, *31*, 157–164. [CrossRef]
17. Lee, J.K.; Kim, H.J.; Lee, S.; Lee, K.C. Design of grasping objects system of robot gripper using artificial neural network. In Proceedings of the 2014 KSMPE Conference, Changwon, Korea, 6–8 November 2014; p. 109.
18. Jaiswal, A.K.; Kumar, B. Vacuum gripper—An important material handling tool. *Int. J. Sci. Technol.* **2017**, *7*, 1–8.
19. Zhang, P.; Kamezaki, M.; Otsuki, K.; He, Z.; Sakamoto, H.; Sugano, S. Development of a vacuum suction cup by applying magnetorheological elastomers for objects with flat surfaces. In Proceedings of the 2020 IEEE/ASME International Conference on Advanced Intelligent Mechatronics (AIM), Boston, MA, USA, 6–9 July 2020; pp. 777–782.
20. Nakamoto, H.; Ohtake, M.; Komoda, K.; Sugahara, A.; Ogawa, A. A gripper for robustly picking various objects placed densely by suction and pinching. In Proceedings of the 2018 IEEE/RSJ International Conference on Intelligent Robots and Systems (IROS), Madrid, Spain, 1–5 October 2018; pp. 6093–6098.
21. Wang, J.; Liang, Y.; Zheng, Y.; Gao, R.X.; Zhang, F. An integrated fault diagnosis and prognosis approach for predictive maintenance of wind turbine bearing with limited samples. *Renew. Energy* **2020**, *145*, 642–650. [CrossRef]
22. Jaiswal, A.K.; Kumar, B. Design constraints of vacuum cup gripper of robots-as a pick and place operation tool. *Int. J. Sci. Technol.* **2016**, *6*, 26–35.
23. Jaiswal, A.K.; Kumar, B. Vacuum cup grippers for material handling in industry. *Int. J. Innov. Sci. Eng. Technol.* **2017**, *4*, 187–194.
24. Papadakis, E.; Raptopoulous, R.; Koskinopoulou, K.; Maniadakis, M. On the use of vacuum technology for applied robotic systems. In Proceedings of the 2020 6th International Conference on Mechatronics and Robotics Engineering, Barcelona, Spain, 12–15 February 2020; pp. 73–77.

25. Valencia, A.J.; Idrovo, R.M.; Sappa, A.D.; Guingla, D.P.; Ochoa, D. A 3D vision based approach for optimal grasp of vacuum grippers. In Proceedings of the 2017 IEEE International Workshop of Electronics, Control, Measurement, Signals and their Application to Mechatronics, Donostia-San Sebastian, Spain, 24–26 May 2017; pp. 1–6.
26. Baek, S.; Kim, D.-O. Effects of Z-axis suction position of vacuum gripper to box pick-up operation. In Proceedings of the 26th Winter Conference of Society for Computational Design and Engineering, Jeju, Korea, 25–27 November 2020; p. 435.
27. Baek, S.; Kim, D.-O.; Lee, S.-J.; Yu, N.-H.; Chea, S.-I. Development of air pressure measurement system of suction cups in a vacuum gripper. In Proceedings of the 21th ACIS International Semi-Virtual Winter Conference on Software Engineering, Artificial Intelligence, Networking and Parallel/Distributed Computing, Ho Chi Minh City, Vietnam, 28–30 January 2021; pp. 1–5.
28. Baek, S.; Jeon, S.H.; Song, E.C. Real-time monitoring of a vacuum gripper pick-up operation in a box packaging process by CNN based the box surface deviation image analysis. *J. Korean Inst. Ind. Eng.* **2020**, *46*, 107–113. [CrossRef]
29. Saufi, S.R.; Ahmad, Z.A.B.; Leong, M.S.; Lim, M.H. Challenges and opportunities of deep learning models for machinery fault detection and diagnosis: A review. *IEEE Access* **2019**, *7*, 122644–122662. [CrossRef]
30. Pillai, D.S.; Blaabjerg, F.; Rajasekar, N. A comparative evaluation of advanced fault detection approaches for PV systems. *IEEE J. Photovolt.* **2019**, *9*, 513–527. [CrossRef]
31. Baek, S. System integration for predictive process adjustment and cloud computing-based real-time condition monitoring of vibration sensor signals in automated storage and retrieval systems. *Int. J. Adv. Manuf. Technol.* **2021**, *113*, 955–966. [CrossRef]
32. Dai, X.; Gao, Z. From model, signal to knowledge: A data-driven perspective of fault detection and diagnosis. *IEEE Trans. Ind. Inf.* **2013**, *9*, 2226–2238. [CrossRef]
33. Bersimis, S.; Psarakis, S.; Panaretos, J. Multivariate statistical process control charts: An overview. *Qual. Reliab. Eng. Int.* **2007**, *23*, 517–543. [CrossRef]
34. Kazemi, P.; Giralt, J.; Bengoa, C.; Masoumian, A.; Steyer, J.-P. Fault detection and diagnosis in water resource recovery facilities using incremental PCA. *Water Sci. Technol.* **2020**, *82*, 2711–2724. [CrossRef]
35. Bencheikh, F.; Harkat, M.F.; Kouadri, A.; Bensmail, A. New reduced kernel PCA for fault detection and diagnosis in cement rotary kiln. *Chemom. Intell. Lab. Syst.* **2020**, *204*, 104091. [CrossRef]
36. Kim, J.-M.; Wang, N.; Liu, Y.; Park, K. Residual control chart for binary response with multicollinearity covariates by neural network model. *Symmetry* **2020**, *12*, 381. [CrossRef]
37. Verron, S.; Tiplica, T.; Kobi, A. Fault detection in a multivariate process with a Bayesian network. *Qual. Assur.* **2007**, *51*, 1–9. [CrossRef]
38. Namgung, K.; Yoon, H.; Baek, S.; Kim, D.Y. Estimating system state through similarity analysis of signal patterns. *Sensors* **2020**, *20*, 6839. [CrossRef] [PubMed]
39. Luo, W.; Hu, T.; Ye, Y.; Zhang, C.; Wei, Y. A hybrid predictive maintenance approach for CNC machine tool driven by Digital Twin. *Rob. Comput. Integr. Manuf.* **2020**, *65*, 101974. [CrossRef]
40. Dini, G.; Gantoni, G.; Failli, F. Grasping leather plies by Bernoulli grippers. *CIRP Ann.* **2009**, *58*, 21–24. [CrossRef]
41. Isermann, R. *Fault-Diagnosis Systems*; Springer: Berlin, Germany, 2006.
42. Montgomery, D.C. *Design and Analysis of Experiments*, 10th ed.; Wiley: Hoboken, NJ, USA, 2019.
43. Kuntoğlu, M.; Sağlam, H. Investigation of progressive tool wear for determining of optimized machining parameters in turning. *Measurement* **2019**, *140*, 427–436. [CrossRef]
44. Shinozaki, H.; Nakao, Y. Equipment comparison analysis using ANOVA of FDC statistics. In Proceedings of the 2018 International Symposium on Semiconductor Manufacturing (ISSM), Tokyo, Japan, 10–11 December 2018; pp. 1–2.
45. Naik, A.B.; Reddy, A.C. Optimization of tensile strength in TIG welding using the Taguchi method and analysis of variance (ANOVA). *Therm. Sci. Eng. Prog.* **2018**, *8*, 327–339. [CrossRef]

processes MDPI

Article

A Systematic Model for Process Development Activities to Support Process Intelligence

Edrisi Muñoz [1,†], Elisabet Capon-Garcia [2,†], Enrique Martinez Muñoz [3,†] and Luis Puigjaner [1,*,†]

1 Chemical Engineering Department, Process and Environment Engineering Centre, Universitat Politecnica de Catalunya, CEPIMA, EEBE - c. Eduard Maristany 10-14, Ed. I-5, 08019 Barcelona, Spain; edrisi.munoz@gmail.com
2 ABB Switzerland Ltd., Segelhofstrasse 1K, 5405 Baden-Dättwil, Switzerland; elisabet.capon@abb.ch
3 Academic Area of Engineering and Architecture, Institute of Basic Sciences and Engineerin, Autonomous University of the State of Hidalgo, UAEH, Ciudad del Conocimiento, Carretera Pachuca-Tulancingo Km. 4.5, 42164 Hidalgo, Mexico; emmunoz@uaeh.edu.mx
* Correspondence: luis.puigjaner@upc.edu
† These authors contributed equally to this work.

Abstract: Process, manufacturing, and service industries currently face a large number of non-trivial challenges ranging from product conception, going through design, development, commercialization, and delivering in a customized market's environment. Thus, industries can benefit by integrating new technologies in their day-by-day tasks gaining profitability. This work presents a model for enterprise process development activities called *the wide intelligent management architecture model* to integrate new technologies for services, processes, and manufacturing companies who strive to find the most efficient way towards enterprise and process intelligence. The model comprises and structures three critical systems: process system, knowledge system, and transactional system. As a result, analytical tools belonging to process activities and transactional data system are guided by a systematic development framework consolidated with formal knowledge models. Thus, the model improves the interaction among processes lifecycle, analytical models, transactional system, and knowledge. Finally, a case study is presented where an acrylic fiber production plant applies the proposed model, demonstrating how the three models described in the methodology work together to reach the desired technology application life cycle assessment systematically. Results allow us to conclude that the interaction between the semantics of formal knowledge models and the processes-transactional system development framework facilitates and simplifies new technology implementation along with enterprise development activities.

Keywords: enterprise process architecture; new technologies integration; process intelligence

Citation: Muñoz, E.; Capon-Garcia, E.; Martinez, E.; Puigjaner, L. A Systematic Model for Process Development Activities to Support Process Intelligence. *Processes* **2021**, *9*, 600. https://doi.org/10.3390/pr9040600

Academic Editor: Mohd Azlan Hussain

Received: 30 January 2021
Accepted: 24 March 2021
Published: 30 March 2021

1. Introduction

Our civilization faces acute and critical challenges, such as climate change, safe drinking water availability, food scarcity, and secure energy supplies, which endanger current and future generations. Therefore, society and industry need to shape their activities based on sustainable principles and to efficiently adopt the rapidly evolving new technologies which can potentially handle the challenges mentioned above. Precisely, this work focuses on the integration of new technologies in the decision-making of process and manufacturing industries. Thus, the proposed methodology applies to the workflow of any productive sector or area where decision-making plays a crucial role.

As for the process and manufacturing industries, complex decision-making lurks at all enterprise levels and the whole product lifecycle, ranging from product conception, design, development, production, commercialization, and delivery. The need to consider highly complex scenarios results in involved and non-trivial decision-making. However, the advent of new technologies supports the successful development and systematization of new structures and frameworks for reaching informed, reasonable, and wise decisions. Therefore, this work aims to integrate new technologies systematically in the decision-making workflow of the process and manufacturing industries. Therefore, this work presents a framework for developing process and product-related activities. Thus, the proposed model allows to unveiling the most efficient way towards integrating enterprise decision support systems and process intelligence into actual enterprise processes.

As discussed in Section 1.1 Decision making in the enterprise, companies have recently adopted decision support systems to handle the complexity of decision-making. However, such systems only tackle part of the complete enterprise structure, and it is necessary to understand the whole picture to reach sensible solutions, as pointed out in Section 1.2 Enterprise Integration. Therefore, this work combines Knowledge management (Section 1.3) and Data management (Section 1.4) to propose a framework for reaching integration of different systems and efficiently apply new technological solutions for decision-making.

1.1. Decision-Making in the Enterprise

Process and manufacturing industries can be regarded as highly involved systems consisting of multiple business and process units. The organization of the different temporal and geographical scales in such units, as well as the other enterprise decision levels, is crucial to understand and analyze their behavior. The key objectives are to gain economic efficiency, market position, product quality, flexibility, or reliability [1]. Recently, indicators related to sustainability and environmental impact have also been included as drivers for decision-making. The basis for solving an enterprise system problem and further implement any action is the actual system representation in a model, which captures the observer's relevant features. Such a model is the basis for decision-making, which is a highly challenging task in these industries due to their inherent complexity.

Therefore, companies have devoted efforts to reach better decisions during the last decades. Indeed, they have invested a large number of resources in exploiting information systems, developing models, and using data to improve decisions. Decision support systems (DSS) are responsible for managing the necessary data and information that allow making decisions. Thus, those systems aim to integrate data transactions with analytical models supporting the decision-making activity at different organizational levels. The work in [2] defines DSS as aiding computer systems at the management level of an organization that combines data with advanced analytical models. The work in [3] presents four components for supporting classic DSS. The components comprise (i) a sophisticated database for accessing internal and external data, (ii) an analytical model system for accessing modeling functions, (iii) a graphical user interface for allowing the interaction of humans and the models to make decisions, and (iv) an optimization-engine based on mathematic algorithms or intuition/knowledge. Traditionally, DSS focus on a single enterprise unit and lack the vision of the boundaries. Thus, DSS rely heavily on rigid data and model structures, and they are difficult to adapt to include new algorithms and technologies.

1.2. Enterprise Integration

Current trends in the process industry outline the importance of being agile and fully integrated to improve decision-making at all scales in the company. Indeed, integration comprises the whole organizational activities from operation to planning and strategic, which differ in physical and temporal scope, but are directly related to each other as decisions made at one level directly affect others. Therefore, companies pursuing integration among different decision levels in the production management environment report substantial economic benefits [4,5]. Therefore, to coordinate and integrate information and decisions among the various functions are crucial for improving global performance.

The use of Standards is the primary conducted method for enterprise integration labor. Groups, commities, and societies have developed those standards in different geopolitical areas where they are of application. Next, the use of some standards serves as well as a brief introduction of their content.

First, the European Committee for Standardization (CEN) and the European Committee for Electrotechnical Standardization (CENELEC) provide standards to characterize, guide, and rule SMEs' activities [6]. CEN standards comprise European Standards (E.N.s), drafts standards (prENs), Technical Specifications (CEN TSs), sDocuments (HDs), Technical Specifications (TSs), Technical Reports (TRs), and CEN Workshop Agreements (CWAs). Finally, CEN work is coordinate with the International Organization for Standardization (ISO) and the International Electrotechnical Commission (IEC).

Next, the National Institute of Standards and Technology (NIST) creates the Integrated Definition Methods (IDEF). IDEF comprises a standard fir function modeling (IDEF 0), Information Modelling (IDEF 1), Data Modelling (IDEF 1X), Process Modelling (IDEF 3), Object-Oriented Design (IDEF 4), and Ontology Description (IDEF 5) currently maintained by the Knowledge-Based Systems, INC. (KBSI) [7]. The standards were funding and are now in use by the United States Air Force and United States Department of Defense agencies. Moreover, many organizations for business process capturing and improvement.

Following, the International Electrotechnical Commission (IEC) develops the International Standards and Conformity Assessment covering areas such as industrial control programming standards (IEC 61131-3) or field devise integration (IEC 61804-2). The standards aim at allowing interoperability, efficiency, the safety of electrical, electronic, and information systems [8].

The International Organization for Standardization (ISO) standards are well-known and widely used standards, covering management systems, quality management, information security management, etc. [9]. Thus, criteria for integration comprises the Enterprise Modeling and Architecture (ISO TEC184 SC5 WG1), Electronic Business Extensible Markup Language (ISO 15000), and the Asset Management System (ISO 55003), among others.

Manufacturing Execution Systems Association (MESA) presents a set of best management practices and information technology aiming to improve business. MESA focuses on asset performance management, lean manufacturing, product lifecycle management, manufacturing performance metrics, quality, regulatory compliance, and return to investment [10].

Next, Machinery Information Management Open Systems Alliance (MIMOSA) association presents the Open Standards for Physical Asset management for information management (I.M.), and information technologies (I.T.) applied to manufacturing environments [11]. MIMOSA standards recently focus on enabling digital twins, big data, industrial internet of things, and analytics specifications.

The Object Management Group (OMG) is dedicated to developing technological standards for enterprise integration and distributed broad-interoperability [12]. The OMG comprises the following standards: Business Process Model and Notation (BPMN), Common Object Request Broker Architecture (CORBA), Common Warehouse Metamodel (CWM), Data-Distribution Service for Real-Time Systems (DDS), Unified Modeling Language (UML), and the Model Driving Architecture (MDA) applied to software visual design, execution, and support.

Next, the Process Industry Practices (PIP) consortium collaborates to define common industry standards and best practices focused on design, maintenance, and procurement activities [13]. Besides, PIP practices facilitate knowledge capturing of process control, mechanical, data management, and Piping and Instrumentation Diagrams.

A key element of integration directly points to enterprise models: computational applications within organizations aiming to represent processes, activities, resources, or physical phenomena. These models are essential for driving design, analysis, management, and prognosis in enterprise functions. Nevertheless, the spread of these models confronts several issues in practice. First of all, the independent creation of systems supporting functions at the enterprise during past years, resulting in heterogeneous enterprise models; that is, the so-called correspondence problem. Different enterprise models refer to the same concept, for example, an activity, each model will probably apply other names, following the example activity, operation, or task. Therefore, most of the time, interpreting agents are necessary to allow communication among those enterprise functions. However, no matter how rational the idea of renaming the concepts is, organizational barriers usually impede it. Furthermore, these representations lack an adequate specification of what the model objects mean; they lack the terminology's actual semantic definition. Instead, concepts are poorly defined, and their interpretations overlap, leading to inconsistent understandings and uses of the knowledge. Finally, the cost of designing, building, and maintaining a model of the enterprise is high. Each model tends to be unique to the enterprise, and objects are enterprise-specific.

Therefore, some efforts have addressed the issues mentioned above using model standardization. On the other hand, the American National Standards Institute (ANSI), developed the Instrumentation, Systems and Automation Society (ISA) standards, known as ANSI/ISA standards for automation and control within the enterprise [14] with wide recognition for process integration. Figure 1 presents the main integration aspects of these standards.

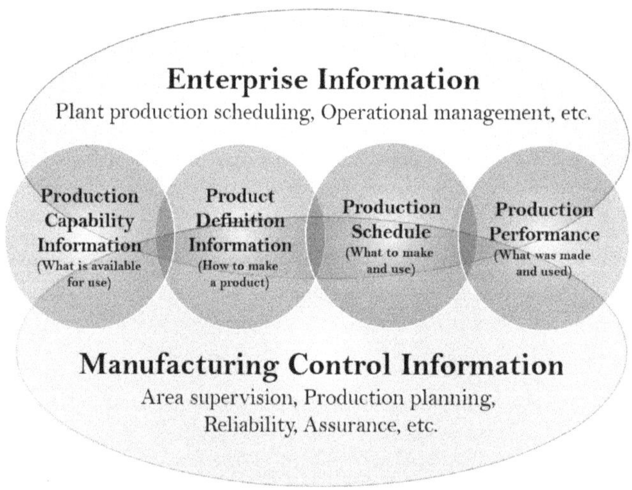

Figure 1. Instrumentation, Systems and Automation Society ISA-95 integration of information schema.

On the other hand, the Purdue reference model provides an "environment" for discrete parts manufacturing and stands for the basis for the other models [15]. In this case, certain activities are identified as directly related to shop floor production and organized in a six-level hierarchical model as depicted in Figure 2. Specific applications may require more or fewer than six levels, but six was deemed sufficient for identifying where integration standards are needed. The following list shows the name of each level and gives its primary responsibility.

- Level 6 Enterprise: Corporate Management (External Influences)
- Level 5 Facility: Planning Production
- Level 4 Section: Material/Resource Supervision
- Level 3 Cell: Coordinate Multiple Machines
- Level 2 Station: Command Machine Sequences
- Level 1 Equipment: Activate Sequences of Motion (Plant Machinery and Equipment)

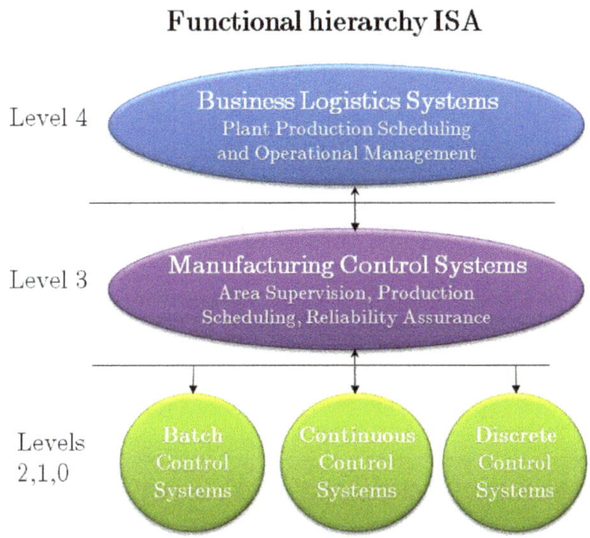

Figure 2. The Purdue Enterprise Reference Architecture (PERA).

These activities apply to manual operations, automated operations, or a mixture of the two at any level. It is worth mentioning the accessible subdivision of the six tasks into control enforcement, systems coordination and reporting, and reliability assurance. In the context of any large industrial plant or an entire industrial company based on one location, the tasks would take place at each level of the hierarchy.

Thus, the Common Information Model (CIM) reference model stands for a reference for computer-integrated manufacturing. It consists of a detailed collection of generic information management and automatic control tasks and their necessary functional requirements for a manufacturing plant. Nevertheless, the CIM reference model scope is limited to the integrated information management and automation system elements. As a result, the company's management, including planning function, financial, purchasing, research, development, engineering, and marketing and sales are all treated as external influences.

The adoption of standard models is the basis for the integration of enterprise processes. Thus, decision-making heavily relies on both the process models and the technologies which tackle the problem. Therefore, this work considers the systematization of data and knowledge management to reach integration in decision-making.

1.3. Knowledge Management

The development of better practices, strategies, and policies is highly related to how organizations use experiences and ideas from customers, suppliers, and employees. Thus, capturing, storing, sharing, and applying knowledge enables the construction of organization intelligence and intellectual assets. Two types of knowledge sources can be generally defined: tangible and intangible. On the one hand, intangible assets are related to skills, expertise, and human resources knowledge. On the other hand, tangible assets are related to data, information, historical records found on databases of customers, suppliers, and employees of the organization [16].

The bases of knowledge management tools can include distributed databases, ontologies, or network maps. This work focuses on formal domain ontologies development as the primary technology for knowledge management. Besides, the use of terms *Semantic Web* or *Web 3.0* can be used to refer to this technology. Ontologies and logic serve as conceptual graphs for knowledge representation in constructing computable models within a specific domain [17]. Additionally, ontologies are defined as formal structures facilitating acquiring, maintaining, accessing, sharing, and reusing information [18,19]. Over the last decades, the Semantic Web pursued theoretical bases for developing knowledge-based applications software: One can communicate

- a shared and common understanding of a domain among people and across application systems and
- an explicit conceptualization that describes the semantics of the data.

Finally, knowledge management systems benefit from ontologies that semantically enrich information and precisely define the meaning of various information artifacts.

1.3.1. Bloom's Cognition Taxonomy

Bloom's Taxonomy is a framework that presents how educational objectives can guide and structure educational goals. This framework's latest work is entitled *A Taxonomy for Teaching, Learning, and Assessment* defining the cognitive processes related to knowledge [20], shown in Figure 3. The framework considers six major categories, with subactivities for better understanding, as follows:

- Remember: Recognizing, Recalling.
- Understand: Interpreting, Exemplifying, Classifying, Summarizing, Inferring, Comparing, Explaining.
- Apply: Executing, Implementing.
- Analyze: Differentiating, Organizing, Attributing.
- Evaluate: Checking, Critiquing.
- Create: Generating, Planning, Producing.

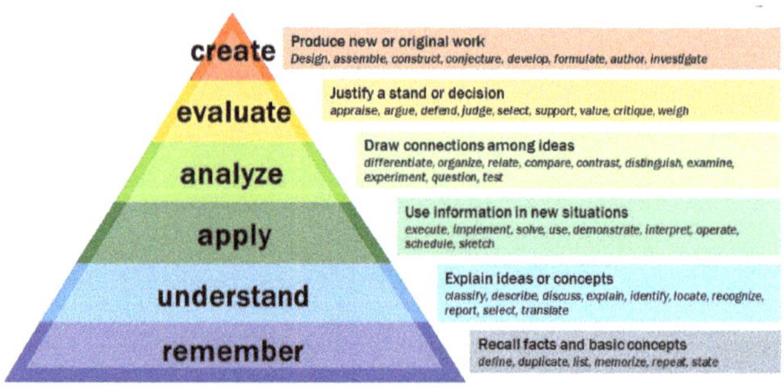

Figure 3. Bloom's taxonomy by Vanderbilt University Center for Teaching.

Finally, the framework defines four types of knowledge used in cognition:

- Factual Knowledge
    - Knowledge of terminology
    - Knowledge of specific details & elements
- Conceptual Knowledge
    - Knowledge of classifications and categories
    - Knowledge of principles and generalizations
    - Knowledge of theories, models, and structures
- Procedural Knowledge
    - Knowledge of subject-specific skills and algorithms
    - Knowledge of subject-specific techniques and methods
    - Knowledge of criteria for determining when to use appropriate procedures

- Metacognitive Knowledge
    - Strategic Knowledge
    - Knowledge about cognitive tasks (appropriate contextual and conditional knowledge)
    - Self-knowledge

1.4. *Transactional System and Data Management*

The performance of enterprise processing activities highly depends on the transactional system's capacity and how well the data is managing.

1.4.1. Transactional System

A transactional system comprises multiple operations that collect, store, modify, and retrieve data transactions within an enterprise. These systems must support a high number of concurrent users and transaction types along the time. Besides, enterprise data are identified by their purpose and type, comprising transactional, analytical, and master data. First, transactional data support the daily operations of an organization. Transactional data refer to data created or modified by the operational systems, such as time, place, number, date, price, payment methods, etc. Next, define analytical data as numerical measurements that support activities, such as decision-making, reporting, query, or analysis. Thus, analytical data are stored and structured as numerical values in some dimensional models. Finally, master data represent the key business entities, involving creating a single view of the data in a master file or master record. Master data comprise data about sites, inventory, levels, demand, products, batches, etc.

1.4.2. Data Management

Enterprise data management aims to govern business data by retrieving, standardizing, storing, integrating, structuring, and disseminating requested data. The transactional system supports data management by enhancing data transaction features for control, analysis, and decision-making. Thus, data management's essential feature comprises communicating all data from different data sources (sensors) and fragmented control systems with all enterprise applications, processes, and entities that require it. Another critical aspect of data management is to store and make data available when needed securely.

2. Materials and Methods

New technologies can accomplish their implementation life cycle with a robust base supported by the proposed architecture, named Wide Intelligence Management Architecture. This presented architecture offers three central systems comprising development activities: *process, knowledge, and transactional systems*, shown in Table 1. First, the process system model introduces seven development activities systematically ordered towards a formalized process maturity process. Thus, the activities range from process definition. The main aspects of how enterprise processes perform are process intelligence, where human and environmental behavior is taken into account to enrich development activities. Next, the knowledge system model aims to strengthen the integration by formalized knowledge from three main perspectives: the domain area, the expertise area (functional activities), and the experience area, enhancing expertise knowledge with success and failure cases. Finally, the third model is related to the transactional data system. This model comprises four main areas: data definition, data improvement, data standardization, and data feeding.

Table 1. Overall Intelligence Management Architecture for technology integration through process activities.

	Process Model	Knowledge Model	Transactional Model
L1	**Process Definition**	**Conceptual**	**Definition**
	Current process matter	Chemical principles	Data definition
		Physics principles	Data collection
		Mechanics principles	
L2	**Improvement**		**Improvement**
	Benchmarking	Good manufacturing practices	Data refining
		Standard operational procedures	Database
L3	**Standardization**		**Standardization**
	Tear levels definition	Process standards	Data metrics
	world-class process	Quality standards	Data language
		Data standards	Structured data
		Security standards	
L4	**Optimization**	**Procedural**	**Integration & Feeding**
	Better performance	Analytic algorithms knowledge	Data to parameters
	Key process variables	Analytical methods knowledge	Data to sets
	Key process parameters		Planning systems
L5	**Automation**	**Expertise**	
	Fixed parameters	Good & bad habits algorithms	Fixed data collection
	Fixed variables		Fixed data structuring
	Key indicators		
	Set values ranges		
L6	**Digitalization**		
	Virtual twin processes	Knowledge-based scenarios	Virtual feeding
	New process scenarios		
L7	**Intelligence**	**Metacognitive**	**Dynamics**
	Problem characterization	Model knowledge characterization	Automated data collection
	Problem classification	Model knowledge classification	Automated data structuring
	Intelligent systems	Knowledge reasoning	Intelligent database
	Intelligent agents	Knowledge creation	
	Autonomous decision-making		

2.1. Process System Model

The Modular Process Reference Model aims to define a coherent and structured manner of process evolution to integrate new technologies and business activities. The reference model comprises seven modules defined by the use of analytical tools and data linked to enterprise activities, represented in Figure 4.

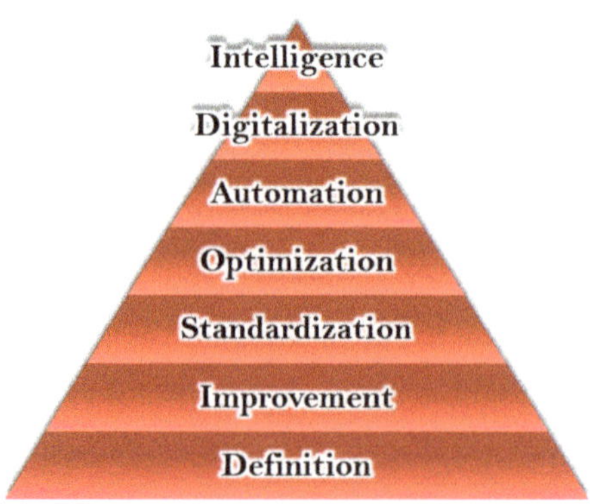

Figure 4. Maturity echelons of the process system model.

2.1.1. Process Definition

This module provides a set of activities aiming to assess enterprise processes performance and to support systematic formalization. On the one hand, this module verifies "if" and measures "how much", existing formalized process follows the current enterprise activities, named as verification phase. Otherwise, this methodology aims to define, design, and standardize enterprise processes, called a definition phase. Thus, a process design phase takes place, considering the followed validation and verification phases. The steps mentioned above (verification and definition) must apply to the enterprise transactional system parallel with the processes.

2.1.2. Process Improvement

The process improvement module makes an exhaustive study of current processes to perform a re-design phase. This re-design phase considers new tendencies on standards, methods, and technologies. Moreover, good manufacturing practices (GMPs) and standard operating procedures (SOPs) are of paramount importance in the improvement task. Moreover, at the same time, a transactional data system must pass through a re-design phase to support the process improvements realized. Finally, updates and documentation regarding enterprise processes, resources, and data improvements must follow (Focus on management).

2.1.3. Process Standardization

This module performs research over standards and models strongly related to main enterprise processes to consider future implementation. Finally, as exposed in the previous module, data and structures from the transactional system are correctly standardized.

2.1.4. Process Optimization

The process optimization module comprises processes by using different rigorous and non-rigorous method approaches for optimization. The process optimization phase aims to provide necessary data and information due to fundamental calculations based on engineering approaches to decide on specific objectives and goals within processes. Thus, as the first step, knowledge, data, and information on processes and systems are crucial to understanding the problem. The development of model design occurs by defining an objective or multi-objective function, a single or multiple purposes, and single or multiple scenarios as a convenience. Finally, WIMa's optimization solutions are enriched by semantics, mathematical, and process semantics model, allowing easier integration within the enterprise.

2.1.5. Process Automation

The process automation module comprises applications such as business process automation (BPA), digital automation (DA), and robotic process automation (RPA). First, BPA makes use of advanced technologies to reduce human intervention in processing tasks across the enterprise. Thus, BPA aims to enhance efficiency by automating (initialize, execute, and complete) the whole or some parts of a complicated process. Next, DA takes the BPA system, aiming to digitalize and improve processes automation, thus meeting the market dynamics customer environment. Finally, software agents carry out RPA, thus pointing to mimic human actions within digital systems to optimize business processes by using artificial intelligence agents.

2.1.6. Process Digitalization

The digitalization module creates a digital integration of all the systems found in business processes. Digitalization encompasses process simulation, industrial augmented reality, predictive systems, proactive systems, industrial internet of things, expert systems, and process virtual twins. Finally, WiiMa's solutions facilitate the digitalization technologies development due to the semantic structure that supports easy access to raw or structured data and processes' formal knowledge.

2.1.7. Process Intelligence

This module aims to understand human behavior principles by reasoning for developing programs for problem solutions by machines, using artificially intelligent tools, computational intelligence systems, and formal knowledge models. One of the first tasks is to manage structured knowledge, which can facilitate and empower the system's understanding. Furthermore, this module is directly affected by the transactional system's efficiency, which considers the collection, structuring, and data communication.

2.2. Data System Model

The transactional system architecture set up data management activity. We have described data management comprising five main activities: data system definition, data system improvement, data standardization, data integration and feeding, and data system dynamics, as shown in Figure 5.

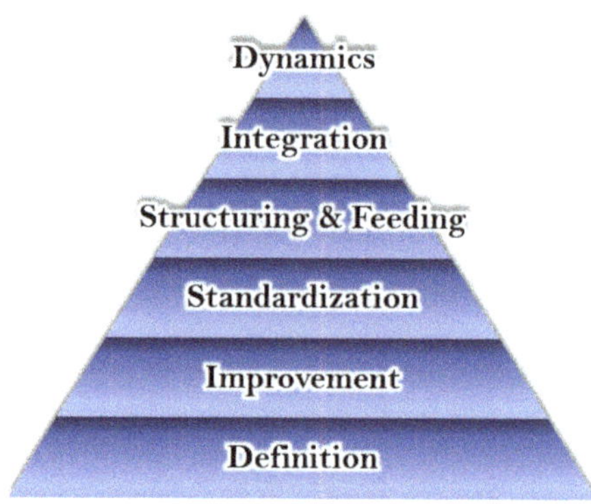

Figure 5. Maturity echelons of the data system model.

2.2.1. Definition

This activity takes into account the process definition (link) to create the data model. The data model establishes the relationship between the process model and the data generated by signals sources, such as process equipment, environment sensors, suppliers, or customers. Even more, transaction data protocols are defined, and the supported physical architecture must be capable of carrying those protocols. Finally, the data management plan is set, providing guidelines and procedures for enhancing security, compliance, quality, efficiency, and access.

2.2.2. Improvement

The data improvement activity refines the relationship between data signals and process models, and explains missing and necessary data. The data system requirements reside in process improvement activity. One can then calculate required data by making analytical computations, adding new technologies, or adding other sensors.

2.2.3. Standardization

The data standardization activity is the process of setting data systems into standard formats. It comprises the selection of common data language, structure, engineering metrics, and time/space Scales. Besides, data conciliation is a crucial task of data integration. Finally, the four language categories data structure comprises data definition language, data query language, data manipulation language, and transaction control language, in compliance with a processes database system.

2.2.4. Integration and Feeding

Data integration and feeding aim to link and integrate the transactional system and analytical systems or models. On the one hand, integration connects data among devices with analytical models, devices with devices, and analytical models with analytical models. On the other hand, data feeding is in charge to structure, send, and deliver data. Thus, integration and feeding tasks focus on support optimization based on decision-making and process action execution that maximize the business's benefit. Besides, material resource planning, distribution requirement planning, or enterprise resource planning systems, as part of the analytical procedure, are setting up if required. As a result, the definition of data-to-parameters and data-to-sets is established, required by transactional models. Finally, at some point, this task considers the feed and integration of virtual systems.

2.2.5. Dynamics

The data system model's last task replaces fixed data parameters and selected data sets by a mechanism for data collecting and structuring. Usually, these mechanisms refer to algorithms, which can become intelligent agents. Finally, the dynamics can enhance the database system, reaching intelligent databases. Intelligent databases are in constant change adapting to the dynamics of process systems and other business features.

2.3. *Knowledge System Model*

The knowledge system model organizes how the cognition process evolves, focusing on knowledge management. This cognition process integrates and adapts Bloom's taxonomy types of knowledge (Section 1.3.1) and ontologies, applied to computational systems. Thus, the resulting model comprises four main modules: conceptual, procedural, expertise, and metacognitive, as shown in Figure 6. Finally, the knowledge model system acts as an integrator between the process and data system models.

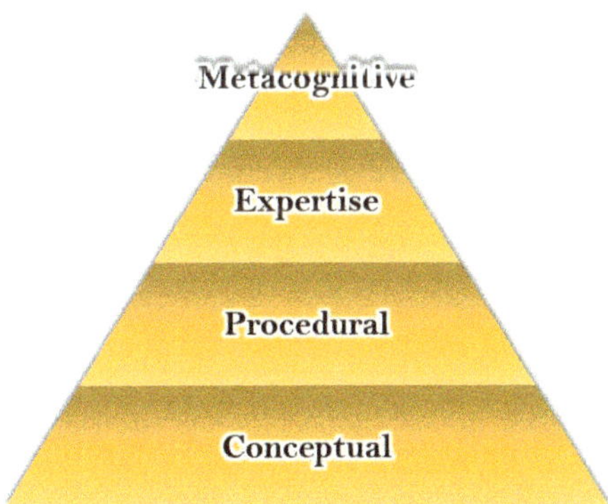

Figure 6. Maturity echelons of the knowledge system model.

2.3.1. Conceptual

The conceptual knowledge aims to develop a formal domain ontology. The ontology considers domain terminology, elements, categories, principles, and generalizations, which can consider chemical, physics, and mechanics phenomena. All those domain information is retrieved from standards, books, handbooks, etc., derived from Sections 2.1.1 and 2.2.1. Additionally, information gathering can consider good manufacturing practices and standard operating procedures derived from Section 2.1.2. Thus, the ontology models those elements in the form of classes (elements involved in the domain area), data-properties (data from specifications and sensors), classes-properties (the relation among elements), axioms (assumptions on behavior), and rules (restrictions on behavior).

2.3.2. Procedural

Procedural knowledge focuses on the interaction that processes and humans have through analytical methods and tools, usually from procedural models, to perform the processing activity. Thus, the primary sources of information are techniques, methods, theories, models, structures, and procedural recipes (general, site, master, and control recipes). Besides, previous knowledge is stored and well identified within the data system. Finally, procedures, analytical models, and analytical tools translate to computational algorithms for the reasoning task.

2.3.3. Expertise

Knowledge expertise aims to collect and manage the criteria that determine the best use of analytics or procedures. This criterion is based on learning on good habits and how to replicate them. Besides, it is built on learning over bad habits and how to avoid them. Finally, expertise knowledge considers creating multiple scenarios to account for the power of knowledge about relations and behavior in the domain based on the conceptual and procedural knowledge activities.

2.3.4. Metacognitive

Metacognitive knowledge aims to go beyond current understanding. The current knowledge system is the one in charge of this metacognitive process. The metacognitive's main comprises the characterization, classification, reasoning, and creation of knowledge to enhance and reach process intelligence.

Finally, Figure 7 represents the interaction among three system models. Thus, the process system model drives the diagonal's general maturing process, reaching from process definition to process intelligence. Next, the X-axis represents how the data system model supports the process and matures, starting from data definition to data dynamics. Last, the Y-axis shows the knowledge system model developing, which interacts with the other two system models and allows an intelligent technological implementation, reaching from conceptual knowledge to metacognitive knowledge.

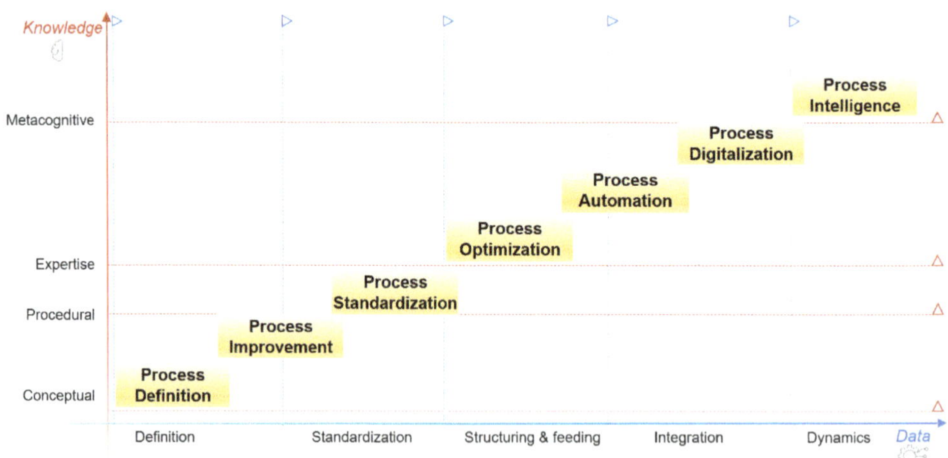

Figure 7. Maturity model system integration: Wide Intelligence Management Architecture's process, data, and knowledge.

3. Results

This section presents the application of the methodology previously described in a process manufacturing case study.

3.1. Case Study

This case study aims to illustrate the WIMa framework's application for the lifecycle assessment (LCA) of an acrylic fiber production plant. The objective is to demonstrate how the three models described in Section 2 work together to reach the desired technology application (in this case, LCA). Life cycle assessment requires a correctly defined process and consistent data to provide sensible results. Furthermore, this is a meaningful example to understand the framework at the process definition level (Section 2.1.1).

The acrylic fibers polymerization process considered in this work is presented initially in [21]. Acrylic fibers' production takes 14 stages in a batch production plant, represented in Figure 8, which involves different material and energetic resources. Two alternative production processes are assessed: acrylic fiber A uses acetone as a solvent in the polymerization, and acrylic fiber B uses benzene. This case study describes the process and data definition required to perform a life cycle assessment according to the WIMa procedure. Still, the actual evaluation is beyond the scope of this contribution. For further details regarding the life cycle assessment, please refer to the work in [21].

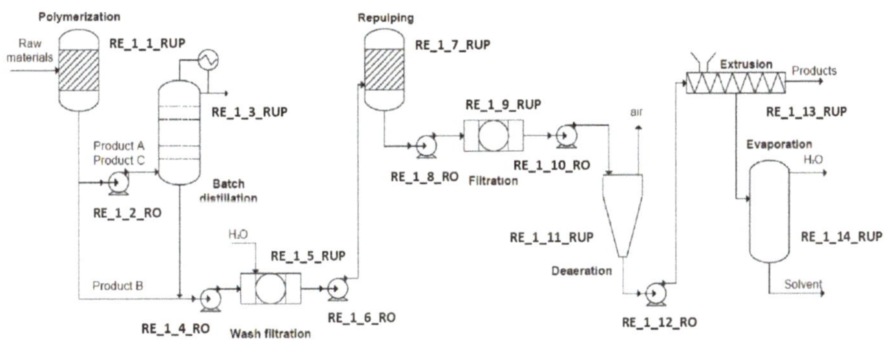

Figure 8. Flowsheet for acrylic fibers' production process (it contains 14 recipe elements, divided into eight recipe unit procedures and six recipe operations).

First of all, define the requirements and objective of the case study. Thus, suppose that "Polymer A.C." wants to develop a high-level decision model agent based on optimization approaches. Besides, they want to standardize their processes to maintain relations with their industrial partners. These requirements comprise the performance of certain levels presented in Table 1:

1. Level one (L1) regarding process definition, conceptual knowledge (process principles, process standards, and data standards), and data definition.
2. Level three (L3) regarding process standardization and data standardization.
3. Level four (L4) regarding process optimization, procedural knowledge, and integration and feeding data.
4. Level seven (L7) regarding process intelligence, metacognitive knowledge, and data dynamics.

Next, present every activity performance in detail structured in each three main models: process model, data model, and knowledge model.

3.1.1. Process Model of Polymer Plant

L1. Process definition: Process definition tackles the formalization of the polymerization process itself according to the technical requirements. On the one hand, the process consists of a complete polymerization plant that produces acrylic fibers using acetone as solvent. The existing documentation comprises the plant flowsheet (Figure 8) and the process recipe, which are the current formalization of the plant process activities. Finally, a characterization of the organization is performed, and the results, presented in Tables 2–4, summarize the features related to the general, tactic, and strategic levels of the organization in which the polymerization plant is installed. Overall, previous information provides clear boundaries of the process and includes all material and energy flows required to perform a life cycle assessment. Therefore, the process considered is according to the process definition model requirements.

Table 2. General features of the organization for system characterization.

General Feature	Value
Production capacity	Medium
Company size	Medium
Supply chain type	Good availability
Production type	Multi-stage
Market competition type	Low
Environmental regulations	Defined
Demand levels	High volume

Table 3. Tactic features of the organization for system characterization.

Tactic Features	Value
Transport type	Land
Supply chain objective	Economic
Production policies	Defined
Customer features	-
Suppliers features	-
Process flow type	Forward
Material storage type	Limited

Table 4. Strategic features of the organization for system characterization.

Strategic Features	Value
Production processing	Sequential
Technology	Multitask
Material storage	Limited
Material resource	Not perishable
Processing resources	Limited
Scheduling objective	Timing
Scheduling mode	On-line

L3. Process standardization: The standardization requires following the ANSI/ISA 88 standard. Thus, a semantic model is based on ANSI/ISA 88 standard, the so-called Batch Process Ontology (BaPrOn). The use is related to the instantiation task, which allows a faster and accurate manner to standardize the process. As a result, formulas of the master process recipes were extracted, as shown in Tables 5 and 6. The production plant considers four stages in batch production mode, eight recipe unit procedures, and six recipe operations, as well as twenty-seven different resources (considering material and energy flows). The master recipe's instantiation results in a set of recipe unit procedures and recipe operations, along with their formula and input, output, and other process parameters. Thus, the environmental performance metrics parameters appear included. Overall, the instantiation results in 934 instances concerning 295 classes, 257 object properties, and 33 data properties. The description logic expressivity of the ontology is $SHIN^{(D)}$, where S refers to attributive language with complement of any concept allowed, not just atomic concepts (ALC); H refers to role hierarchy (subproperties); I refers to inverse properties; N refers to cardinality restrictions a special case of counting quantification; and $^{(D)}$ refers to the use of data-type properties, data values or data types. As an example of class instantiation, the RawMaterial class has Input1_1 (Acrylonitrile), Input1_2 (MethylMethacrilate), Input1_3 (VinilChloride), Input1_4 (Solvent-Acetone) as instances.

Table 5. The formula for the master recipe of acrylic fiber A production process 1/2.

Recipe ID	Recipe Type	Element ID	Procedural Element Type	Parameter ID	Parameter Name
MR-01	Master	RE-3	Unit procedure	I15	Output3_2
MR-01	Master	RE-14	Unit procedure	I56	Output14_1
MR-01	Master	RE-14	Unit procedure	I57	Output14_2
MR-01	Master	RE-2	Operation	I61	CleaningWater_total
MR-01	Master	RE-1	Unit procedure	I62	CoolingWater_total
MR-01	Master	RE-1	Unit procedure	I63	Electricity_total
MR-01	Master	RE-13	Unit procedure	I50	Output13_1
MR-01	Master	RE-1	Unit procedure	I1	Input1_1
MR-01	Master	RE-1	Unit procedure	I2	Input1_2
MR-01	Master	RE-1	Unit procedure	I3	Input1_3
MR-01	Master	RE-1	Unit procedure	I4	Input1_4
MR-01	Master	RE-1	Unit procedure	I5	Input1_5
MR-01	Master	RE-7	Unit procedure	I26	Output7_2
MR-01	Master	RE-3	Unit procedure	I65	Steam_total

Table 6. The formula for the master recipe of acrylic fiber A production process 2/2.

Resource Type	Subtype	Resource Name	Procedural Information	Value	Unit
Material	By-product	ByProduct1	Output Parameter	1750	kg
Material	By-product	ByProduct3	Output Parameter	1217	kg
Material	By-product	ByProduct4	Output Parameter	1734	kg
Material		CleaningWaterT1	Process Parameter	240,000	kg
Energetic	Cooling water	CoolingWaterT1	Process Parameter	8,613,983	kg
Energetic	Electricity	ElectricityT1	Process Parameter	458,979	kWh
Material	Final product	FinalProduct1	Output Parameter	1000	kg
Material	Raw material	RawMaterial1	Input Parameter	100	kg
Material	Raw material	RawMaterial2	Input Parameter	50	kg
Material	Raw material	RawMaterial3	Input Parameter	25	kg
Material	Raw material	RawMaterial4	Input Parameter	0	kg
Material	Raw material	RawMaterial5	Input Parameter	0	kg
Material	Residue	Residue1	Output Parameter	1974	kg
Energetic	Steam	SteamP1	Process Parameter	441,323	kg

L4. Process optimization: This case study tackles the optimization of multistage batch plants' scheduling problem under sequence-dependent changeovers presented by Capon et al. (2011, 2012). The problem can be defined as follows: given a set of process operations planning data, including (i) time horizon, (ii) set of product recipes, (iii) equipment technologies for processing stages, (iv) product demands, (v) changeover methods, (vi) economic data related to costs and prices, and (vii) environmental data related to raw material, equipment, and product manufacturing environmental interventions; all of them provided by the data model. Four objective functions are relevant for decision-making: productivity (P), total environmental impact (TEI), makespan (M), and total profit (TP). The problem's modeling uses an immediate precedence mathematical formulation, managing the possible use of different product changeover cleaning methods; multiple alternative pieces of equipment at each stage; limited storage policies; and product batching, allocation, and timing constraints. Such problem representation is suitable for applying any of the three different optimization strategies considered in this case study. Specifically, we will solve the multi-objective problem using mathematical programming with a normalized constraint method (MP), a genetic algorithm (GA), and a hybrid optimization approach (HA). Each solution method's suitability depends on the combination of problem features and objective function, and will be further explored in level 7 related to process intelligence. At this level, the solution techniques are considered independently, and the knowledge model supports the implementation of the optimization providing adequate data from the data model.

L7. Process intelligence: This level comprises the development of intelligent agents for the scheduling problem. Indeed, one can use different problem representations for optimization purposes, and which is the most suitable depends on the issue features. That is precisely the function provided by the process intelligence framework using agents. The agents perform different functions and are integrated, allowing communication among them. The agent's functions include communication, search, classification, and solution. A common vocabulary is necessary to achieve communication, and all the agents rely on the ontologies described in the knowledge model.

The solution mechanism consists of the following steps: (i) problem definition, (ii) modeling process for reaching a problem model, (iii) model analysis, (iv) model solutions, and (v) problem implementation. Based on the answers in (iv), we can make inferences and reach decisions about the problem (v). The assessment of decisions goodness feeds back to the intelligent system to enable learning.

Next, the communication agent prepares the classification agent for analyzing the problem using a knowledge-driven classification procedure. Next, a solution strategy is proposed based on a similitude measure resulting from the problem instance compared with (i) existing problems tackled in the past and stored in the database and (ii) existing problem approaches from the state-of-the-art. The problem instances solved in the original papers are the basis for the database of this problem. A total of 415 problem solutions are included with different problem descriptions and objective function values. As a result of the similitude measure, a set of ranked solution approaches is proposed to the decision-maker. Finally, a solution agent uses the solution algorithms to reach the optimal solution for the problem instance. The solution agent also sends problem solutions to the decision-maker and stores them in the future reasoning database.

The framework testing with new problem instances and results are shown in Table 7. The first column describes the problem size (number of batches for each problem). The second column specifies the objective function. The third column presents the solution implementation method's selection, while the fourth column includes the objective function's Value. Finally, the fifth column stands for the distance to the best optimal solution found. For small problem instances, the rigorous mathematical programming approach has been selected, whereas problem instances with many variables are solved using a genetic algorithm. Thus, the objective function also has an essential role in the selection of the solution strategy. Indeed, for productivity maximization, the hybrid approach is selected. In most cases, the solution proposed by the agent-based system is close to 5% to the optimal solution. Overall, this framework stands for a systematic approach to scheduling model selection and solution implementation, thus supporting the engineers' high-level decision, who do not need to have a thorough understanding of advanced optimization techniques.

Programming the different agents uses Jython because it combines Python as a programming language and uses Java APIs for communicating with the ontological models.

Table 7. Results for different problem instances from the agent-based framework at Polymer plant.

Number Batches (A/B/C)	Objective Function	Solution Approach	Solution Value	Optimal Solution Value
4/4/4	Productivity	Mathematical programming	2170	2174
4/4/4	Environmental impact	Mathematical programming	48,200	48,200
4/4/4	Makespan	Mathematical programming	48,000	48,000
17/10/13	Productivity	Hybrid approach	1301	1302
17/10/13	Environmental impact	Mathematical programmin	218,814	217,236
17/10/13	Makespan	Genetic algorithm	199,507	197,686
20/18/15	Productivity	Hybrid approach	1354	1356

3.1.2. A Knowledge Model of the Polymer Plant

L1. Conceptual knowledge: A knowledge system harmonizes and manages sets of valuable information, making them accessible for their use considering specific purposes. In this case study, knowledge conceptualization uses the Enterprise Ontology Project (EOP) [22]. EOP is an ontology containing three active ontologies: batch process ontology, environmental ontology, and enterprise ontology.

First, batch process ontology (BaPrOn) tackles features such as physical, procedural, recipe, and process models based on the ANSI/ISA 88 standard. It focuses on the production operation management of batch processes. Next, environmental ontology (EVO) considers life cycle assessment and environmental impact categories features, which allows the trace and calculation of environmental impact produced by product or processes activities. Finally, the enterprise ontology project (EOP) is based on the ANSI/ISA 95 standard. It considers the integration of enterprise activities, such as quality, maintenance, and inventory management. Additionally, EOP also considers financial features to tackle supply chain management activities.

Besides, the EOP model takes into account and models the following key knowledge:

- Production system characterization: comprising physical, procedural, and recipe (site and general) models.
- Products contemplated: according to the processing order activity in the industry recipes defining the production requirements and production path for the products in the physical, process and recipe (master and control) models.
- Resource availability and plant status: provided by the process management and production information management activities.

Finally, Figure 9 shows the first classes found in the taxonomy of the enterprise ontology project.

L4. Procedural knowledge: Mathematical programming has been choose as strategy for making optimization knowledge explicit and available. Thus, this case study makes use of the mathematical modeling ontology (MMO) [23,24] and the operation research ontology (ORO) [25].

On the one hand, MMO aims to represent knowledge of mathematical domain based on mathematical structures comprising elements, terms, and operations. Thus, the mathematical term is the atomic part of a mathematical expression. Mathematical elements and expressions are related through mathematical operations, which can be logic or algebraic types. An element or expression can define specific conceptual meanings, such as processing time, the opening Value of the valve, the effort calculation equation, etc. In the same manner, an element or expression has a behavior that is related to variables, constants values, etc. Finally, MMO allows the definition of object-oriented mathematical modeling relating mathematical elements and expression with concepts from other semantic representations. In this case study, MMO is integrated with EOP. That allows linking mathematical models and equations with instances of the acrylic fiber process. Figure 10 shows the first classes found in the taxonomy of the mathematical modeling ontology.

On the other hand, ORO aims to capture the knowledge of operation research area that is a branch of mathematics. This ontology structures mathematical expressions fed by MMO in the form of mathematical programming. That allows a formal study and solution of complex problems for decision-making activity. As a result, an enriched semantic structure is obtaining. It considers the main parts of mathematical programming, such as the objective function in the form of an equation, a set of constraints in the form of mathematical equations. In the same manner, logic and algebraic operations are supported by MMO. Figure 11 shows the first classes found in the taxonomy of the operation research ontology.

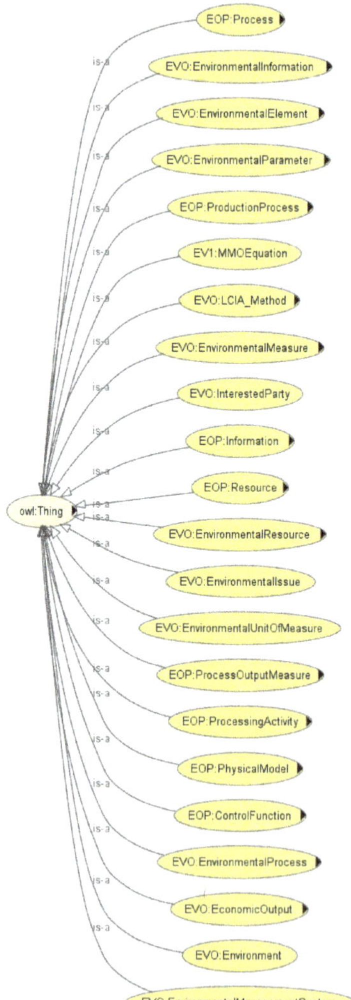

Figure 9. First taxonomical representation of the Enterprise Ontology Project classes.

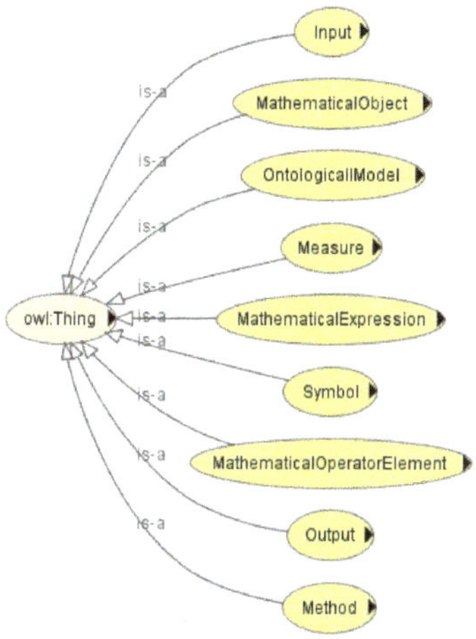

Figure 10. First taxonomical layer representation of the Mathematical Modeling Ontology classes.

L7. Metacognitive knowledge: This level aims to create an autonomous problem definition agent to construct a semantically enriched problem statement [26]. The agent works in a semantic environment where machines can access explicit knowledge codified in Python and Jython. The strategy comprises the following: (1) Semantic definition of the system. (2) Recognition of current situation. (3) The setting of key process features and variables. (4) The setting of confidential intervals for monitoring task. (5) Searching for relation to key features. (6) Definition problem statement.

First, the system's semantic definition refers to Tables 2–4 presented in Section 3.1.1. Based on the system instantiation, the current process is introduced semantically, indicators, related key features, and engineering metrics are set, such as resource availability, energy consumption, demand un-accomplishment, and cleaning overtimes desired. Table 8 shows a brief example of the resulting process of system setup for monitoring. This table performs a SWOT analysis defining strengths (S), weaknesses (W), opportunities (O), or threats (T). The following two rows show indicators and related features coming from classes representing the process domain concepts. The next row shows the engineering metrics associated with indicators. Finally, the last two rows refer to the upper and lower bounds values defined for evaluating the current performance.

Table 8. The semantic search of key features and confidential intervals, where SWOT refers to a strength (S), weakness (W), opportunity (O), or threat (T) and EOP refers to Enterprise Ontology Project.

SWOT	Indicator	Related Feature	Metric	UBV	LBV
DataString	*EOP_Class*	*EOP_Class*	*EOP_DataProperty*	*EOP_DataProperty*	*EOP_DataProperty*
W	Time delivered hauler	Tardiness finish orders	u/month	15	45
W	Demand unacomplishment	Unfinished orders	u/month	5	15

Next, the intelligent agent defines optimization goal statements (maximization or minimization). Then, using previous decision variables definitions, the system is ready for construct or semantic problem statement definition. Finally, the agent fills the problem statement template automatically to present it in the form of natural language, as follows:

Empty template. "Taking into account -*EOP classes found as key variables*- variables, and -*EOP classes found as key parameters*- parameters; -*Goal statement*- - *EOP class defined as decision variable*- related to -*EOP class defined as an indicator*- indicator."

Filled template: "Taking into account *Processing start time, Storage level* variables, and *Maximum storage capacity, Batch processing time, Batch due date* parameters; *Minimize Makespan* related to *Number of late jobs* indicator."

At this level, intelligent agents provide additional capabilities for reasoning using semantic technologies. The main task focuses on decision-support for industry 4.0 microenvironments.

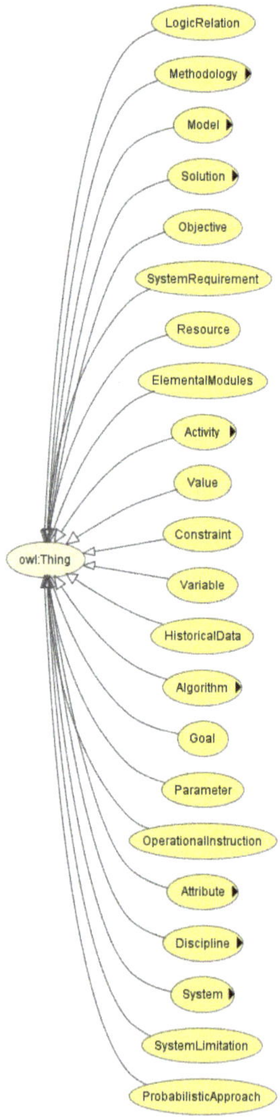

Figure 11. First taxonomical layer representation of the Operations Research Ontology classes.

3.1.3. The Data Model of the Polymer Plant

L1. Data definition: The data definition is concerned with the data collection and data system architecture. Accordingly, the transactional system needs to be verified. In this case study, the information related to the recipe, namely, the energy and material flows, is stored in aStructured Query Language (SQL) database. All flows are listed, identified, and quantified in the database, and their sign (positive or negative) indicates whether they enter or leave the process boundaries. The data relating to the material and energy flows stems from the factory floor, and only process engineers have access and permission to modify the data.

L3. Data standardization: This level aims to standardize data properties. Data properties comprise data metrics, data language, and data structure. In general, properties are related to specifications within the processing system. Many of the systems are ruled by the technology implemented. In contrast, this methodology pursues defined, choose and consensus data properties desired for the processing system. In this particular case study, ANSI/ISA standards define those properties, standardizing the data through the instantiation process. The data model provides specifications for developed interfaces, software requirements. Table 9 presents properties, detail of Boolean, and Direction Type data from EOP (based on ANSI/ISA standards).

Table 9. ANSI/ISA standard properties for enumeration members.

Enumeration Set	Enumeration Value	Enumeration String	Description
Boolean	0	FALSE	Definition of a Boolean value.
	1	TRUE	
Direction Type	0	Invalid	Entry not valid
	1	Internal	Identifies how a parameter is handled. Internal = only available within the Recipe Element. Defined at creation or created as an intermediate value.
	2	Input	The Recipe Element receives the Value from an external source.
	3	Output	The Recipe Element creates the Value and makes it available for external use.
	4	Input/Output	The Recipe Element and external element exchange the Value, and may change its Value.
	5–99		Reserved
	100+		User defined

Next, Table 10 presents data details from the polymer process.

Table 10. Data properties from EOP within the Polymer process system.

Object/Data Property	Range
hasParameterSource	Resource
hasID_ParameterID	ParameterID
parameter_type	constant; variable
hasEquationAsReferenceValue	MathematicalElement
value	float
engineering_units	string
description	string
scaled	float

L4. Data integration and feeding: This level performs the definition of data and data sets required by the optimization software or other software. We consider software specialized in mathematical programing and solving by optimization software, which contains strict and non-strict approaches. Thus, Tables 11–13 show some structure data or data sets required by the polymer process plant's optimization activity. Besides, optimization software can call for single data at any time.

Table 11. Capacity data set of the Polymer plant, structuring two columns: Unit_ID and Data value.

ID_R1	4000
ID_P1	13,000
ID_C1	4000
ID_P2	5000
IN001B	5000
IN002A	5000
IN002B	11,000
ID-FP00A	36,000
ID_FP00B	36,000
ID_R001	36,000
ID_R002	36,000

Table 12. Subtasks time of product A of the Polymer plant, structuring seven columns: Task number, Unit_ID, Preparation time, Load time, Operation time, Unload time, and Cleaning time.

1	ID_R1	0.20	0.00	2.00	0.30	0.25
2	ID_P1	0.20	0.00	0.30	0.75	0.25
3	ID_C1	0.50	0.30	2.50	0.00	0.75
4	ID_P2	0.20	0.00	0.75	0.00	0.25
5	IN001B	0.50	0.00	0.75	0.00	0.50
6	IN002A	0.20	0.00	0.75	0.75	0.25
7	IN002B	0.30	0.75	1.00	0.00	0.25
8	ID-FP00A	0.20	0.00	0.75	0.00	0.25
9	ID_FP00B	0.30	0.00	0.75	0.00	0.25
10	ID_R001	0.20	0.00	0.75	0.00	0.50
11	ID_R002	0.20	0.00	0.75	0.00	0.25

Table 13. Subtasks time of product B of the Polymer plant, structuring seven columns: Task number, Unit_ID, Preparation time, Load time, Operation time, Unload time, and Cleaning time.

1	ID_R1	0.20	0.00	3.00	0.75	0.25
2	ID_P1	0.00	0.00	0.00	0.00	0.00
3	ID_C1	0.00	0.00	0.00	0.00	0.00
4	ID_P2	0.50	0.00	0.75	0.00	0.25
5	IN001B	0.20	0.00	0.75	0.00	0.50
6	IN002A	0.30	0.00	0.75	0.00	0.25
7	IN002B	0.20	0.75	0.74	0.74	0.25
8	ID-FP00A	0.20	0.00	0.74	0.00	0.25
9	ID_FP00B	0.50	0.00	0.74	0.00	0.25
10	ID_R001	0.20	0.00	0.74	0.00	0.50
11	ID_R002	0.20	0.00	0.74	0.00	0.25

L7. Data dynamics: This level aims to develop an algorithm based on Jython being capable of structuring data. For this specific case study, the algorithm was under construction. The strategy focuses on queries and the structure of triples from the semantic models, which can dynamically define data sets as shown in L4. Data structuring and feeding. Finally, we want to point out that ontologies have a database structure but are semantically enriched and supported by knowledge.

4. Discussion

This work introduces the wide intelligent management architecture and the application to acrylic fiber production as a case study. As a result, the production process first creates a process definition (system characterization and flowsheet of the plant), data definition (database based on the semantic model), and knowledge conceptualization (a semantic model for representing concepts and data of the process and system). The standardization of data and concepts has been done using the semantic model to represent the process (ANSI/ISA standards). Thus, using the architecture for data structuring and feeding facilitates the integration of the Life Cycle Assessment approach improvement. From this point, the acrylic fiber production plant has the basis for developing process automation, developing process digitalization, or developing intelligent agents for decision-making. Figure 12 shows the process activities performed in the case study regarding process definition, standardization, optimization, and intelligence (yellow boxes and blue arrows path). Moreover, the acrylic fiber company can perform process improvement, automation, or digitalization based on the current plant status (green arrows pointing gray boxes). Finally, using a comprehensive intelligent management architecture model can be adapted to any company necessity by choosing how to evolve their processes.

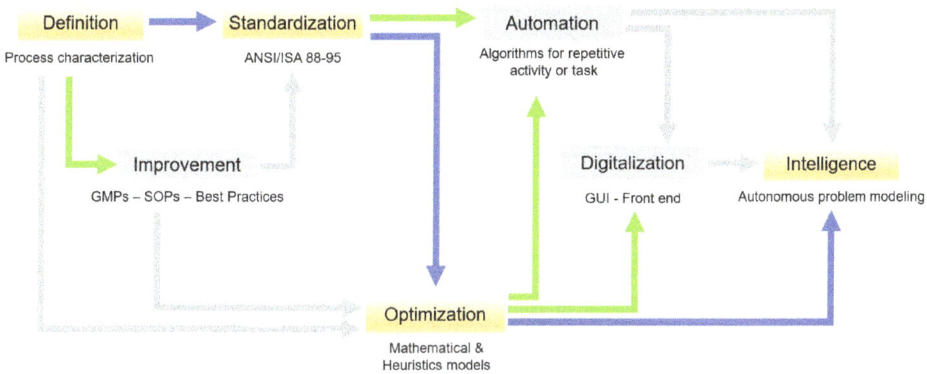

Figure 12. Potential activities derived from the case study, where GMPs refers to Good Manufacturing Practices, SOPs refers to Standard Operating Procedures, and GUI refers to Graphical User Interface.

5. Conclusions

This work presents a novel architecture for technology integration on process activities and systematically manages processes and their related data using formal knowledge. Knowledge models provide additional reasoning capabilities to support decision support systems for the wide optimization and industry 4.0 approaches. The architecture comprises and structures three critical systems: process system, knowledge system, and transactional system. As a result, analytical tools belonging to process activities and transactional data systems can be guided by a systematic development framework consolidated with formal knowledge models. Thus, the model improves the interaction among process life cycles, analytical models, transactional systems, and knowledge. A comprehensive intelligent management architecture model can be seen as an ordered, adaptable, and configurable tool for integrating technologies and processes maturity. A critical aspect of this method regards formal knowledge models that become usable and reusable in technology integration and maturity. Finally, this method is a new alternative supporting companies when they need to decide "what to do" by explaining to them "why to do" and "how to do it", taking as starting point their processes and characteristics.

Author Contributions: Conceptualization, E.M. and E.C.-G.; methodology, E.M. and E.C.-G.; software, E.C.-G. and E.M.; validation, E.M., L.P. and E.C.-G.; formal analysis, E.M., E.C.-G. and E.M.M.; investigation, E.M., E.C.-G. and L.P.; data curation, E.C.-G.; writing-original draft preparation, E.M., E.C.-G. and L.P.; writing-review and editing, E.M. and L.P; visualization, E.C.-G. and E.M.M.; supervision, L.P.; project administration, E.M.; funding acquisition, L.P. All authors have read and agreed to the published version of the manuscript.

Funding: This research was funded by Ministerio de Economia, Industria y Competitividad and the European Regional Development Fund, both with grant numbers DPI2017-87435-R and PCIN-2015-001/ELAC2014/ESE0034.

Institutional Review Board Statement: Not applicable.

Informed Consent Statement: Not applicable.

Data Availability Statement: The data presented in this study are available on request from funding institution (see Funding).

Acknowledgments: The technical support in this work from Enterprise WISDOM Company is fully acknowledge.

Conflicts of Interest: The authors declare no conflict of interest.

References

1. Venkatasubramanian, V.; Zhao, C.; Joglekar, G.; Jain, A.; Hailemariam, L.; Suresh, P.; Akkisetty, P.; Morris, K.; Reklaitis, G. Ontological informatics infrastructure for pharmaceutical product development and manufacturing. *Comput. Chem. Eng.* **2006**, *30*, 1482–1496. [CrossRef]
2. Simon, F.; Murray, T. Decision support systems. *Commun. ACM* **2007**, *50*, 39–40.
3. Shim, J.P.; Warkentin, M.; Courtney, J.F.; Power, D.J.; Sharda, R.; Carlsson, C. Past, present, and future of decision support technology. *Decis. Support Syst.* **2002**, *33*, 111–126. [CrossRef]
4. Shobrys, D.E.; White, D.C. Planning, scheduling and control systems: Why cannot they work together. *Comput. Chem. Eng.* **2002**, *26*, 149–160. [CrossRef]
5. Park, J.; Du, J.; Harjunkoski, I.; Baldea, M. Integration of Scheduling and Control Using Internal Coupling Models. *Comput. Chem. Eng.* **2014**, *33*, 529–534.
6. CEN Standards. European Committee for Standardization (CEN). Available online: https://www.cen.eu/about/Pages/default.aspx (accessed on 15 March 2021).
7. Integrated Definition Methods (IDEF) Standards. Available online: https://www.idef.com/ (accessed on 15 March 2021).
8. International Electrotechnical Commission (IEC) Standards. Available online: https://www.iec.ch/homepage (accessed on 15 March 2021).
9. International Organization for Standardization (ISO). Available online: https://www.iso.org/standards.html (accessed on 15 March 2021).
10. Manufacturing Execution Systems Association (MESA). Available online: http://www.mesa.org/en/modelstrategicinitiatives/MSI.asp (accessed on 15 March 2021).
11. Machinery Information Management Open Systems Alliance MIMOSA. Available online: https://www.mimosa.org/mimosa-osa-eai/ (accessed on 15 March 2021).
12. Object Management Group (OMG). Available online: https://www.omg.org/about/index.htm (accessed on 15 March 2021).
13. Process Industry Practices (PIP). Available online: https://www.pip.org/ (accessed on 15 March 2021).
14. International Society for Measurement and Control. ISA-88/95 technical report: Using ISA-88 and ISA-95 together. In *ISA The Instrumentation, Systems, and Automation Society 2007*; Technical Report; ISA, Durham, NC, USA, 2017.
15. Williams, T.J. A Reference Model for Computer Integrated Manufacturing from the Viewpoint of Industrial Automation. *IFAC Proc. Vol.* **1990**, *23*, 281–291. [CrossRef]
16. Apostolou, D.; Mentzas, G.; Abecker, A. Ontology-enabled knowledge management at multiple organizational levels. In Proceedings of the 2008 IEEE International Engineering Management Conference, Estoril, Portugal, 28–30 June 2008; pp. 1–6.
17. Sowa, J.F. *Conceptual Structures: Information Processing in Mind and Machine*; Addison-Wesley: Reading, MA, USA, 1984.
18. Gruber, T.R. A translation approach to portable ontology specifications. *Knowl. Acquis.* **1993**, *5*, 199–220. [CrossRef]
19. Fensel, D. A Silver Bullet for Knowledge Management and Electronic Commerce. In *Ontologies*; Springer: Berlin/Heidelberg, Germany, 2004.
20. Anderson, L.W.; David, R.K.; Benjamin, S.B. *A Taxonomy for Learning, Teaching, and Assessing: A Revision of Bloom's Taxonomy of Educational Objectives*; Lorin, W., Anderson, D.K., Eds.; Longman: New York, NY, USA, 2001.
21. Munoz, E.; Capon-Garcia, E.; Puigjaner, L. Supervised Life-Cycle Assessment Using Automated Process Inventory Based on Process Recipes. *ACS Sustain. Chem. Eng.* **2018**, *6*, 11246–11254. [CrossRef]
22. Munoz, E.; Capon-Garcia, E.; Espuna, A.; Puigjaner, L. Ontological framework for enterprise-wide integrated decision-making at operational level. *Comput. Chem. Eng.* **2012**, *42*, 217–234. [CrossRef]
23. Munoz, E.; Capon-Garcia, E.; Lainez, J.; Espuna, A.; Puigjaner, L. Integration of enterprise levels based on an ontological framework. *Chem. Eng. Res. Des.* **2012**, *91*, 1542–1556. [CrossRef]
24. Munoz, E.; Capon-Garcia, E. Intelligent Mathematical Modelling Agent for Supporting Decision-Making at Industry 4.0. In *Trends and Applications in Software Engineering. Advances in Intelligent Systems and Computing*; Mejia, J., Munoz, M., Rocha, A., Pena, A., Perez-Cisneros, M., Eds.; Springer: Cham, Switzerland, 2019; Volume 865.
25. Munoz, E.; Capon-Garcia, E.; Lainez-Aguirre, J.M.; Espuna, A.; Puigjaner, L. Operations Research Ontology for the Integration of Analytic Methods and Transactional Data. In *Trends and Applications in Software Engineering. Advances in Intelligent Systems and Computing*; Mejia, J., Munoz, M., Rocha, A., Calvo-Manzano, J., Eds.; Springer: Cham, Switzerland, 2018; Volume 45.

26. Munoz, E.; Capon-Garcia, E.; Puigjaner, L. Advanced Model Design Based on Intelligent System Characterization And Problem Definition. In *Computer-Aided Chemical Engineering*; Anton, A.K., Edwin, Z., Richard, L., Leyla, Ö., Eds.; Elsevier: Amsterdam, The Netherlands, 2019; Volume 46, pp. 1045–1050.

processes

MDPI

Article

MINLP Model for Operational Optimization of LNG Terminals

Zhencheng Ye, Xiaoyan Mo and Liang Zhao *

Key Laboratory of Smart Manufacturing in Energy Chemical Process, Ministry of Education, East China University of Science and Technology, Shanghai 200237, China; yzc@ecust.edu.cn (Z.Y.); Y30180850@mail.ecust.edu.cn (X.M.)
* Correspondence: lzhao@ecust.edu.cn; Tel.: +86-21-6425-2755

Abstract: Liquefied natural gas (LNG) is a clear and promising fossil fuel which emits less greenhouse gas (GHG) and has almost no environmentally damaging sulfur dioxide compared with other fossil fuels. An LNG import terminal is a facility that regasifies LNG into natural gas, which is supplied to industrial and residential users. Modeling and optimization of the LNG terminals may reduce energy consumption and GHG emission. A mixed-integer nonlinear programming model of the LNG terminal is developed to minimize the energy consumption, where the numbers of boil-off gas (BOG) compressors and low-pressure (LP) pumps are considered as integer variables. A case study from an actual LNG terminal is carried out to verify the practicality of the proposed method. Results show that the proposed approach can decrease the operating energy consumption from 9.15% to 26.1% for different seasons.

Keywords: LNG terminal; operational optimization; BOG compressor; MINLP

Citation: Ye, Z.; Mo, X.; Zhao, L. MINLP Model for Operational Optimization of LNG Terminals. *Processes* **2021**, *9*, 599. https://doi.org/10.3390/pr9040599

Academic Editor: Luis Puigjaner

Received: 2 March 2021
Accepted: 24 March 2021
Published: 30 March 2021

1. Introduction

In recent years, environmental protection and the reduction of carbon dioxide emissions have become a hot spot worldwide [1,2]. Compared with other fossil fuels, natural gas (NG) is considered a sustainable and potential source of energy in the future [3–5]. Considering that the volume of liquefied natural gas (LNG) is 600 times smaller than the gaseous state of NG [6,7], LNG is considered as an economic transportation approach when the gas transportation pipeline is longer than 1500 km [8–10].

The traditional LNG supply chain includes NG liquefaction plants, ship transportation, and LNG import terminals [11,12]. Natural gas is first exploited and purified in liquefied facilities and then cooled to −162 °C for transportation [13]. Then, LNG is transported to the demand region by LNG carriers. Once the LNG ship arrived at the terminals, the LNG is unloaded and kept in cryogenic storage tanks. LNG is regasified through evaporation, and NG is provided to different users [14,15].

In the whole supply chain, the LNG terminal is an important part, which connects LNG resources and end users. It is responsible for receiving LNG from vessels, storing LNG in insulated tanks, vaporizing the liquid, and then delivering NG into the gas pipeline network [16]. The storage capacity of LNG is primarily affected by seasonal variations of requirements and the unloading cycles. LNG terminals are the regasification-to-end-user section of the supply chain, and they can be operated for the whole year. LNG can be transported further from the terminals to customers by the pipe network or by LNG trucks.

The cryogenic operations in an LNG import terminal consume considerable power for driving devices, such as compressors and pumps [13,17]. Energy consumption in LNG import terminals can be reduced in two ways. The first one refers to the LNG cold energy recovery. In the past decades, the recovery of cold energy from the regasification process has become a research hotspot. Around 830 kJ of cold energy is generally stored in per kilogram LNG [18]. Thus, the larger the system, the more cold energy is wasted [19]. Researches introduced different LNG cold energy utilization systems and discussed other potential directions beyond electric power generation [11,20,21]

The second way refers to the modeling and optimization of the boil-off gas (BOG) handling process. Due to the low bubble point of LNG, the BOG always arises at terminals and can cause damages [22]. Specifically, the heat will leak to LNG through the tank and the shell of the circling pipeline. Thus, the timely removal of the BOG is important to ensure the safe operation of the storage tank under the absolute pressure. An excessive amount of the BOG in a tank can result in safety issues, whereas a scant amount of the BOG causes an unnecessary waste of energy [23]. Accordingly, these two issues are important to address in the design and optimization of an LNG terminal.

BOG compressors are used to remove extra gas and ensure the safety of tanks. They have intensive and high-energy properties. Thus, they are the first target for energy saving. The minimization of the total compression energy is the general objective function of the LNG terminals, although many mathematical models of the compressors have been developed and applied in the simulation and optimization of LNG terminals [24–26]. Terminals normally used several multi-stage compressors in parallel to keep the BOG flow rate in a specific range. Several investigators have studied BOG compressor systems. Shin et al. proposed a mixed-integer linear programming (MILP) model for optimizing the BOG compressors [27]. A simplified tank model was then proposed to predict the pressure when failure occurred [28]. To improve the accuracy of the model, they lately used the rigorous model developed by Aspen Dynamic simulation [29].

Some researchers focused on the issues of multi-stage compression, multi-stage condensation, and cooling before or after a compressor in an LNG terminal. For example, Rao et al. used the Nonlinear Optimization by Mesh Adaptive Direct Search (NOMAD) algorithm to prove that the two-stage recondensation is superior to other structures [30]. Tak et al. investigated the influences of multi-stage compression on single-mixed refrigerant processes [31]. Yuan et al. analyzed the parameters in four types of BOG recondensation systems. They compared the power consumptions between the integrated and the non-integrated systems considering the conditions of different BOG components [18].

Various researches recover the LNG cold energy for utilization [11,12,19–21]. Many studies investigate the design optimization of BOG handling process to improve the energy efficiency while ensuring the system safety [32–35]. Studies on BOG compressor systems have also been done [24–29]. However, there is only a little focused on the recirculation operations. Park et al. determined the optimal recirculation flow rate to reduce operating costs in LNG terminal [15]. Wu et al. built a dynamic simulation model to optimize the recirculation and branch flow rate [34]. However, there is no literature that considers the scheduling optimization of LP pumps related to the send-out and recirculation flow rate, to the best of our knowledge. Additionally, a mixed-integer nonlinear programming model was first employed to solve the scheduling optimization problem of an LNG terminal. For estimating the generation rate of BOG, a nominal boil-off ratio of 0.05%-1% for the LNG tank capacity per day is used [34,36]. Besides, an empirical equation corrected by the data from the LNG storage tank manufacturers is proposed [28]. In this work, the HYSYS dynamic model of the industrial LNG terminal was developed to generate the data of BOG generation, and the regression model was obtained by the data. Therefore, the model is more suitable for LNG terminal optimization than the methods in the literature.

In this work, a typical LNG terminal was studied, which consists of tanks, pumps, recondensers, compressors, and vaporizers. The contributions of this work are given as follows.

- An MINLP model is developed for the operational optimization of the LNG terminal.
- A regression model of BOG generation is proposed considering both model accuracy and computational complexity.
- An industrial case study in an actual LNG terminal is employed to indicate the effectiveness of the proposed method.

2. Problem Statement

The schematic of an actual LNG terminal, which is composed of various devices, such as pumps, tanks, a recondenser, and vaporizers, is illustrated in Figure 1. As shown in Figure 1, the BOG produced in the LNG storage tanks is compressed into the recondenser with compressors, and the LNG is pumped into the recondenser by in-tank LNG pumps. When the BOG is completely condensed by the subcooled LNG in the recondenser, the BOG and the subcooled LNG are mixed into one stream. Then, the HP LNG pumps send the stream into an open rack vaporizer (ORV) or submerged vaporizer (SCV), which converts LNG to NG for commercial and household users. In some cases, the NG demands are low, and thus LNG cannot recondense all the BOG. Consequently, the HP compressors are employed to send the BOG to the NG pipes. This BOG handling process is simple, but the operating energy consumption is higher than the recondensation way [32].

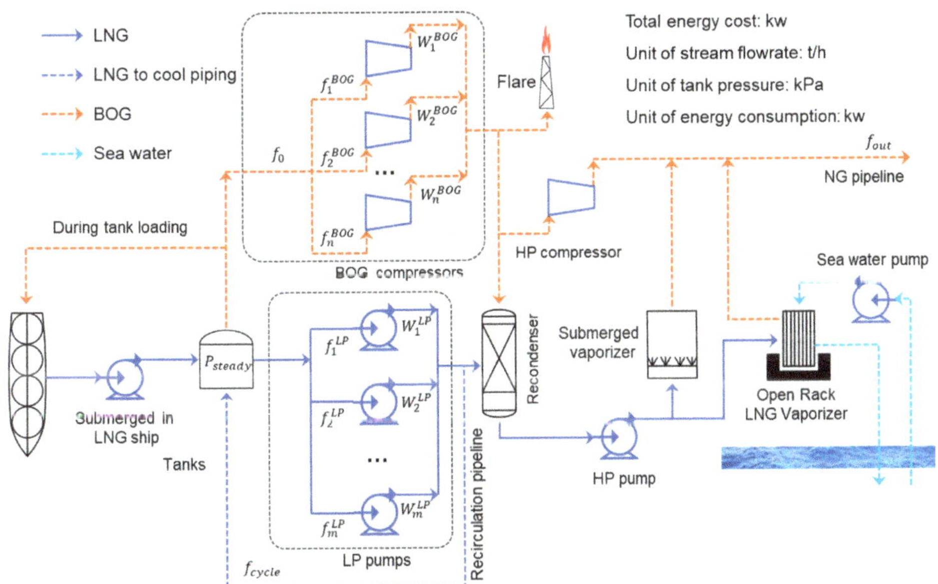

Figure 1. Structure of the liquefied natural gas (LNG) terminal with decision variables.

This work aims to minimize the energy consumption by optimizing the recirculation flow rate and scheduling the LP pumps and BOG compressors according to natural gas demands.

As shown in Figure 1, P_{steady} is the steady pressure of the tank, and f_0 is the flow rate of the total BOG removed from the tank. f_i^{BOG} and W_i^{BOG} denote the BOG flow rate and energy consumption of compressor i, respectively. f_j^{LP} and W_j^{LP} denote the LNG load and energy consumption of the LP pump j, respectively. f_{cycle} is the flow rate of recirculating LNG. f_{out} is the flow rate of the output NG.

The following assumptions are made to develop the operational optimization model of the LNG terminal:

(1) The terminal has n BOG compressors, whose load is divided into l levels;
(2) the terminal has m fixed speed pumps, whose power consumption and flow rate load are the same;
(3) the status of each pump or compressor is identical;
(4) the recondensation method is used to handle BOG.

The binary variables are introduced to indicate whether the compressors or pumps are operated. Furthermore, many constraints are considered in the model.

3. Model Formulation

3.1. Basic Component Models

The models of basic components such as the storage tank, BOG compressor, LP pump, and circulating pipeline are developed as follows.

3.1.1. Tank Model

LNG storage tanks play a vital role in the terminal [37], which serve primarily as a buffer to balance the LNG supplies from ships and NG demands from local users [38]. Given the continuous heat leaking into the storage tanks, the BOG is produced inevitably [22,39]. Although the cryogenic tanks are heavily insulated from the sides and proof, external heat leakage into the LNG is unavoidable [40].

In the research of design and optimization for LNG terminals, a normal parameter is used for predicting boil-off rate generated by heat transfer from the surroundings to the tank [41]. The quantity of BOG is normally expressed as the percentage of total volume of LNG in the tank. The boil-off rate can be calculated by the following expression:

$$f = B_s \frac{V_L \rho_L}{24}, \tag{1}$$

where B_s is the boil-off rate on specification ranging from 0.05%–0.1% per day [36]; V_L is the volume of LNG in tank, and ρ_L is the density of LNG.

In addition, a corrected empirical equation is widely used in recent years [28]:

$$f = \frac{C_R B_s \rho_L V_L}{K_1 K_2 K_3}, \tag{2}$$

where the coefficient C_R is the rollover effected by the flow rate of circulating LNG, and its value is usually set as 1.2. K_1, K_2, and K_3 are the correction factors for the offset of the tank pressure (P) from the LNG vapor pressure (P_v), LNG temperature (T_L), and ambient temperature (T_a), respectively.

In this work, the HYSYS dynamic model of the LNG tank was used to generate the data of BOG generation rate varying with the operations. For convenience, the simulation data were used to regress the parameters of Equation (3), by which the total BOG generation can be calculated.

$$f = \beta_1(P - P_v) + \beta_2 T_L + \beta_3 T_a + \beta_4, \tag{3}$$

where $P - P_v$, T_L, and T_a are the differences between the pressure of the gas phase in the tank and the vapor pressure of the LNG, the temperature of LNG, and the ambient temperature, respectively. β_1, β_2, and β_3 are the correction factors for $P - P_v$, T_L, and T_a, respectively. β_4 is the boil-off rate of BOG on specific conditions. The parameters can be derived from the simulation data by multi-linear regression.

3.1.2. Compressor Model

The BOG compressors are used to remove excess BOG, which may damage the infrastructure and operations of the tanks. In most LNG terminals, the optimization of compressors is the primary goal for reducing the consumption of energy, as they are highly energy intensive [26]. Industrial compressors have several types, such as reciprocating, rotary, axial, and centrifugal. In this study, the two-stage reciprocating compressors are used, and the total power consumption can be calculated as follows:

$$W^{BOG} = \sum\nolimits_{i=1}^{n} W_i^{BOG}, \tag{4}$$

where W_i^{BOG} is the power consumption of compressor i and defined as $W_i^{BOG} = \sum_{z=0}^{l} c^z t_i^z$. The superscript z is the load level number of compressors, and c^z is the power consumption of level z. t_i^z is the fraction of the operation period for compressor i to run at level z.

3.1.3. Pump Model

In the LNG terminal, LP pumps are used to transfer the LNG of tanks to a recondenser for cooling BOG and carry out the cold LNG to the recirculation pipeline. Therefore, the power consumption of the LP pumps is related to the send-out and recirculation flow rate. In this study, the total energy consumption (W^{LP}) can be calculated as follows:

$$W^{LP} = \sum_{j=1}^{m} W_j^{LP},\ j = 1, \cdots, m, \tag{5}$$

where j is the index of pumps, and W_j^{LP} is the power consumption of the LP pump j.

3.1.4. Recirculation Pipeline Model

A stream of recirculating LNG is used to keep the unloading arms in a low temperature to prevent the flow rate of the produced BOG from increasing rapidly, which may damage the devices and disturb the normal operations [22]. The heat (Q) transfers from the air to the recirculation pipeline, whose relationship with mass flow rate of recirculation pipeline is shown as follows:

$$Q = f_{cycle} c_p \Delta T, \tag{6}$$

where f_{cycle} is the mass flow rate of recycling LNG, c_p is the specific heat capacity, and $\Delta T = T_o - T_{in}$ is the temperature difference between inlet and outlet of recirculation pipeline. Q can also be calculated as follows:

$$Q = KA\Delta T_m, \tag{7}$$

$$\Delta T_m = \frac{(T_o - T_a) - (T_{in} - T_a)}{ln\frac{T_o - T_a}{T_{in} - T_a}}, \tag{8}$$

According to Equations (6)–(8), T_o can be calculated as follows:

$$T_o = T_a - \frac{T_a - T_{in}}{e^{\frac{KA}{f_{cycle} c_p}}}, \tag{9}$$

where K is the total transfer coefficient; A is the heat transfer area; T_o is the outlet temperature, and T_{in} is the inlet temperature of LNG. ΔT_m is log mean temperature difference.

The power consumption of LP pumps can be reduced by low f_{cycle}. However, low f_{cycle} also leads to an increase of power consumption of BOG compressors simultaneously. Therefore, f_{cycle} must be optimized.

3.2. Operational Optimization Model of the LNG Terminal

3.2.1. Objective Function

This work aims to obtain the optimal operation condition by minimizing the total energy consumption of the BOG compressors and LP pumps. Based on the developed basic component models, the objective function is defined as follows:

$$min \quad Energy\ Consumption = \sum_{i=1}^{n} W_i^{BOG} + \sum_{j=1}^{m} W_j^{LP} + \sigma \sum_{i=1}^{n} \sum_{z=0}^{l} u_i^z, \tag{10}$$

where the item $\sum_{i=1}^{n} W_i^{BOG}$ and $\sum_{j=1}^{m} W_j^{LP}$ are the electricity consumptions of compressors and LP pumps, respectively. The third one is the penalty item for the complicated operations of compressors, where σ is a small positive penalty coefficient. u_i^z is the binary integer variable that indicates whether the operation mode of compressor i at level z is used. For example, using a small number of compressors is better than using several compressors. The index i and j represent the compressor and pump number, respectively, and z is the compressor load level.

3.2.2. Compressor Constraint

In order to remove the generated BOG in time, the mass flow balance for compressor i can be expressed as follows:

$$\sum_{i=1}^{n} f_i^{BOG} = f_0, \tag{11}$$

$$f^z = \delta_z f_{max}^{BOG},\ z = 0, \cdots, l, \tag{12}$$

$$f_i^{BOG} = \sum_{z=0}^{l} \theta_i^z f^z,\ i = 1, \cdots, n, \tag{13}$$

where δ_z is the load fraction at level z; f_{max}^{BOG} is the mass flow rate of the compressor in the load fraction of 100%. f^z is the mass flow rate of level z. The operation time constraint is given as follows:

$$x_i = \sum_{z=0}^{l} \theta_i^z,\ i = 1, \cdots, n, \tag{14}$$

$$u_i^z \geq \theta_i^z,\ i = 1, \cdots, n,\ z = 0, \cdots, l, \tag{15}$$

where x_i is a binary integer variable indicating whether compressor i is to be used; θ_i^z is the fraction of the operation period for compressor i to run at level z; u_i^z indicates whether the operation mode of compressor i at level z is used. The following constraint is used to avoid multiple equivalent solutions for compressors:

$$f_i^{BOG} \geq f_{i+1}^{BOG},\ i = 1, \cdots, n. \tag{16}$$

3.2.3. Pump Constraint

The total load stream supply for pumps must satisfy the stream demand of customers (f_{out}) when considering the mass flow of BOG and circular LNG, which can be expressed as follows:

$$f_{LNG} = f_{out} - f_0 + f_{cycle}, \tag{17}$$

$$\sum_{j=1}^{m} y_j f_j^{LP} \geq f_{LNG}, \tag{18}$$

where y_j is a binary variable that denotes whether pump j is running or not; f_j^{LP} is the load of pump j, and the index j is the pump number. f_{LNG} is the minimum flow rate for LP pumps.

The following constraint is used to avoid multiple equivalent solutions for pumps:

$$f_j^{LP} \geq f_{j+1}^{LP},\ j = 1, \cdots, m. \tag{19}$$

3.2.4. Recirculation Pipeline Constraint

The temperature difference (ΔT) between the inlet and outlet of recirculation pipeline is primarily influenced by the ambient temperature and flow rate of recirculating LNG. When the ambient temperature is fixed, the ΔT is decided by the flow rate of recirculating LNG. When the flow rate increases, the ΔT will decrease accordingly, otherwise, ΔT will increase. The temperature difference constraints of the recirculation pipeline are expressed as follows:

$$\Delta T_{min} \leq \Delta T \leq \Delta T_{max}, \tag{20}$$

where ΔT_{min} and ΔT_{max} are the lower and upper bounds of ΔT [42].

The operational optimization model for the LNG terminal (LNGT-OOM) is an MINLP model, which is formally cast as follows:

$$\begin{array}{c} \min\limits_{f_{cycle},\ x_i,\ u_i^z,\ \theta_i^z,\ y_i} \quad Energy\ Consumption\ defined\ in\ (10) \\ s.t.\ Compressors\ constraints\ (11)-(16) \\ Pumps\ constraints\ (17)-(19) \\ Recirculation\ pipeline\ constraint\ (20) \end{array} \quad . \text{(LNGT} - \text{OOM)}$$

3.3. Modeling the Backup Compressors

The backup compressor must always be kept in hot standby mode to start from standby mode immediately under some sudden failures. The hot standby mode of BOG compressors also consumes energy. This operation condition is discussed in this study. Starting up a backup compressor unnecessarily is a waste of energy.

Since the load of compressors is greatly influenced by the vaporized gas of tanks, an appropriate equation of state is necessary for the sufficiently accurate description of the BOG. Considering that BOG is primarily composed of methane and nitrogen, the Soave–Redlich–Kwong (SRK) equation is used to describe the gas phase in the tank, which is calculated as follows [43,44]:

$$P = \frac{RT}{V_m - b} - \frac{a(T)}{V_m(V_m - b)}, \tag{21}$$

$$V_m = \frac{V}{N}, \tag{22}$$

$$b = \psi_b \frac{RT_c}{P_c}, \tag{23}$$

$$a(T) = \psi_a \frac{(RT_c)^2}{P_c \alpha_T}, \tag{24}$$

$$\alpha(T) = \left[1 + k_e\left(1 - T_{re}^{0.5}\right)\right]^2, \tag{25}$$

$$k_e = \psi_{k1} + \psi_{k2} w_e - \psi_{k3} w_e^2, \tag{26}$$

where P is the system pressure, and R is the ideal gas constant. T is the system temperature, and V_m is the molar volume. V is the system volume, and N is the moles of the system. a and b are the correction factor for intermolecular attraction and volume repulsion, respectively. w_e is the acentric factor, and the subscripts e, c, and r represent the components, critical properties, and contrast nature, respectively. α and k_e are used to make a key function of temperature and improve the accuracy of the equation [45]. Considering that the value of P changes a little with variables except for N, it can be assumed as a function of N. Among the variables, $\gamma_{k1} = 0.48$; $\psi_a = 0.42747$; $\psi_b = 0.08664$; $\psi_{k1} = 0.48$; $\psi_{k2} = 1.574$, and $\psi_{k3} = 0.176$.

The accumulation of molar flow rate (*dN/dt*) can be calculated as follows:

$$\frac{dN}{dt} = \frac{f - f_0}{M}, \tag{27}$$

where f is the mass flow rate of BOG generation caused by heat leak from tanks; f_0 is the total mass load of BOG compressors; M is the molecular weight of BOG, and t is the operation time.

The operation time when moles change can be estimated as follows:

$$\Delta t = \frac{\Delta n}{\left(\frac{dN}{dt}\right)}, \tag{28}$$

where the symbol Δ represents the differences.

If the pressure of the tank can still be kept below the flare pressure during the startup time while an operating compressor fails, then the backup compressor can be shut down during the normal operation.

4. Case Study

4.1. Case Description

A case study on energy optimization of an actual LNG terminal in China is presented to demonstrate the effectiveness of the proposed approach. The parameters of the original condition are shown in Figure 2, and the variables and related process parameters are listed in Table 1. Table 2 shows the regression parameters for calculating the BOG generation rate f. Figure 3 shows the comparison between the simulated and predicted values. The average of the simulated value is 2.39 t/h, that is 0.09% for B_s (Equation (1)).

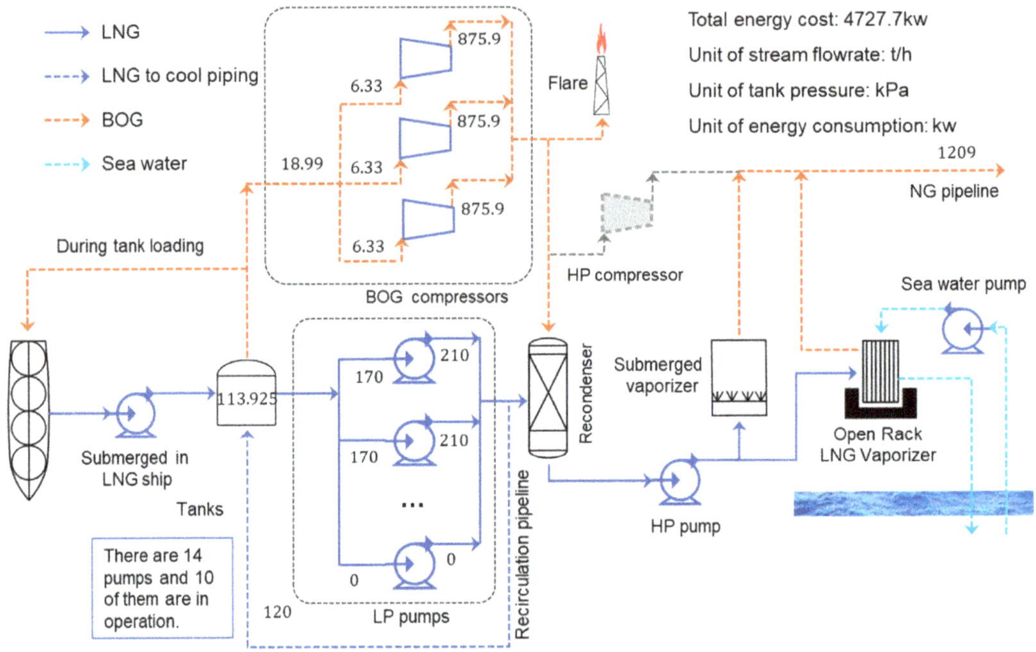

Figure 2. Structure of the original condition.

Table 1. Environmental variables and related process parameters of optimization.

Parameters	Values	Units
Tank number	4	/
Tank volume	16,000	m^3
Tank liquid level	85	%
LNG temperature	−159.8	°C
Length of the LNG unloading pipeline	2909	m
Diameter of the LNG unloading pipeline	1.487	m
Length of the LNG cooling cycle pipeline	2942	m
Diameter of the LNG cooling cycle pipeline	0.574	m
Total heat transfer coefficient of the pipeline	0.38476	$W/(m^2{\cdot}K)$
Average ambient temperature	5	°C
Send-out flow rate	1209	t/h

Table 2. Regression parameters for calculating f.

Parameters	Values
β_1	−0.12161
β_2	−2.1252
β_3	0.053183
β_4	−332.666

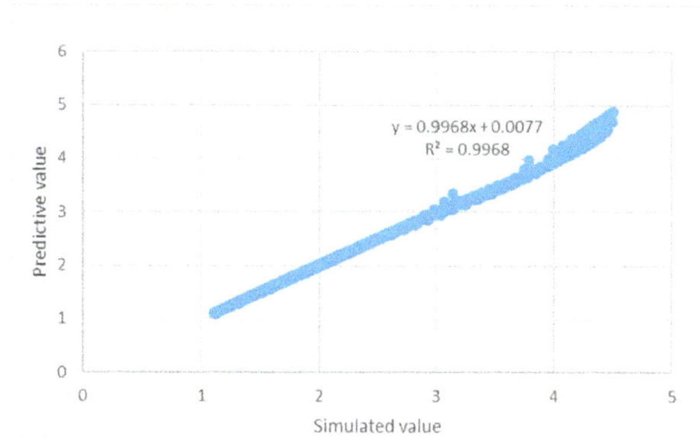

Figure 3. Comparison between simulated and predicted values.

4.2. Parameters of the Proposed Models

The tanks are equipped with a cold insulation layer to ensure that the tank's daily maximum evaporation rate does not exceed 0.1%. The flare pressure of the storage tank is 25 kPaG. Table 3 shows the compositions of lean and rich LNG. Table 4 lists the basic thermodynamic parameters of each component in NG. Table 5 shows the binary interaction parameters of the SRK equation of state.

Table 3. Compositions of different types of LNG.

	Lean LNG		Rich LNG	
	Mass%	Mole%	Mass%	Mole%
Methane	99.84	99.91	72.33	84.23
Ethane	0.04	0.02	20.65	12.83
Propane	0	0	6.33	2.68
i-Butane	0	0	0.31	0.1
n-Butane	0	0	0.28	0.09
Nitrogen	0.012	0.07	0.1	0.07
Total	100	100	100	100

Table 4. Basic thermodynamic parameters of the NG components. SRK: Soave–Redlich–Kwong.

Component	Chemical Formula	Molecular Weight	SRK Acentric	Critical Temperature (°C)	Critical Pressure (kPa)
Methane	CH_4	16.043	0.00740	−82.45	4641
Ethane	C_2H_6	30.07	0.09830	32.28	4884
Propane	C_3H_8	44.097	0.15320	96.75	4257
i-Butane	C_4H_{10}	58.123	0.18250	134.9	3648
n-Butane	C_4H_{10}	58.123	0.20080	152	3797
Nitrogen	N_2	28.013	0.03580	−147.0	3394

Table 5. Binary interaction parameters of the SRK equation of state.

	Methane	Ethane	Propane	i-Butane	n-Butane	Nitrogen
Methane	/	0.00224	0.00683	0.01311	0.0123	0.03120
Ethane	0.00224	/	0.00126	0.00457	0.00410	0.03190
Propane	0.00683	0.00126	/	0.00104	0.00082	0.08860
i-Butane	0.01311	0.00457	0.00104	/	0.00001	0.13150
n-Butane	0.01230	0.00410	0.00082	0.00001	/	0.05970
Nitrogen	0.0312	0.03190	0.08860	0.13150	0.05970	/

As shown in Figure 2, the LNG terminal has three BOG compressors and ten LP pumps operating in the process, whose stream flow rates and power consumption are provided in Table 6. The operating characteristics of compressors are presented in Table 7.

Table 6. Original operation condition of the LNG terminal.

	Variables	Original Value	Energy Consumption (kw)
Boil-off gas (BOG) compressors	x_1	1	875.9
	x_2	1	875.9
	x_3	1	875.9
	u_1^3	1	/
	u_2^3	1	/
	u_3^3	1	/
BOG load (t/h)	f_0	19	/
LP pumps	y_1	1	210
	y_2	1	210
	y_3	1	210
	y_4	1	210
	y_5	1	210
	y_6	1	210
	y_7	1	210
	y_8	1	210
	y_9	1	210
	y_{10}	1	210
	y_{11}	0	0
	y_{12}	0	0
	y_{13}	0	0
	y_{14}	0	0
Recirculation flow rate (t/h)	f_{cycle}	120	/
Steady pressure (kPa)	P_{steady}	113.925	/
Objective function	*Energy Consumption*	/	4727.7

Table 7. Operating characteristic of BOG compressors.

Property	Unit	Variable	Value				
Road Levels	/	z	0	1	2	3	4
Mass load	t/h	f^z	0	2.11	4.22	6.33	8.44
Load fraction	%	δ	0	25	50	75	100
Power consumptions	kw	W_c	448.3	586.2	793.1	875.9	1000
Startup time	min	Δt_s			30		

5. Results and Discussion

The flowchart of the proposed optimization modeling framework is illustrated in Figure 4. It was programmed and performed in MATLAB R2019a on a computer with an Intel I Core (TM) i9-9900 CPU @ 3.10 GHz and 32 GB RAM. The deterministic model (LNGT-OOM) was programmed in GAMS 24.1.2 and solved by the Discrete and Continuous Optimizers (DICOPT 24.1.2).

In the proposed operational optimization framework, the steady-state pressure is first presented to determine whether the results are optimal or not. Meanwhile, the SRK equation of state is selected for the physical property calculation. First, MATLAB provides the initial variables based on the actual operating condition and minimum compressor load. Additionally, then the variables are input to GAMS to obtain the optimal recirculation flow rate and number of LP pumps in operation by solving the model (LNGT-OOM). The obtained operation strategy will be sent back to MATLAB and steady-state pressure of the tank can be calculated. If the steady-state pressure is higher than the flare pressure, the compressor load must be increased, and then a new steady-state pressure is calculated. After the termination condition is achieved, whether a standby compressor needs to be turned on or not must be decided. Finally, the total power consumption of the LP pumps and BOG compressors is obtained.

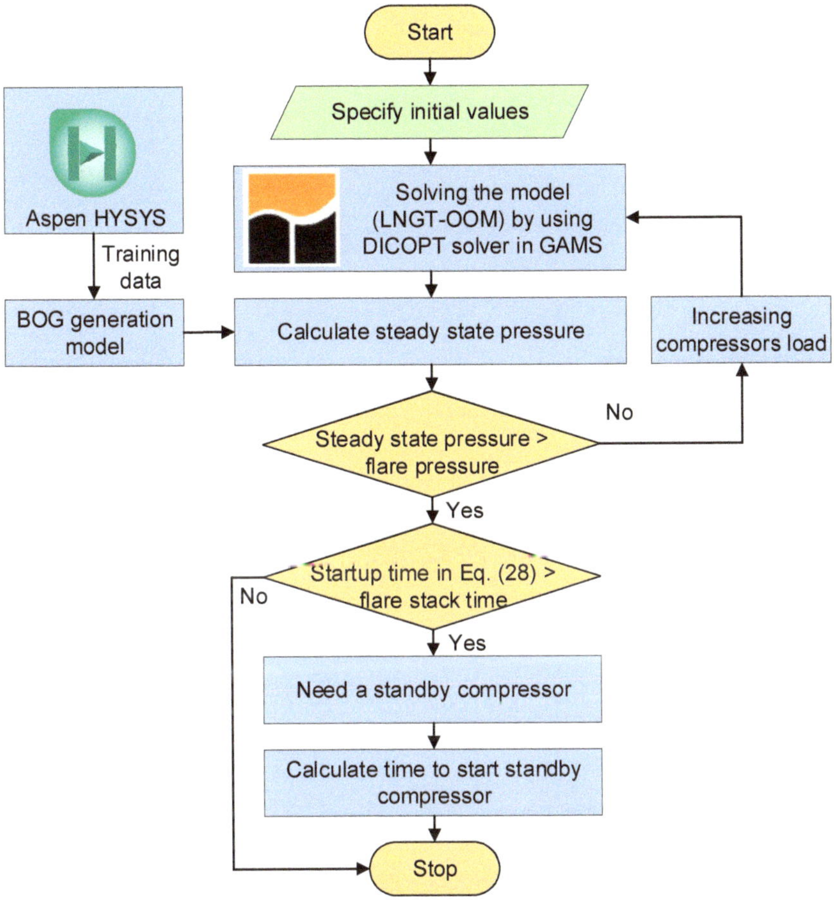

Figure 4. Schematic diagram of the proposed optimization framework.

The problem sizes and the computation time of the proposed MINLP model for the LNG terminal are shown in Table 8.

Table 8. Problem sizes and computation time.

	Value
Number of continuous variables	57
Number of binary variables	32
Constraints	34
Number of iterations	19
Computation time (s)	0.017

The optimized results are shown in Figure 5 and Table 9. As shown in Table 9, the total energy consumption is 2680 kw, and the steady pressure of the tank is 124.49 kPa. Two BOG compressors and two LP pumps are turned off from running. Therefore, the recirculation flow rate of LNG is increased to 122.58 t/h, and the energy consumption is reduced by 13.31%.

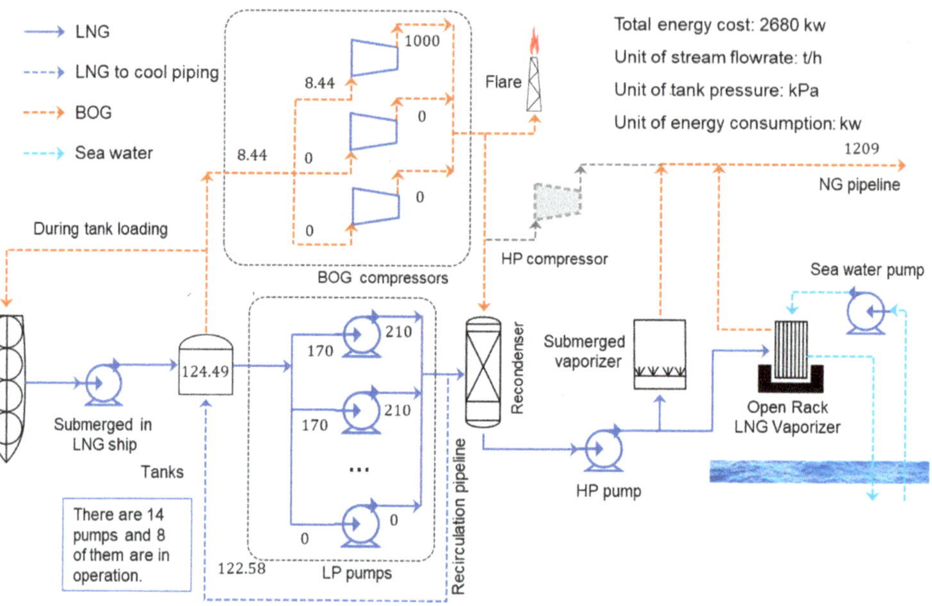

Figure 5. Optimized system configuration determined by the model (operational optimization model for the LNG terminal (LNGT-OOM)).

Table 9. Optimized results and energy consumption of the model (LNGT-OOM).

	Variables	Optimized Value	Energy Consumption (kw)
	x_1	1	1000
	x_2	0	0
BOG compressors	x_3	0	0
	u_1^4	1	/
	u_2^z	0	/
	u_3^z	0	/
BOG load (t/h)	f_0	8.44	/
	y_1	1	210
	y_2	1	210
	y_3	1	210
	y_4	1	210
	y_5	1	210
	y_6	1	210
LP pumps	y_7	1	210
	y_8	1	210
	y_9	0	0
	y_{10}	0	0
	y_{11}	0	0
	y_{12}	0	0
	y_{13}	0	0
	y_{14}	0	0
Recirculation flow rate (t/h)	f_{cycle}	122.58	/
Steady pressure (kPa)	P_{steady}	124.49	/
Objective function	*Energy Consumption*	/	2680

An operating compressor can possibly fail, therefore, the mass flow rate of BOG generation is more than the output flow rate, which leads to the accumulation of BOG and the increased pressure of the tank. The time consumed for changing from steady pressure to flare pressure (Δt_f) is 5.68 min, which can be calculated by Equation (28). It is smaller than the startup time. Therefore, a backup compressor must be turned on all the time.

The optimal operation condition is shown in Figure 6, and the energy consumption comparisons between the original and optimized operation conditions are presented in Table 10. The energy consumption is reduced by 33.83% compared with the original condition. The energy saving results from the reduction in the number of LP pumps and the increase of the tank pressure. Moreover, the safety of the LNG tanks is ensured by the operation strategy of the backup compressors.

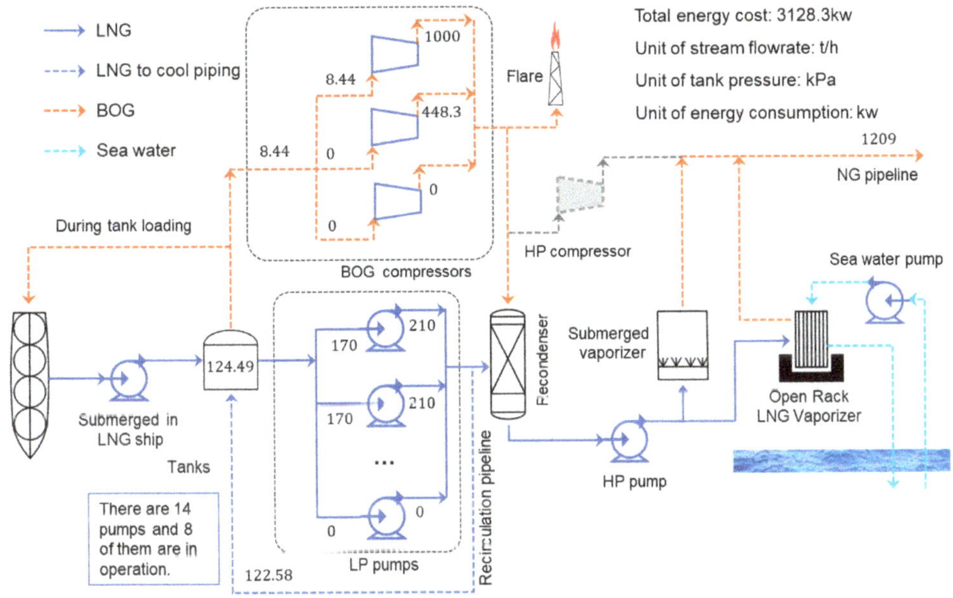

Figure 6. Optimized operation condition of the proposed method.

Table 10. Energy consumption comparisons.

	Original Condition	Optimized Condition
Compressor loads (t/h)	6.33, 6.33, 6.33	8.44, 0, 0
Target pressure (kPa)	113.93	124.49
Circulation flow (t/h)	120	122.58
Pump number	10	8
Power consumption (kw)	4727.70	3128.30
Energy save (%)	/	33.83

Furthermore, the NG demands of the end users and the ambient temperature vary all the time. Two typical scenarios in different months are implemented to indicate the effectiveness of the proposed method. The comparisons among ambient temperatures, user demands, decision variables, and optimization results for the two scenarios are summarized in Table 11. For the given LNG terminal, the average ambient temperature is 30 °C, and the NG demand is 555.56 t/h from April to October. Energy consumption is reduced by 9.15%. The average ambient temperature is 5 °C, and the NG demand is 1388.89 t/h from November to March. For this scenario, 26.1% energy saving is achieved. The optimal operating variables obtained vary due to different ambient temperatures and flow rates of send-out NG.

Table 11. Comparisons of data of the operating variables and results.

	April to October		November to March	
	Original	Optimized	Original	Optimized
Average ambient temperature (°C)	30	30	5	5
Send-out flow rate (t/h)	555.56	555.56	1388.89	1388.89
Compressor loads (t/h)	19	14.77	19	8.44
Need a standby compressor or not	No	Yes	No	Yes
Target pressure (kPa)	113.93	122.98	113.93	124.49
Circulation flow (t/h)	120	139.21	120	122.58
Pump number	4	4	9	9
Power consumption (kw)	3467.7	3150.4	4517.7	3338.3
Energy save (%)	9.15		26.1	

6. Conclusions

This work proposed an operational optimization model of the LNG terminal to minimize the energy consumption of BOG compressors and LP pumps. An MINLP model was formulated, which determined whether the pumps were running or on standby, and the number of compressor level was selected as a binary variable. Operating strategies for varied flow rates of the send-out rate and the ambient temperature can be proposed using the model. An actual case study on the LNG terminal was presented to indicate the effectiveness of the proposed approach. The minimum energy consumption was determined by using the optimization model, and the corresponding decision variables were obtained.

One BOG compressor and two pumps can be turned off after optimization. The energy consumption can be reduced from 4727.70 kw to 3128.30 kw and 33.83% energy saving was obtained for the given operating condition. Furthermore, the scenarios of different months were analyzed. From April to October, when the compressor load changed from 19 t/h to 14.77 t/h and the recirculation flow rate increased from 120 t/h to 139.21 t/h, the energy consumption can be reduced by 9.15%. From November to March, the optimal operating pressure rose to 124.49 kPa due to the decrease of ambient temperatures. The optimized compressor load and recirculation flow rate were 8.44 t/h and 122.58 t/h, respectively. Compared with the previous period, 26.1% of energy can be saved after optimization. About 16.21% of energy consumption can be saved annually.

The proposed optimization method would significantly contribute to the existing LNG terminals. However, the research was on the condition that the LNG was not unloading and the LNG terminal used a recondenser instead of HP compressors to handle BOG. The other working condition will also be studied in the future. Besides, average temperatures of the months were used in this work, which is not very realistic since the ambient temperature changes all the time.

Author Contributions: Conceptualization, Z.Y. and L.Z.; methodology, L.Z. and Z.Y.; software, X.M.; validation, Z.Y. and L.Z.; investigation, Z.Y.; writing—original draft preparation, X.M.; writing—review and editing, L.Z. and Z.Y.; supervision, Z.Y.; funding acquisition, Z.Y. and L.Z. All authors have read and agreed to the published version of the manuscript.

Funding: This research was funded by National Natural Science Foundation of China (Basic Science Center Program: 61988101, 61873092), International (Regional) Cooperation and Exchange Project (61720106008), National Natural Science Fund for Distinguished Young Scholars (61725301) and Fundamental Research Funds for the Central Universities (222202017006).

Institutional Review Board Statement: Not applicable.

Informed Consent Statement: Not applicable.

Data Availability Statement: Not applicable.

Conflicts of Interest: The authors declare no conflict of interest.

Acronyms

The following acronyms are used in this manuscript:

BOG	Boil-off gas
GHG	Greenhouse gas
HP	High-pressure
LNG	Liquefied natural gas
LNGT-OOM	Operational optimization model for the LNG terminal
LP	Low-pressure
MILP	Mixed-integer linear programming
MINLP	Mixed-integer nonlinear programming
NG	Natural gas
NOMAD	Nonlinear Optimization by Mesh Adaptive Direct Search
SRK	Soave–Redlich–Kwong

References

1. He, T.; Lin, W. A novel propane pre-cooled mixed refrigerant process for coproduction of LNG and high purity ethane. *Energy* **2020**, *202*, 117784. [CrossRef]
2. Arteconi, A.; Brandoni, C.; Evangelista, D.; Polonara, F. Life-cycle greenhouse gas analysis of LNG as a heavy vehicle fuel in Europe. *Appl. Energy* **2010**, *87*, 2005–2013. [CrossRef]
3. Al-Haidous, S.; Al-Ansari, T. Sustainable Liquefied Natural Gas Supply Chain Management: A Review of Quantitative Models. *Sustainability* **2020**, *12*, 243. [CrossRef]
4. Qi, M.; Park, J.; Kim, J.; Lee, I.; Moon, I. Advanced integration of LNG regasification power plant with liquid air energy storage: Enhancements in flexibility, safety, and power generation. *Appl. Energy* **2020**, *269*, 115049. [CrossRef]
5. Agarwal, R.; Rainey, T.J.; Steinberg, T.; Rahman, S.M.A. LNG regasification-Effects of project stage decisions on capital expenditure and implications for gas pricing. *J. Nat. Gas Sci. Eng.* **2020**, *78*, 12. [CrossRef]
6. Kurle, Y.M.; Wang, S.; Xu, Q. Dynamic simulation of LNG loading, BOG generation, and BOG recovery at LNG exporting terminals. *Comput. Chem. Eng.* **2017**, *97*, 47–58. [CrossRef]
7. Wahl, P.E.; Løvseth, S.W.; Mølnvik, M.J. Optimization of a simple LNG process using sequential quadratic programming. *Comput. Chem. Eng.* **2013**, *56*, 27–36. [CrossRef]
8. Qyyum, M.A.; Qadeer, K.; Lee, S.; Lee, M. Innovative propane-nitrogen two-phase expander refrigeration cycle for energy-efficient and low-global warming potential LNG production. *Appl. Therm. Eng.* **2018**, *139*, 157–165. [CrossRef]
9. Abdul Qyyum, M.; Qadeer, K.; Lee, M. Closed-Loop Self-Cooling Recuperative N2 Expander Cycle for the Energy Efficient and Ecological Natural Gas Liquefaction Process. *ACS Sustain. Chem. Eng.* **2018**, *6*, 5021–5033. [CrossRef]
10. Kumar, S.; Kwon, H.-T.; Choi, K.-H.; Lim, W.; Cho, J.H.; Tak, K.; Moon, I. LNG: An eco-friendly cryogenic fuel for sustainable development. *Appl. Energy* **2011**, *88*, 4264–4273. [CrossRef]
11. He, T.; Chong, Z.R.; Zheng, J.; Ju, Y.; Ling, P. LNG cold energy utilization: Prospects and challenges. *Energy* **2019**, *170*, 557–568. [CrossRef]
12. Atienza-Marquez, A.; Bruno, J.C.; Coronas, A. Cold recovery from LNG-regasification for polygeneration applications. *Appl. Therm. Eng.* **2018**, *132*, 463–478. [CrossRef]
13. Lim, W.; Choi, K.; Moon, I. Current Status and Perspectives of Liquefied Natural Gas (LNG) Plant Design. *Ind. Eng. Chem. Res.* **2013**, *52*, 3065–3088. [CrossRef]
14. Economides, M.J.; Wood, D.A. Natural gas and LNG trade-a global perspective. *Hydrocarb Process.* **2006**, *85*, 39–45.
15. Park, C.; Lee, C.-J.; Lim, Y.; Lee, S.; Han, C. Optimization of recirculation operating in liquefied natural gas receiving terminal. *J. Taiwan Inst. Chem. Eng.* **2010**, *41*, 482–491. [CrossRef]
16. Liu, C.; Zhang, J.; Xu, Q.; Gossage, J.L. Thermodynamic-Analysis-Based Design and Operation for Boil-Off Gas Flare Minimization at LNG Receiving Terminals. *Ind. Eng. Chem. Res.* **2010**, *49*, 7412–7420. [CrossRef]
17. Hatcher, P.; Khalilpour, R.; Abbas, A. Optimisation of LNG mixed-refrigerant processes considering operation and design objectives. *Comput. Chem. Eng.* **2012**, *41*, 123–133. [CrossRef]
18. Le, S.; Lee, J.-Y.; Chen, C.-L. Waste cold energy recovery from liquefied natural gas (LNG) regasification including pressure and thermal energy. *Energy* **2018**, *152*, 770–787. [CrossRef]
19. Kanbur, B.B.; Xiang, L.; Dubey, S.; Choo, F.H.; Duan, F. Cold utilization systems of LNG: A review. *Renew. Sustain. Energy Rev.* **2017**, *79*, 1171–1188. [CrossRef]
20. Lee, I.; Park, J.; You, F.; Moon, I. A novel cryogenic energy storage system with LNG direct expansion regasification: Design, energy optimization, and exergy analysis. *Energy* **2019**, *173*, 691–705. [CrossRef]
21. Kochunni, S.K.; Chowdhury, K. LNG boil-off gas reliquefaction by Brayton refrigeration system–Part 1, Exergy analysis and design of the basic configuration. *Energy* **2019**, *176*, 753–764. [CrossRef]
22. Rao, H.N.; Wong, K.H.; Karimi, I.A. Minimizing Power Consumption Related to BOG Reliquefaction in an LNG Regasification Terminal. *Ind. Eng. Chem. Res.* **2016**, *55*, 7431–7445. [CrossRef]

23. Park, C.; Lim, Y.; Lee, S.; Han, C. BOG Handling Method for Energy Saving in LNG Receiving Terminal. In *Computer Aided Chemical Engineering*; Pistikopoulos, E.N., Georgiadis, M.C., Kokossis, A.C., Eds.; Elsevier: Amsterdam, The Netherlands, 2011; pp. 1829–1833.
24. Lee, I.; Moon, I. Total Cost Optimization of a Single Mixed Refrigerant Process Based on Equipment Cost and Life Expectancy. *Ind. Eng. Chem. Res.* **2016**, *55*, 10336–10343. [CrossRef]
25. Park, J.; Lee, I.; You, F.; Moon, I. Economic Process Selection of Liquefied Natural Gas Regasification: Power Generation and Energy Storage Applications. *Ind. Eng. Chem. Res.* **2019**, *58*, 4946–4956. [CrossRef]
26. Reddy, H.V.; Bisen, V.S.; Rao, H.N.; Dutta, A.; Garud, S.S.; Karimi, I.A.; Farooq, S. Towards energy-efficient LNG terminals: Modeling and simulation of reciprocating compressors. *Comput. Chem. Res.* **2019**, *128*, 312–321. [CrossRef]
27. Shin, M.W.; Shin, D.; Choi, S.H.; Yoon, E.S.; Han, C. Optimization of the Operation of Boil-Off Gas Compressors at a Liquified Natural Gas Gasification Plant. *Ind. Eng. Chem. Res.* **2007**, *46*, 6540–6545. [CrossRef]
28. Shin, M.W.; Shin, D.; Yoon, C.E.S. Optimal operation of the boil-off gas compression process using a boil-off rate model for LNG storage tanks. *Korean J. Chem. Eng.* **2008**, *25*, 7–12. [CrossRef]
29. Jang, N.; Shin, M.W.; Choi, S.H.; Yoon, E.S. Dynamic simulation and optimization of the operation of boil-off gas compressors in a liquefied natural gas gasification plant. *Korean J. Chem. Eng.* **2011**, *28*, 1166–1171. [CrossRef]
30. Nagesh Rao, H.; Karimi, I.A. Optimal design of boil-off gas reliquefaction process in LNG regasification terminals. *Comput. Chem. Res.* **2018**, *117*, 171–190. [CrossRef]
31. Tak, K.; Lee, I.; Kwon, H.; Kim, J. Comparison of Multistage Compression Configurations for Single Mixed Refrigerant Processes. *Ind. Eng. Chem. Res.* **2015**, *54*, 9992–10000. [CrossRef]
32. Park, C.; Song, K.; Lee, S.; Lim, Y.; Han, C. Retrofit design of a boil-off gas handling process in liquefied natural gas receiving terminals. *Energy* **2012**, *44*, 69–78. [CrossRef]
33. Yuan, T.; Song, C.; Junjiang, B.; Zhang, N.; Zhang, X.; He, G. Minimizing power consumption of boil off gas (BOG) recondensation process by power generation using cold energy in liquefied natural gas (LNG) regasification process. *J. Clean. Prod.* **2019**, *238*, 117949. [CrossRef]
34. Wu, M.; Zhu, Z.; Sun, D.; He, J.; Tang, K.; Hu, B.; Tian, S. Optimization model and application for the recondensation process of boil-off gas in a liquefied natural gas receiving terminal. *Appl. Therm. Eng.* **2019**, *147*, 610–622. [CrossRef]
35. Li, Y.; Li, Y. Dynamic optimization of the Boil-Off Gas (BOG) fluctuations at an LNG receiving terminal. *J. Nat. Gas Sci. Eng.* **2016**, *30*, 322–330. [CrossRef]
36. Li, Y.; Chen, X. Dynamic simulation for improving the performance of boil-off gas recondensation system at lng receiving terminals. *Chem. Eng. Commun.* **2012**, *199*, 1251–1262. [CrossRef]
37. Khan, M.S.; Effendy, S.; Karimi, I.A.; Wazwaz, A. Improving design and operation at LNG regasification terminals through a corrected storage tank model. *Appl. Therm. Eng.* **2019**, *149*, 344–353. [CrossRef]
38. Trotter, I.M.; Gomes, M.F.M.; Braga, M.J.; Brochmann, B.; Lie, O.N. Optimal LNG (liquefied natural gas) regasification scheduling for import terminals with storage. *Energy* **2016**, *105*, 80–88. [CrossRef]
39. Ferrín, J.L.; Pérez-Pérez, L.J. Numerical simulation of natural convection and boil-off in a small size pressurized LNG storage tank. *Comput. Chem. Res.* **2020**, *138*, 106840. [CrossRef]
40. Effendy, S.; Khan, M.S.; Farooq, S.; Karimi, I.A. Dynamic modelling and optimization of an LNG storage tank in a regasification terminal with semi-analytical solutions for N2-free LNG. *Comput. Chem. Res.* **2017**, *99*, 40–50. [CrossRef]
41. Dobrota, Đ.; Lalić, B.; Komar, I. Problem of Boil-off in LNG Supply Chain. *Trans. Marit. Sci.* **2013**, *2*, 91–100. [CrossRef]
42. Lee, C.-J.; Lim, Y.; Han, C. Operational strategy to minimize operating costs in liquefied natural gas receiving terminals using dynamic simulation. *Korean J. Chem. Eng.* **2012**, *29*, 444–451. [CrossRef]
43. Yuan, Z.; Cui, M.; Song, R.; Xie, Y. Evaluation of prediction models for the physical parameters in natural gas liquefaction processes. *J. Nat. Gas Sci. Eng.* **2015**, *27*, 876–886. [CrossRef]
44. Ramirez-Jimenez, E.; Justo-Garcia, D.N.; Garcia-Sanchez, F.; Stateva, R.P. VLL Equilibria and Critical End Points Calculation of Nitrogen-Containing LNG Systems: Application of SRK and PC-SAFT Equations of State. *Ind. Eng. Chem. Res.* **2012**, *51*, 9409–9418. [CrossRef]
45. Soave, G. Equilibrium constants from a modified Redlich-Kwong equation of state. *Chem. Eng. Sci.* **1972**, *27*, 1197–1203. [CrossRef]

processes

MDPI

Article

Ontology-Based Process Modelling-with Examples of Physical Topologies

Heinz A Preisig

Department of Chemical Engineering, Norwegian University of Science and Technology, 7491 Trondheim, Norway; Heinz.Preisig@chemeng.ntnu.no

Abstract: Reductionism and splitting application domain into disciplines and identify the smallest required model-granules, termed "basic entity" combined with systematic construction of the behaviour equations for the basic entities, yields a systematic approach to process modelling. We do not aim toward a single modelling domain, but we enable to address specific application domains and object inheritance. We start with reductionism and demonstrate how the basic entities are depending on the targeted application domain. We use directed graphs to capture process models, and we introduce a new concept, which we call "tokens" that enables us to extend the context beyond physical systems. The network representation is hierarchical so as to capture complex systems. The interacting basic entities are defined in the leave nodes of the hierarchy, making the overall model the interacting networks in the leave nodes. Multi-disciplinary and multi-scale models result in a network of networks. We identify two distinct network communication ports, namely ports that exchange tokens and ports that transfer information of tokens in accumulators. An ontology captures the structural elements and the applicable rules and defines the syntax to establish the behaviour equations. Linking the behaviours to the basic entities defines the alphabet of a graphical language. We use this graphical language to represent processes, which has proven to be efficient and valuable. A set of three examples demonstrates the power of the graphical language. The Process Modelling framework (ProMo) implements the ontology-centred approach to process modelling and uses the graphical language to construct process models.

Keywords: model-based; computational engineering; process simulation; digital twin

Citation: Preisig, H.A. Ontology-Based Process Modelling-with Examples of Physical Topologies. *Processes* **2021**, *9*, 592. https://doi.org/10.3390/pr9040592

Academic Editor: Luis Puigjaner

Received: 1 February 2021
Accepted: 22 March 2021
Published: 29 March 2021

1. Introduction

The title refers to modelling, implying the generation of what is often referred to as a digital twin, namely, a digital reproduction of a process' behaviour. The term "model" is used in very different contexts and, correspondingly, has many facets. However, people always model processes for a specific purpose, and most often, the main objective is to generate a simulation with given inputs and a specified process characteristic. Many applications use simulation as an inner loop, with the outer loop defining the main objective, such as optimisation. For example, process design seeks optimal process parameters and optimising control finds an input that optimises an objective. From this perspective, modelling naturally sits at the very front end of the overall process, associated with solving a simulation-related problem, which makes modelling a core activity; therefore, errors in the model are the most costly ones. Thus, it is a natural request to have a systematic approach to modelling that is safe in terms of generating structurally sound models.

We claim that chemical engineering lacks such a systematic model-building process. The closest we have is flowsheeting software which uses customised unit-operation building blocks as its base entities. Readers interested in the historical aspects are referred to [1]. These traditional modelling environments suffer from a couple of problems, which provides the motivation for seeking an alternative, new approach. Let us give a couple of observation-based reasons.

1.1. Why We Need a Systematic Modelling Approach

Incompleteness: Published models tend to be incomplete, in the sense that, usually, the authors do not provide all relevant information leading to the given input/output behaviour. As an example, the publication [2] describes a non-trivial decanting two-liquid-phase reactor. On page two is a simple figure that shows the process schematically. On page three, one finds the mathematical model, which consists of eleven equations. It is left to the knowledgable reader to interpret these equations and determine which one describes which element in the process. If the exercise is used to analyse these equations, close to one hundred equations are obtained, which explain the nature of the description and assumptions in detail. The number of equations is somewhat puzzling, considering the low structural complexity, and a strong indication that the modeller hid many details or is not aware of them. The model is not particularly complex. It describes the reactor's contents as two lumped systems, each with component balances for five species, an energy balance for the two lumps together and a description of the reactor's holdup.

Flowsheeting-balance closure: Above, we mentioned flowsheeting. The typical supplier provides libraries, often designed for particular users, and thus subdiscipline-specific process units or component models. A model builder or composer then utilises the library to construct process models. Next, a solver kernel is attached, providing the implementation of the appropriate solution methods. Examples are gPROMS®, ASPEN® and UNISYM®. Within parts of the user community, it seems to be well known that these simulators do not guarantee the conserved quantities' closure, which represents a fundamental problem for most applications.

Documentation: Process-unit-oriented libraries document the unit's behaviour on the unit level. The details of the models are, therefore, mostly hidden. There is little detail on the actual implementation, as it is considered unnecessary information for the typical user. This information is seen as only relevant for the expert implementing the software module. The result of this is that the individual models are black boxes that provide input/output behaviour. Consequently, the software users must work on the corresponding level of insight; they do not have access to the lower modelling level, through either the code or the documentation.

1.2. An Alternative

The author's process modelling project ProMo takes an alternative route. While we also build the models using a set of building blocks, our basic model entities are below the unit-operation level. They are in the context of the model application on the smallest granularity level; we call them "basic entities". It is essential to acknowledge that these entity blocks are basic on a particular granularity level, with the granularity level depending on the intended application.

While macroscopic physics was the starting point, the concept has been expanded to other disciplines, including control and lower-scale physics, like molecular dynamics. The project constructs its building blocks only from the fundamental concepts defining discipline-basic entities. These entities form the foundation for the discipline-mechanistic construction of the models. The underlying philosophy is therefore reductionism. Since they are deductive, mechanistic models cannot necessarily adequately capture real-world behaviours; a practical modelling system must allow for both glass-box (white-box) models and holistic surrogates (black-box models).

State-space representation: The approach is state-space based. For macroscopic physical systems, the conserved quantities serve as the state. Consequently, any numerical method, primarily integration, closes the respective balance to the defined accuracy. We can thus always guarantee the closure of the fundamental quantities once the algorithm has converged.

Implementation: The definition space, namely, the part associated with defining the basic building blocks, is ontology-driven. Using ontologies at the front end of the definition process imposes a structure onto the process and yields overall consistency. On a more technical level, the equations are in the "pure" form, meaning neither the user nor the software performs any transformations or substitutions. Each equation appears only once in the code, in contrast to unit-operation-based modelling, where each module typically represents the complete model for the mimicked unit. All model-related code is centralised and appears only once. Any composite model uses indexing to refer to the block representations, which renders modelling a bookkeeping problem.

Project Process Modelling ProMo's main objective: The project, which we named ProMo, aims to construct an environment that steers the modelling process as tightly as possible, without imposing constraints, which we achieve by using an ontology. The domain-specific ontology defines the structure of the information underlying the models which will be constructed in the domain captured by the ontology. The construction of the ontology is thus essential for the project's software. The following section is devoted to defining the elements of the ontology.

This paper: The goal of the paper is to define the basis for the graphical representation of process models. In the first step, we discuss the reductionism-based approach, seeking a minimal set of entities that enable us to construct the models for a class of modelling domains. Next, we discuss an abstraction that allows us to extend the modelling beyond physical systems. An ontology tailored to specific application domains provides the fundamental structures for the behaviour description and represents the construction codes for complex models. Defining the entity behaviour in terms of mathematical equations, namely as input/output functions, and linking them to the graphical objects sets the stage for the graphical representation. The latter forms the basis of a graphical language, which we demonstrate the use of in a set of applications. The visual language is handy when discussing a paper-and-pencil model design, and it is also directly used in the ProMo model design tool.

We will leave out two main parts from ProMo. The first is the definition of the behaviour equations, which we will have to present on another occasion. The core is the definition of a formal language and the ability to compile it into different target codes [3]. The second is the generation of the actual simulation code, which is similarly technical in terms of computer science issues, including the realisation of the automatic generation of application interfaces and the semantic network for interoperability. Both sections are full of technical details, which are not relevant to the current exposition.

2. Reductionism

The objective is to generate a mathematical representation of the modelled system, confining the description to a mathematical input/output representation. This behaviour description will be the result of a network of interacting basic entities. For these reasons, we apply reductionism to identify the smallest underlying relevant entities in the context of an intended application. Applying reductionism in model construction places the base entities' definition, the base building blocks, into the centre of development.

From the perspective of mathematics, describing the dynamic results in time-dependent differential equations for time-discrete systems uses time-difference equations, and for event-discrete systems, automata are used, and thus state-discrete difference equations. All the models must satisfy a fundamental condition: Independent of their nature, all equations must represent realisable systems, which, for physics, translates to *causal* or *nonanticipative systems*.

2.1. Continuous Macroscopic Physical Systems

Reductionism, when applied to continuous macroscopic systems, recursively splits the physical system into smaller volumes. The dividing process stops once one reaches a "suitable" granularity. The resulting base entities are three-dimensional dynamical systems that live in four-dimensional spacetime. Partial differential equations describe their behaviour and the applicable fundamental conservation principles. They describe the evolution of the state of the modelled smallest volume; the identified entity, namely, the state change is the consequence of the interaction with the environment. As a mechanism to establish the conservation and balance equations, one "walks" the surface of the volume and accounts for all interactions, and thus the conserved quantity crossing the boundary, which gives rise to the term "control volume". Overall, an accounting operation, that is based on the system and the conserved quantity. The assembly of extensive quantities then represents the fundamental state of the volume. The network of elementary control volumes, the base entities, and their interactions across common boundaries describe the behaviour of a modelled fundamental entity.

2.2. Particulate Physical Systems

In many cases, the modelling is not focused on a complexity described by a hierarchy of systems, but complexity arises from having many objects on the same level of granularity. Examples include molecular dynamics or large quantities of particles or models that approximate continuous systems with many particles, such as smoothed-particle hydrodynamics [4] or the like. One of the main problems with these systems is the formulation of boundary conditions and the forces acting between them. The fundamental entity in such systems is the particle.

Dynamic state equations describe the behaviour of the particles. With the particles' capacities, this results in ordinary differential equations, with population balances being a typical representation. As each particle has a state, the size of the equation set is an apparent computational problem.

2.3. Control Systems

Control is an enabling technology. It allows for steering and maintaining the state of the process. Processes must be driven; there must be driving forces acting on the process to keep it from its natural steady state. Thus, processes must be embedded in an environment that is not in equilibrium with the process. The parts that make up the environment are also not in equilibrium with each other, thereby enabling the process to "run" like a water wheel between an elevated water reservoir, ejecting it to the lower-level reservoir. This Carnot-kind of viewpoint is very generic.

The manipulation of the flows between the different constituent parts of the system makes it possible to move the process into any place in the attainable region defined by the environment (Refers to controllability.). Thus, the state of the environment determines the attainable region, and it is the main controls that act on the flows between the environment and the system that control the overall state of the process.

A control system implements the externally provided objectives in terms of target values for states or state-dependent quantities. Analysing the ideal controller, namely, requesting that the process instantaneously follows the given setpoint, shows that the controller would have to invert the plant. Since the plant exhibits capacity behaviour, the inversion is not causal, and therefore not feasible. Consequently, all controllers implement an approximate inverse plant, and their states mirror the plant's states.

The nature of the equations depends on the nature of the controlled system. In the case of a sampled system, time-discrete difference equations are used, while for event-observed systems, an automaton is used, and thus also a difference equation. Continuous control implies analogue controllers, which are physical systems and may also be modelled in the physical domain.

2.4. Other Relevant Subjects

The simulation of technical and natural systems is often augmented with analytical tools. For technical systems, these may be techno-economical or ecological analytical tools, such as Life-Cycle Analysis or financial return. Statistical tools also belong to this class of extension.

3. Networks & Tokens

Overall, reductionism describes the modelled physical system as a network of interacting capacities that partially project onto the control, and all other relevant subjects or disciplines.

If we attach the term "discipline" to physics and control, it is natural to subdivide each discipline into more specific "subdisciplines". For physics, this leads to a "tree" of disciplines, which captures continuous, macroscopic, particulate, small-scale, atomic, etc. For control, the control pyramid is used, with the event-discrete layers on top, planning, scheduling, optimisation and sampled systems below, as well as model-predictive control, optimising control, unit-level control and low-level control.

The subdivision into disciplines and subdisciplines for physics is somewhat involved. Carrying out subdivision on a global level leads to a very complex and large framework. The members of the European Material Modelling Council are working on such a global classification, called the European Materials Modelling Ontology (EMMO) (For information, see https://emmc.info/emmo-info/, accessed on 27 March 2021 and for the ontology https://github.com/emmo-repo/EMMO, accessed on 27 March 2021). ProMo enables the definition of subtrees of the EMMO discipline classification, aiming at the construction of smaller EMMO-related ontologies, which are more focused on specific application domains. Also, ProMo aims to handle large-scale models and is not limited to materials, although it has a substantial section that is associated with materials.

3.1. *Higher-Level Abstraction*

3.1.1. Tokens

The multidisciplinary nature of the problem requires higher-level abstraction. The network is the first element which we lift by adding the concept "tokens". The domain-specific "tokens" are the items living in the specific "networks". The "networks", have directed graphs as node capacities for the "tokens", whilst the edges are transporting "tokens". The "tokens" within a capacity, a graph's node, define the state of the node.

For physical systems, we use the conserved quantities as "tokens", primarily mass, energy and momenta, but also items that make up a population. In control, the token is a signal, and a monetary measure in economic systems. Thus, the abstraction of a (sub)-discipline is a network of capacities for tokens communicating tokens.

3.1.2. Intra-Faces and Inter-Faces

The definition of networks and tokens provides the basis for the definition of the two types of interaction: "Intraface" is a communication path for tokens, while "interface" communicates state-dependent information.

The "intraface" thus couples similar networks, where *similar* implies that the two coupled "networks" contain the same "tokens". An "intraface" thus communicates "tokens". In contrast, if one couples two networks that are not similar, but have different "tokens", the link will transfer information, usually state-related information. To give an example, in the first case, one may transfer mass from a liquid to a gas phase. Mass is transferred through an "intraface". In contrast, for the second case, state information is passed to the controller through an "interface". The generated value for a manipulated variable in the physical system is the information given back to the physical system.

3.1.3. Nodes & Arcs

"Networks" are directed graphs. The "nodes" or vertices represent the ***capacities for the tokens***, and the "arcs" or edges represent the ***flow of tokens***. The directed arcs define a reference coordinate for each flow of tokens. Tokens of the same type may be grouped and transported by one arc.

Figure 1 shows all elements that make up the graph of graphs in a minimal example. The physical part of the plant, labelled with "plant", shows two networks, a liquid phase and a gas phase. Those two communicate with each other mass through a common "intraface". The liquid phase seeks material information from the material model, providing state information, typically species, pressure, temperature, and asking for a physical property, like density. Both channels operate through an "interface". The liquid phase also provides state information to a control network, again through "interfaces".

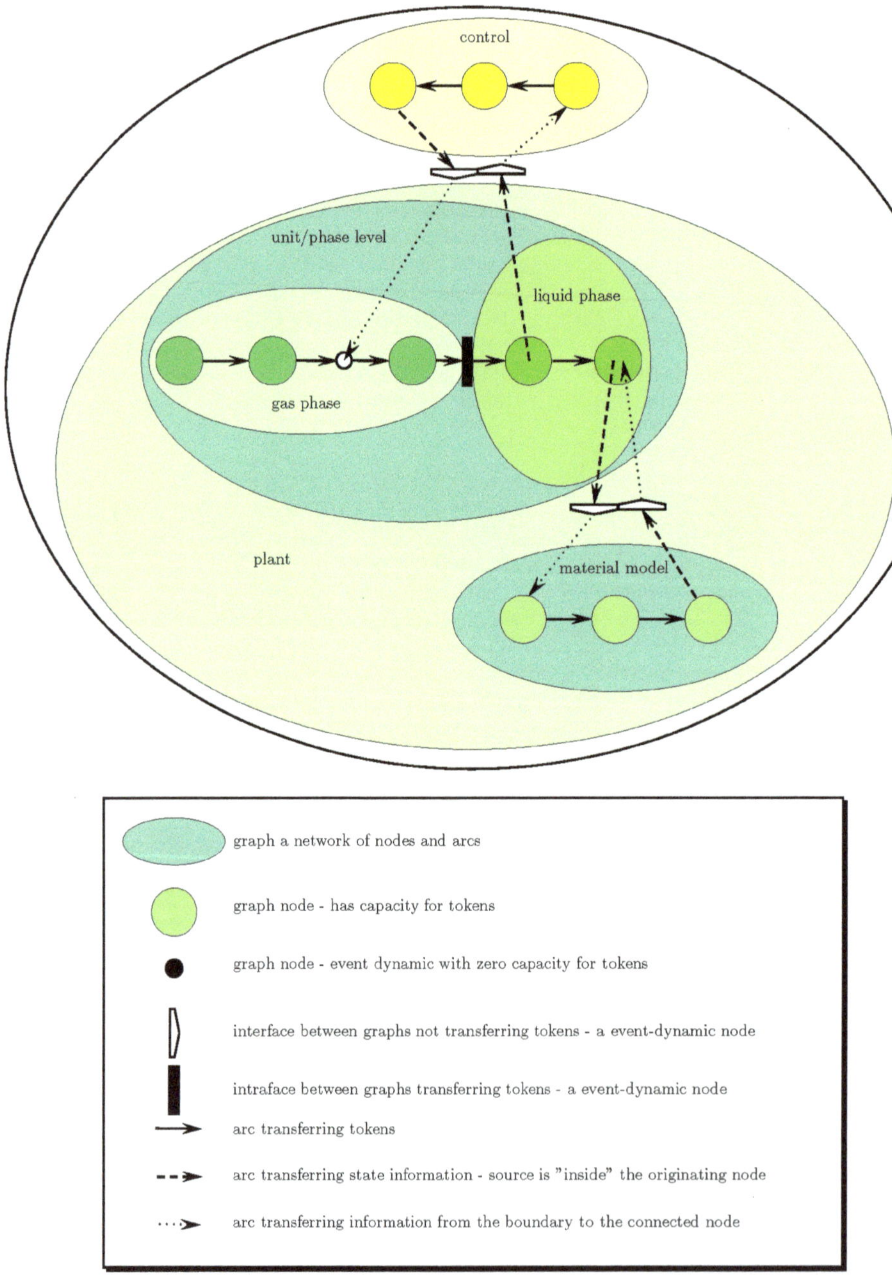

Figure 1. Network of networks representation of multi-disciplinary model graphs.

On the smallest scale, the "nodes" also represent basic entities, which result from applying reductionism.

3.2. Formal Definitions—A Summary

The discussion in this section yields the definition of a couple of items:

Network: *"directed graph" representing the structured modelled system on a given granularity level.*

Token: *"item" living in the "network".*

Node: *is a component of the "network", which exhibits the ability to store "tokens".*

Arc: *is the component of the "network" representing the transfer of "tokens" between "nodes".*

Basic or fundamental entity: *is the smallest building block required to establish models in a given domain – the smallest basic item in a hierarchical granular system (European Committee for Committee for Standardisation CEN defines the term* ***entity*** *in the Workshop agreement MODA [5]). In other words, a basic entity is context-dependent and defined as the smallest granule that describes the group of composite systems. It is the application of the resulting model that determines the granule resolution.*

Intraface: *communicates bidirectionaly tokens between nodes of two intra-related networks.*

Interface: *communicates unidirectionally state-related information from a node of one network to a node in the inter-related network.*

4. Basic Entities

"Entities" are discipline and application-specific, but also have some common properties, namely, the dependency on the free variables' time and the three spatial variables, and thus the four-dimensional spacetime.

4.1. Time Aspects

Any model of a dynamic system embraces three time scales: (i) "constant"—the parts that are not changing with time, (ii) "dynamic"—those parts that exhibit dynamic capacity behaviour, and (iii) "event-dynamic"—those parts that change instantaneously.

4.2. Spatial Aspects

For "physical entities", objects live in the four-dimensional spacetime. Continuous domains may then be categorised into classes.

Lumped systems: *are finite-dimensional volumes where the relevant intensive properties* ***are not*** *a function of the spatial coordinates, thus a spatial domain in which the relevant intensive properties are uniform in terms of the spatial distribution.*

In contrast:

Distributed systems: *are finite-dimensional volumes where the relevant intensive properties* ***are*** *a function of the spatial coordinates, thus a spatial domain in which the relevant intensive properties are not uniform in terms of the spatial distribution.*

Both "lumped systems" and "distributed systems" are basic building blocks (entities) when modelling macroscopic systems.

4.3. Deterministic, Stochastic & Ergodic

"Entities" may exhibit deterministic or stochastic behaviour. Given a deterministic input and a fixed entity, then the output is also deterministic. If any either the input or the entity's properties are stochastic, then the output will also be stochastic.

"Egodicity" is when the time average is identical to the state average of identical processes at one point in time, or in other terms: "all accessible microstates are "equiprobable" over a long period of "time" (Source wikipedia).

4.4. Continuous Physical Systems

The nature of "basic entities" is described above. The objective is to generate equations for numerical computations. Therefore, we need to define the input/output behaviour as a set of mathematical equations.

We create the equations on the background of system theory [6], leaning towards Willhelm's behaviour theory [7]. A state-space view serves the purpose of analysing and controlling the structure. The "state" takes a centre position. The definition of the term "state" is hard to trace, but one can find an early version in Caratheodory [8]: "... ein "Zustand" des Systems S, und wir wollen für die Zahlen x_i selbst den Namen "Zustandskoordinaten" einführen." We shall have a look at the two essential domains, namely, continuous systems and control.

4.4.1. Thermodynamic Systems

The domain of macroscopic, thermodynamic systems builds on the concepts of conserved mass, energy and momentum, while Newton's laws govern mechanical systems. Focusing on thermodynamic systems, one would typically also allow for the conversion of chemical or biological species, which, building on atomic mass conservation, gives rise to species balances.

Balances and conservations follow the same principle: one defines them by walking the system's boundary, accounting for all the quantities crossing the surface. The change occurring inside the system is then equated with the transfer, and, in the case of species conversion, it is augmented with the species' transposition, yielding a "species balance". Expressing these concepts verbally: "accumulation = net flow across the surface + net consumption". Conservation principles do not include the last term; it only appears as a consequence of allowing for conversion and inducing the atomic mass conservation, all of which turn into a "balance".

4.4.2. Mechanical Systems

Mechanical systems are handled in much the same way, except that one balances momenta, and the energy analysis focuses mainly on kinetic and potential contributions. However, fields other than gravitational are also considered depending on the application.

4.5. Particulate Physical Systems

In many cases, reductionism yields particles as the smallest entities required to capture the system behaviour. Molecular dynamics is one example, and fluid models building on particles is another. It is particles that characterise these systems, and momentum balances with forces, representing the momentum flow, form the core of the description.

4.6. A First Classification of Variables

The generic conservation/balancing operation enables us to define a first set of variable classes, namely, the classes "state", "transport", "conversion". The "accumlation" term is represented by the spacetime derivative of the state variables.

If one draws up the scheme defining the model, one quickly recognises that one part is missing. For these reasons, the missing piece is often called "closure", which, by their nature, are state variable transformations. Looking at a diagrammatic representation, Figure 2 shows the mathematical representation of a generic deterministic physical system, with $\underline{\mathbf{x}}$ being the "state" of the system in the form of a block diagram also showing the connections for a control system. The scheme has four external connections: the initial conditions $\underline{\mathbf{x}}(0)$, the boundary conditions $\underline{\mathbf{y}}_b$ and the two connections to the control system, namely, the manipulated variables $\underline{\mathbf{u}}$, the measurement $\underline{\mathbf{y}}$.

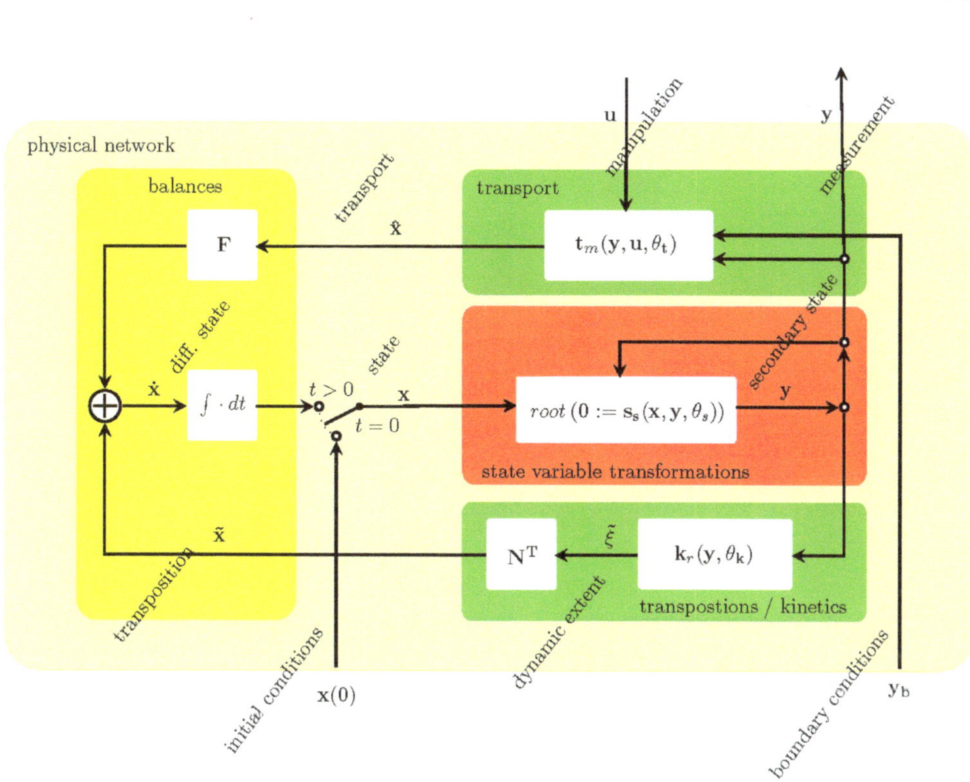

Figure 2. A generic dynamic physical system.

The dark-yellow box on the left indicates the balances and conservations. If the state is limited to the conserved quantities, then the yellow box is the conservation, and otherwise there is a balance. The green box on the top represents the transport laws for the extensive quantities, and the lower green box the reactions or transposition. Both are a function of a class of variables, which we call "secondary states". Typical members are the quantities driving the transport of extensive quantities, namely temperature, pressure and chemical potential, while concentration defines the probability function in reaction kinetics. It is the red box that closes the gap between the state and the secondary state. The red box's main components are thermodynamic models relating the extensive state to the intensive properties and geometrical relations that link geometrical variables to volume. The main physical–mathematical underlying frameworks are Hamiltonian systems for mechanical and contact geometry for the thermodynamic parts. Both define a configuration space. While classical mechanics is well established, contact geometry does not enjoy similar popularity. The energy formulation reads

$$U\left(S, V, \underline{\mathbf{n}}, \frac{\partial U}{\partial S}, \frac{\partial U}{\partial V}, \frac{\partial U}{\partial \underline{\mathbf{n}}}\right) \tag{1}$$

with the last three elements being the temperature T, the negative pressure p and the chemical potential μ. The thermodynamic configuration space is the assembly, including $U, S, V, \underline{\mathbf{n}}, \frac{\partial U}{\partial S}, \frac{\partial U}{\partial V}, \frac{\partial U}{\partial \underline{\mathbf{n}}}$. An early introduction can be found in [9], followed by [10–12], which provides an extensive exposition of the subject.

Processes are controlled. A single-layer control scheme may be of the form shown in Figure 3. Some of the variable classes are different from the ones in the physical system. There is still the state and the differential state. One now has to classify outputs and inputs, setpoints and control error. Figure 3 also shows the interfaces transferring state information in one direction, with control settings in the other direction, with the latter also being a function of the controller state via the output function.

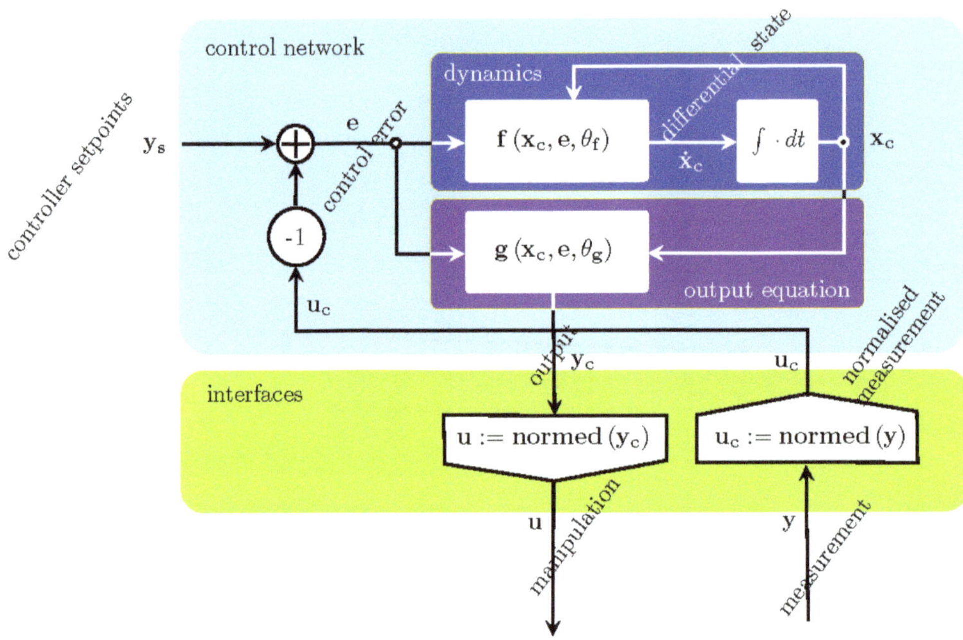

Figure 3. A generic control system.

The two examples demonstrate that the variable classes are domain-specific. While one can streamline things to some extent by using a state-space approach, one needs to adapt the variable classes to the disciplines to effectively support the process of defining the variable and equation system.

5. ProMo Ontology

The previous sections provide the main elements used to define the structure of the ProMo ontology. The top definition is the tree of disciplines. For each discipline, we must define elements associated with the structural elements enabling efficient handling of the building blocks and the elements that provide the framework for defining each building block's behavioural equations.

We do not aim at generating a global ontology for all processes. Even if this was a feasible undertaking, the argument here is the improved targeting of groups of modellers (see also the discussion in Section 3), at least not in the first instance. Another argument is flexibility. An ontology is a living structure and will change with time, continuously aligning with new requirements. It is for these reasons that ProMo has its own ontology structure and a corresponding editor.

The organisation of the disciplines follows a couple of simple rules:

- Networks that house the same tokens are in an intra-relation;
- Networks that do not house the same tokens are in an inter-relation;
- Each discipline has its own branch;
- Disciplines may have subdisciplines.

The sample ontology Figure 4 shows the discipline tree of an ontology designed for modelling a large group of macroscopic processes. The green disciplines are in an **inter-relation**, while the blue ones are in an **intra-relation**. On the top level, the ontology includes a physical and a control branch. We have the generic macroscopic systems, the material descriptions, and an example of a mixing domain devoted to empirical mixing models on the physical level.

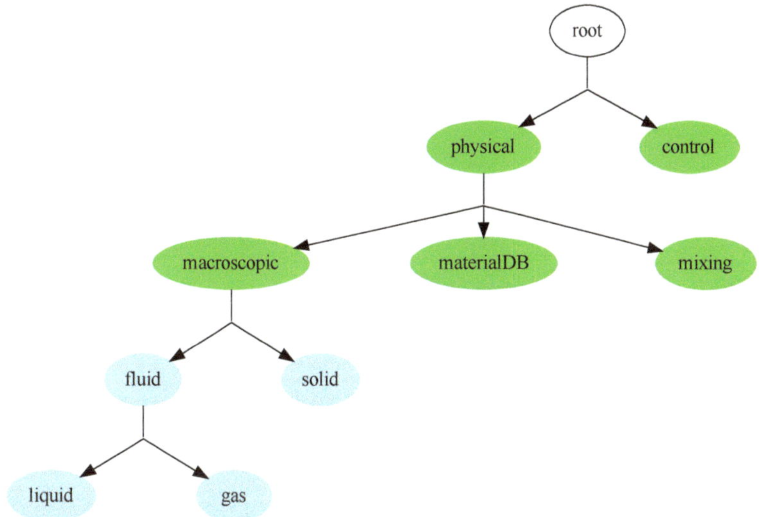

Figure 4. A sample discipline tree. It is designed to capture a very wide range of processing plant models. The green nodes are in an inter-relation, and thus transfer tokens, while the blue ones are in an intra-relation, and thus transfer state information.

Each node has two branches of definitions: the branch associated with defining the process models' structure and the branch with the definition of the variable classes. Figure 5 shows the definition of the root of the discipline tree.

The overall system is defined as dynamic with basic entities belonging to all three time-scale domains. For the representation of the network structures, we declare the variable class "network" and "projection". Both are associated with the "graph", the network. For the system theoretical definitions of the model equations, one defines the variable classes "frame" for the free variables, "state" and "constant". The two additional classes are currently explicitly defined, namely, classes associated with the spaces generated by differentiating the states and frame variables. These definitions are required because of the design choice of uniquely identifying the dimensions of all mathematical objects. The issue of indexing has been introduced and discussed in [1].

Moving down the discipline tree, one can extend and augment the definitions. Figure 6 shows the example of the node labelled with "physical". Thid is the parent to the nodes "macroscopic", "materialDB" and "mixing". It augments variable classes with "secondary state" and "effort". The former appears in the generic representation of the dynamic physical system, Figure 2, while the term "effort" is a class that captures the effort variables [13], also called the conjugates to the thermodynamic potentials or generalised forces. In the context of contact geometry Equation (1), the temperature, the pressure and the chemical potential are three elements of the odd configuration space [9,10].

The structure includes the extension of the nodes, which applies to all nodes in the tree. It captures the time-scale property of capacitive elements and details it with the distribution effect.

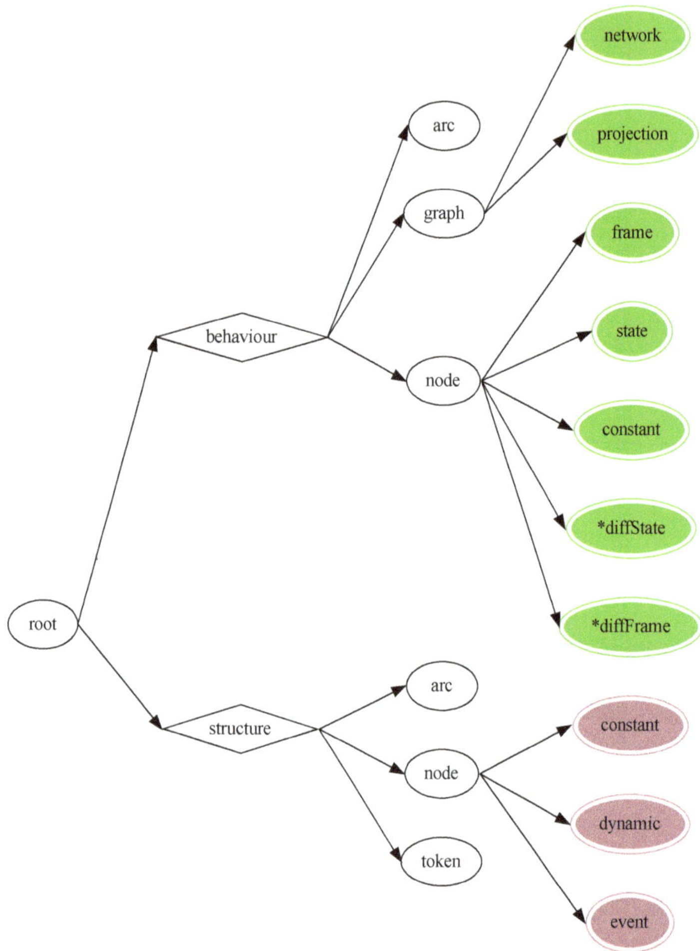

Figure 5. The **root** node in the discipline tree. It has two branches: the structure branch is used to define the ontology items used for designing the items associated with capturing the model as a directed graph; the behaviour branch serves the purpose of defining the variable classes used to capture the mathematical description of the entity behaviour models.

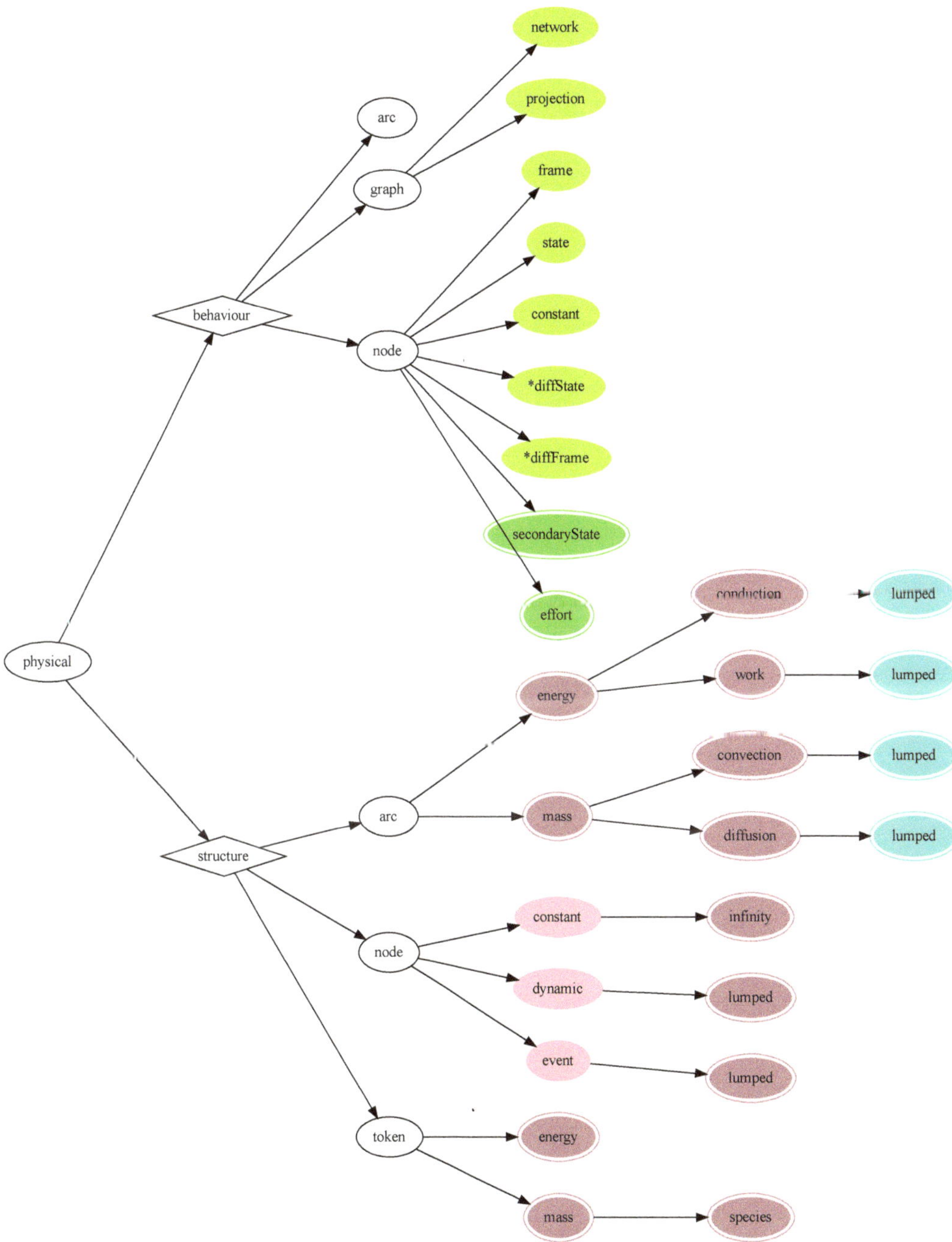

Figure 6. The physical node in the discipline tree. Both branches are expanded with additional items. In the structure branch, the basic entities are populated with tokens and specified with the modelled distribution effects.

6. Entity Behaviour

6.1. Equations

The ontology intrinsically defines the basic entities in the discipline-specific frame. For physical systems, this is the frame spacetime, while for control, the frame is usually limited to time.

ProMo's equation editor can enter and edit the mathematical behaviour description. As mentioned above, we start with a set of "port" variables: the constants, the frame variables, and the variables defining the configuration spaces' base. We define a small, tailored language to enter the equations, namely, a variable defined by an expression. The description of the language and the mechanisms associated with implementing the details is beyond this paper and will have to be reported separately. Further information can be found in [3]. Thus, in very brief terms: The equation editor implements a parser and a template machine. The parser generates an abstract syntax tree, and the template machine uses the abstract syntax tree and generates different compiled versions of the equations, including LaTex rendering of online documentation. The main features of the parser are that it implements the index (For the discussion on the indexing, we refer to [1]) and rigorous unit handling.

The variable/expression bipartite multi-graph: We construct the equation system starting with the "port" variables: frame, constants and the state variables. For physical systems, the base state variables are the ones defining the configuration spaces. The state variables appear in the block diagrams Figures 2 and 3 after the integral. We then follow the paths in the respective block diagram. For physical systems, shown in Figure 2, the next step is to define the equations representing the red box, labelled with " state variable transformations", whereby each equation can only be a function of already defined variables. For the control system, these are the nonlinear dynamic function and the nonlinear static output function.

We allow for the definition of more than one equation for the same variable. This approach allows us to implement the basic thermodynamic variables, for example, temperature as a partial derivative of internal energy with respect to entropy. Variables with an implicit model equation must first be defined with an explicit model equation. Only then can the implicit model be added. This enables the proper handling of indexing and unit handling. Further details can be found in [3].

6.2. Graphical

The ontology defines the base entities, and the equations describe the behaviour of these entities. For the physical systems, we have combinations of the time-scale behaviour, the distribution effects and the present tokens. By linking the mathematical behaviour description to graphical symbols, we generate a visual modelling language, to which the remaining paper shall be devoted. The graphical language uses a small number of visual symbols. For each discipline, a few capacity components and a few transport components remain to be defined. Appendix A shows a table of symbols for macroscopic physical systems and control.

The structure of the model, in terms of its composition of basic building blocks, is the main model-design issue, and not the equations. We can define the basic blocks' behaviour, except for the specific material models, which have to be injected at the initialisation stage. A very elegant solution has been suggested and realised by Bjørn-Tore Løvfall's PhD documented in his thesis [14]. He constructs one of the energy functions, usually Helmholtz, from two state equations, which serve as models of the specific material's behaviour. Legendre transformation provides the additional energy funtions. The properties then are defined by derivatives of the surface with respect to state-dependent variables. Automatic differentiation makes the sytem tick.

The graphical language is an excellent model design tool. We not only teach it in our modelling course, we also use it extensively in projects to discuss the model structure. We shall now discuss a few examples.

7. Graphical Model-Design—Three Examples

This section discusses three examples demonstrating the power of the graphical model design and their use in ProMo's model composer. An explanation for graphical symbols is in Appendix A.

7.1. Stirred Jacketed Tank

First, we look at a standard piece of equipment, the stirred tank reactor. Figure 7 shows a possible configuration.

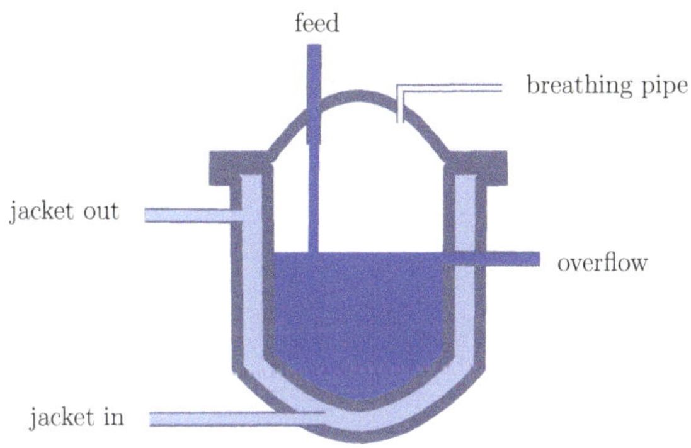

Figure 7. The configuration of a jacket tank with an inflow, an overflow and a breathing pipe for pressure compensation.

The tank is connected to one liquid feed, has an overflow, and is "breathing" to the outside. The jacket controls the temperature of the contents. Figure 8 shows a relatively detailed model for energy flows but a relatively simple model for the reaction fluid.

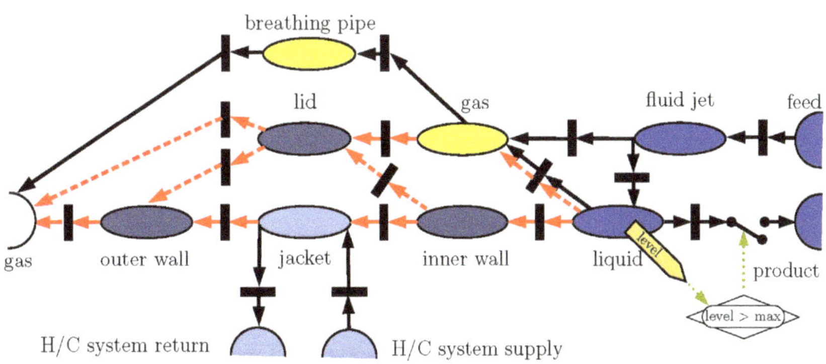

Figure 8. A relatively complex model without condensation on the lid.

The overflow is of interest: an imaginary controller switches the overflow on when the level in the tank has reached the maximum. In the following examples, we use this approach of "control" to handle events that change the configuration of the model. The controller shown in Figure 8 is not real, but implements the event of reaching the fluid level in the tank when the overflow begins to become active. The graphical language also enables model reduction or model simplification. Different applications require different details, and rougher or finer granularity. We can apply a series of assumptions:

- The shell is well insulated from the room and heat loss through the outer shell is insignificant;
- The fluid jet is very fast;
- The lid is well insulated from the lower part of the tank;
- The lid, gas and fluid are at about the same temperature as the room. The heat loss through the lid is insignificant;
- The inner wall conducts very well;
- The material exchange of the gas phase with the room is insignificant;
- The fluid jet is event-dynamic.

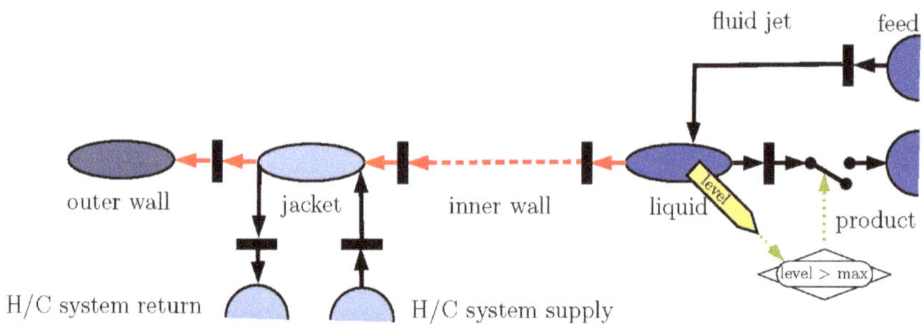

Figure 9. A simplified tank model. No heat loss through the outer shell, lid and gas phase; "breathing" is neglected, jet is event-dynamic.

This helps us to achieve a significant simplification, shown in Figure 9.

One can simplify the model further by assuming:

- The outer shell has a negligible capacity;
- The contents is ideally mixed—thus, mixing is very fast, resulting in uniform conditions;
- The fluid flow in the jacket is extremely fast—yielding uniform temperature in the jacket.

This yields a much more simple model, as shown in (Figure 10).

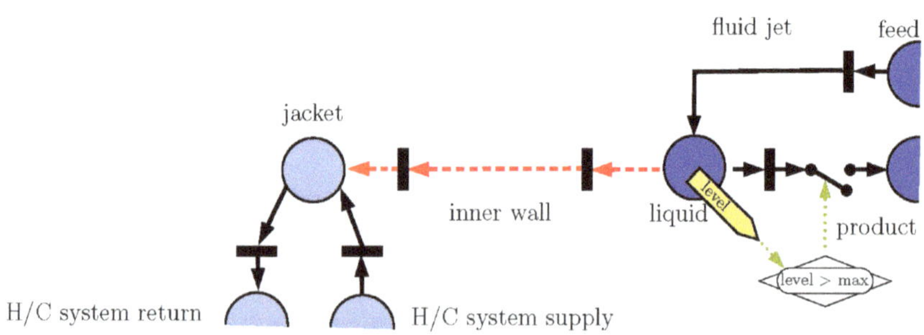

Figure 10. An even more simplified tank model. The outer wall is neglected and jacket, and the contents are lumped.

7.2. Melting Process

This second example focuses on "model control". It shows the melting of a solid as shown in Figure 11. The process itself is rather straightforward: a solid is exposed to a heat source and heated up. Once the melting temperature at the hot surface has been reached, a liquid phase is generated. The model switches to describe two phases, assuming that the solid has no direct contact with the hot surface, and a liquid film is in between the solid and the surface. The liquid's thermal distribution effect can be quite complex, and is not shown at this level of description.

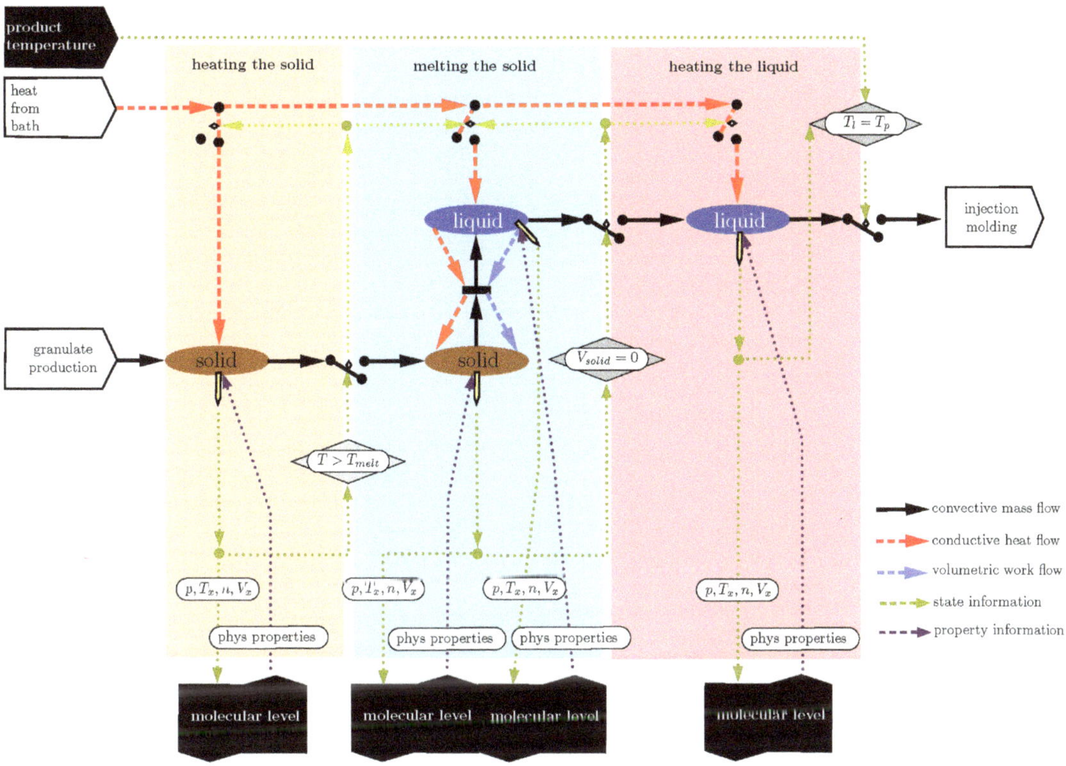

Figure 11. A generic melting process demonstrating how the model switching is included in the model, leading to a surprisingly complicated overall model.

The topology indicates a connection to molecular modelling, which, as shown, implies close coupling, thus running the molecular modelling task at every point in time. Molecular codes are notoriously computationally intensive and slow. There is at least one unique method, COSMO [15], which works for some configurations well and is fast. Close coupling is not an option for the codes that minimise a vast number of molecules' overall energy. In these cases, the approach used replaces the molecular modelling module with a surrogate model that spans the variable space sufficiently and is of a simple structure. The molecular modelling task is replaced by an input/output function with the property function in a subspace of the fundamental variables, such as p, T, n.

7.3. Moving Boundary Problem

Many processes are characterised by one part of the material growing at the cost of another. We take the example of a corrosion process: an iron-rod-reinforced concrete pillar in water. The problem is well known and of broad interest. Water, carbon dioxide, chlorine anions and oxygen mainly diffuse into the concrete and cause the iron to oxidise, resulting in a loss of strength over time. Since it is not the reaction that is of interest, Figure 12 shows a model representing the process with a simplistic abstract reaction. The main issue, in this case, is to demonstrate the approach used to model moving boundaries.

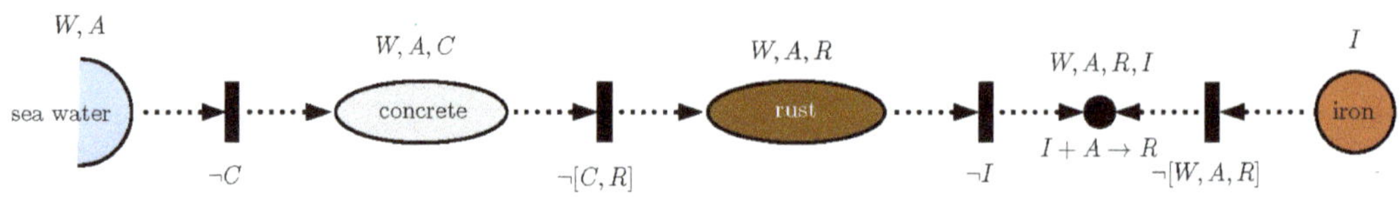

Figure 12. Rusting iron in concrete is an example of a moving boundary problem. The top row shows the species present in the respective capacities, while, in the lower row, the transfer constraints for the intrafaces and the reactions in the point capacity are shown.

The reaction is placed into an infinitely small reaction front, represented by an infinitely small volume where the reaction occurs, and the point capacity is combined with restrictive intrafaces to the left and right. The rust is transported to the left, while iron comes from the right. The species water, active component and rust do not transfer into the iron $\neg[W, A, R]$, and iron is not transferred to the rust $\neg[I]$. Rust and concrete are not transferred between the rust and the concrete domain, and concrete is not transferred into the water.

The Figures 13 and 14 show two pictures taken from ProMo's model composer. Figure 13 shows the top layer, while Figure 14 shows the model of the pillar with the iron embedded in concrete and the reaction front receiving iron from the iron bar, converting it into rust, which is transported to the rust node, representing the rust capacity.

The ProMo software uses simple graphical objects limited to ellipses and rectangles for visual representation. Decorators are used to indicate the membership to networks—liquid and solid in the example—in the form of circles. The membership to specialised networks, like concrete, rust or iron, is indicated by coloured circles. ProMo provides an editor that allows the user to design the graphical object. Arcs are directional, with the tail as a small circle and the head a larger and darker circle. Intrafaces are the black squares.

Figure 13. Rusting iron in concrete, a view of the ProMo's graphical modeller showing the top of the hierarchy with the pillar in the seawater. The seawater is modelled as a reservoir system (large mauve circle) with two circular indicators attached, as seawater is a liquid. The black dot indicates mass, and the "W" and "A" stand for the species water and active component.

7.4. Other Applications

The reader may be interested to learn that we used this graphical approach for very many different processes. In process modelling lectures, we discuss tanks, mixing models, heat exchangers, distillation, chicken coop and greenhouses and wood drying and fruit transport, life-support systems, fermentation processes, biological cells, water treatment plants, solar reactor, mammal blood flow, moving bed reactor, decaffeination plant, a methanol plant, laboratory equipment like Soxhlet, mokka maker, crystaliser, bubble column, etc.

We used the same approach in European projects, including the modelling of polyurethane foams, production of high-precision ceramic products, biorefining processes, wastewater cleaning plants, transport of fruit and vegetables, life-support systems for space travelling, membrane processes, catalytic bed reactors, coating processes, etc.

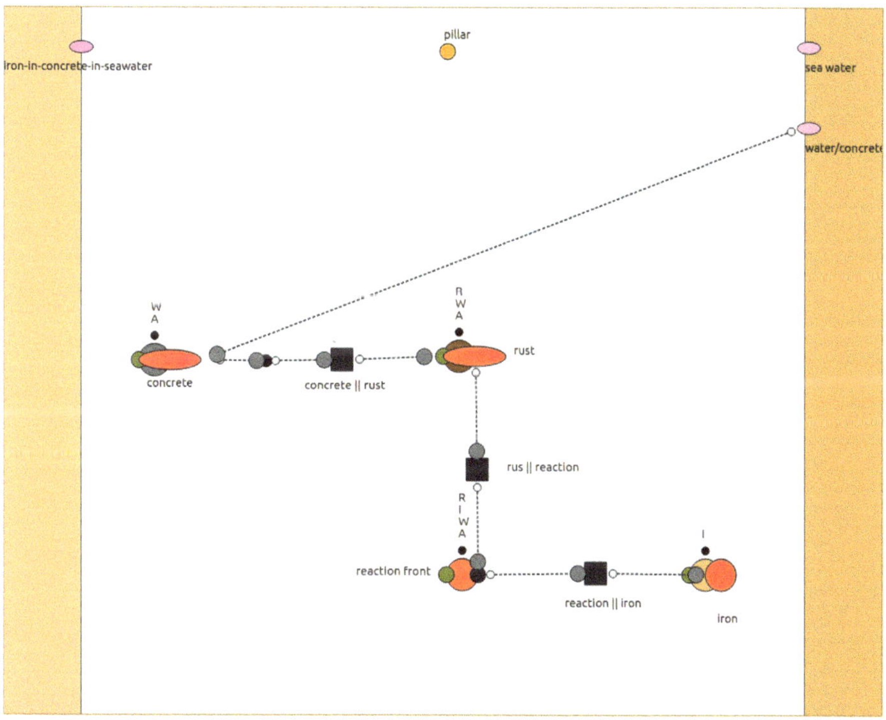

Figure 14. Rusting iron in concrete, a view of the ProMo's graphical modeller showing the pillar model. The concrete and rust are modelled as distributed systems (ellipses). The reaction front is modelled as an infinitely small lumped system (black dot), decorated with the large red dot indicating the reaction front and the green dot indicating the solid. The iron is modelled as a lumped system. "R", "I" stand for the species rust and iron. The left strip elements show the parents, while the ones on the right are the siblings in the hierarchy.

8. Conclusions

Motivated by reductionism, models are seen as directed "networks", where abstract "tokens" live in nodes and move about through arcs. The concept of "tokens" allows us to apply the network concept to different disciplines, thereby enabling the multi-disciplinary and multi-scale model building process. The nodes are then capacities for the tokens, and the arcs' transport tokens.

For physics, the tokens are first the conserved quantities that form the basis of the Hamiltonian's and the contact geometry's configuration space. Since we also model reactive systems, the tokens are extended to refining mass with species, consequently augmenting the representation with balances. This approach thus captures the physical system on all levels, including electronic and atomic, and particulate and macroscopic systems. For some particulate processes, the behaviour description is augmented with population balances. For financial systems, the tokens are monetary values. The node represents an account, with the monetary value being the state, and interest plays the role of a production term similar to the role that reactions take in component mass balances.

Reductionism is applied recursively, resulting in a hierarchical representation of the modelled process. The subdivision is continued until a basic level is reached, where "basic" implies that it can be considered a basic building block. The definition of "basic" is the smallest granules in the decomposition process, such that the resulting model serves the intended purpose.

ProMo implements an ontology, which captures the knowledge of a user-defined application domain, which, in turn, reflects the hierarchy of disciplines considered for the application domain. For each discipline, the ontology defines the infrastructure for the definition of the model structure and the behaviour description, with the latter being a multi-bipartite graph of variables and their defining relations. This bipartite graph captures the behaviour of each basic building block for all involved disciplines. The variable/expression set is lower triangular in the sense that one begins by defining the state that reflects the relevant tokens in a node. We allow for multiple definitions of the variables, thereby supporting the use of principle definition equations as well as more practical versions, which ultimately enables the substitution of a complex with a simple model, also termed the "surrogate".

The ProMo suite resolves a number of issues in process modelling:

Incompleteness and consistancy: The systematic approach of constructing the variable/expression system guarantees the completeness of the model equations. The fundamental base used to describe the entity's behaviour models is tiny, namely, the thermodynamic and mechanical system's configuration space. Physical units and dimensionality is defined only for the fundamental quantities. They automatically propagate when defining new variables.

Code generation: is automated and does not require manual intervention. The entity behaviour code is centralised and not distributed over unit models, as this is commonly the case in chemical engineering software.

Documentation: The expressions defining a variable are compiled as part of the definition process. Consequently, the documentation is complete and available during definition time.

Closure: Defining the fundamental behaviour equations as the conservation equations has two main advantages: (i) the equations are not substituted, and (ii) the closure of the balances can be guaranteed for the defined accuracy.

The graphical model-design language: enables us to capture the structure of the modelled process and generate alternatives quickly. The simplicity makes it an efficient discussion tool for exploring functionality, required detail, involved disciplines, and complexity to generate alternatives.

Funding: ProMo research was funded in part by: (i) Bio4Fuels Research Council of Norway (RCN) project 257622 (ii) MARKETPLACE H2020-NMBP-25-2017 project 760173. (iii) VIPCOAT H2020-NMBP-TO-IND-2020 project 952903T (iv) MODENA FP7-NMP- Specific Programme "cooperation": Nanosciences, Nanotechnologies, materials and new Product Technologies Grant agreement ID: 604271.

Acknowledgments: I gratefully acknowledge the contributions of the doctoral students who worked on this project over the years: Tae-Yeong Lee (TAMU), Mathieu Westerweele (TUE), Sigve Karolius (NTNU), Arne Tobias Elve (NTNU), and Robert Pujan (NTNU, DBFZ) and several master students and PostDoc Niloufar Abtahi.

Conflicts of Interest: The author declare no conflict of interest.

Appendix A. Graphical Symbols

The number of graphical items required to discuss controlled, physical processes is small. The following set was sufficient in the cases where we applied the method to applications including basic units: tanks, mixing models, heat exchangers, distillation, chicken coop and greenhouses and wood drying and fruit transport, life-support systems, fermentation processes, biological cells, water treatment plants, solar reactor, mammal blood flow, moving bed reactor, decaffeination plant, methanol plant, laboratory equipment like Soxhlet, mokka maker, crystaliser, bubble column, polyurethane foam, pressure distributions in plants, molecular modelling of polymer/ceramic powder mix, melting processes, and more.

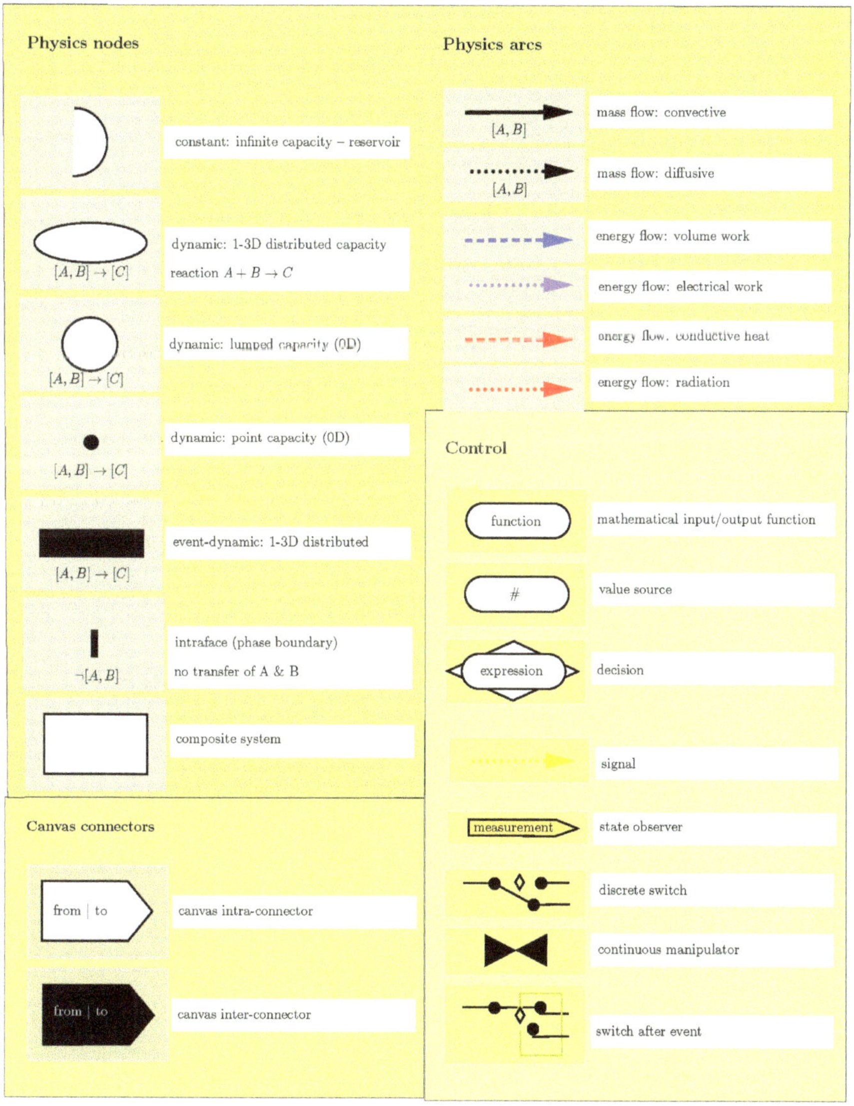

Figure A1. A starter set of graphical symbol defining the graphical modelling language.

References

1. Preisig, H.A. Constructing and maintaining proper process models. *Comput. Chem. Eng.* **2010**, *34*, 1543–1555. [CrossRef]
2. Harold, M.P.; Ostermaier, J.J.; Drew, D.W.; Lerou, J.J.; Luss, D. The continuoulsy stirred decanting reactor: Stead state and dynamic features. *Chem. Eng. Sci.* **1996**, *51*, 1777–1786. [CrossRef]
3. Elve, A.T.; Preisig, H.A. From ontology to executable program code. *Comput. Chem. Eng.* **2018**, [CrossRef]
4. Lind, S.; Rogers, B.; Stansby, P. Review of smoothed particle hydrodynamics: Towards converged Lagrangian flow modelling. *Proc. R. Soc. A* **2020**, *476*. [CrossRef] [PubMed]
5. CEN Workshop Committee. *CEN/WS MODA-Materials Modelling-Terminology, Classification and Metadata (CWA 17284)*; CEN Workshop Committee: Brussels, Belgium, 2018.
6. Bernhard, P.; Deschamps, M. Kalman 1960: The birth of modern system theory. *Math. Popul. Stud.* **2019**, *26*, 123–145. [CrossRef]
7. Willems, J.C. The behavioral approach to systems theory. In Proceedings of the Conference on New Approaches in System and Control Theory, Istanbul, Turkey, 2005; pp. 59–71. Available online: https://homes.esat.kuleuven.be/~sistawww/smc/jwillems/Articles/ConferenceArticles/2005/7.pdf (accessed on 17 March 2021).
8. Caratheodory, C. Untersuchung ueber die Grundlagen der Thermodynamik. *Math. Ann.* **1909**, *67*, 355–386. [CrossRef]
9. Peterson, M.A. Analogy betwween thermodynamics and mechanics. *Am. J. Phys.* **1979**, *47*, 488–490. [CrossRef]
10. Rajeev, S.G. A Hamilton–Jacobi formalism for thermodynamics. *Ann. Phys.* **2007**, *323*, 2265–2285. [CrossRef]
11. Bravetti, A. Contact Hamiltonian Dynamics: The Conceptand Its Use. *Entropy* **2017**, *19*, 535. [CrossRef]
12. Bravetti, A. Contact geometry and thermodynamics. *Int. J. Geom. Methods Mod. Phys.* **2019**, *16*. [CrossRef]
13. Breedveld, P.C.; Rosenberg, R.C.; Zhou, T. Bibliography of Bond Graph Theory and Application. *J. Frankl. Inst.* **1991**, *328*, 1067–1109. [CrossRef]
14. Løvfall, B.T. Computer Realisation of Thermodynamic Models Using Algebraic Objects. Ph.D. Thesis, NTNU, Trondheim, Norway, 2008.
15. Klamt, A. The COSMO and COSMO-RS solvation models. *WIREs Comput. Mol. Sci.* **2018**, *8*, e1338, doi:10.1002/wcms.1338. [CrossRef]

processes

MDPI

Article

Algorithmic Approaches to Inventory Management Optimization

Hector D. Perez, Christian D. Hubbs, Can Li and Ignacio E. Grossmann *

Department of Chemical Engineering, Carnegie Mellon University, Pittsburgh, PA 15123, USA; hperezpa@andrew.cmu.edu (H.D.P.); chubbs@andrew.cmu.edu (C.D.H.); canl1@andrew.cmu.edu (C.L.)
* Correspondence: grossmann@cmu.edu; Tel.: +1-412-268-3642

Abstract: An inventory management problem is addressed for a make-to-order supply chain that has inventory holding and/or manufacturing locations at each node. The lead times between nodes and production capacity limits are heterogeneous across the network. This study focuses on a single product, a multi-period centralized system in which a retailer is subject to an uncertain stationary consumer demand at each time period. Two sales scenarios are considered for any unfulfilled demand: backlogging or lost sales. The daily inventory replenishment requests from immediate suppliers throughout the network are modeled and optimized using three different approaches: (1) deterministic linear programming, (2) multi-stage stochastic linear programming, and (3) reinforcement learning. The performance of the three methods is compared and contrasted in terms of profit (reward), service level, and inventory profiles throughout the supply chain. The proposed optimization strategies are tested in a stochastic simulation environment that was built upon the open-source OR-Gym Python package. The results indicate that, of the three approaches, stochastic modeling yields the largest increase in profit, whereas reinforcement learning creates more balanced inventory policies that would potentially respond well to network disruptions. Furthermore, deterministic models perform well in determining dynamic reorder policies that are comparable to reinforcement learning in terms of their profitability.

Keywords: inventory management; supply chain; multi-echelon; stochastic programming; reinforcement learning

Citation: Perez, H.D.; Hubbs, C.D.; Li, C.; Grossmann, I.E. Algorithmic Approaches to Inventory Management Optimization. *Processes* **2021**, *9*, 102. https://doi.org/10.3390/pr9010102

Received: 1 December 2020
Accepted: 30 December 2020
Published: 6 January 2021

1. Introduction

Modern supply chains are complex systems that interconnect the globe. Efficient supply chains are able to control costs and ensure delivery to customers with minimal delays and interruptions. Inventory management is a key component in achieving these goals. Higher inventory levels allow for suppliers to maintain better customer service levels, but they come at a higher cost, which often gets passed on to their customers and, ultimately, to the end consumers. This is particularly the case for perishable items that have a limited shelf life and can go to waste if the inventory exceeds demand. Thus, every participant in the supply chain has an incentive to find the appropriate balance in inventory levels to maximize profitability and maintain market competitiveness. Efficient supply chains are able to coordinate material flows amongst its different stages to avoid the "bullwhip effect", whereby over corrections can lead to a cascading rise or fall in inventory, having a detrimental impact on the supply chain costs and performance [1].

Extensive literature exists in supply chain and inventory management. Relevant review papers in the area of inventory optimization include those of Eruguz et al. [2] and Simchi-Levi and Zhao [3]. The inventory management problem (IMP) that is presented in this work is built upon the problem structure presented in Glasserman and Tayur [4], which presents a single-product, multi-period, serial capacitated supply chain with production and inventory holding locations at each echelon. In their work, Glasserman and Tayur [4] use infinitesimal perturbation analysis (IPA) in order to determine optimal base stock levels in an order-up-to policy by optimizing over a sample path of the system.

Other approaches for solving the IMP have been reported in the literature. Chu et al. [5] use agent-based simulation-optimization on a multi-echelon system with an (r, Q) inventory policy. Expectations are determined via Monte Carlo simulation. Improvements are only accepted after passing a hypothesis test to mitigate the effect of noise on the improvement. Two-stage stochastic programming (2SSP) is used to optimize small supply chains in the works by Dillon et al. [6], Fattahi et al. [7], and Pauls-Worm et al. [8]. The studied supply chains are either single or two-echelon chains with centralized or decentralized configurations, a single perishable or unperishable product, and (r, S) or (s, S) policies. Zahiri et al. [9] present a multi-stage stochastic program (MSSP) for a four-level blood supply network with uncertain donation and demand. The model is reformulated and solved while using a hybrid multi-objective meta-heuristic. Bertsimas and Thiele [10] apply robust optimization to both uncapacitated and capacitated IMP. However, production capacity is not explicitly included. Their models are solved with linear programming (LP) or mixed-integer linear programming (MILP), depending on the usage of fixed costs. The reader is referred to Govindan and Cheng [11] for a review of robust optimization and stochastic programming approaches to supply chain planning.

Additionally, there have been a number of efforts to optimize multi-echelon supply chain problems via dynamic programming (DP). A neuro-dynamic programming approach was developed by Roy et al. [12] in order to solve a two-stage inventory optimization problem under demand uncertainty to reduce costs by 10% over the benchmarked heuristics. Kleywegt et al. [13] formulate a vendor managed inventory routing problem as a Markov Decision Process (MDP) and develop an approximate dynamic programming (ADP) method to solve it. Topaloglu and Kunnumkal [14] develop a Lagrangian relaxation-based ADP to a single-product, multi-site system to manage inventory for the network that outperforms a linear programming method used in the benchmark. Kunnumkal and Topaloglu [15] use ADP to develop stochastic approximation methods to compute optimal base-stock levels across three varieties of inventory management problems: a multi-period news vendor problem with backlogs and lost sales, and an inventory purchasing problem with uncertain pricing. Cimen and Kirkbride [16] apply ADP to a multi-factory and multi-product inventory management problem with process flexibility. They find that, in most scenarios, the ADP approach finds a policy within 1% of the optimal DP solution in approximately 25% of the time. Additional resources on supply chain management with DP and ADP is provided by Sarimveis et al. [17].

Reinforcement learning has also been applied to IMPs in recent years. Mortazavi et al. [18] use Q-learning for a four-echelon IMP with a 12 week cycle and non-stationary demand. Oroojlooyjadid et al. [19] train a Deep Q-Network in order to play the Beer Game—a classic example of a multi-echelon IMP—and achieve near optimal results. Kara and Dogan [20] use Q-learning and SARSA to learn stock-based replenishment policies for an IMP with perishable goods. Sultana et al. [21] use a hierarchical RL model to learn re-order policies for a two-level multi-product IMP with a warehouse and three retailers.

We extend the problem in Glasserman and Tayur [4] to general supply networks with tree topologies. Our focus is not on finding optimal parameters for static inventory policies, but rather to determine and compare different dynamic policy approaches to the IMP. We build on the previous works in the literature by exploring the IMP while using different approaches and discuss their relative merits and drawbacks. The approaches studied include

1. A deterministic linear programming model (DLP) that uses either the rolling horizon or shrinking horizon technique in order to determine optimal re-order quantities for each time period at each node in the supply network. Customer demand is modeled at its expectation value throughout the rolling/shrinking horizon time window.

2. A multi-stage stochastic program (MSSP) with a simplified scenario tree, as described in Section 2.7. Shrinking and rolling horizon for the MSSP model are both implemented to decide the reorder quantity at each time period.
3. A reinforcement learning model (RL) that makes re-order decisions based on the current state of the entire network.

We build off of the work of Hubbs et al. [22] by extending the IMPs presented therein in order to address multi-echelon problems with multiple suppliers at each echelon, and contribute new environments to the open-source OR-Gym project (See https://www.github.com/hubbs5/or-gym). The intial version of the IMP in the OR-Gym project was limited to serial multi-echelon systems and it did not include multi-stage stochastic programming models for reorder policy optimization. The library was thus generalized in order to simulate and optimize supply networks with tree topologies under uncertain demand, while using the dynamic reorder policies mentioned above.

2. Materials and Methods

2.1. Problem Statement

In this work, we focus on the multi-echelon, multi-period, single-product, and single-market inventory management problem (IMP) in a make-to-order supply network with uncertain stationary demand. The base case supply network has a tree topology with four echelons, as shown in Figure 1. The different sets that are used for the nodes in the base case network are designated in the figure's legend (raw material, J^{raw}; main, J; retail, J^{retail}; distributor, J^{dist}; producer, J^{prod}; and, market nodes, J^{market}).

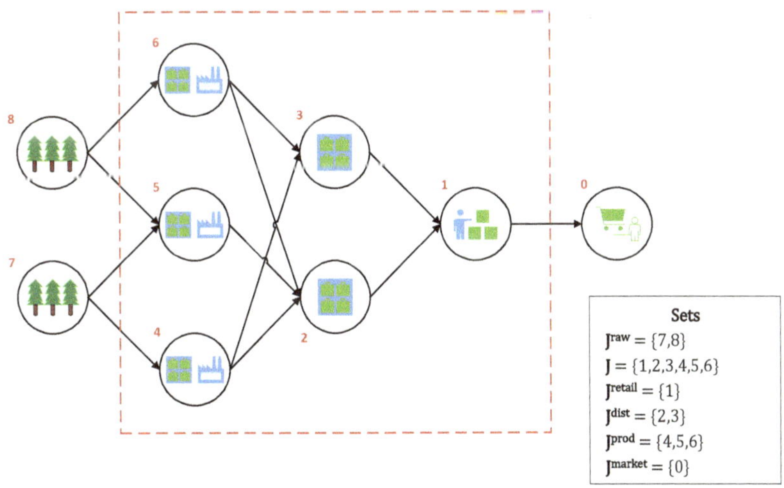

Figure 1. Supply Chain Network Schematic.

2.2. Sequence of Events

The sequence of events in each period of the IMP simulation environment occurs, as follows,

1. Main network nodes (retailer, distributors, and producers) place replenishment orders to their respective suppliers. Replenishment orders are filled according to available production capacity and available feedstock inventory at the respective suppliers. The supply network is assumed to be centralized, such that replenishment orders never exceed what can be provided by the suppliers to each node.
2. The main network nodes receive incoming feedstock inventory replenishment shipments that have made it down the product pipeline (after the associated lead times

have transpired). The lead times between stages include both production times and transportation times.

3. Single-product customer demand occurs at the retail node and it is filled according to the available inventory at that stage.
4. One of the following occurs at the retailer node,
   (a) Unfulfilled sales are backlogged at a penalty. Backlogged sales take priority in the following period.
   (b) Unfulfilled sales are lost and a goodwill loss penalty is levied.
5. Surplus inventory is held at each node at a holding cost. Inventory holding capacity limits are not included in the present formulation, but they can be easily added to the model, if needed. The IMP that is presented here is capacitated in the sense that manufacturing at production nodes is limited by both the production capacity and the availability of feedstock inventory at each node. Because the supply network operates as a make-to-order system, only feedstock inventories are held at the nodes. All of the product inventory is immediately shipped to the downstream nodes upon request, becoming feedstock inventory to those nodes (or simply inventory if the downstream node is a distributor/retailer). A holding (e.g., transportation) cost is also placed on any pipeline inventory (in-transit inventory).
6. Any inventory remaining at the end of the last period (period 30 in the base case) is lost, which means that it has no salvage value.

2.3. Key Variables

Table 1 describes the main variables used in the IMP model. All of the key variables are continuous and non-negative.

Table 1. Main variables in the inventory management problem (IMP) Model.

Variable	Description
$a_{t,j,k}$	The reorder quantity requested to supplier node j by node k at the beginning of period t (the amount of material sent from node j to node k)
$S^d_{t,j,k}$	The amount retailer j sells to market k in period $t-1$. Note: Retail sales are indexed at the next period since these occur after demand in the current period is realized.
$S^o_{t,j}$	The on-hand inventory at node j just prior to when the demand is realized in period t.
$S^p_{t,j,k}$	The in-transit (pipeline) inventory between node j and node k just prior to when the demand is realized in period t.
$u_{t,j,k}$	The unfulfilled demand at retailer j associated with market k in period $t-1$. Note: indexing is also shifted since any unfulfilled demand occurs after the uncertain demand is realized.
$R_{t,j}$	The profit (reward) in node j for period t.

2.4. Objective Function

The objective of the IMP optimization is to maximize the time-averaged expected profit of the supply network (R, Equation (1)). The uncertain parameter vector that is associated with the demand in period t is given by $\boldsymbol{\xi}_t$. A specific realization of the uncertain parameter is denoted with ξ_t. The sequence of uncertain parameters from period t through t' is represented with $\boldsymbol{\xi}_{[t,t']}$, with $\xi_{[t,t]}$ being a specific realization of that sequence (note: $\xi_{[1,1]}$ refers to stage 1, which is deterministic). The present formulation assumes a single retailer–market link with a single demand in each period. The profit in each period is the sum of the profits in the main network nodes ($R_1 = \sum_{j\in J} R_{1,j}$ and $R_t = \sum_{j\in J} R_{t,j}(\xi_t)\ \forall t \in T$). The main network nodes include the production/manufacturing, distribution, and retail nodes.

$$\max R = \frac{1}{|T|}\cdot\left(R_1 + \mathbb{E}_{\boldsymbol{\xi}_{[2,|T|]}|\xi_{[1,1]}}\left[\max R_2(\xi_2) + \ldots + \mathbb{E}_{\boldsymbol{\xi}_{[|T|,|T|]}|\xi_{[1,|T|-1]}}\left[\max R_{|T|}(\xi_{|T|})\right]\right]\right) \quad (1)$$

2.5. IMP Model

The dynamics of the IMP under demand uncertainty are modeled as a Linear Programming (LP) problem using the linear algebraic constraints that are given below.

2.5.1. Network Profit

The profit in period t at node j ($R_{t,j}$, Equations (2a) and (2b)) is obtained by subtracting procurement costs ($PC_{t,j}$, Equations (4a) and (4b)), operating costs ($OC_{t,j}$, Equations (5a) and (5b)), unfulfilled demand penalties ($UP_{t,j}$, Equation (6)), and inventory holding costs ($HC_{t,j}$, Equations (7a) and (7b)) from the sales revenue ($SR_{t,j}$, and Equations (3a) and (3c)). The operating costs refer to production costs and, hence, do not apply to distribution nodes, as no manufacturing occurs at these nodes. There is also no unfulfilled demand at non-retail nodes, as all inter-network requests must be feasible.

In the equations shown below, the parameters $p_{j,k}$, $b_{j,k}$, and $g_{k,j}$ refer to the material unit price, the unfulfilled unit demand penalty, and the unit material pipeline holding cost (transportation cost) for the link going from node j to node k, respectively. o_j, v_j, and h_j refer to the unit operating cost, production yield (0 to 1 range), and on-hand inventory holding cost at node j, respectively. The sets J_j^{in} and J_j^{out} are the sets of predecessors and successors to node j, respectively.

$$R_{1,j} = SR_{1,j} - PC_{1,j} - OC_{1,j} - UP_{1,j} - HC_{1,j} \quad \forall j \in J \tag{2a}$$

$$R_{t,j}(\xi_t) = SR_{t,j}(\xi_t) - PC_{t,j}(\xi_t) - OC_{t,j}(\xi_t) - UP_{t,j}(\xi_t) - HC_{t,j}(\xi_t) \quad \forall t \in \{2, ..., |T|\}, j \in J \tag{2b}$$

$$SR_{1,j} = \sum_{k \in J_i^{out}} p_{j,k} \cdot a_{1,j,k} \quad \forall j \in J^{prod} \cup J^{dist} \tag{3a}$$

$$SR_{t,j}(\xi_t) = \sum_{k \subset J_i^{out}} p_{j,k} \cdot a_{t,j,k}(\xi_t) \quad \forall t \in \{2, ..., |T|\}, j \in J^{prod} \cup J^{dist} \tag{3b}$$

$$SR_{t,j}(\xi_t) = \sum_{k \in J_i^{out}} p_{j,k} \cdot S^d_{t,j,k}(\xi_t) \quad \forall t \in \{2, ..., |T|\}, j \in J^{retail} \tag{3c}$$

$$PC_{1,j} = \sum_{k \in J_i^{in}} p_{k,j} \cdot a_{1,k,i} \quad \forall j \in J \tag{4a}$$

$$PC_{t,j}(\xi_t) = \sum_{k \in J_i^{in}} p_{k,j} \cdot a_{t,k,j}(\xi_t) \quad \forall t \in \{2, ..., |T|\}, j \in J \tag{4b}$$

$$OC_{1,j} = \frac{o_j}{v_j} \cdot \sum_{k \in J_j^{out}} a_{1,j,k} \quad \forall j \in J^{prod} \tag{5a}$$

$$OC_{t,j}(\xi_t) = \frac{o_j}{v_j} \cdot \sum_{k \in J_i^{out}} a_{t,j,k}(\xi_t) \quad \forall t \in \{2, ..., |T|\}, j \in J^{prod} \tag{5b}$$

$$UP_{t,j}(\xi_t) = \sum_{k \in J_j^{out}} b_{j,k} \cdot u_{t,j,k}(\xi_t) \quad \forall t \in \{2, ..., |T|\}, j \in J^{retail} \tag{6}$$

$$HC_{1,j} = h_j \cdot S^o_{1,j} + \sum_{k \in J_j^{in}} g_{k,j} \cdot S^p_{1,k,j} \quad \forall j \in J \tag{7a}$$

$$HC_{t,j}(\xi_t) = h_j \cdot S^o_{t,j}(\xi_t) + \sum_{k \in J_j^{in}} g_{k,j} \cdot S^p_{t,k,j}(\xi_t) \quad \forall t \in \{2, ..., |T|\}, j \in J \tag{7b}$$

2.5.2. Inventory Balances

The on-hand inventory at each node is updated while using material balances that account for incoming and outgoing material, as shown in Equations (8a)–(9c). The inventory levels at each node are updated by adding any incoming inventory and subtracting any outgoing inventory to the previously recorded inventory levels at the respective nodes. The parameter $S^o_{0,j}$ is the initial inventory at node j. The variable $a'_{t,k,j}$ is the pipeline inventory that arrives at node j from node k in period t (Equation (10)). Outgoing inventory is the inventory that is transferred to downstream nodes, $a_{t,j,k}$, or sold to the market at the retailer node, $S^d_{t,j,k}$. At production nodes, the sales quantities are adjusted for production yields (ν_j). For distribution nodes, ν_j is set to 1.

$$S^o_{1,j} = S^o_{0,j} + \sum_{k \in J^{in}_j} a'_{1,k,j} - \frac{1}{\nu_j} \cdot \sum_{k \in J^{out}_j} a_{1,j,k} \quad \forall j \in J^{prod} \cup J^{dist} \tag{8a}$$

$$S^o_{2,j}(\xi_2) = S^o_{1,j} + \sum_{k \in J^{in}_j} a'_{2,k,j} - \frac{1}{\nu_j} \cdot \sum_{k \in J^{out}_j} a_{2,j,k}(\xi_2) \quad \forall j \in J^{prod} \cup J^{dist} \tag{8b}$$

$$S^o_{t,j}(\xi_t) = S^o_{t-1,j}(\xi_{t-1}) + \sum_{k \in J^{in}_j} a'_{t,k,j} - \tfrac{1}{\nu_j} \cdot \sum_{k \in J^{out}_j} a_{t,j,k}(\xi_t) \quad t \in \{3, ..., |T|\} \forall j \in J^{prod} \cup J^{dist} \tag{8c}$$

$$S^o_{1,j} = S^o_{0,j} + \sum_{k \in J^{in}_j} a'_{1,k,j} \quad \forall j \in J^{retail} \tag{9a}$$

$$S^o_{2,j}(\xi_2) = S^o_{1,j} + \sum_{k \in J^{in}_j} a'_{2,k,j} - \sum_{k \in J^{out}_j} S^d_{2,j,k}(\xi_2) \quad \forall j \in J^{retail} \tag{9b}$$

$$S^o_{t,j}(\xi_t) = S^o_{t-1,j}(\xi_{t-1}) + \sum_{k \in J^{in}_j} a'_{t,k,j} - \sum_{k \in J^{out}_j} S^d_{t,j,k}(\xi_t) \quad \forall t \in \{3, ..., |T|\}, j \in J^{retail} \tag{9c}$$

$$a'_{t,k,j} = \begin{cases} 0, & \text{if } t - L_{k,j} < 1 \\ a_{1,k,j}, & \text{if } t - L_{k,j} = 1 \\ a_{t-L_{k,j},k,j}(\xi_{t-L_{k,j}}), & \text{if } t - L_{k,j} > 1 \end{cases} \quad \forall t \in T, j \in J^{retail}, k \in J^{in}_j \tag{10}$$

Equations (11a)–(11c) provide the pipeline inventory balances at each arc. Once again, inventories are updated by deducting delivered inventory downstream and adding new inventory requests to the previously recorded pipeline inventory levels. It is assumed that, at $t = 0$, there is no inventory in the pipeline.

$$S^p_{1,k,j} = -a'_{1,k,j} + a_{1,k,j} \quad \forall j \in J, k \in J^{in}_j \tag{11a}$$

$$S^p_{2,k,j}(\xi_2) = S^p_{1,k,j} - a'_{2,k,j} + a_{2,k,j}(\xi_2) \quad \forall j \in J, k \in J^{in}_j \tag{11b}$$

$$S^p_{t,k,j}(\xi_t) = S^p_{t-1,k,j}(\xi_{t-1}) - a'_{t,k,j} + a_{t,k,j}(\xi_t) \quad \forall t \in \{3, ..., |T|\}, j \in J, k \in J^{in}_j \tag{11c}$$

2.5.3. Inventory Requests

Upper bounds on the replenishment orders are set, depending on the type of node. For the production nodes, downstream replenishment requests are limited by the production capacity, c_j, as given in Equations (12a) and (12b). The requests are also limited by the available feedstock inventory at the production nodes that is transformed into products with a yield of ν_j, as stated in Equations (13a) and (13b). Because distribution-only nodes do not have manufacturing areas, the upper bounds on any downstream replenishment requests are set by the available inventory at the distribution nodes, which is equivalent to setting ν_j to 1 in Equations (13a) and (13b). These sets of constraints ensure that the reorder quantities are always feasible, which means that the quantities requested are quantities that can be sold and shipped in the current period.

$$\sum_{k \in J^{out}_j} a_{1,j,k} \le c_j \quad \forall j \in J^{prod} \tag{12a}$$

$$\sum_{k \in J^{out}_j} a_{t,j,k}(\xi_t) \le c_j \quad \forall t \in \{2, ..., |T|\}, j \in J^{prod} \tag{12b}$$

$$\sum_{k\in J_j^{out}} a_{1,j,k} \leq S_{1,j}^{o} \cdot \nu_j \quad \forall j \in J^{prod} \cup J^{dist} \tag{13a}$$

$$\sum_{k\in J_j^{out}} a_{t,j,k}(\xi_t) \leq S_{t,j}^{o}(\xi_t) \cdot \nu_j \quad \forall t \in \{2,...,|T|\}, j \in J^{prod} \cup J^{dist} \tag{13b}$$

2.5.4. Market Sales

The retailer node sells up to its available on-hand inventory in each period when the demand is realized, as given in Equations (14a) and (14b). This includes the start-of-period inventory plus any reorder quantities that arrive at the beginning of the period (before the markets open). Sales at the retailer nodes do not exceed the market demand, $d_{t,j,k}(\xi_t)$, as shown in Equation (15a). If backlogging is allowed, then any previous backlogged orders are added to the market demand, as shown in Equation (15b). If unfulfilled orders are counted as lost sales, then u is removed from Equation (15b).

$$\sum_{k\in J_j^{out}} S_{2,j,k}^{d}(\xi_2) \leq S_{1,j}^{o} \quad \forall j \in J^{retail} \tag{14a}$$

$$\sum_{k\in J_j^{out}} S_{t,j,k}^{d}(\xi_t) \leq S_{t-1,j}^{o}(\xi_{t-1}) \quad \forall t \in \{3,...,|T|\}, j \in J^{retail} \tag{14b}$$

$$S_{2,j,k}^{d}(\xi_2) \leq d_{2,j,k}(\xi_2) \quad \forall j \in J^{retail}, k \in J_j^{out} \tag{15a}$$

$$S_{t,j,k}^{d}(\xi_t) \leq d_{t,j,k}(\xi_t) + u_{t-1,j,k}(\xi_{t-1}) \quad \forall t \in \{3,...,|T|\}, j \in J^{retail}, k \in J_j^{out} \tag{15b}$$

Unfulfilled demand at the retailer is the difference between the market demand and the actual retail sale in the current period (Equations (16a) and (16b). If the network operates under the lost sales mode, then the u term in the right-hand side of Equation (16b) is removed.

$$u_{2,j,k}(\xi_2) = d_{2,j,k}(\xi_2) - S_{2,j,k}^{d}(\xi_2) \quad \forall j \in J^{retail}, k \in J_j^{market} \tag{16a}$$

$$u_{t,j,k}(\xi_t) = d_{t,j,k}(\xi_t) + u_{t-1,j,k}(\xi_{t-1}) - S_{t,j,k}^{d}(\xi_t) \quad \forall t \in \{3,...,|T|\}, j \in J^{retail}, k \in J_j^{market} \tag{16b}$$

2.5.5. Variable Domains

$$R_{1,j} \in \mathbb{R}^1 \quad \forall j \in J \tag{17a}$$

$$R_{t,j}(\xi_t) \in \mathbb{R} \quad \forall t \in \{2,...,|T|\}, j \in J \tag{17b}$$

$$S_{1,j}^{o} \geq 0 \quad \forall j \in J \tag{18a}$$

$$S_{t,j}^{o}(\xi_t) \geq 0 \quad \forall t \in \{2,...,|T|\}, j \in J \tag{18b}$$

$$a_{1,k,j},\ S_{1,k,j}^{p} \geq 0 \quad \forall j \in J, k \in J_j^{in} \tag{19a}$$

$$a_{t,k,j}(\xi_t),\ S_{t,k,j}^{p}(\xi_t) \geq 0 \quad \forall t \in \{2,...,|T|\}, j \in J, k \in J_j^{in} \tag{19b}$$

$$S_{t,j,k}^{d}(\xi_t),\ u_{t,j,k}(\xi_t) \geq 0 \quad \forall t \in \{2,...,|T|\}, j \in J^{retail}, k \in J_j^{out} \tag{20}$$

2.6. Scenario Tree for Multistage Stochastic Programming

Equations (1)–(20) describe a multistage stochastic inventory management problem. In principle, continuous or discrete probability distributions can be used to model the uncertain demands in the stochastic process $(\xi_1, \xi_2, \ldots, \xi_t, \ldots, \xi_{|T|})$. However, in most applications, a scenario-based approach is assumed for ease of computation, i.e., there are a finite number of realizations of the uncertain parameter. For illustration purposes, Figure 2 shows a scenario tree that corresponds to a three-stage stochastic programming problem. At stage one, the decision-maker does not know the realizations of the uncertain parameters in the future time periods. At stage two, there are two different realizations of the uncertain parameters. The decision-maker can take different actions, depending on the realization of the uncertainty at stage two. For each realization at stage two, there are two different realizations at stage three. Therefore, the scenario tree that is presented in Figure 2 has four scenarios in total.

It is easy to observe that the number of scenarios grows exponentially with respect to the number of stages. For example, in the IMP, if we consider three realizations of the demand per stage, the total number of scenarios will be 3^{29} for a 30 period problem. Moreover, a Poisson distribution is assumed for the distribution of the demand. In principle, there is an infinite number of realizations per stage. We introduce an approximation of the multistage stochastic programming problem in the next subsection in order to reduce the computational complexity.

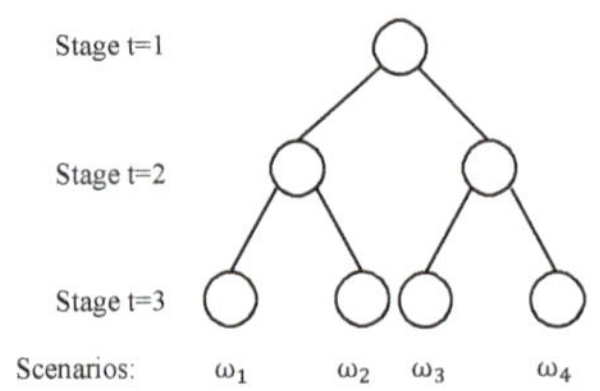

Figure 2. The scenario tree for a three-stage stochastic program with two realizations per stage.

2.7. Approximation for the Multistage Scenario Tree

In order to make the multistage stochastic IMP tractable, we make the following two simplifications.

First, the Poisson(λ) distribution (as shown in Equation (21)) is approximated by a discrete distribution with three realizations.

$$p(x = k) = \frac{\lambda^k e^{-\lambda}}{k!} \tag{21}$$

Note that the mean and variance of Poisson (λ) are both λ. The values of the three realizations are chosen to be $\lambda - \lceil\sqrt{\lambda}\rceil, \lambda, \lambda + \lceil\sqrt{\lambda}\rceil$. The probabilities of the three realizations are chosen, such that the Wasserstein-1 distance to the original Poisson(λ) is minimized. In other words, for all $k = 1, 2, \ldots, \infty$, the probability of $x = k$ in Poisson (λ) is assigned to the realization of the new distribution that is closest to k. The probabilities for this new distribution are given in Equations (22a)–(22c).

$$p(x = \lambda - \lceil\sqrt{\lambda}\rceil) = \sum_{k=1}^{\lambda - \lceil\frac{\sqrt{\lambda}}{2}\rceil} \frac{\lambda^k e^{-\lambda}}{k!} \tag{22a}$$

$$p(x = \lambda) = \sum_{k=\lambda - \lceil\frac{\sqrt{\lambda}}{2}\rceil + 1}^{\lambda + \lceil\frac{\sqrt{\lambda}}{2}\rceil - 1} \frac{\lambda^k e^{-\lambda}}{k!} \tag{22b}$$

$$p(x = \lambda + \lceil\sqrt{\lambda}\rceil) = \sum_{k=\lambda + \lceil\frac{\sqrt{\lambda}}{2}\rceil}^{+\infty} \frac{\lambda^k e^{-\lambda}}{k!} \tag{22c}$$

This scenario generation approach, where the values of the realizations are fixed and the probabilities of each realization are chosen to minimize the Wasserstein-1 distance between the probability distribution of the scenario tree from the "true distribution", has been reported in [23].

Second, even with three realizations per stage, the number of scenarios for $T = 30$ becomes 3^{29}. Because the decisions in the later periods have a smaller impact on the decisions here-and-now, we only consider three realizations per stage until stage 6. After stage 6, the demands are assumed to be deterministic and they take the mean value λ.

With these two simplifications, Figure 3 shows the scenario tree for the approximate multistage stochastic IMP. The size of the scenario tree grows exponentially until stage 6 and it only grows linearly after stage 6. In total, $3^5 = 243$ scenarios are considered. With slight abuse of notation, we still denote the problem that is shown in Figure 3 as MSSP.

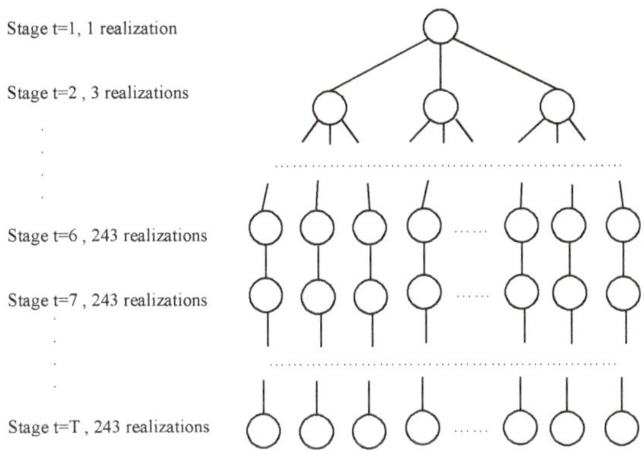

Figure 3. An approximation of the multistage stochastic program.

2.8. Perfect Information and Deterministic Model

We benchmark the MSSP model with a perfection information model and a deterministic model. In the perfect information model, it is assumed that the demand realizations from $t = 1$ to $t = T$ are known beforehand. Equation (1) reduces to Equation (23) in a perfect information model, where we assume that we know the realization of $(\xi_2, \ldots, \xi_t, \ldots, \xi_{|T|})$ and optimize over this given realization.

$$\max R(\xi) = \max \frac{1}{|T|} \cdot \left(R_1 + \sum_{t \in \{2,\ldots,|T|\}} R_t(\xi_t) \right) \tag{23}$$

In the deterministic model (DLP), we optimize over the expected value of ξ_t, which is denoted as $\bar{\xi}_t$. This means that the demand is assumed to be λ (the mean of the Poisson distribution) for all periods.

2.9. Reinforcement Learning Model

Reinforcement learning (RL) is a machine learning method whereby an *agent* learns to maximize a *reward* via interactions with an *environment*. The feedback the agent receives from the reward allows it to learn a *policy*, which is a function that directs the agent at each step throughout the environment. Because of the data-intensive and interactive nature of RL, agents are typically trained by interacting with Monte Carlo simulations to make decisions at each time step.

To this end, we formulate the IMP as a Markov Decision Process (MDP)—a stochastic, sequential decision making problem. At each time period t, the agent observes the current state of the system (S_t), and then selects an action (a_t) that is passed to the system. The simulation then advances to the next state (S_{t+1}) based on the realization of the random variables and the selected action, and it returns the associated reward (R_t), which is simply the profit function that is given in Equation (2a) summed over all nodes j ($R_t = \sum_{j \in J} R_{t,j}$).

The state consists of a vector with entries for the current demand at the retail node, the inventory levels at each node in the network, and the inventory in the pipeline along every edge in the network. In terms of the model notation from the previous subsection, $S_t = \{d_{t,j,k}, S^o_{t,j}, S^p_{t,k',j} | j \in J, k \in J^{out}_j, k' \in J^{in}_j\}$. The action at each period is a vector with all of the reorder quantities in the network ($a_t = \{a_{t,k,j} | j \in J, k \in J^{in}_j\}$).

In this work, we rely on the Proximal Policy Optimization (PPO), as described in Schulman et al. [24]. This has become a popular algorithm in the RL community, because it frequently exhibits stable learning characteristics. PPO is an actor-critic method that uses two neural networks that interact with one another. The actor is parameterized by θ_π and it learns the policy, while the critic is parameterized by θ_v and it learns the value function. The policy the actor learns is probabalistic in nature and yields a probability distribution over actions during each forward pass. For our IMP, the action space consists of re-order values for each node in the network. These are discrete values that range from 0 to the maximum order quantity at each individual node. If the actor chooses an action that is greater than the quantity that can be supplied – for example, if the maximum re-order quantity is 100, but only 90 units are available – then the minimum of these two values will be supplied.

The critic learns the value function, which allows it to estimate the sum of the discounted, future rewards available at each state. The difference between the actual rewards and the estimated rewards supplied by the critic is known as the *temporal difference error* (TD-error, or δ_T). This difference is summed and discounted in order to provide the advantage estimation of the state, $\hat{A}_k$, as given in Equation (24),

$$\hat{A}_k = \sum_{T=1}^{t} \gamma^{t-T+1} \delta_T \tag{24}$$

where γ is the discount factor used to prioritize current rewards over future rewards. This value is then used in the loss function, $\mathcal{L}(\theta)$, whereby the paramaters of the networks are updated while using stochastic gradient descent to minimize the loss function. PPO has shown to be effective in numerous domains, exhibiting stable learning features, as discussed in Schulman et al. [24]. PPO achieves this by penalizing large policy updates by optimizing a conservative loss function given by Equation (25), where $r_k(\theta)$ is the probability ratio between the new policy $\pi_k(\theta)$ and the previous policy, $\pi_{k-1}(\theta)$. Here, we use k to denote each policy iteration, since the parameters have been initialized. The clip function reduces the incentive for moving $r_k(\theta)$ outside the interval $[1-\epsilon, 1+\epsilon]$. The hyperparameter ϵ limits the update of the policy, such that the probability of outputs does not change more than $\pm\epsilon$ at each update. For more detail, see the work by Schulman et al. [24].

$$\mathcal{L}(\theta) = \min\left(r_k(\theta)\hat{A}_k,\ \text{clip}\left(r_k(\theta), 1-\epsilon, 1+\epsilon\right)\hat{A}_k\right) \tag{25}$$

In the present work, we rely on the implementation of the PPO algorithm found in the Ray package [25]. A two-layer, 256-node feed-forward network is trained with over 70,000 episodes—simulated 30-day periods, whereby the agent learns to maximize the expected reward by interacting with the environment. Given that this is a model-free approach, the agent must learn through this trial-and-error approach. The initial policy consists of randomly initialized weights and biases, so the output actions are on par with random decisions. After each episode, the results are collected and the weights and biases are updated in order to minimize the loss function according to the PPO algorithm, as discussed in Section 2.9. As one would expect, this initial policy performs poorly, with the agent losing roughly $300 per episode. However, as shown in the training curve presented in Figure 4, the agent is able to improve on the policy with additional experience and learn a very effective policy to control the inventory across the network.

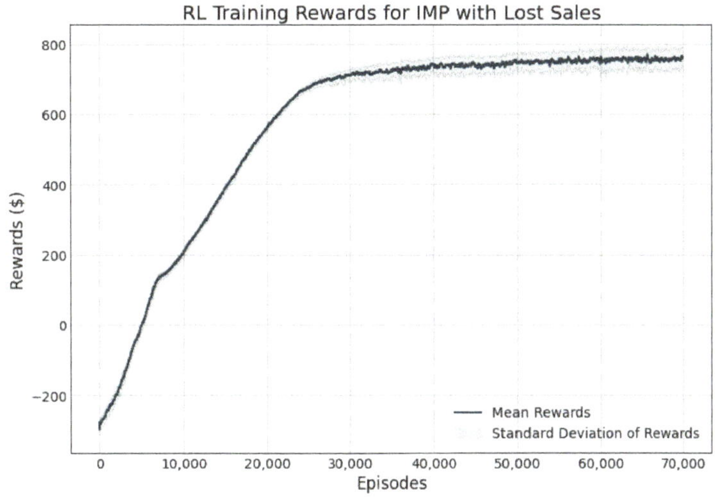

Figure 4. Training curve for the reinforcement learning (RL) model.

2.10. Case Study

A 30-period system is used for the case study in order to represent one month's worth of inventory management in a supply network. A time step of one day is used, such that demand is received on a daily basis. Figure 5 depicts the network structure of the base case system used. Major parameters and their values are included in the figure. Parameters that are next to nodes are specific to the node (initial state or on-hand inventory, S_0; unit holding costs, h; unit operating costs, o; production yield v; and, production capacity, c), whereas those next to links are specific to that link (unit sales price, p; unfulfilled unit demand penalty, b; market demand distribution, d; unit pipeline holding costs, g; and, lead times, L). Node and link subscripts in the schematic are dropped for clarity. In order to compare the performance of the three modeling approaches (DLP, MSSP, and RL), 100 unique sample paths were generated. Each sample path consists of 30 demand realizations, one for each period in the simulation horizon, sampled from a Poisson distribution with a mean of 20.

The execution of the 100 simulations follows the sequence of events for each time period that is described in Section 2.2. At the beginning of each period and prior to Step 1 in the event sequence, each optimization model is called to obtain the reorder quantities for each node in that period. The models have no knowledge of the demand realizations that will occur in the current and future periods, but they can rely on the variables/states from previous periods. The reorder quantities for the current period obtained by the models are then passed as the actions in Step 1 of the sequence of events. Subsequently, the subsequent events for that period unfold, with the demand realization for that given period being taken from the respective sample path that is assigned to that simulation instance. The process is repeated for the next period, re-solving the models at the beginning of each time period, until the 30 periods are complete.

The three modeling approaches are benchmarked against a perfect information model (also referred to as the *Oracle*). 10-period windows are used for the rolling horizon (RH) modes in the DLP and MSSP models. However, towards the end of the simulation, the RH becomes a shrinking horizon (SH), since the window does not roll past period 30.

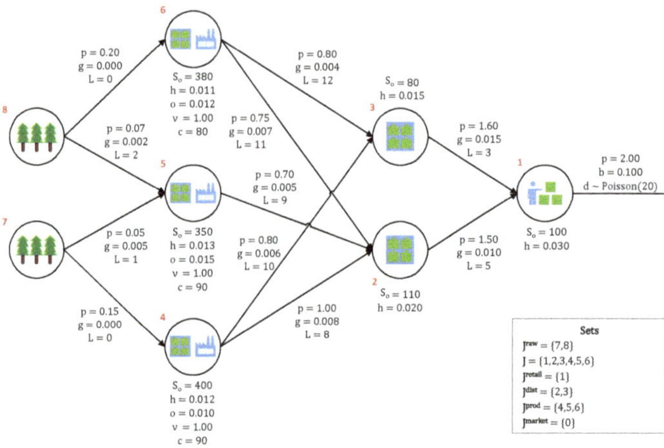

Figure 5. Supply Chain Network Schematic with Network Parameters used in the Case Study.

3. Results

The DLP and MSSP models were solved while using Gurobi (version 9.1). The DLP models are quite small, with the largest one being the DLP-SH solved at $t = 1$ in the simulations (1231 constraints and 1291 variables). The DLP-RH with a 10-period window size has 411 constraints and 431 variables. The shrinking and rolling horizon DLP models both have CPU solve times that are below 0.3 s on average. The largest MSSP model solved is the MSSP-SH solved at $t = 1$ (263,958 constraints and 299,941 variables), which has an average CPU solve time of 119 s. The MSSP-RH with a 10-period window size has 64,698 constraints and 71,521 variables. The average CPU time to solve the MSSP-RH model is 12 s.

Table 2 summarizes the performance results from each solution method. Figure 6 shows the total inventory profiles at each node for the lost sales case. The total inventory includes both on-hand inventory and pipeline inventory incoming from a node's suppliers. Similar results (not shown) were observed for the backlogging case. Figures 7 and 8 show sample network flow plots for the RL and MSSP-RH models, respectively. The cumulative network flow plots for both DLP instances and MSSP-SH are not shown as they are similar to that of MSSP-RH. The edge thickness is proportional to the average total amount of material requested through that link. These network flows indicate the suppliers that are prioritized by the different model policies. Figure 9 shows the average unfulfilled market demand at the retailer node (lost sales), which gives an indication of the service levels of the supply network. A similar result is obtained for the backlogging case.

Table 2. Total reward comparison for the various models used to solve the IMP. *Performance Ratio* is defined as the ratio of the final cumulative profit of the perfect information model to that of the model used. *DLP* = Deterministic linear program; *MSSP* = Multi-stage stochastic program; *RL* = Reinforcement Learning; *RH* = rolling horizon; *SH* = shrinking horizon; *Oracle* = perfect information LP.

	DLP-RH	DLP-SH	MSSP-RH	MSSP-SH	RL	Oracle
Backlog						
Mean Profit	791.6	825.3	802.7	847.7	737.2	861.3
Standard Deviation	52.5	37.0	56.3	49.4	24.8	56.4
Performance Ratio	1.09	1.04	1.07	1.02	1.17	1.00
Lost Sales						
Mean Profit	735.8	786.9	790.6	830.6	757.8	854.9
Standard Deviation	31.2	30.8	47.8	37.7	33.1	49.9
Performance Ratio	1.16	1.09	1.08	1.03	1.13	1.00

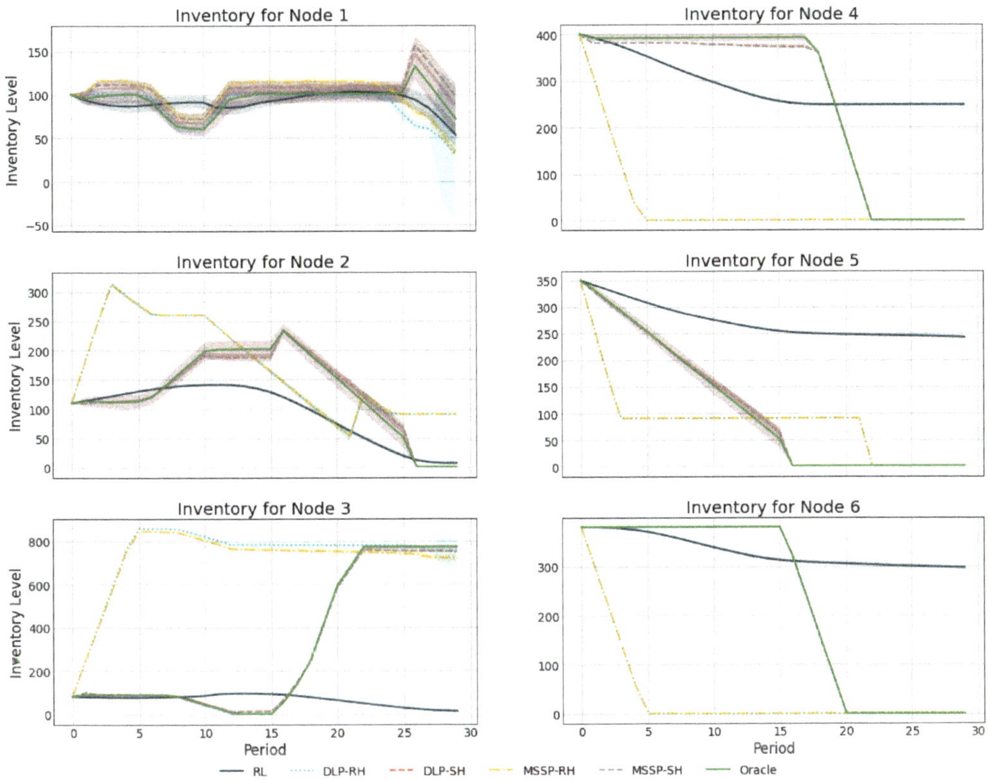

Figure 6. Average total inventory at each main network node (lost sales mode). Shaded areas denote ±1 standard deviation of the mean value.

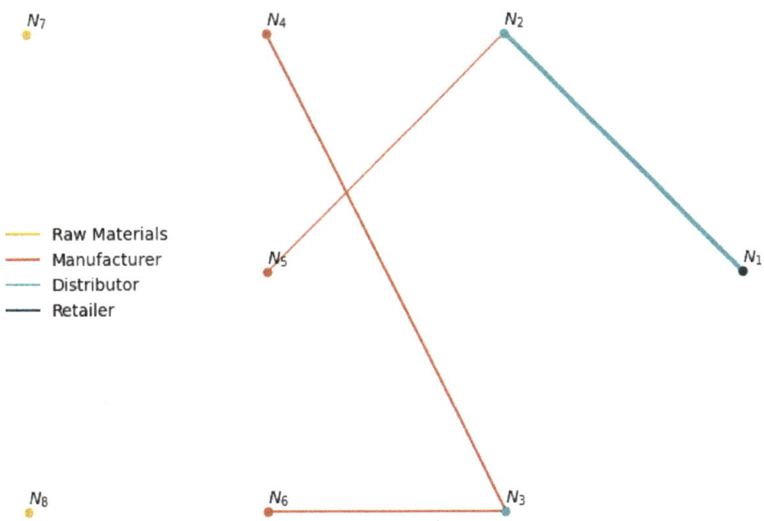

Figure 7. Average network flow with the RL policy (lost sales mode). Total flow is proportional to the edge thickness.

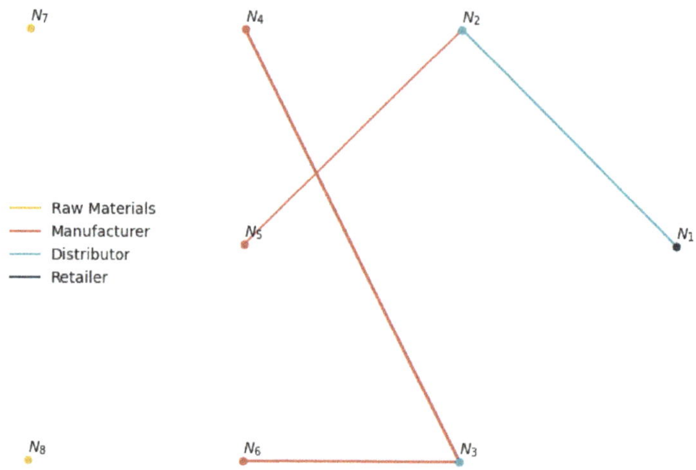

Figure 8. Average network flow with the MSSP-RH policy (lost sales mode). Total flow is proportional to the edge thickness.

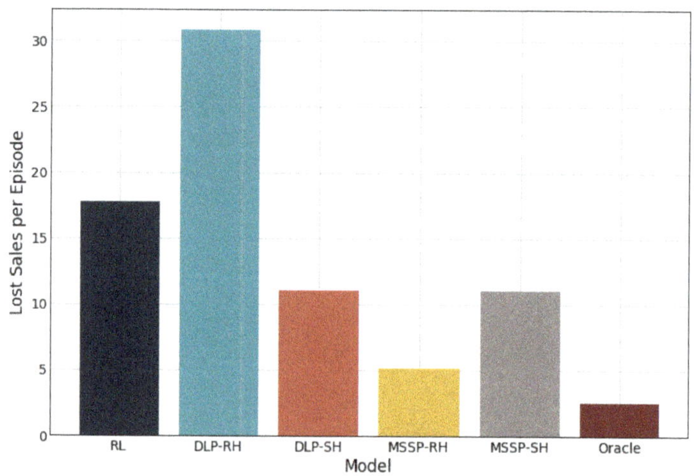

Figure 9. Average unfulfilled demands at the retailer node (lost sales mode).

4. Discussion

The results shown in Table 2 indicate that the rolling horizon DLP model outperforms the RL model when backlog is included, but it is outperformed by the latter when unfulfilled demands become lost sales. When backlogging is allowed, unfulfilled demand can be satisfied at a later period with a penalty, which reduces the need for high service levels. However, the service levels become more important in the lost sales case, where, not only is a goodwill penalty assessed, but potential profit from the sales is lost. Because the RL does a better job at maintaining on-hand inventories it displays the higher service levels shown in Figure 9 and superior performance in the lost sales case. It should be noted that the differences between the two approaches are rather small (7% and 3%, respectively), and within 15% of the perfect information model.

As expected, the shrinking horizon DLP exhibits superior performance relative to its rolling horizon counterpart, because it looks further ahead in time during the optimization. In the rolling horizon approach, the short-sighted model tends to drop inventory levels at the top tier suppliers (nodes 4–6) sooner in an attempt to reduce the inventory holding costs towards the end of the optimization window. However, since the simulation horizon extends beyond the 10-period optimization window, that inventory ends up accumulating in the medium tier suppliers (nodes 2–3), driving up holding costs overall. From a service level standpoint, the shrinking horizon DLP maintains higher inventory levels at the retailer than its rolling horizon counterpart, allowing it to achieve higher service levels (see Figure 9). However, it is interesting to note that the opposite is observed in the MSSP model, which, despite having a higher profit, has lower service levels in the shrinking horizon case (higher unfulfilled demand). The greater profit is a result of the shrinking horizon reducing holding costs by 13% overall, which has a greater impact on profit than the unfulfilled demand penalties. Just at the retailer node, the holding cost to demand penalty ratio is 3:1, which incentivizes the model to sacrifice some demand satisfaction to reduce the holding costs. Overall, the MSSP model yields superior performance in all cases, coming in within 8% and 3% of the best possible outcome (*Oracle*), on average, for the rolling horizon and shrinking horizon modes, respectively.

From an operational standpoint, the Oracle and shrinking horizon models prioritize inventory flow to the retailer via nodes 5 and 2, which have a lower holding cost than the alternatives as shown in the timing of inventory transfers in Figure 6 and the flow patterns in Figure 8. The transportation cost for this path is 0.015, with a lead time of 14 days, whereas the other paths have transportation costs in the range 0.017–0.021, with lead times in the 13–16 day range. Once inventory at node 5 is depleted, the other top level suppliers (nodes 4 and 6) begin to send inventory downstream. On the other hand, the rolling horizon models send inventory from all of the top level suppliers from the start due to the myopic effects of the reduced optimization window. In general terms, the inventory profiles that are shown in Figure 6 for the rolling horizon models are similar to their shrinking horizon counterparts, except that the inventory changes are shifted to earlier times. All of the mathematical programming models also take advantage of the fact that the pipeline inventory costs are lower than holding costs at the supplier nodes. Therefore, they trigger sending more inventory to node 3 from nodes 4 and 6 than is needed so as to reduce costs. This additional inventory ends up accumulating in node 3 for the most part, as it is cheaper to source the retailer from node 2 than node 3. Although the DLP and MSSP models exhibit similar inventory profiles, the superiority of the MSSP model arises from the fact that, unlike the DLP model, it accounts for uncertainty in the demand, which enables it to target superior service levels and reduce holding costs.

In contrast to the mathematical programming models, the RL model avoids drastic changes in the inventory positions, maintaining levels throughout the simulation. This is supported not only by the inventory levels in Figure 6, but also by the flow pattern shown in Figure 7, which indicate that, contrary to the mathematical programming models, the RL model distributes requests more evenly amongst the suppliers of each node. This conservative approach explains why the profits obtained with the RL model are lower than those that were obtained by most of mathematical programming models. In practice, the policy from the RL model is preferred as it reduces shocks to the inventory levels. Furthermore, the RL policy manages the supply network with potentially greater resiliency to disruptions as a result of the balanced load distribution within the network. Unlike the other models that have virtually no flow between the raw material nodes to the top tier suppliers and rely solely on the initial inventory at these nodes, the RL model gradually replenishes inventories at the top tier nodes in order to avoid their depletion. This conservative behavior of the RL is observed as a result of the PPO algorithm used, which penalizes large policy changes.

A drawback from the current implementation of the supply network is that all of the models exhibit end-of-simulation effects, in which the inventory drops to zero or near zero at the end of the simulation to avoid excess holding costs. In a real application, this could be avoided by imposing penalties on the models in order to avoid depleting inventories near the end of the simulation, adding terminal inventory constraints (Lima et al. [26]), or running the models for longer simulation horizons, since most of the applications extend beyond 30 periods. The latter option would not be viable for the stochastic programming models as it would affect their tractability. Despite these limitations, the three approaches show promise in obtaining dynamic reorder policies that improve the supply network performance to within 3% to 15% of perfect information dynamic policies, which do not exist in practice.

5. Conclusions

The present work extends to the open-source package OR-Gym for general single-product, multi-period make-to-order supply networks with production and inventory holding sites throughout the network. The work introduces additional tools for solving inventory management problems within the OR-Gym framework (e.g. multi-stage stochastic programming and rolling horizon implementations for deterministic and stochastic models). The inventory management policies that are obtained via deterministic linear programming, stochastic linear programming, and reinforcement learning are compared and contrasted in the context of a four echelon supply network with uncertain stationary demand. The results show that the stochastic model yields superior results in terms of supply network profitability. However, the reinforcement learning model manages the network in a way that is potentially more resilient to network disruptions, showing promise in using AI for supply chain applications. Although deterministic linear models ignore the stochastic nature of the supply network, they rapidly solve in fractions of a second, while providing solutions that are comparable to the profitability of the reinforcement learning policies. Extensions to this work may include studying the effects of non-stationary demand on the models used and mitigating the end-of-simulation effects that have been discussed previously.

Author Contributions: conceptualization, H.D.P., C.D.H. and C.L.; software, H.D.P., C.D.H. and C.L.; formal analysis, H.D.P. and C.D.H.; writing, H.D.P., C.D.H. and C.L.; visualization, H.D.P. and C.D.H.; supervision, I.E.G. All authors have read and agreed to the published version of the manuscript.

Funding: This research was funded in part by Dow Inc.

Acknowledgments: The authors acknowledge the support from the Center for Advanced Process Decision-making (CAPD) at Carnegie Mellon University.

Conflicts of Interest: The authors declare no conflict of interest.

Abbreviations

The following abbreviations are used in this manuscript:

LP	Linear Programming
DLP	Deterministic Linear Programming
MILP	Mixed-integer Linear Programming
2SSP	Two-stage Stochastic Programming
MSSP	Multi-stage Stochastic Programming
RL	Reinforcement Learning
AI	Artificial Intelligence
PPO	Proximal Policy Optimization
MDP	Markov Decision Process
IMP	Inventory Management Problem

References

1. Lee, H.L.; Padmanabhan, V.; Whang, S. Information distortion in a supply chain: The bullwhip effect. *Manag. Sci.* **1997**, *43*, 546–558. [CrossRef]
2. Eruguz, A.S.; Sahin, E.; Jemai, Z.; Dallery, Y. A comprehensive survey of guaranteed-service models for multi-echelon inventory optimization. *Int. J. Prod. Econ.* **2016**, *172*, 110–125. [CrossRef]
3. Simchi-Levi, D.; Zhao, Y. Performance Evaluation of Stochastic Multi-Echelon Inventory Systems: A Survey. *Adv. Oper. Res.* **2012**, *2012*, 126254. [CrossRef]

4. Glasserman, P.; Tayur, S. Sensitivity analysis for base-stock levels in multiechelon production-inventory systems. *Manag. Sci.* **1995**, *41*, 263–281. [CrossRef]
5. Chu, Y.; You, F.; Wassick, J.M.; Agarwal, A. Simulation-based optimization framework for multi-echelon inventory systems under uncertainty. *Comput. Chem. Eng.* **2015**, *73*, 1–16. [CrossRef]
6. Dillon, M.; Oliveira, F.; Abbasi, B. A two-stage stochastic programming model for inventory management in the blood supply chain. *Int. J. Prod. Econ.* **2017**, *187*, 27–41. [CrossRef]
7. Fattahi, M.; Mahootchi, M.; Moattar Husseini, S.M.; Keyvanshokooh, E.; Alborzi, F. Investigating replenishment policies for centralised and decentralised supply chains using stochastic programming approach. *Int. J. Prod. Res.* **2015**, *53*, 41–69. [CrossRef]
8. Pauls-Worm, K.G.; Hendrix, E.M.; Haijema, R.; Van Der Vorst, J.G. An MILP approximation for ordering perishable products with non-stationary demand and service level constraints. *Int. J. Prod. Econ.* **2014**, *157*, 133–146. [CrossRef]
9. Zahiri, B.; Torabi, S.A.; Mohammadi, M.; Aghabegloo, M. A multi-stage stochastic programming approach for blood supply chain planning. *Comput. Ind. Eng.* **2018**, *122*, 1–14. [CrossRef]
10. Bertsimas, D.; Thiele, A. A robust optimization approach to inventory theory. *Oper. Res.* **2006**, *54*, 150–168. [CrossRef]
11. Govindan, K.; Cheng, T.C. Advances in stochastic programming and robust optimization for supply chain planning. *Comput. Oper. Res.* **2018**, *100*, 262–269. [CrossRef]
12. Roy, B.V.; Bertsekas, D.P.; North, L. A neuro-dynamic programming approach to admission control in ATM networks. In Proceedings of the IEEE International Conference on Acoustics, Speech, and Signal Processing, Munich, Germany, 21–24 April 1997.
13. Kleywegt, A.J.; Non, V.S.; Savelsbergh, M.W. Dynamic Programming Approximations for a Stochastic Inventory Routing Problem. *Transport. Sci.* **2004**, *38*, 42–70. [CrossRef]
14. Topaloglu, H.; Kunnumkal, S. Approximate dynamic programming methods for an inventory allocation problem under uncertainty. *Naval Res. Logist.* **2006**, *53*, 822–841. [CrossRef]
15. Kunnumkal, S.; Topaloglu, H. Using stochastic approximation methods to compute optimal base-stock levels in inventory control problems. *Oper. Res.* **2008**, *56*, 646–664. [CrossRef]
16. Cimen, M.; Kirkbride, C. Approximate dynamic programming algorithms for multidimensional inventory optimization problems. In Proceedings of the 7th IFAC Conference on Manufacturing, Modeling, Management, and Control, Saint Petersburg, Russia, 19–21 June 2013; Volume 46, pp. 2015–2020. [CrossRef]
17. Sarimveis, H.; Patrinos, P.; Tarantilis, C.D.; Kiranoudis, C.T. Dynamic modeling and control of supply chain systems: A review. *Comput. Oper. Res.* **2008**, *35*, 3530–3561. [CrossRef]
18. Mortazavi, A.; Arshadi Khamseh, A.; Azimi, P. Designing of an intelligent self-adaptive model for supply chain ordering management system. *Eng. Appl. Artif. Intell.* **2015**, *37*, 207–220. [CrossRef]
19. Oroojlooyjadid, A.; Nazari, M.; Snyder, L.; Takáč, M. A Deep Q-Network for the Beer Game: A Reinforcement Learning algorithm to Solve Inventory Optimization Problems. *arXiv* **2017**, arXiv:1708.05924.
20. Kara, A.; Dogan, I. Reinforcement learning approaches for specifying ordering policies of perishable inventory systems. *Expert Syst. Appl.* **2018**, *91*, 150–158. [CrossRef]
21. Sultana, N.N.; Meisheri, H.; Baniwal, V.; Nath, S.; Ravindran, B.; Khadilkar, H. Reinforcement Learning for Multi-Product Multi-Node Inventory Management in Supply Chains. *arXiv* **2020**, arXiv:2006.04037.
22. Hubbs, C.D.; Perez, H.D.; Sarwar, O.; Sahinidis, N.V.; Grossmann, I.E.; Wassick, J.M. OR-Gym: A Reinforcement Learning Library for Operations Research Problems. *arXiv* **2020**, arXiv:2008.06319.
23. Hochreiter, R.; Pflug, G.C. Financial scenario generation for stochastic multi-stage decision processes as facility location problems. *Ann. Oper. Res.* **2007**, *152*, 257–272. [CrossRef]
24. Schulman, J.; Moritz, P.; Levine, S.; Jordan, M.I.; Abbeel, P. High-dimensional continuous control using generalized advantage estimation. *arXiv* **2016**, arXiv:1506.02438.
25. Moritz, P.; Nishihara, R.; Wang, S.; Tumanov, A.; Liaw, R.; Liang, E.; Elibol, M.; Yang, Z.; Paul, W.; Jordan, M.I.; et al. Ray: A Distributed Framework for Emerging AI Applications. In Proceedings of the 13th USENIX Symposium on Operating Systems Design and Implementation (OSDI 18), Carlsbad, CA, USA, 8–10 October 2018.
26. Lima, R.M.; Grossmann, I.E.; Jiao, Y. Long-term scheduling of a single-unit multi-product continuous process to manufacture high performance glass. *Comput. Chem. Eng.* **2011**, *35*, 554–574. [CrossRef]

processes

Article

Improving Transactional Data System Based on an Edge Computing–Blockchain–Machine Learning Integrated Framework

Zeinab Shahbazi and Yung-Cheol Byun *

Department of Computer Engineering, Jeju National University, Jeju 63243, Korea; zeinab.sh@jejunu.ac.kr
* Correspondence: ycb@jejunu.ac.kr

Abstract: The modern industry, production, and manufacturing core is developing based on smart manufacturing (SM) systems and digitalization. Smart manufacturing's practical and meaningful design follows data, information, and operational technology through the blockchain, edge computing, and machine learning to develop and facilitate the smart manufacturing system. This process's proposed smart manufacturing system considers the integration of blockchain, edge computing, and machine learning approaches. Edge computing makes the computational workload balanced and similarly provides a timely response for the devices. Blockchain technology utilizes the data transmission and the manufacturing system's transactions, and the machine learning approach provides advanced data analysis for a huge manufacturing dataset. Regarding smart manufacturing systems' computational environments, the model solves the problems using a swarm intelligence-based approach. The experimental results present the edge computing mechanism and similarly improve the processing time of a large number of tasks in the manufacturing system.

Keywords: smart manufacturing; edge computing; machine learning; blockchain; Industrial Internet of Things

Citation: Shahbazi, Z.; Byun, Y.-C. Improving Transactional Data System Based on an Edge Computing–Blockchain–Machine Learning Integrated Framework. *Processes* **2021**, *9*, 92. https://doi.org/10.3390/pr9010092

Received: 24 October 2020
Accepted: 30 December 2020
Published: 4 January 2021

1. Introduction

The progress of industrialization has been changed and transformed from automation to digitalization. Similarly, Industry 4.0 in Germany faces the same problems that originated in different countries, such as the Internet industry in the United States made by China, Japan Industry 4.1, and South Korea manufacturing Industry Innovation 3.0. The connection of entities is based on two main features. Digitalization and identification are important features for entity connection. From another perspective, the Internet of Things (IoT) is determined for managing the identification problems, which mostly happen in the Industrial Internet of Things (IIoT). The cyber-physical system is defined to solve the entities' connection problems.

In a recent development, smart manufacturing was named a core of modern production in the manufacturing industry's digitalization. Similarly, it is the smart factory's foundation [1]. The smart manufacturing process uses information technology (IT) to connect the facilities and terminal devices that are digitalized [2]. The interactions between the devices produce massive amounts of data, which causes multiple requirements for the processing of data, e.g., unstructured, able to handle massive amounts, and less time delay. Big data techniques, cloud computing techniques, and artificial intelligence techniques are presented to simplify data processing, which is part of data technology (DT). Furthermore, operational technology (OT) achievement is based on the combination of detailed control machines and data computation, e.g., a distributed control system, programmable logic controller, data acquisition, and supervisory control. Cloud manufacturing services are applied for further processes of the inner performance of smart manufacturing. This section presents a brief explanation of smart manufacturing and related techniques. There are three main topics discussed in this section—edge computing; blockchain; IoT, Industrial Internet of Things (IIoT), Industry 4.0, and cyber-physical systems (CPS).

1.1. Edge Computing

In recent years, many researchers have focused on the edge computing issue regarding intelligent manufacturing. To address some of the low latency and limited resources of this system, Yin et al. [3] proposed a novel visualization service for task scheduling based on fog computing and explored a new approach to the task scheduling algorithm based on a container role. The proposed system is able to reduce the delay rate of the tasks and improve the concurrent tasks on fog nodes. Lei et al. [4] presented the architecture of adaptive transmission containing edge computing and software-defined network (SDN) to solve the problem of data exchanging in IIoT and intelligent devices. Suganuma et al. [5] proposed the Flexible and Advanced Internet of Things (FLEC) to overcome the integration of traditional Internet of Things and edge computing problem that focuses on user positioning adapting to the environment. Lin et al. [6] presented the swarm optimization algorithm connected with a genetic algorithm to overcome the load balancing problem in traditional data placement based on optimizing the transfer time. To achieve detailed control of smart manufacturing systems, communication latency and a reliable environment are required. The multi-access edge computing (MEC) provides all the mentioned requirements. Similarly, cloud computing's capabilities and information technology provide environmental services on the edge network, despite the access technology [7]. Chen et al. [8] proposed a multi-micro-controller structure, which is the gateway for the Industrial Internet and combines the array-based programmable gateway of hardware with multiple scalable micro-controllers. Li et al. [9] proposed adaptive transmission architecture based on the centralized global support for am IIoT edge computing network. Another approach presented by Yu et al. [10] is the survey of edge computing performance on IoT applications—smart cities, smart farms, smart transportation, etc. Porambage et al. [11] showed an MEC overview for IoT applications realization and synergy.

1.2. Blockchain

Blockchain technology is one of the famous areas for trust and safety, which can apply in any related topics to keep the information and data private. Similarly, it is a novel technology for decentralized and distributed computing architecture that keeps the dataset with encrypted blocks in a chain [12–14]. Digital information related to transactions, date and time, amount, etc., which are elaborated in the transaction process, is all stored in blocks. The saved data are available within the distributed network, containing nodes' participants to validate the transaction. All nodes throughout blockchain are linked with each other and support the crypto and transaction codes. Another important feature in blockchain technology is the mathematical algorithms, which are very strong in this network. It provides block validation to minor nodes without any effect on data through the blockchain network, which is why blockchain is secure and transparent [15–23]. There are many of research requirements for addressing the security problems and recommendation systems based on blockchain and knowledge discovery technology [24–29]. This process needs to carry out the integration of blockchain and IoT. Similarly, the security issues which are mentioned by many authors specify the blockchain as a good solution. In [30], blockchain's key features are defined as trust, security, programmability, etc. A blockchain can be one of three different types—a public blockchain, a private blockchain, or a consortium blockchain. The public blockchain is famous for digital currencies. The main objective of a consortium blockchain is to combine the stakeholder and service trading entities. Li et al. [31] presented the energy trading system based on a consortium blockchain. Min [32] proposed to leverage blockchain methods to enhance supply chain flexibility in risky situations. In a business trading system, blockchain technology can be assumed for IoT applications for implementing private blockchains. In [33], an IoT-oriented data exchange system was designed based on the Hyperledger Fabric to overcome the automatic maintenance of a distributed management system problem.

1.3. Internet of Things, Industrial Internet of Things, Industry 4.0, and Cyber-Physical Systems

The growth of the IoT system provides substantial support for the digitalization environment. Furthermore, the IoT applications cover different perspectives—smart farms, smart cities, traffic monitoring, etc. Similarly, the machine-to-machine (M2M) techniques are also covered by IoT systems, which is a way forward of digitalizing the manufacturing system [34]. The abstraction of Industry 4.0 becomes apparent when IIoT meets the cyber-physical system (CPS), which is the best solution for improving the efficiency of productivity in smart manufacturing. Yang et al. [35] presents the IoT applications and issues in the smart manufacturing system. The conclusion of the proposed work shows that IoT visualized the interconnection of the physical world and cyberspace. On the other hand, in [36], a cyber-physical production system (CPPS) was proposed to authorize the dataset efficiency transferring based on the intelligent network and trustworthy communication technology. The Industrial Internet Consortium (IIC) is one of the most famous techniques launched in US top five companies—GE, AT&T, Cisco, Intel, and IBM. This technique mainly points to the standardization of network innovations, applications, and constructions; data circulation growth; and industrial digital transformation. The IIoT sub-concept was first launched in Germany by the name of Industry 4.0 and globally partial CPS facts based on artificial intelligence in smart manufacturing. In short, CPS shows the relationships between information and the physical world, relying on the interconnection of things. The IoT technology selects the interconnections between physical address objects to check if they are related to the industry or not. Table 1 shows the studies related to smart manufacturing systems. Ten studies are compared based on the industry sector, internal equipment, external equipment, and concept of creation.

Table 1. A taxonomy of smart manufacturing applications.

#	Authors	Industry Sectors	Internal Equipment	External Equipment	Creation Concept (Design, Production, Test, Service)
1	Chen et al. (2018) [37]	Automotive industry	No	No	Yes
2	Zhou et al. (2017) [38]	Energy industry	No	No	Yes
3	Dutta et al. (2018) [39]	Transportation equipment manufacturing	Yes	No	No
4	Weissenblock et al. (2014) [40]	Chemical fibers manufacturing	No	No	Yes
5	Chen et al. (2017) [41]	Food processing industry	No	No	Yes
6	Amirkhanove et al. (2014) [42]	Ordinary machinery industry	No	No	Yes
7	Zhou et al. (2011) [43]	Iron and steel industry	Yes	No	No
8	Wu et al. (2018) [44]	Chemical industry	No	No	Yes
9	Coffey et al. (2013) [45]	Specialized equipment manufacturing	No	No	Yes
10	Millette et al. (2016) [46]	Electronic equipment manufacturing	No	Yes	No

Table 2 presents the recent challenges on the integration of blockchain and IoT technology in the smart manufacturing industry. The comparison shows the techniques applied in this research, the main contributions of the presented methods, the usage of blockchain and IoT, the challenges of the proposed systems, and the limitations of the research.

Table 2. Challenges of blockchain and IoT integrated methods.

#	Authors	Applied Technique	Contribution	Blockchain	Internet of Things	Challenges	Limitations
1	Asutosh et al. [47]	Decentralized and cryptographical platform	Avoiding the central authority usage in decentralized and cryptographical platform for verification and connection	Yes	Yes	No	There is no improvement on data confidentiality
2	Marco et al. [48]	The technology of full-stack and view-point of system level	Choosing 6G technology based on view-point of system-level in communication models	No	No	Yes	No verification for security enhancement
3	Emanuel et al. [49]	Transaction Model	Improving the IoT privacy based on blockchain operations	Yes	Yes	No	No reduction on computational cost
4	Chao et al. [50]	Structure of Blockchain	Identifying the process between IoT and Blockchain	Yes	Yes	No	No changes in level of security
5	Bong et al. [51]	IoT devices security modul	Limit hacking based on usage of blockchain	Yes	Yes	Yes	Verification didn't improve the security level
6	Yueyue et al. [52]	Secure and intelligent architecture	Applying deep reinforcement learning to increase the effectiveness of system based on secure and intelligent architecture	Yes	No	Yes	No improvement on privacy level
7	Maroufi Mohammad et al. [53]	IoT and Blockchain	Managing short comings and limitations based on high-level solution technology	Yes	Yes	Yes	Exact issue not designed with the proposed architecture
8	Alfonso et al. [54]	Integration of IoT and Blockchain	Testing the related researches to IoT and Blockchain	Yes	Yes	Yes	The level of complexity didn't minimized
9	Lei et al. [55]	Blockchain and IoT integrated method	Integrated method secure the sensing data.	Yes	Yes	Yes	No reduction on overheard communication
10	Ishan et al. [56]	Centralized architecture	Reducing the over-head computational based on centralized architecture	Yes	Yes	Yes	Reduction of computational over-head has no effect on energy consumption changes

The development of smart manufacturing underpins integrating information technology, data technology, and operational systems. The ever-increasing facilities and devices are leading to data processing and application challenges in existing technology. To reduce this issue's effectiveness, multi-access edge computing was extracted from cloud technology as a solution for the mentioned problems and for its ability to simplify the data processing in the Industrial Internet of Things and industrial cloud computing [57]. Another issue in the smart manufacturing system is the transmission of data and business transactions. Blockchain technology is a suitable answer to overcome this issue, which stabilizes data transmission and business transactions by using the distributed control mechanism [58]. Smart manufacturing systems' immense data processing causes the issues mentioned in [59,60]—high dimensionality, feature space, etc. Deep learning allows the data processing to automatically go through complex feature abstraction using multiple layers, and similarly provides advanced data analysis for smart manufacturing. The challenges mentioned above are being analyzed using state-of-the-art machine learning techniques and smart manufacturing applications. Figure 1 shows the data-driven role in the smart manufacturing system. The data-driven process is divided into three main layers named data-driven, manufacturing system, and benefits. The data-driven layer contains machine learning, deep learning, artificial intelligence, the Internet of Things, big data, and cloud computing techniques. After data-driven, the manufacturing system layer contains three main steps, named technology in manufacturing, network, and advanced analysis. This step's important information includes the design, process, equipment, records, customers, suppliers, parts, and workforce information. The last layer of the data-driven system has the manufacturing system's benefits: quality, energy, cycle time, etc.

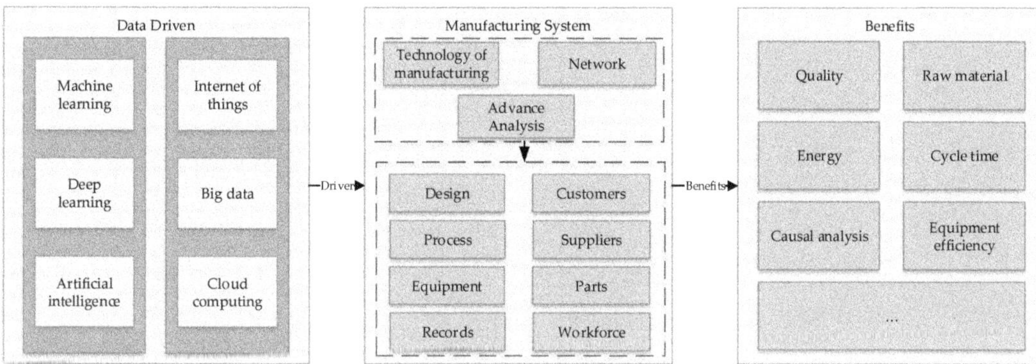

Figure 1. Smart manufacturing's data-driven roles.

The main contributions of this paper are:

- Investigating the multi-access edge computing potential problems, blockchain, and machine learning in the smart manufacturing system.
- The proposed approach's conceptual scenario is the integration of multi-access edge computing, blockchain, and machine learning.
- The multi-access edge computing changed the smart manufacturing architecture from centralized management to decentralized style.
- Addressing the terminal device's task assignment issue.
- Representing the allocation issue between the edge servers.
- Providing an optimization process by applying the swarm intelligence to the presented smart manufacturing system.
- The main objectives of applying machine learning in this system are reducing the manufacturing environment's predicted values and improving the productivity rate.
- Securing the information of stored data in blocks based on blockchain technology.
- Improving the productivity and cost reduction using blockchain technology.

The rest of this paper is divided up as follows: Section 2 presents the proposed integrated model's conceptual scenario in smart manufacturing. Section 3 presents the final result and validation of the system's performance, and we conclude this paper in the conclusion section.

2. System Architecture of the Proposed Smart Manufacturing Environment

The integration of edge computing, blockchain, and machine learning can simplify data processing and transactions in s smart manufacturing system. The following steps present the details of the proposed method in a smart manufacturing system.

2.1. Prototype System Based on Edge Computing

The edge computing system's main concept is to apply the computing technique as close to a data source as possible. Figure 2 presents the edge computing architecture in the smart manufacturing system. The local infrastructure is used to process the data in an edge-computing system, and it takes the cloud server to the hardware. There are three main layers in the edge computing system named the physical layer, network layer, and application layer. The physical layer consists of sensors, robots, actuators, etc., organizing the physical layer's main components. The second layer contains the various edge servers, which process the terminal devices for the third layer's input. Unlike a cloud server, an edge server provides a computational service limited to capacity and resources. The root of enterprise-level applications is IIoT cloud server data processing, all done in the application layer. Enterprise information systems (EIS), supply chain managements (SCM), and smart contracts (SC) are some application layer examples. Applying edge computing in smart manufacturing is far greater than cloud server supplementary resources. Edge computing's prosperity is highly based on virtualization technologies. Virtualization technology contains virtual machines and containers. The main differences between them are the implementation and level of isolation; in the virtual machine, the implementation needs hardware visualization. In the virtual container, the performance is based on light-weight visualization.

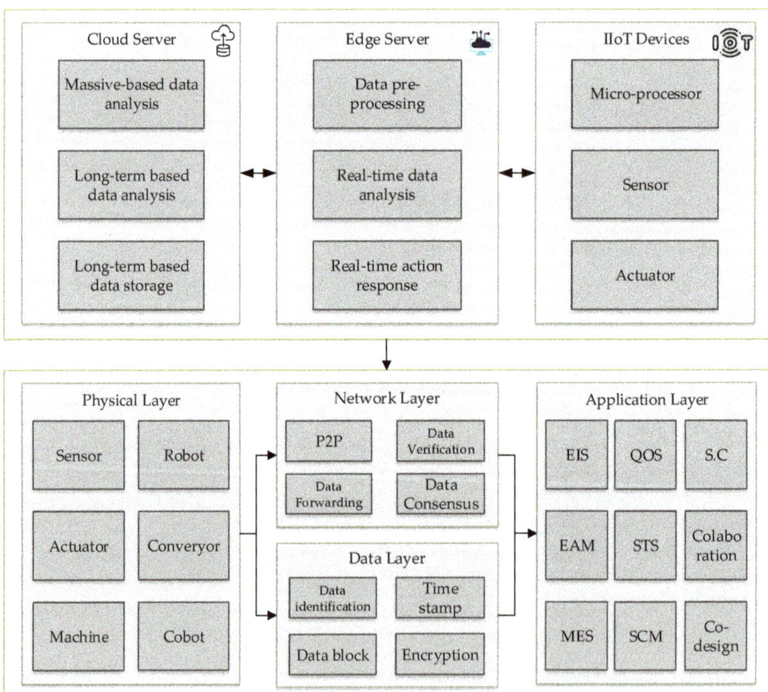

Figure 2. Overview of edge computing architecture.

2.2. Service Validation Based on Blockchain

The blockchain technology in smart manufacturing consists of two main contributions. The first one is IIoT, and edge computing servers' smart manufacturing changes from cloud-centered to the distributed system architecture. In this process, the blockchain system is applied to strengthen data integrity and decrease data transmission risk to authorize the validation key and identification in a distributed manner. To avoid operation defectiveness, the data transactions should be time-stamped through the hash code and refrain from positioning the fake data in the linked chain. The second one is the consensus mechanism, which is used to decide whether adding a validated block into blockchain is possible or not. Smart manufacturing digitalization recommends manufacturing virtualization, leading the cloud manufacturing service from another point of view.

Figure 3 presents the manufacturing system based on two main fields, contents and metadata: identify the unique service and give a detailed description of the process. The service block was created based on the manufacturing system abstraction, and similarly broadcasting the distributed manufacturing in-network service to further validate network entities. The service transaction block creation is based on purchasing and querying the manufacturing service. The transaction block is in the same manufacturing system network, and validates based on the other peer-to-peer entities. Similarly, the transaction block adds to the blockchain transaction system too. In contrast, blockchain's transaction process organizes the smart contract between the business partners, facilitates the inner protocols, and verifies a contract's performance.

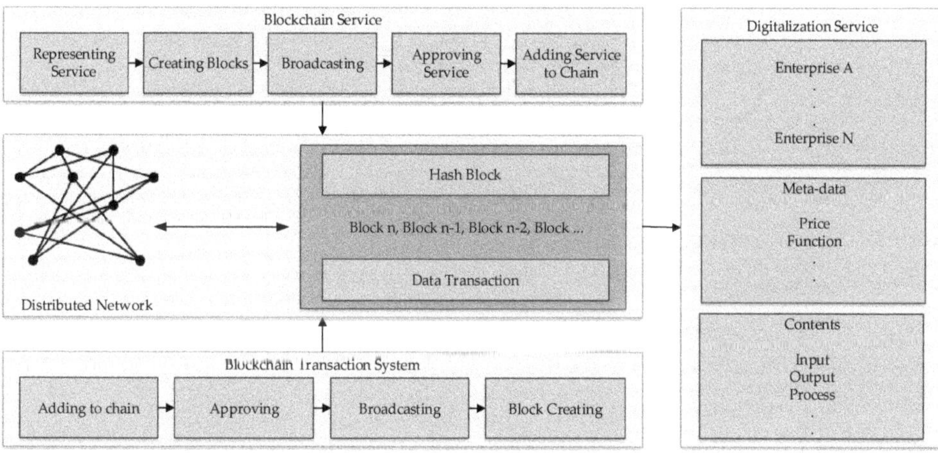

Figure 3. Overview of a blockchain service.

2.3. Machine Learning-Based Smart Manufacturing

Based on the recent new technologies—big data, IoT, etc.—smart facilities are positively developing intelligence manufacturing to impact the cross-organization in smart manufacturing systems. The manufacturing system is experiencing an unexampled data extension based on the data collection from sensors in various formats, structures, and semantics. Data collection is based on the multiple manufacturing systems, e.g., lines of product, manufacturing equipment, processes, etc. Hug data in the manufacturing system need data modeling and analysis to handle the high-volume dataset growth and support the real-time data-processing. Machine learning techniques contain some advantages for improving smart manufacturing: cost reduction, security, fault reduction, increasing production, operator safety, etc. These advantages include a great and strong bond for the operating procedure. Furthermore, the system's fault detection is one of the decisive components for predictive preservation, and it is essential in the case of industry. Figure 4 presents the overall architecture of smart manufacturing based on the integration of edge computing, blockchain, and machine learning. Each of these methods is well-known, but the integration between them has a huge effect on the manufacturing industry regarding safety, cost reduction, increasing production, etc. The edge computing section is based on the physical, network, and application layers. The physical layer provides the smart sensors connected to the IoT platform for real-time data collection and monitoring. Similarly, in this layer, the ability to check the condition of machines is also available. The network layer updates the information and tracks the dataset over time. The application layer corresponds and reviews the data quality, and finally measures and reports the monitoring results. The edge computing process's final report moves to a blockchain service for securing the collected information in blocks. This information is in terms of assets, design, and block security. The process moves to the machine learning section to control the quality of the service and fault rate prediction. In this section, there is a various level of data analysis. This process contains predictive analysis, diagnostic analysis, descriptive analysis, and prescriptive analysis. The main goal of descriptive analysis is to give the product manufacturing process and operation information, capturing the environmental conditions and parameters. If the product's performance decreases, the diagnosis analysis examines the issue and presents the reason for the problem. The predictive analysis operates the statistical models and predicts the possible future equipment and products based on a historical dataset. The final analysis is the prescriptive analysis, which further recommends actions and measures the identification to improve the rates of outcomes, solve the problems, and present each final decision outcome. Based on the advanced machine learning analysis, the smart facilities are highly optimized. This process's benefits are reducing the costs of operation, meeting changing consumer demands, improving productivity, and reducing downtime.

Equations (1) and (2) present the evaluation of cost reduction in manufacturing industry based on machine learning prediction process. In the first step is a derivation function applied to decrease the error of cost function. The cost function is evaluated below:

$$B = \frac{1}{m} \sum_{n=0}^{m} (g_n - (xh_n + d))^2 \tag{1}$$

where g_n is the predicted value and xh_n is the actual value of the cost prediction process. $\frac{\alpha}{\alpha x}$ represent the partial derivative values. d and e are representing the intercept, and x represents the slope of the evaluation.

$$\frac{\alpha}{\alpha x} = \frac{2}{M} \sum_{n=1}^{M} -h_n(g_n - (xh_n) + e) \tag{2}$$

The predictive accuracy evaluation is based on two main metrics: mean absolute prediction error (MAPE) and normalized root mean square error (NRMSE). Equations (3) and (4) present the MAPE and NRMSE evaluations.

$$MAPE = \frac{1}{m} \sum_{n=1}^{m} |\frac{g_n - \hat{g}_n}{g_n}| \tag{3}$$

$$NRMSE = \frac{1}{m} \sqrt{\sum_{n=0}^{m} (\frac{g_n - \hat{g}_n}{g_n})^2} \tag{4}$$

The MAPE evaluates the prediction's total error compared with initial values, and NRMSE evaluates the normalized squared errors.

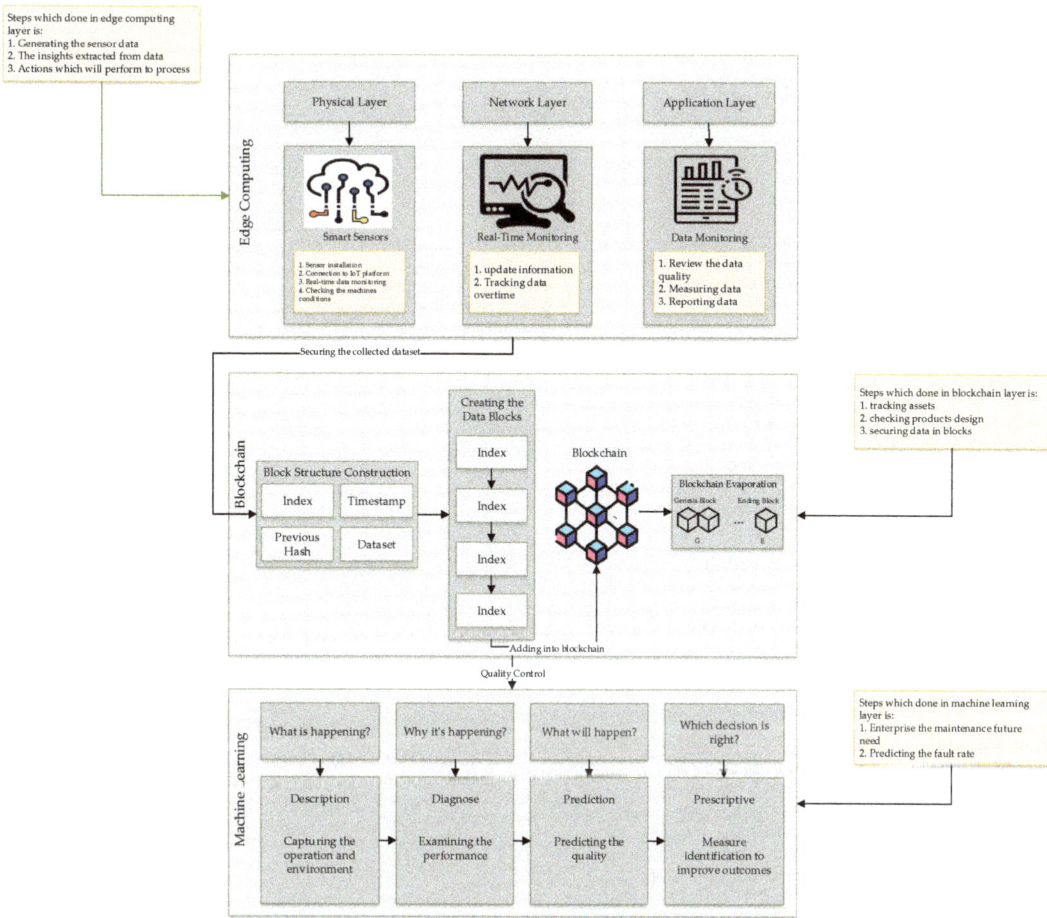

Figure 4. Smart manufacturing overall architecture based on an integrated system.

2.4. Fault Assessment Diagnostic Analysis

Generally, the manufacturing system faces failures based on abnormal and degradation operations. The failing causes high costs, disqualifies the product, and causes lower productivity. Based on the implementation of a smart manufacturing system, it is necessary for smart factories to monitor the condition of machines, identify the primary defects, recognize the root causes of failures, and finally combine the information for manufacturing system production [61]. Based on the data collected from sensors, there are many machine learning algorithms to investigate the fault diagnosis and classification [62]. The convolutional neural network (CNN) combines feature learning and identification into one model and has been applied in many sectors— wind generator [63], rotor [64], bearing [65–68], etc.

3. Results

In this section, a brief explanation of the results is provided. Section 3.1 presents the process of data collection and dataset information. Section 3.2 presents the performance evaluation of the proposed system. Section 3.3 presents smart manufacturing challenges and opportunities.

3.1. Implementation Environment

The implementation of the proposed system structure and environment is presented in this section. Table 3 summarizes the purposed system experimental setup. All the experiments were done on an Intel(R) Core(TM) i7-8700 @3.20 GHz processor with 32 GB memory. Moreover, the docker environment was processed in the 18.06.1-ce version, and the container configuration in the virtual machine was processed based on the docker composer 1.13.0 version. The Hyperledger Fabric framework project is from the Linux Foundation.

Table 3. Development environment of the proposed system.

Component	Description
IDE	Composer-Playground
Memory	32 GB
CPU	Intel(R) Core(TM) i7-8700 @3.20 GHz
Python	3.6.2
Operating System	Ubuntu Linux 18.04.1 LTS
Docker Engine	Version 18.06.1-ce
Docker Composer	Version 1.13.0
Hyperledger Fabric	V1.2
CLI Tool	Composer REST Server
Node	V8.11.4

Figure 5 shows the operation of the transaction process function. For improving the assets and participants, create, delete, update, and other functions were defined in the blockchain network. The functions of the transaction processor were implemented in JavaScript and defined as a smart contract. The specified ShareRecord function is used to update the manufacturing records based on the events and registry.

```
/**
* share Manufacturing Repositry Record with Manger
* @param {composers.ManufacturingRepositry.shareRepositryRecordsWithManger} newDetail - the update DrugRepositry transaction
* @transaction
*/
async function shareRepositryRecordWithManager(record) {
    //payBill.user.balanceDue = payBill.bill.amount;
    return getAssetRegistry('composers.ManufacturingRepositry.Repositry')
        .then(function(assetRegistry){
        record.manufacturingRepositry.manager = record.mangerID;
        console.log(record.manufacturingRepositry.manager);
        let factory = getFactory();
        let shareRecordEvent = factory.newEvent('composer.ManufacturingRepositry', 'shareRepositryRecorWithManagernotification');
        shareRecordEvent.manufacturingRepositry = record.manufacturingRepositry;
        emit (shareRecordEvent);
        return assetRegistry.update(record.manufacturingRepositry);
    })
    .catch(function (error) {

    });
}
```

Figure 5. Transaction processor function in a manufacturing blockchain platform in the proposed manufacturing system.

To control the domain model elements, the access control language (ACL) is needed. ACL provides the ability to define rules to specify the roles and users, which are authorized to make changes in the business network domain. Figure 6 shows the ACL rules defined in this network that give participants access to make changes in the network.

```
//6
rule MangerUpdateManufacturingOrderTransaction{
    description: "Only Manager can update Manufacturing Order transaction"
    participant: "composers.participant.Manager"
    operation: READ, CREATE, UPDATE
    resource: "composers.ordersdetails.UpdateorderStatus"
    action: Allow
}

//7
rule MangerUpdateTransactionRecords{
    description: "Only Manager can update Transaction records"
    participant: "composers.participant.Manager"
    operation: READ, CREATE, UPDATE
    resource: "composers.prescription.UpdateTransactionRecords"

//8
rule MangerViewTransactions{
    description: "Only Manager can view Transactions"
    participant: "composers.participant.Manager"
    operation: READ, CREATE, UPDATE
    resource: "composers.prescription.ViewTransactions"
```

Figure 6. Access control definition in the proposed manufacturing system.

3.2. Dataset Management

The smart manufacturing system's data increase in volume based on the traditional algorithms' ability, mostly when the user wants to extract useful information from the collected dataset. High sample volume in a large dataset, when the records are not similar, needs the consolidation and isolation algorithms for implementation and knowledge utilization. In this research, the data were collected from various sources related to IoT; the production equipment was collected from various sensors to monitor the product in real-time—e.g., the built-in sensors measured, monitored, and reported the status of manufacturing equipment and product based on the temperature, humidity, pressure, etc. Figure 7 shows the data driven process in smart manufacturing.

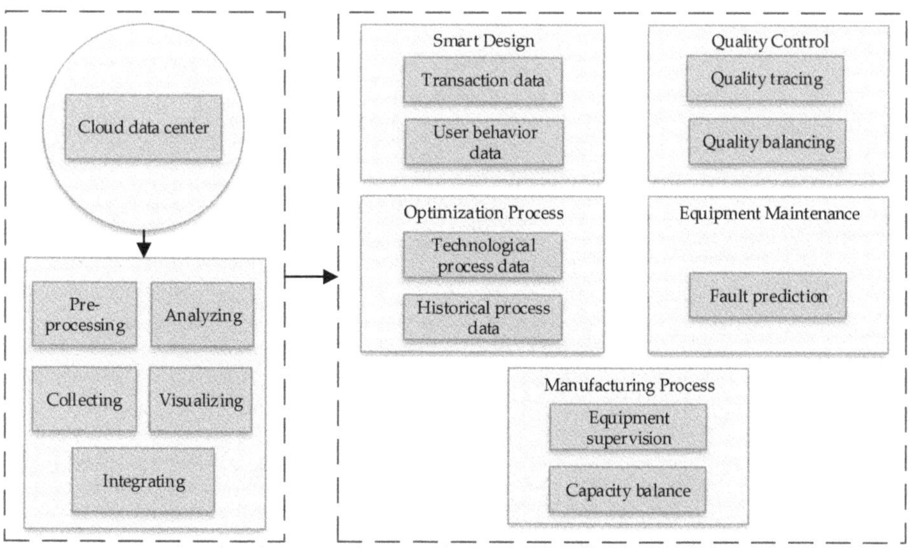

Figure 7. Data-driven process.

Table 4 presents the configuration of IoT devices and sensors for real-time data collection. During the smart manufacturing (SM) cycle, the IoT devices are located in the main areas of manufacturing resources at various levels, e.g., machines, factories, etc. The radio-frequency identification (RFID) tags are mainly configured enough in practical documents to report important machines' quality, design, and production procedures in the manufacturing process.

Table 4. IoT device configuration information for data usage in smart manufacturing.

IoT Devices	Type of Device	Monitoring Resources	Purpose
Smart Sensors	Temperature	SM machine	Temperature data monitoring
Smart Sensors	Humidity	SM machine	Humidity data monitoring
Smart Sensors	Pressure	SM machine	Pressure data monitoring
RFID Tags	Ultra high frequency	Drawing, model, material	Trace and monitor real-time data
RFID Tags	Ultra high frequency	Operate, product, etc	Trace and monitor real-time data
RFID Reader	Ultra high frequency	Material, maintenance	Identify and track components

3.3. Optimization

Smart manufacturing based on the edge computing system has high scalability and huge IIoT devices, which is suitable based on the expansion potentiality. Data analysis and transmission are considered computational tasks. They are supposed to allocate data on an edge server or cloud to recognize the suitable task assignment for reducing the process time of the incoming task. There are X defined device terminals, Y edge nodes, and one industrial server for the cloud to design this issue in smart manufacturing. Within the manufacturing process, the requests from terminal devices are managed by a cloud or edge server. The process timing for the tasks in edge server requires two main components called computation time $\theta^{y}_{i,c}$ and data transmission time $\theta^{y}_{i,d}$; see Equation (5).

$$\beta^{y}_{i} = \theta^{y}_{i,c} + \theta^{y}_{i,d} \tag{5}$$

The task computation time in an edge server is evaluated based on Equation (6).

$$\theta^{y}_{i,c} = L_i / \Sigma^{n^{i}_{max}}_{n=1} a^{y}_{i,n} \tag{6}$$

y is defined as the edge server, i represents the task computation time, and $a^{y}_{i,n}$ represents the edge server's computational resources through the n period of maintaining tasks. L is defined as the length of the tasks. The task processing time in a cloud server is evaluated as it was presented in Equation (7):

$$\beta^{t}_{i} = \theta^{t}_{i,c} + \theta^{y}_{i,d} + \theta^{y,t}_{i,d} \tag{7}$$

The computation time in the cloud is defined as $\theta^{t}_{i,c}$. Data transmission between the edge and device is defined as $\theta^{y}_{i,d}$. The transmission time from the edge server to cloud is defined as $\theta^{y,t}_{i,d}$. The task assignment's presented issue is the deployment of a parallel mechanism and heterogeneous units' processing in the computational task assignment issue. Accordingly, the swarm intelligence approach is applied in this process.

Swarm Intelligence

Generally, the swarm intelligence (SI) approach is a famous process among artificial intelligence algorithms. Two main strategies follow based on this algorithm named approximate and non-deterministic to consider and utilize the searching spaces to find the near-optimal solutions [69]. SI contains various approaches; among them, the artificial bee colony (ABC) algorithm demonstrates SI's classic features. The importance and required process for intelligence performance, self-organization, collective behavior, and decentralization of SI are sufficient [70]. Moreover, the mentioned three features contain the simple mechanism control, which is tuned with only two parameters. The bee colony's size determines whether the solution can be dropped or whether there is no need to drop it. Figure 8 shows the process of solving the computational task problem based on the artificial bee colony workflow.

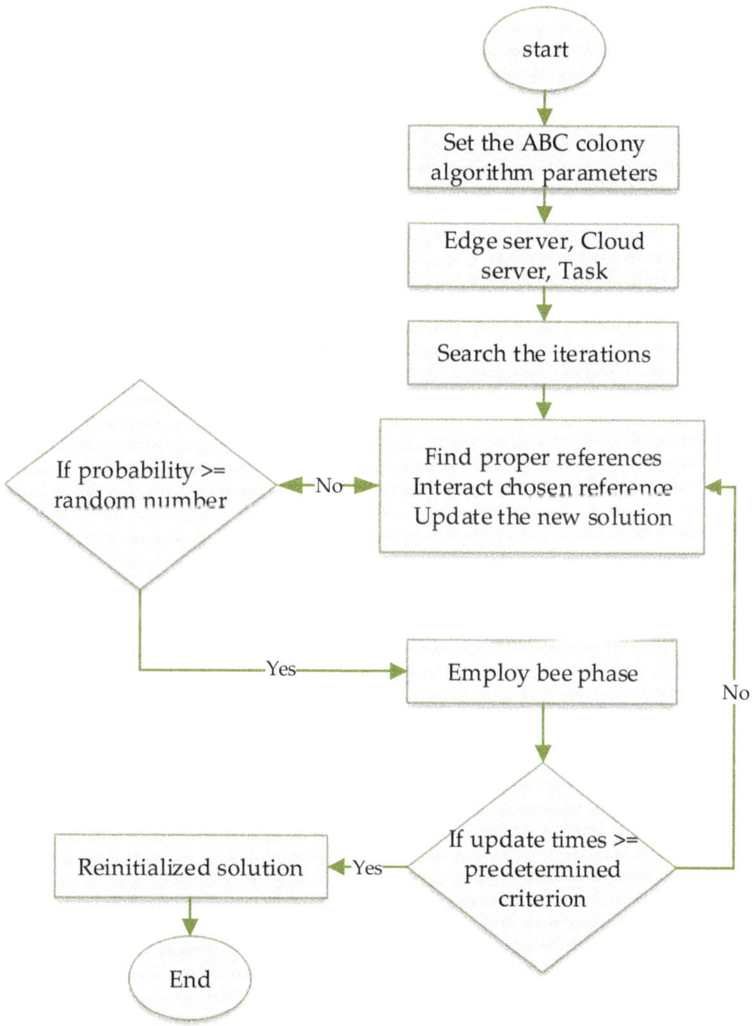

Figure 8. Artificial bee colony workflow.

3.4. Performance Evaluation

The test models are generated based on the following patterns. The task length follows a uniform distribution in the range of [1, 10] million specifications. The data volume is defined as 100 KB to 10 MB. The time delay is 100 milliseconds to 10 s. The average processing performance based on the edge server is defined as 10 million instructions per second (MIPS). The cloud volume is 1000 MIPS. Edge server and device connections work through wireless communication. The edge server and cloud connections go through broadband. Tasks are specified to the evident edge server, which forwards the information to the cloud. This causes the edge server to be limited to processing enough resources for the under-processed task during the delay time. Figures 9 and 10 present the analysis of parameters for abandonment and solution number (SN) criteria of α for incoming tasks (200).

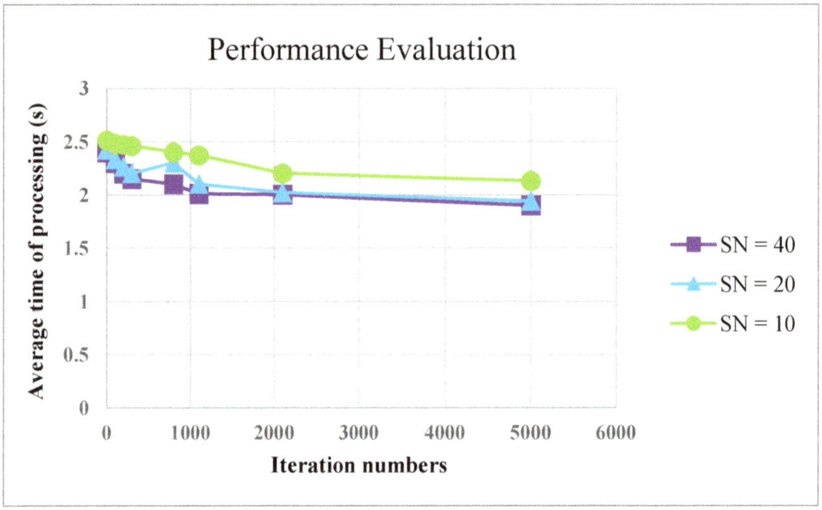

Figure 9. Performance evaluation of various solution number (SN) settings.

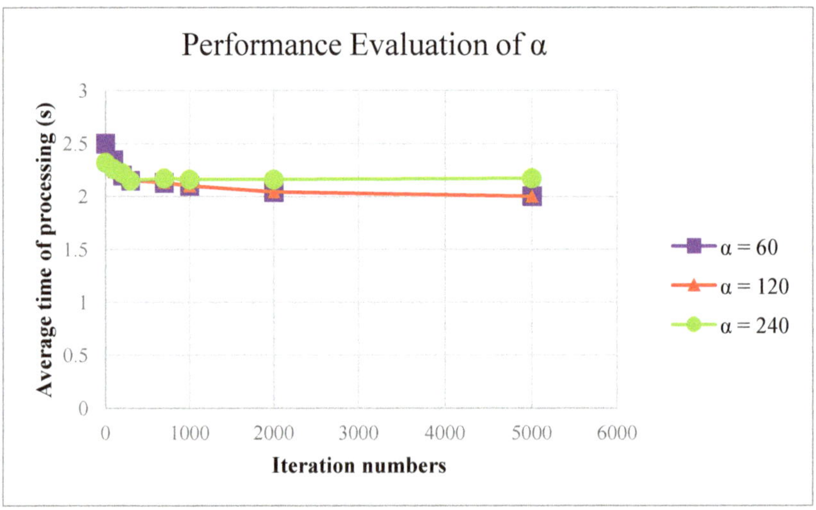

Figure 10. Performance evaluation of various α settings.

To show edge computing's effectiveness in the system of smart manufacturing, Figure 11 shows a various number of incomes based on three main frameworks, i.e., cloud, edge, and mixed-mode. The meanings of these three scenarios show the computational task between them. As shown in Figure 11, the mixed-mode shows the combined outperformance of edge and cloud. Similarly, it is increasing the number of tasks along with the cloud mode's average processing time. When the tasks are less than the cloud server's capacity, there is a decrease at a certain level; on the other hand, if the number of functions increases, then the edge server does not modify the processing time appropriately.

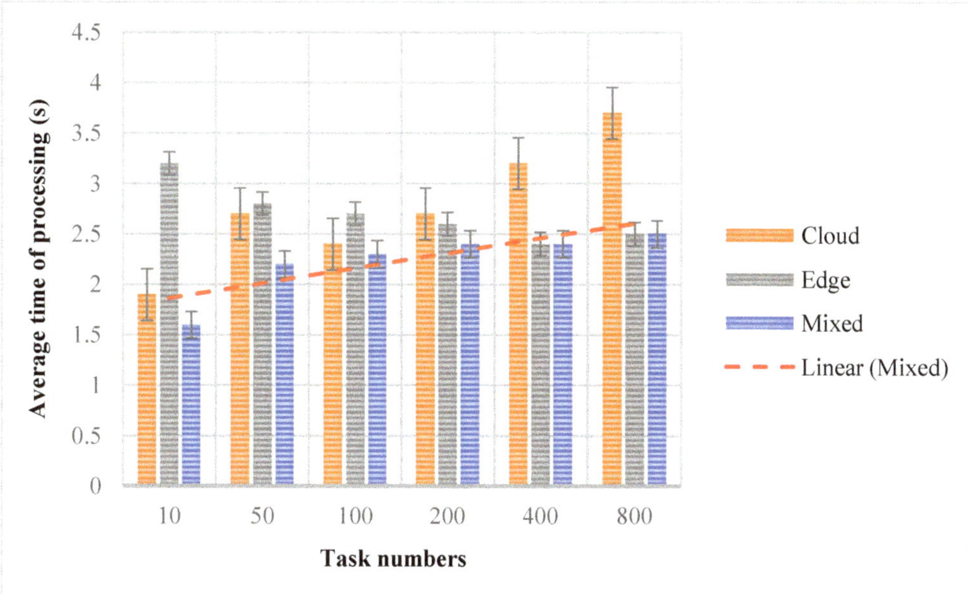

Figure 11. Performance evaluation of various scenarios.

Figures 12 and 13 present the machine learning techniques applied in this process: k-means clustering algorithm (IKCD), k-means clustered deployment (KCD), and random deployment (RD). Figure 12 shows each algorithm's delay rate in different edge computing nodes (ECNs). When the number of nodes increases in the edge computing system, the amount of equipment production also reduces due to the network's delay reduction. Based on the presented results, the network delay in the IKCD process is the shortest one, and RD is the worst among the compared methods. Based on the ECNs in the system, when there are between 1 and 3, the IKCD method is better than KCD, and when the number of ECNs is more than three, based on the network latency, the differences between IKCD and KCD decrease.

Figure 13 presents the system cost deployment differences for the ECNs. We can see the ECNs incurred greater costs based on the system node increases. Based on the deployment of ECNs for the higher costs, the costs for all three methods increased. Comparing the three methods, IKCD had the highest and most outstanding performance. The RD method's deployment used a number of ECNs randomly and did not deploy any node in the production node. This process caused the node to be chosen without consideration and constraints. The deployed nodes recorded in the KCD process are based on the Euclidean distance between the devices, which is not sensible and causes the network delay and access time communication for data processing in real-time requirements.

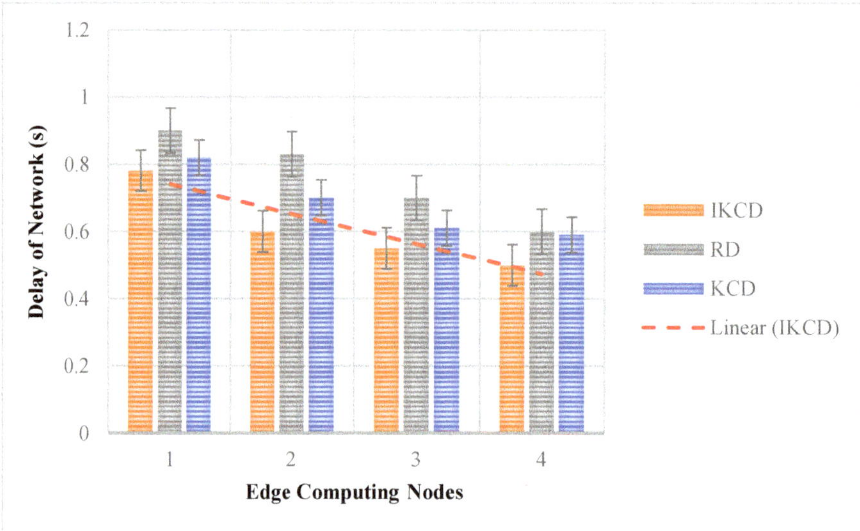

Figure 12. Changes of network nodes based on edge computing.

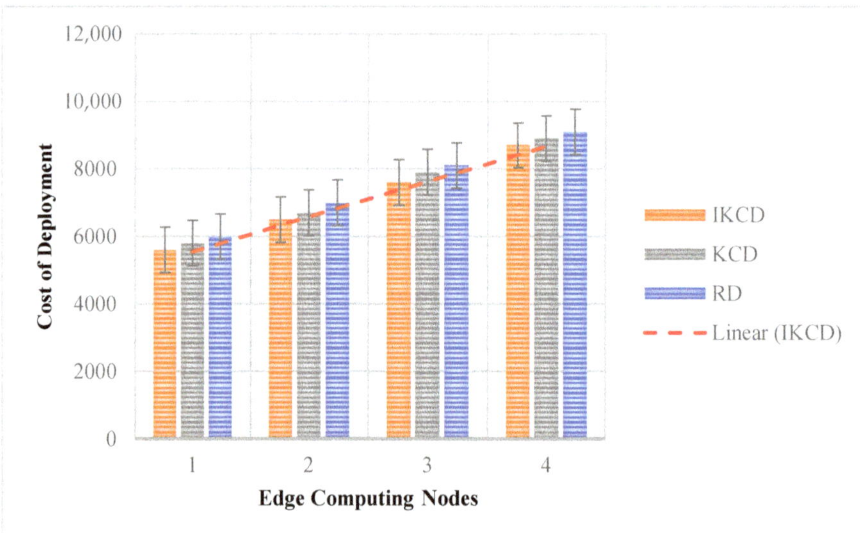

Figure 13. Cost dependency deployment on edge computing nodes.

Figure 14 presents the relationship between system cost and edge computing nodes based on the compared methods.

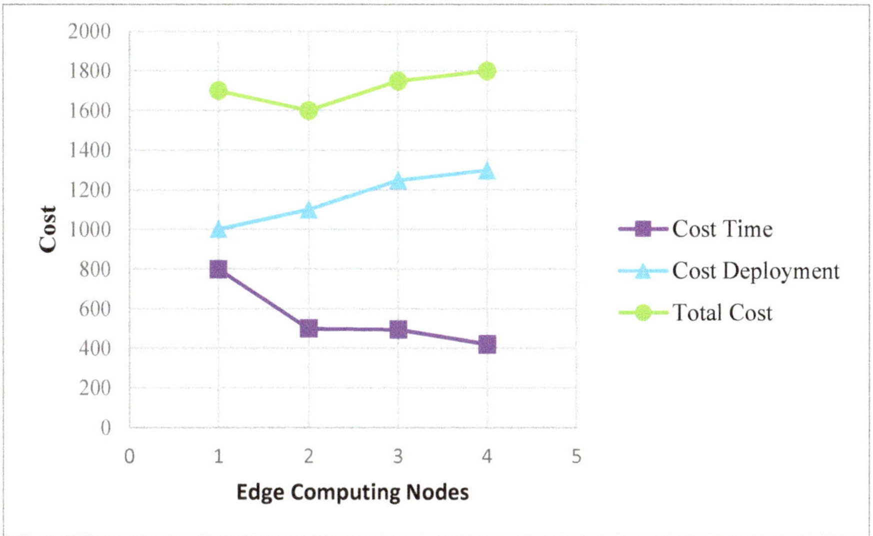

Figure 14. Total cost of IKCD and edge computing node relationship.

The results show the advantages of the IKCD method, and similarly, show the reduction of network delay based on the ECNs and increasing the computing cost of ECNs. The sum of the process decreases at the start and then increases.

3.5. Challenges and Opportunities of the Smart Manufacturing System

Data management, performance evaluation, and standardization of edge computing in the IIoT system are briefly explained in the above sections. Analyzing the proposed system based on the integrated methods reveals great opportunities and challenges for edge networks, data processing, security, etc., in IIoT technology related to edge computing. The below information explains several challenges and opportunities for smart manufacturing systems.

- Data offloading and load balancing: The IIoT system having various devices, which are important in data offloading among the large servers and devices. The IIoT system, based on edge computing, reflecting on data processing, increases this process's difficulty.
- Edge intelligence: In a recent IIoT system designed based on edge computing, the devices could only accomplish the light-weight tasks. To make the system intelligent, edge intelligence (EI) must be applied to the process.
- Data sharing security: One of the IIoT system's advantages is the huge amount of data in real-time devices, websites, etc., which is efficient to improve industrial production.

4. Conclusions and Future Research

Smart manufacturing is a favorable movement for the evolution of the manufacturing industry and production in a new industry. The manufacturing system's implementation causes the support of data technology, information technology, and operational technology, surrounded by the development of integrated edge computing, blockchain, and machine learning based on the Industrial Internet for operational processes in the manufacturing environment. This paper's proposed system was designed based on integrating edge computing, blockchain technology, and machine learning to support the manufacturing system's design. The assignment problem of the system was formulated based on the optimization model. Unlike other research in edge computing and IIoT, the presented method's stresses illuminate the integration method's importance in future developments.

The future research plan is to improve the manufacturing system more and analyze it in more detail. The blockchain system's applications can be quantified and further analyzed. Other technologies can be incorporated to enhance the development of the manufacturing system. The experiments and results can be analyzed with an on-site dataset to identify the possible impact factors and regulate the proposed model's configured parameters.

Author Contributions: Conceptualization, Z.S.; Formal analysis, Z.S.; Funding acquisition, Y.-C.B.; Methodology, Z.S.; Writing–review and editing, Z.S.; Investigation, Z.S.; Resources, Y.-C.B.; Methodology, Z.S.; Project administration, Y.-C.B.; Supervision, Y.-C.B. All authors have read and agreed to the published version of the manuscript.

Funding: This research was financially supported by the Ministry of Small and Medium-sized Enterprises(SMEs) and Startups(MSS), Korea, under the "Regional Specialized Industry Development Program(R&D, S2855401)" supervised by the Korea Institute for Advancement of Technology(KIAT).

Conflicts of Interest: The authors declare no conflict of interest.

References

1. Tao, F.; Qi, Q.; Liu, A.; Kusiak, A. Data-driven smart manufacturing. *J. Manuf. Syst.* **2018**, *48*, 157–169. [CrossRef]
2. Cheng, J.; Chen, W.; Tao, F.; Lin, C.L. Industrial IoT in 5G environment towards smart manufacturing. *J. Ind. Inf. Integr.* **2018**, *10*, 10–19. [CrossRef]
3. Yin, L.; Luo, J.; Luo, H. Tasks scheduling and resource allocation in fog computing based on containers for smart manufacturing. *IEEE Trans. Ind. Inform.* **2018**, *14*, 4712–4721. [CrossRef]
4. Li, X.; Li, D.; Wan, J.; Liu, C.; Imran, M. Adaptive transmission optimization in SDN-based industrial Internet of Things with edge computing. *IEEE Internet Things J.* **2018**, *5*, 1351–1360. [CrossRef]
5. Suganuma, T.; Oide, T.; Kitagami, S.; Sugawara, K.; Shiratori, N. Multiagent-based flexible edge computing architecture for IoT. *IEEE Netw.* **2018**, *32*, 16–23. [CrossRef]
6. Lin, B.; Zhu, F.; Zhang, J.; Chen, J.; Chen, X.; Xiong, N.N.; Mauri, J.L. A time-driven data placement strategy for a scientific workflow combining edge computing and cloud computing. *IEEE Trans. Ind. Inform.* **2019**, *15*, 4254–4265. [CrossRef]
7. Reznik, A.; Arora, R.; Cannon, M.; Cominardi, L.; Featherstone, W.; Frazao, R.; Giust, F.; Kekki, S.; Li, A.; Sabella, D.; et al. Developing software for multi-access edge computing. *ETSI White Pap.* **2017**, *20*.
8. Chen, C.H.; Lin, M.Y.; Liu, C.C. Edge computing gateway of the industrial internet of things using multiple collaborative microcontrollers. *IEEE Netw.* **2018**, *32*, 24–32. [CrossRef]
9. Khan, P.W.; Byun, Y.C.; Park, N. A Data Verification System for CCTV Surveillance Cameras Using Blockchain Technology in Smart Cities. *Electronics* **2020**, *9*, 484. [CrossRef]
10. Yu, W.; Liang, F.; He, X.; Hatcher, W.G.; Lu, C.; Lin, J.; Yang, X. A survey on the edge computing for the Internet of Things. *IEEE Access* **2017**, *6*, 6900–6919. [CrossRef]
11. Porambage, P.; Okwuibe, J.; Liyanage, M.; Ylianttila, M.; Taleb, T. Survey on multi-access edge computing for internet of things realization. *IEEE Commun. Surv. Tutor.* **2018**, *20*, 2961–2991. [CrossRef]
12. Wang, S.; Ouyang, L.; Yuan, Y.; Ni, X.; Han, X.; Wang, F.Y. Blockchain-enabled smart contracts: Architecture, applications, and future trends. *IEEE Trans. Syst. Man Cybern. Syst.* **2019**, *49*, 2266–2277. [CrossRef]
13. Khan, P.W.; Byun, Y.C.; Park, N. IoT-Blockchain Enabled Optimized Provenance System for Food Industry 4.0 Using Advanced Deep Learning. *Sensors* **2020**, *20*, 2990. [CrossRef] [PubMed]
14. Khan, P.W.; Byun, Y. Secure Transactions Management Using Blockchain as a Service Software for the Internet of Things. In *Software Engineering in IoT, Big Data, Cloud and Mobile Computing. Studies in Computational Intelligence* ; Springer: Cham, Switzerland, 2021; pp. 117–128.
15. Jamil, F.; Hang, L.; Kim, K.; Kim, D. A novel medical blockchain model for drug supply chain integrity management in a smart hospital. *Electronics* **2019**, *8*, 505. [CrossRef]
16. Khan, P.W.; Byun, Y. A Blockchain-Based Secure Image Encryption Scheme for the Industrial Internet of Things. *Entropy* **2020**, *22*, 175. [CrossRef]
17. Jamil, F.; Iqbal, M.A.; Amin, R.; Kim, D. Adaptive thermal-aware routing protocol for wireless body area network. *Electronics* **2019**, *8*, 47. [CrossRef]
18. Jamil, F.; Ahmad, S.; Iqbal, N.; Kim, D.H. Towards a Remote Monitoring of Patient Vital Signs Based on IoT-Based Blockchain Integrity Management Platforms in Smart Hospitals. *Sensors* **2020**, *20*, 2195. [CrossRef]
19. Jamil, F.; Kim, D.H. Improving Accuracy of the Alpha–Beta Filter Algorithm Using an ANN-Based Learning Mechanism in Indoor Navigation System. *Sensors* **2019**, *19*, 3946. [CrossRef]
20. Jamil, F.; Iqbal, N.; Ahmad, S.; Kim, D.H. Toward Accurate Position Estimation Using Learning to Prediction Algorithm in Indoor Navigation. *Sensors* **2020**, *20*, 4410. [CrossRef]
21. Ahmad, S.; Jamil, F.; Khudoyberdiev, A.; Kim, D. Accident risk prediction and avoidance in intelligent semi-autonomous vehicles based on road safety data and driver biological behaviours. *J. Intell. Fuzzy Syst.* **2020**, *38*, 4591–4601. [CrossRef]

22. Jamil, F.; Kim, D. Payment Mechanism for Electronic Charging using Blockchain in Smart Vehicle. *Korea* **2019**, *30*, 31.
23. Shahbazi, Z.; Byun, Y.C. Towards a Secure Thermal-Energy Aware Routing Protocol in Wireless Body Area Network Based on Blockchain Technology. *Sensors* **2020**, *20*, 3604. [CrossRef] [PubMed]
24. Shahbazi, Z.; Jamil, F.; Byun, Y. Topic modeling in short-text using non-negative matrix factorization based on deep reinforcement learning. *J. Intell. Fuzzy Syst.* **2020**, *39*, 753–770. [CrossRef]
25. Shahbazi, Z.; Hazra, D.; Park, S.; Byun, Y.C. Toward Improving the Prediction Accuracy of Product Recommendation System Using Extreme Gradient Boosting and Encoding Approaches. *Symmetry* **2020**, *12*, 1566. [CrossRef]
26. Shahbazi, Z.; Byun, Y.C. Product Recommendation Based on Content-based Filtering Using XGBoost Classifier. *Int. J. Adv. Sci. Technol.* **2019**, *29*, 6979–6988.
27. Shahbazi, Z.; Byun, Y.C. Toward Social Media Content Recommendation Integrated with Data Science and Machine Learning Approach for E-Learners. *Symmetry* **2020**, *12*, 1798. [CrossRef]
28. Shahbazi, Z.; Byun, Y.C. Analysis of Domain-Independent Unsupervised Text Segmentation Using LDA Topic Modeling over Social Media Contents. *Int. J. Adv. Sci. Technol.* **2020**.
29. Shahbazi, Z.; Byun, Y.C.; Lee, D.C. Toward Representing Automatic Knowledge Discovery from Social Media Contents Based on Document Classification. *Int. J. Adv. Sci. Technol.* **2020**.
30. Yuan, Y.; Wang, F.Y. Blockchain: The state of the art and future trends. *Acta Autom. Sin.* **2016**, *42*, 481–494.
31. Li, Z.; Kang, J.; Yu, R.; Ye, D.; Deng, Q.; Zhang, Y. Consortium blockchain for secure energy trading in industrial internet of things. *IEEE Trans. Ind. Inform.* **2017**, *14*, 3690–3700. [CrossRef]
32. Min, H. Blockchain technology for enhancing supply chain resilience. *Bus. Horiz.* **2019**, *62*, 35–45. [CrossRef]
33. Yang, J.; Lu, Z.; Wu, J. Smart-toy-edge-computing-oriented data exchange based on blockchain. *J. Syst. Archit.* **2018**, *87*, 36–48. [CrossRef]
34. Lee, C.; Zhang, S.; Ng, K. Development of an industrial Internet of things suite for smart factory towards re-industrialization. *Adv. Manuf.* **2017**, *5*, 335–343. [CrossRef]
35. Yang, C.; Shen, W.; Wang, X. The internet of things in manufacturing: Key issues and potential applications. *IEEE Syst. Man Cybern. Mag.* **2018**, *4*, 6–15. [CrossRef]
36. Wan, J.; Chen, B.; Imran, M.; Tao, F.; Li, D.; Liu, C.; Ahmad, S. Toward dynamic resources management for IoT based manufacturing. *IEEE Commun. Mag.* **2018**, *56*, 52–59. [CrossRef]
37. Chen, Y.; Xu, P.; Ren, L. Sequence synopsis: Optimize visual summary of temporal event data. *IEEE Trans. Vis. Comput. Graph.* **2017**, *24*, 45–55. [CrossRef]
38. Zhou, F.; Lin, X.; Luo, X.; Zhao, Y.; Chen, Y.; Chen, N.; Gui, W. Visually enhanced situation awareness for complex manufacturing facility monitoring in smart factories. *J. Vis. Lang. Comput.* **2018**, *44*, 58–69. [CrossRef]
39. Dutta, S.; Shen, H.W.; Chen, J.P. In situ prediction driven feature analysis in jet engine simulations. In Proceedings of the 2018 IEEE Pacific Visualization Symposium (PacificVis), Kobe, Japan, 10–13 April 2018; pp. 66–75.
40. Weissenböck, J.; Amirkhanov, A.; Li, W.; Reh, A.; Amirkhanov, A.; Gröller, E.; Kastner, J.; Heinzl, C. Fiberscout: An interactive tool for exploring and analyzing fiber reinforced polymers. In Proceedings of the 2014 IEEE Pacific Visualization Symposium, Yokohama, Japan, 4–7 March 2014; pp. 153–160.
41. Chen, Y.; Du, X.; Yuan, X. Ordered small multiple treemaps for visualizing time-varying hierarchical pesticide residue data. *Vis. Comput.* **2017**, *33*, 1073–1084. [CrossRef]
42. Amirkhanov, A.; Fröhler, B.; Kastner, J.; Gröller, E.; Heinzl, C. InSpectr: Multi-Modal Exploration, Visualization, and Analysis of Spectral Data. *Comput. Graph. Forum* **2014**, *33*, 91–100. [CrossRef]
43. Zhou, C.Q.; Lopez, P.R.; Moreland, J.; Wu, B. Visualizing the future in steel manufacturing. *Iron Steel Technol.* **2011**, *8*, 37–50.
44. Wu, W.; Zheng, Y.; Chen, K.; Wang, X.; Cao, N. A visual analytics approach for equipment condition monitoring in smart factories of process industry. In Proceedings of the 2018 IEEE Pacific Visualization Symposium (PacificVis), Kobe, Japan, 10–13 April 2018; pp. 140–149.
45. Coffey, D.; Lin, C.L.; Erdman, A.G.; Keefe, D.F. Design by dragging: An interface for creative forward and inverse design with simulation ensembles. *IEEE Trans. Vis. Comput. Graph.* **2013**, *19*, 2783–2791. [CrossRef] [PubMed]
46. Millette, A.; McGuffin, M.J. DualCAD: Integrating augmented reality with a desktop GUI and smartphone interaction. In Proceedings of the 2016 IEEE International Symposium on Mixed and Augmented Reality (ISMAR-Adjunct), Merida, Mexico, 19–23 September 2016; pp. 21–26.
47. Biswal, A.K.; Maiti, P.; Bebarta, S.; Sahoo, B.; Turuk, A.K. Authenticating IoT Devices with Blockchain. In *Advanced Applications of Blockchain Technology*; Springer: Warsaw, Poland 2020; pp. 177–205.
48. Giordani, M.; Polese, M.; Mezzavilla, M.; Rangan, S.; Zorzi, M. Toward 6g networks: Use cases and technologies. *IEEE Commun. Mag.* **2020**, *58*, 55–61. [CrossRef]
49. Jesus, E.F.; Chicarino, V.R.; de Albuquerque, C.V.; Rocha, A.A.D.A. A survey of how to use blockchain to secure internet of things and the stalker attack. *Secur. Commun. Netw.* **2018**, *2018*, 9675050. [CrossRef]
50. Qu, C.; Tao, M.; Zhang, J.; Hong, X.; Yuan, R. Blockchain based credibility verification method for IoT entities. *Secur. Commun. Netw.* **2018**, *2018*, 7817614. [CrossRef]
51. Choi, B.G.; Jeong, E.; Kim, S.W. Multiple Security Certification System between Blockchain Based Terminal and Internet of Things Device: Implication for Open Innovation. *J. Open Innov. Technol. Mark. Complex.* **2019**, *5*, 87. [CrossRef]

52. Dai, Y.; Xu, D.; Maharjan, S.; Chen, Z.; He, Q.; Zhang, Y. Blockchain and deep reinforcement learning empowered intelligent 5G beyond. *IEEE Netw.* **2019**, *33*, 10–17. [CrossRef]
53. Maroufi, M.; Abdolee, R.; Tazekand, B.M. On the convergence of blockchain and Internet of Things (IoT) technologies. *arXiv* **2019**, arXiv:1904.01936.
54. Panarello, A.; Tapas, N.; Merlino, G.; Longo, F.; Puliafito, A. Blockchain and iot integration: A systematic survey. *Sensors* **2018**, *18*, 2575. [CrossRef]
55. Hang, L.; Kim, D.H. Design and implementation of an integrated iot blockchain platform for sensing data integrity. *Sensors* **2019**, *19*, 2228. [CrossRef]
56. Mistry, I.; Tanwar, S.; Tyagi, S.; Kumar, N. Blockchain for 5G-enabled IoT for industrial automation: A systematic review, solutions, and challenges. *Mech. Syst. Signal Process.* **2020**, *135*, 106382. [CrossRef]
57. Taleb, T.; Samdanis, K.; Mada, B.; Flinck, H.; Dutta, S.; Sabella, D. On multi-access edge computing: A survey of the emerging 5G network edge cloud architecture and orchestration. *IEEE Commun. Surv. Tutor.* **2017**, *19*, 1657–1681. [CrossRef]
58. Li, Z.; Barenji, A.V.; Huang, G.Q. Toward a blockchain cloud manufacturing system as a peer to peer distributed network platform. *Robot. Comput.-Integr. Manuf.* **2018**, *54*, 133–144. [CrossRef]
59. Helu, M.; Libes, D.; Lubell, J.; Lyons, K.; Morris, K.C. Enabling smart manufacturing technologies for decision-making support. In Proceedings of the International Design Engineering Technical Conferences and Computers and Information in Engineering Conference. American Society of Mechanical Engineers, Charlotte, NC, USA, 21–24 August 2016; Volume 50084, p. V01BT02A035.
60. Wuest, T.; Weimer, D.; Irgens, C.; Thoben, K.D. Machine learning in manufacturing: Advantages, challenges, and applications. *Prod. Manuf. Res.* **2016**, *4*, 23–45. [CrossRef]
61. Park, J.K.; Kwon, B.K.; Park, J.H.; Kang, D.J. Machine learning-based imaging system for surface defect inspection. *Int. J. Precis. Eng. Manuf.-Green Technol.* **2016**, *3*, 303–310. [CrossRef]
62. Zhao, R.; Yan, R.; Chen, Z.; Mao, K.; Wang, P.; Gao, R.X. Deep learning and its applications to machine health monitoring: A survey. *arXiv* **2016**, arXiv:1612.07640.
63. Dong, H.; Yang, L.; Li, H. Small fault diagnosis of front-end speed controlled wind generator based on deep learning. *WSEAS Trans. Circuits Syst.* **2016**, *15*, 64.
64. Wang, J.; Zhuang, J.; Duan, L.; Cheng, W. A multi-scale convolution neural network for featureless fault diagnosis. In Proceedings of the 2016 International Symposium on Flexible Automation (ISFA), Cleveland, OH, USA, 1–3 August 2016; pp. 65–70.
65. Janssens, O.; Slavkovikj, V.; Vervisch, B.; Stockman, K.; Loccufier, M.; Verstockt, S.; Van de Walle, R.; Van Hoecke, S. Convolutional neural network based fault detection for rotating machinery. *J. Sound Vib.* **2016**, *377*, 331–345. [CrossRef]
66. Lu, C.; Wang, Z.; Zhou, B. Intelligent fault diagnosis of rolling bearing using hierarchical convolutional network based health state classification. *Adv. Eng. Inform.* **2017**, *32*, 139–151. [CrossRef]
67. Guo, X.; Chen, L.; Shen, C. Hierarchical adaptive deep convolution neural network and its application to bearing fault diagnosis. *Measurement* **2016**, *93*, 490–502. [CrossRef]
68. Verstraete, D.; Ferrada, A.; Droguett, E.L.; Meruane, V.; Modarres, M. Deep learning enabled fault diagnosis using time-frequency image analysis of rolling element bearings. *Shock Vib.* **2017**, *2017*. [CrossRef]
69. Blum, C.; Merkle, D. Swarm intelligence. In *Swarm Intelligence in Optimization*; Blum, C., Merkle, D., Eds.; Springer: Berlin/Heidelberg, Germany, 2008; pp. 43–85.
70. Zhang, S.; Lee, C.K.; Chan, H.K.; Choy, K.L.; Wu, Z. Swarm intelligence applied in green logistics: A literature review. *Eng. Appl. Artif. Intell.* **2015**, *37*, 154–169. [CrossRef]

processes

Article

Analysis and Optimization of Two Film-Coated Tablet Production Processes by Computer Simulation: A Case Study

Stefanie Hering [1,2], Nico Schäuble [2], Thomas M. Buck [2], Brigitta Loretz [3], Thomas Rillmann [2], Frank Stieneker [4] and Claus-Michael Lehr [1,3,*]

[1] Department of Pharmacy, Saarland University, 66123 Saarbrücken, Germany; hering@riemser.com
[2] SW Pharma GmbH, 66578 Schiffweiler, Germany; schaeuble@riemser.com (N.S.); buck@riemser.com (T.M.B.); rillmann@riemser.com (T.R.)
[3] Helmholtz Institute for Pharmaceutical Research Saarland (HIPS), 66123 Saarbrücken, Germany; brigitta.loretz@helmholtz-hzi.de
[4] Free Consultant and Qualified Person According to German Law, 79410 Badenweiler, Germany; fs@stieneker.com
* Correspondence: claus-michael.lehr@helmholtz-hips.de; Tel.: +49-681-98806-1000

Abstract: Increasing regulatory demands are forcing the pharmaceutical industry to invest its available resources carefully. This is especially challenging for small- and middle-sized companies. Computer simulation software like FlexSim allows one to explore variations in production processes without the need to interrupt the running process. Here, we applied a discrete-event simulation to two approved film-coated tablet production processes. The simulations were performed with FlexSim (FlexSim Deutschland—Ingenieurbüro für Simulationsdienstleistung Ralf Gruber, Kirchlengern, Germany). Process visualization was done using Cmap Tools (Florida Institute for Human and Machine Cognition, Pensacola, FL, USA), and statistical analysis used MiniTab® (Minitab GmbH, Munich, Germany). The most critical elements identified during model building were the model logic, operating schedule, and processing times. These factors were graphically and statistically verified. To optimize the utilization of employees, three different shift systems were simulated, thereby revealing the advantages of two-shift and one-and a half shift systems compared to a one-shift system. Without the need to interrupt any currently running production processes, we found that changing the shift system could save 50–53% of the campaign duration and 9–14% of the labor costs. In summary, we demonstrated that FlexSim, which is mainly used in logistics, can also be advantageously implemented for modeling and optimizing pharmaceutical production processes.

Keywords: 3D simulation modeling and analysis; model implementation; bottleneck analysis; production costs; resource conservation

Citation: Hering, S.; Schäuble, N.; Buck, T.M.; Loretz, B.; Rillmann, T.; Stieneker, F.; Lehr, C.-M. Analysis and Optimization of Two Film-Coated Tablet Production Processes by Computer Simulation: A Case Study. *Processes* **2021**, *9*, 67. https://doi.org/10.3390/pr9010067

Received: 30 November 2020
Accepted: 26 December 2020
Published: 30 December 2020

1. Introduction

The pharmaceutical industry is known to be prosperous but inflexible. Regulatory authorities expect increasing standards for medicinal products, e.g., during clinical trials or production [1–3], which makes the industry less prosperous and even more inflexible. Multinational pharmaceutical companies react with various strategies, such as outsourcing or mergers and acquisitions [4], while small- and middle-sized companies need to compensate for their losses differently. For both, the strategy of addressing production costs is promising because of this strategy's generally low equipment utilization [5] and because of production's high costs (production makes up to 30% of the overall costs) [6]. Instead of real-world experiments and tests, computer simulations enable one to test different scenarios without any interruptions or threats to daily business.

The chosen processes for this case study involve the production of two film-coated tablets for the treatment of tuberculosis. As of 2020, about one-quarter of the world's population is infected with latent tuberculosis. Ending this epidemic by 2030 is one of the health targets of the United Nations Sustainable Development Goals, so the incidence of the disease is decreasing by 2% each year. This disease is mainly caused by acid-fast rod-like Mycobacterium tuberculosis. Most patients suffer from a curable lung infection that is treated with a combination of different agents [7]. According to the World Health Organization, isoniazid, rifampicin, pyrazinamide, and ethambutol are the most essential first-line anti-tuberculosis drugs [8]. The main active pharmaceutical ingredients of the two investigated products are isoniazid and ethambutol; thus, the products are abbreviated as P_{INA} and P_{EMB}.

Background information about simulations can be found in Banks (2005), who defined them as the "imitation of the operation of a real-world process or system over time" [9], while a model can be described as a "representation of a [...] process intended to enhance our ability to understand, predict, or control its behavior" [10]. The link between both is that "the exercise or use of a model to produce a result" is a simulation [11]. Simulating different scenarios, thereby changing the parameters in a model and running it, enables one to test and evaluate the effects of certain parameters. Hence, diverse fields of application are possible in a pharmaceutical context. These applications could support decision making for pipeline management [12–14] and optimize supply chain management [15–17]. Simulations of production processes are also of interest. They can be divided into simulations of single process steps, such as computational fluid dynamics for mixing steps [18], and simulations of multiple production steps [19] or even of entire continuous production processes [20]. The models of Sundaramoorthy et al. [16] and Matsunami et al. [21] were used to investigate a mixture of the abovementioned factors. Their prospective models targeted the planning of production capacities when a product is still in its developmental stage. Matsunami et al. compared a batch with continuous production considering various uncertainties, prices, and market demands for one product. Sundaramoorthy et al., however, address the production capacities of multiple products. Habibifar et al. recently published a study on the optimization of an existing production line, including a sensitivity analysis, the design of multiple scenarios, and a data envelopment analysis. Other comparable work was also examined intensively—11 references from 2007–2019 were investigated and compared [5]. The applied techniques (simulations, mathematical modeling, and statistical techniques) differed, as did the focuses of the studies. Some papers concentrated on the optimization of specific process steps [22], while others pursued a more holistic approach [23]. The high variability in this small population demonstrates that there are many different approaches and even more available software solutions for optimizing pharmaceutical production. One described attempt involved a discrete-event simulation, in which the variables of a model changed due to defined events. Recent work on discrete-event simulation addressed room occupancies in a hospital [24], a flow shop [25], and manufacturing scheduling [26].

In contrast to the above-mentioned studies (i.e., prospective simulations for future medicinal products), this study addressed two already existing production processes. The motivation of this work was to optimize the validated and approved production processes of P_{INA} and P_{EMB} via discrete-event simulations without the need to interrupt or interfere with the continuing production processes themselves. Initially, it was investigated whether it is possible to establish a discrete-event simulation approach with limited resources and simulation knowhow. Meanwhile, the three most critical steps in model building (implementing model logic, operating schedule, and processing times) were determined, verified, and partly validated. Based on the generated as-is models of the P_{INA} and P_{EMB} production processes, bottlenecks in the production process were identified and different production scenarios designed to find the optimal one under existing conditions. The optimizations focused on process efficiency, not pharmaceutical or validation questions. The market authorization of these products limited the possible changes to only organizational ones. Therefore, the shift systems of the created as-is models for P_{INA} and P_{EMB} were changed from the existing one-shift system to a one-and-a-half shift system and a two-shift system to optimize employee utilization. The campaign duration, labor costs, and used resources were calculated to generate comparable outcomes.

2. Materials and Methods

For discrete event simulation of these pharmaceutical production processes, the software package FlexSim was chosen since it is easy to use and already widely implemented in various industrial sectors for logistics or production, as well as in national institutions. To the best of our knowledge, this is the first time that FlexSim has been used to optimize pharmaceutical bulk production in its entirety. This report suggests a possible implementation path for a batch production. It starts in a semi-automated facility to obtain data, continues with the creation of a representing simulation model and ends with a case study optimizing the capacity utilization of the investigated batch production. Where the available FlexSim software was not sufficient, additional software, such as Cmap Tools (Florida Institute for Human and Machine Cognition, Pensacola, FL, USA), Microsoft® Excel (Microsoft, Seattle, WA, USA) and Minitab® (Minitab GmbH, Munich, Germany), was used.

Since our literature research yielded few standards for creating a discrete-event simulation in the context of pharmaceutical processes, an intuitive attempt was pursued and implemented. At the beginning, basic decisions about the model were made for model description. Afterwards, information about the investigated production processes was collected. All information was clustered into numerical and logical information. The numerical information covers the collection of historical processing times, their analysis as well as the selection of the most representative distribution for each process step. The logical information was used for model building. The most meaningful information sources were historical batch data, official validation and qualification data of the process owner, instruction manuals, on site observations, and work experiences. The gathered process information was firstly depicted in a flow chart and afterwards transferred to a simulation model. These elements of the methodological approach were addressed separately, but simultaneously. Together, they resulted in an as-is model of the production process, which was later verified and partly validated. It furthermore served as a process analysis and optimizing tool. This methodological approach is depicted in Figure 1.

Figure 1. Methodological approach of this case study: Seven steps (blue boxes) were implemented from model description until its application as optimizing tool. The according sub-steps are listed underneath and include further information, such as applied software and cross references to other figures and tables.

2.1. Employed Software

Different software was used during this case study on a Windows (Microsoft, Seattle, WA, USA)-based laptop equipped with 12.0 GB RAM and a 64-bit processor. Initially, the historical processing times and their deviations, as well as the new data for model verification and application, were collected in Microsoft® Excel. Moreover, the basic questions were analyzed statistically, such as the calculations of standard deviations and the minimum and maximum values.

An extensive flow chart was created in Cmap Tools, which is freely available from the Florida Institute for Human and Machine Cognition (Pensacola, FL, USA), to understand and take stock of the entire process. This program is ideal for creating flow charts or visualizing logical relations and features easy drag and drop handling. The knowledge gained from the program facilitates the steps for later model building in FlexSim [27].

Further statistical analyses were performed using Minitab® (Minitab GmbH, Munich, Germany, version 18.0), a commercial statistics package that is widely used in Lean Six Sigma projects [28]. Some basic Lean Six Sigma approaches were integrated into the data handling of this case study. Data analysis, such as hypothesis tests, and graphical evaluations, such as boxplots or individual moving range charts, can be easily performed in Minitab®.

Discrete-event simulations were conducted in FlexSim (FlexSim Deutschland—Ingeni eurbüro für Simulationsdienstleistung Ralf Gruber, Kirchlengern, Germany), a commercially available 3D simulation software designed for modeling production and logistic processes. FlexSim provides discrete-event simulations that are object-oriented, which means to implement all components as objects, to assign specific attributes and methods to them for their characterization as well as for manipulating the overall system. In addition to a graphical 3D click-and-drag simulation environment, programming in C++ is offered. The user can choose between different views and methods for representing data [9]. The applied, non-configured student version is FlexSim 19.0.0. Here, dynamic process flows can be captured on a functional level, plainly visualized, and extensively analyzed. This program gives decision makers the opportunity to forecast the outcomes of possible changes in their processes, such as changes in product flow, resource utilization (staff, money, and machinery), or plant design.

The production processes were captured by implementing and connecting all single components of the system by their procedural functions and attributes. Additional critical parameters (set-up times, staff) and logic (priorities, random events) made the models as close to reality as possible. The 3D visualization of the process provided an intuitive understanding of the current state of the system and future possibilities. The results of the simulations were analyzed via performance and output statistics. Capacity utilization, transport time, and state statistics are examples of the metrics of interest.

2.2. Production Processes

Two approved coated tablet production processes were selected for this case study because they were similar but different in their ingredients and product properties, such as tablet size. Their obviously different punches require different processes and cleaning times, but their other equipment and operations are similar. Since both processes are comparable, comparable results were expected. Comparable results would enable the transfer of optimizations to other processes and thereby increase the impact of this study. The processes consisted of 47 sub-steps that were merged to the following 13 super-ordered steps:

- Setting up the scales
- Weighing the granule and granulation liquid
- Dissolution of the solid components to finish the granulation liquid
- Compulsory mixing
- Fluid bed granulation
- In-process controls

- Sieving
- Tumble blending
- Compaction
- Weighing the coating
- Dissolution of the solid components to finish the coating
- Coating
- Bulk packaging

Most machines (in FlexSim referred to as "processors") are about 25 years old and largely still operated manually. The two products, P_{INA} and P_{EMB}, are produced batch-wise in variable campaign sizes from three to 18 batches. The simulations were chosen to represent the average campaign sizes to identify bottlenecks and increase productivity. Therefore, simulations for P_{INA} covered four batches and those for P_{EMB} covered ten batches.

Since the intention of this work was to describe standard production campaigns of these products, deviations as machine breakdowns, personnel shortages, and other human failures were excluded. During data preparation, we assessed whether the observed deviations influenced any of the historical processing times. In addition, outlier tests were performed to exclude deviating historical batch data, so the resulting data pool solely represented standard processing times. Further excluded data are product-dependent cleaning times as well as times for setting up. The cleaning and set-up efforts before a product campaign vary widely, since the production of analogous products containing the same APIs at different doses lowers the necessary effort dramatically. Thus, this study only includes product-independent daily routine cleaning times, as well as the daily times for setting up the machines.

2.3. *Statistical Data Processing*

The processing times are the most important since the campaign duration endpoint in this study influences a second endpoint: the labor costs. Therefore, the collection and handling of processing times are of great importance. We distinguished between the initial historical batch data and the FlexSim-generated verification data. The collection of historical batch data strongly depends on the production equipment. While data is easily available in automated production lines, most semiautomated production plants do not have automated tracking and data generation. For these two production processes, no digitally workable data were available. Hence, the processing times of each process step were collected manually and batch-wise before they were transferred into digital files. The overall statistical data process is depicted in Figure 2.

During *data preparation*, we investigated in Minitab® whether the data were under statistical control with individual moving range charts (I-MR chart). Even though I-MR charts are, strictly speaking, only to be used for normally distributed data, they nevertheless indicate shifts, trends, and process variations. Therefore, the first impressions of the process stability were obtained. Secondly, we tested if the data were normally distributed using probability plots for further data handling. Additional statistical tests for the outliers, data pooling possibilities, and visualization were also performed in Minitab®.

FlexSim contains a tool named ExpertFit, which helps to find the best fit distribution based on raw data. FlexSim distinguishes two main types of probability distributions: discrete and continuous distributions. Continuous distributions are subdivided into non-negative, unbounded, and bounded distributions. Historical batch data were entered as raw data in ExpertFit during *model building*. The results were extracted, and the suitability of all distributions were evaluated. Afterwards, several graphical comparisons between different distributions were performed to select the best-fitting distribution. The last step in ExpertFit was to transfer the selected distribution and its parameters into the according process step in the FlexSim model. If none of the available distributions provided a satisfying evaluation, the usage of an empirical table was recommended and implemented. The results of this analysis can be found in the Supplementary Materials (Table S1). Interestingly, even though some data were earlier shown to be normally distributed, ExpertFit never evaluated a normal distribution as the best choice.

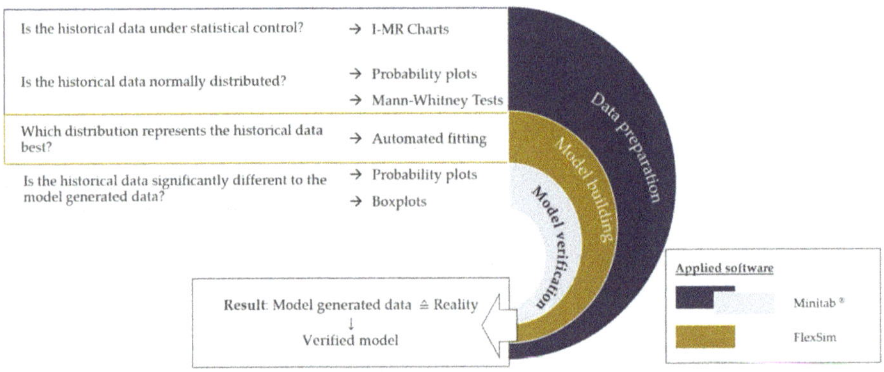

Figure 2. Chronological workflow of statistical data processing: Minitab® was used during data preparation and model verification to analyze the historical batch data and to later compare it to the FlexSim-generated data. The ExpertFit tool of FlexSim performed the automated fitting of the historical batch data to over 20 distributions to identify the best representation for each process step during model building.

During *model verification*, the FlexSim-generated data were compared to the historical batch data. Since the probability tests during data preparation proved that most processing times were not normally distributed, Mann–Whitney tests were performed in Minitab®.

Choosing the best way of *model validation* was challenging. Initially, a face validity was made. The head of production investigated the models, their behavior, and logics. For a historical data validation, the data pool of historical processing times was statistically too small. Therefore, an additional, predictive validation was attempted. Therefore, FlexSim-generated data were compared to processing times of new campaigns.

3. Results

3.1. Model Development: Design and Building

Model design started with the creation of a detailed flow chart in Cmap tools, resulting in a conceptual model. The flow chart includes all production steps and sub-steps, the operators, the average historical processing times, and the relevant machines. Connections and cross references in the overall process become evident here. Information on merging multiple process steps, the elimination or integration of process steps, and the required model logic was collected. The correctness of the flow chart was approved by the process owner—the head of production. Altogether, this step of model design ensured deep process knowledge and identified all the dependencies and necessary components for the subsequent FlexSim models. Since the manufacturing process is confidential, this process flow chart cannot be shown. However, the complexity of such production processes makes it highly recommendable to start with such depiction and its approval by the head of production.

After the basic information was determined, *model building* in FlexSim began (Figure 3) following the common guiding idea to include only crucial attributes in the model and to keep the model as simple as possible [29]. Additionally, in the models, we assumed that no machine breakdowns or other major deviations would occur.

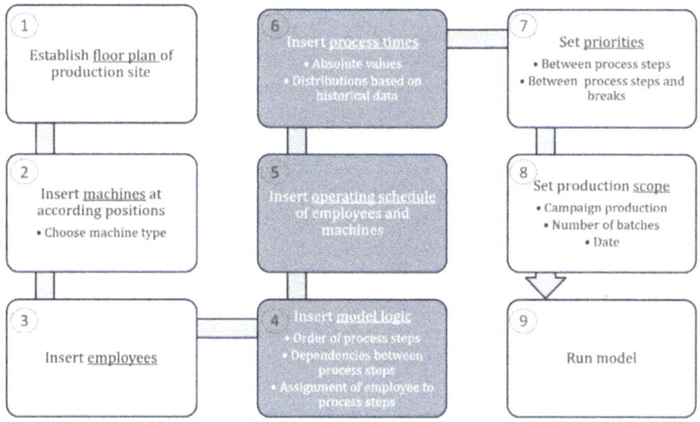

Figure 3. Steps for building a process model using FlexSim: In the first third, simple facts established the foundation of the model. After inserting the model logic, the working schedule and the processing times became more complex and included the most critical attributes of the model to be verified (colored blue).

Initially, the floor plan (step 1) and the machines (step 2) were transferred out of official documents into FlexSim (Figure 4). The machines are represented as so-called processors. There are three processor types: one to process the item, one to combine multiple items, and one to separate an item into multiple items.

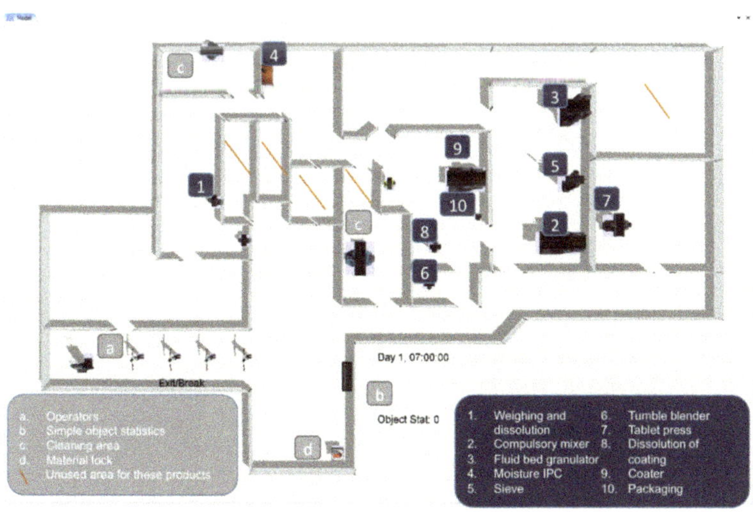

Figure 4. Simplified floor plan in FlexSim: The production area is depicted, including processors (blue boxes: number represent the chronological order of the process steps) and further details (grey boxes). Brown bars represent areas that are not used for the production of the investigated products.

The employees (step 3) are called operators. The headcount in the model, symbolized with the four males on the left, corresponds to the headcount of the original process.

Inserting the model logic (step 4) is one of the most complex parts in building the simulation model. Its basis is the batch documentation, which gives the order of the process steps. Some steps can be connected easily. More complex sequences occur when certain events condition other steps, e.g., the weighing of a new batch cannot start until certain steps of the previous batch are finished, even if the scales and operators are not occupied. Such conditions are considered in the model by triggers like opening and closing the processor ports, sending messages, or placing certain information on the items. It must be determined which and how many operators fulfill a process step.

Next, the operating schedule (step 5) and the times off are entered. Working hours are from 7:00 a.m. to 3:45 p.m., with a breakfast break from 9:15 to 9:30 a.m. and a lunch break from 12:15 to 12:45 a.m. For some processors, it must be guaranteed that the entire process can be finished at once. Therefore, the operating schedules of the processors must be considered as well.

Processing times (step 6) are then established as described in the statistical data processing. If a pairwise comparison revealed no statistical significance (Mann–Whitney test), the times for the different process steps of the two products, P_{INA} and P_{EMB}, were pooled (see Table S1 in Supplementary Materials: Weighing granulation liquid on the table and on the floor scale, dissolution of granulation liquid, mixing in a tumble blender, and weighing the coating on the table and on a floor scale).

Some working steps and times are not explicitly part of the batch documentation, which only provides the daily routine. These times are instead based on employees' experiences and were also implemented in the model. As these times do not vary depending on the product, the implemented times were identical in all models (Table S2 in the Supplementary Material).

Not all process steps have the same importance. Some process steps can be stopped, while others cannot. Only some of the process steps require the attendance of an operator for the entire processing time. Therefore, the priorities (step 7) have to be defined.

Additionally, the scope (step 8) of the production must be implemented. The prevalent conditions are the campaign production with a certain number of batches. FlexSim also offers the implementation of actual dates.

In this case study, the most important model parameters to verify were the processing times, model logic, and operating schedules. Model verification is unique and strongly depends on the model itself. Thus, finished components are rarely available. Sometimes, additional checkpoints and workarounds, such as labeling the so-called flow items with informational stickers, had to be implemented. FlexSim offers different statistical analysis modules that must be transformed to enable the verification of these parameters. Some of these analysis modules can be added via drag-and-drop and do not need further changes. To track the processed flow items and operators, each must be individually in C++. The implementation of these elements for model verification is the last step of model building.

Testing of the models (step 9) can now be performed in single runs that are started and stopped manually. Alternatively, it is possible to use the FlexSim module Experimenter. Experimenter offers the opportunity to predefine the amount of replications (runs), the statistics for evaluation, and the variables to compare, as well as subsequently perform the necessary replications.

3.2. Model Verification

The blue colored boxes (steps 4–6) in Figure 3 highlight the most important parameters to verify, including simple but productive model logic, an accurate operating schedule for operators and processors, and the correct processing times for each process step. To establish a sound foundation for the FlexSim-generated data, the Experimenter module was used. The amount of model replications was determined as identical to the amount of historical batch data (P_{INA}: 25 runs; P_{EMB}: 45 runs). While using the Experimenter, FlexSim accesses the deposited statistical distribution of each process step and thereby generates different processing times for each run. Afterwards, an interactive report and a performance measure report can be exported. This interactive report includes data on the model logic in item-trace Gantt charts and all data on the working schedules in state charts. These charts are produced replication-wise. The performance measure report includes a statistical summary, a replication plot, a frequency histogram, and the single values of all replications for each process step.

3.2.1. Model Logic

Verification of the model logic is complex to integrate. The goal of this step is to prove that all process steps run in the correct order and that the conditions and dependencies between different process steps are correctly implemented.

Therefore, all items run through the model are tagged, and triggers are programmed for all processors to leave information on these tags. This type of information is best captured using item-trace Gantt charts. Figure 5 shows a schematic item-trace Gantt chart for one batch. One batch consists of six items, with two each for the granule, the granulation liquid, and the coating. The items are pictured as one bar. Each process step is represented as one colored square of the bar. Following from top to bottom and left to right, the process order and dependencies of the different process steps become evident. As an example, the processing step for coating (light blue squares) is provided. This process requires the previously formed tablets (green square) and the newly dissolved coating (light yellow squares). Coating then combines these two items on one bar for the coated tablets, which are packed in the subsequent step (light pink). Hence, the process order and processor type of the coater, a combiner, are verified. During model building, a trigger was set to withhold weighing the coating until compaction is almost finished to prevent long holding times for the liquid parts of the coating. This chart verifies the implementation of this trigger since weighing (purple) starts shortly before the end of compaction (green). Overall, the item-trace Gantt charts were able to verify the overall model logic. Additionally, the overall processing time became evident, which is important for the subsequent comparison of different optimization scenarios.

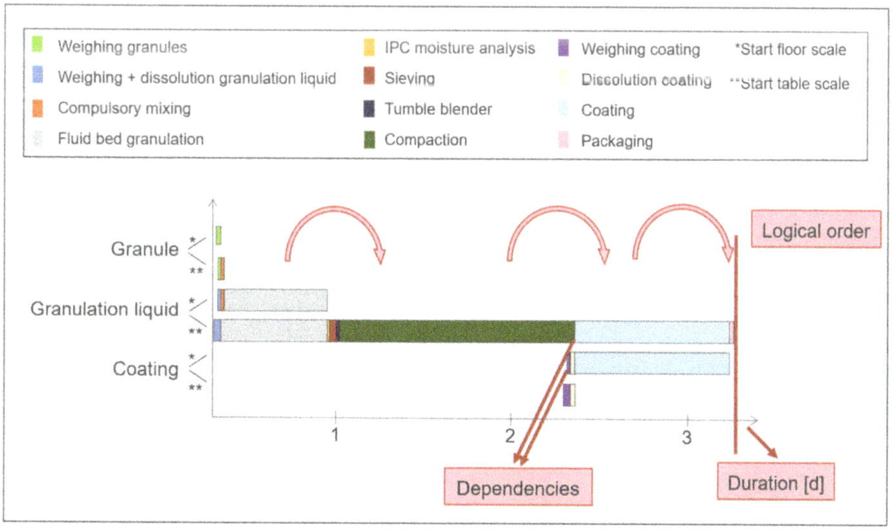

Figure 5. Example of the FlexSim-generated item-trace Gantt chart for model verification: This chart represents the production of one batch including the most important process steps. The granule, granulation liquid, and coating are necessary to produce film-coated tablets. Floor and table scales are used for each of these components, as different mass ranges are weighted. One bar symbolizes one of these components depending on the initially used scale. Hence, one batch consists of six bars. The different colored subdivisions of the bars show the finished process steps of each component. This type of item-trace Gantt chart allows to (i) control the correct logical order of the process steps; (ii) check if the dependencies are correct (e.g., the weighing of coating can only start shortly before the end of the compaction to avoid long holding times); and (iii) determine the total campaign duration.

3.2.2. Operating Schedules

The operating schedules are pictured in the state charts for the operators and processors. Each operator and each processor is represented by one bar, which is divided into several states, including utilize, idle, or scheduled down. The x-coordinate shows the duration in days.

Figure 6 shows a schematic state operator chart (top) and a schematic processor chart (bottom). These charts enable the modeler to verify break times (duration and fixed moments). Moreover, the capacity utilization of each operator and processor becomes evident in this chart. The process steps are arranged in chronological order. This way, one can follow a batch by starting with the utilized part of the first process step and continue watching it work downstream through the processors. The purple parts indicate processing times at which the process becomes stuck because of too few operators, while the light-yellow sections indicate blocked processors. FlexSim can export processing times and states in a table form. This way, export for further data analysis is easy to handle. An evaluation of the state charts verified the break times and process order.

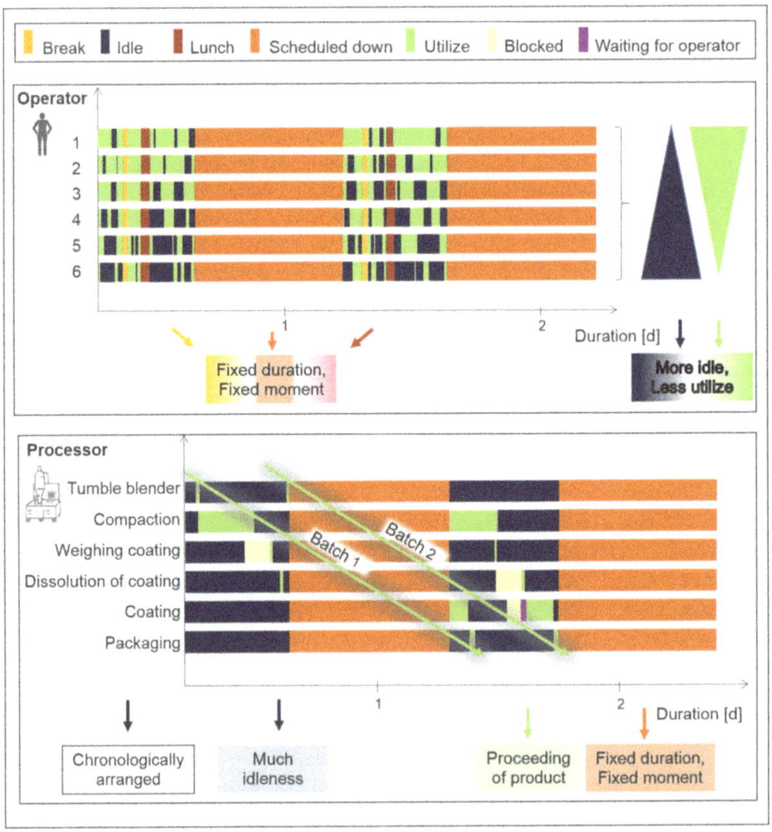

Figure 6. Example of the FlexSim-generated state operator chart (**top**) and state processor chart (**bottom**): The production of two work days is depicted, including the most important states for the six operators and the most important states for the last process steps. Each operator is symbolized by one bar. The different-colored subdivisions of the bars represent the operators' states over time. The processors are also represented by bars divided into different-colored states. The major outcomes of both charts include verification of break, lunch, and after-work hours, as well as capacity utilization (idle/utilize). This enables one to i) test the model (chronological order) and to identify bottlenecks (capacity utilization of operators and processors).

3.2.3. Processing Times

The last parameter to verify is the processing time. For this purpose, the performance measure report was used. Single FlexSim-generated values were transferred into Minitab® where these values were compared to the processing times of the historical batch data. Mann–Whitney tests with a confidence interval of 0.95 were performed, as most data are not normally distributed. None of the generated processing times were significantly different to the historical processing times (Table 1).

Table 1. Statistical analysis: Results of the probability plots of each process step and of the Mann–Whitney tests comparing historical batch data to the FlexSim-generated data for the products P_{INA} and P_{EMB}. There was no significant difference ($p < 0.05$) between the historical batch and FlexSim-generated data.

Process Step	Normally Distributed?		p-Value	
	P_{INA}	P_{EMB}	P_{INA}	P_{EMB}
Dissolution of granulation liquid	Yes	No	0.760	0.243
Compulsory mixer	Yes	No	0.594	0.479
Fluid Bed Granulation	No	No	0.993	0.127
IPC Moisture Analysis	No	No	0.806	0.527
Sieving	No	Yes	0.767	0.602
Tumble blender	No	No	0.319	0.110
Compaction	Yes	No	0.331	0.107
Dissolution of coating	Yes	No	0.399	0.561
Coating	Yes	Yes	0.679	0.246
Packaging	No	Yes	0.886	0.086

3.3. Model Validation

To prove the reproducibility of the simulation models, a model validation was conducted. There are different options to validate computer models. Initially, it was chosen to prove the correctness and reasonability of our models by face validity. Also, a predictive validation was added for which future production campaigns of P_{INA} and P_{EMB} were picked and specifications about the process flow, processing times, and campaign durations were defined. The campaigns, covering four batches for P_{INA} and ten for P_{EMB}, were run under normal conditions. Non-standard conditions, changes, and deviations were additionally monitored and documented. Afterwards, the relevant data was collected from batch documentation, transferred into Minitab®, and compared to FlexSim-generated data. Analyzing the new data showed a valid process flow. The other parameters, the processing times and the campaign duration, however did not meet the specifications caused by severe deviations during both campaigns. Due to confidential issues, detailed explanations are limited. Some of the deviations, such as machine breakdowns, were intentionally excluded during model description and therefore not considered in the FlexSim-generated data. Another very influential deviation was personnel shortage; for none of the production days, the full head count was available. On top, urgent, non-campaign work cut the already minimized work capacities. Besides these issues, the predictive model validation gathered additional important information for the process owner and validated the model logic. The fact that not meeting the specifications was at least partly caused by a lack of the necessary personnel which, however, at least indirectly validates the models.

3.4. Model Application: Optimization and Evaluation of Fictive Shift Systems

After the successful verification and partly validation of the as-is model, realistic and meaningful changes in the real production processes were discussed with the head of production. As Figure 6 (bottom) illustrates, processors have long idle times with little utilization. This raised the question whether the system could run more profitably under different shift systems. Profitability was investigated as the total duration for one campaign and the labor costs. Labor costs included not only the salaries of the employees but also the costs for the machine's run times.

3.4.1. Establishment of Models with Different Shift Systems

The shift systems of interest were one-shift (OS), one-and-a-half-shift (OHS), and two-shift (TS). The related operating schedules can be found in Table 2.

Table 2. Operating schedules: Operating schedules of the optimization scenarios classified into the shift models of one-shift (OS), one-and-a-half-shift (OHS), and two-shift (TS) systems. The working hours for the one-shift system were adjusted according to an in-company agreement. * Different weekly hours for some operators in the past.

	One-Shift *	One-and-a-Half-Shifts	Two-Shifts
Operating schedule	07:00 a.m.–03:15/03:45 p.m.	06:00 a.m.–02:15 p.m. 09:15 a.m.–05:30 p.m.	06:00 a.m.–02:15 p.m. 02:00 p.m.–10:15 p.m.

The different parameters between the different scenarios were operating schedule, product type, and the number of operators. These variations were implemented manually without any optimization algorithm. In addition to simply changing the schedules and headcounts, other aspects must be considered. The campaign duration strongly depends on the weighing operations of the different batches. Therefore, it is important to identify the best times after the start of the campaign for each batch. This is done by supervising and evaluating the interactive report of the Experimenter module. The shift system and number of operators for the OS system of P_{INA} influence these times. The impact of this parameter on the overall campaign duration is visualized in Figure 7. The campaign of P_{INA} includes four batches (left side), and the campaign of P_{EMB} includes ten batches (right side). The last batch of P_{INA} can be weighted in after 8 h under a TS system compared to 48 h using an OS system. For P_{EMB}, the time could be reduced from 9 d to only 3 d after campaign start. Hence, a significant reduction in campaign duration was already expected when building the optimization scenarios.

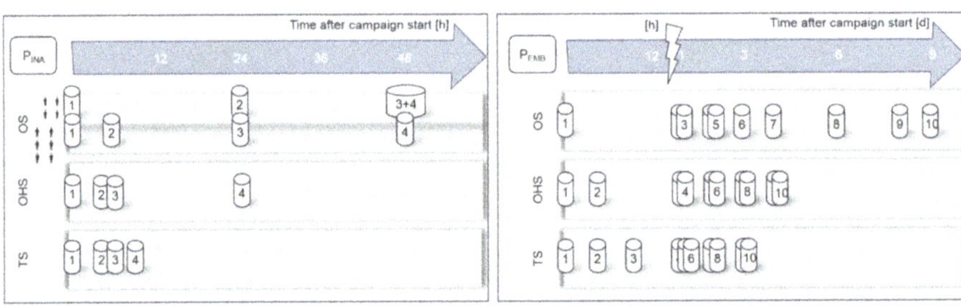

Figure 7. Summary of the optimal starts for weighing the granules of all batches after campaign start for P_{INA} (**left side**) and P_{EMB} (**right side**): The cylinders represent batches, the numbers indicate the batch number in the campaign. Therefore, the campaign of P_{INA} includes four numbered cylinders and the one of P_{EMB} includes ten cylinders. The start depends on the applied shift system and (only for the one-shift system of P_{INA}) also on the number of operators. This indicates that the processors limit the weighing strategies for all other cases. The start times have a significant influence on the overall campaign time.

3.4.2. Results of the Shift Systems

The implemented shift systems were evaluated by the number of successful replications, the utilization degree of the operators, and the campaign duration. A well-established model with suitable logic and processing times can complete the replication. Therefore, the presence of several successful replications indicates a harmonious model that can be used to evaluate model optimization. The predefined number of replications aligns with the replications used during model verification and thus to the number of available historical batch documents. As already mentioned, the schematic operator state charts (Figure 6, top) visualize the utilization degree of the operators. The mean campaign duration and headcounts were used as the basis for the labor cost calculations.

Originally, the overall headcount of the case study was always four operators. Hence, in the first step, the optimization scenarios for both products with OS, OHS, and TS systems were built. The four operators worked simultaneously in the OS system and at staggered intervals for the OHS and TS systems. For some combinations, some replications did not finish (Table 3). It was also impossible to build a running TS model with four operators for P_{EMB}, since only two operators were available for monitoring up to four simultaneous running processors. Obviously, such real-life limits of this process were also reflected by the computational models. The operator and processor state charts also indicated that too many processors needed an operator at the same time. As a result, alternative models were built featuring an increased headcount of six, with three operators always working simultaneously. A summary of the results for P_{INA} and P_{EMB} can be found in Table 3.

Table 3. Results of the different shift models for P_{INA} and P_{EMB}: Generally, the campaigns of P_{INA} consisted of four batches, and the campaigns for P_{EMB} consisted of ten batches, yielding the resulting model scope. The number of successful replications indicates whether the model is stable. The arrows symbolize the utilization degree of the operators (↓ = some idleness, ↓↓ = much idleness, ↓↓↓ = operator is barely working, and ↑↑↑ = work overload, ✔ =appropriate work load). Additional metrics of interest are the campaign duration, including the standard error of the mean, and labor costs. The best scenarios are highlighted with grey boxes.

	One-Shift (OS)		One-and-a-Half-Shift (OHS)		Two-Shift (TS)	
			P_{INA}			
Operators $_{total}$	4	6	4	6	4	6
Replications	30/30	30/30	30/30	30/30	28/30	30/30
Operator Utilization	✔	Op5 + Op6 ↓↓	✔	Op6 ↓↓	↑↑↑	✔
Duration $_{mean}$ [d]	3.2 ± 0.00	3.2 ± 0.00	2.4 ± 0.02	2.3 ± 0.00	2.3 ± 0.02	1.6 ± 0.02
Labor costs [€]	7488	8928	7128	7866	8798	6840
			P_{EMB}			
Operators $_{total}$	4	6	4	6	4	6
Replications	41/45	41/45	45/45	44/45	–/45	42/45
Operator Utilization	✔	Op4 – Op6 ↓↓	✔	Op1 – Op6 ↓	↑↑↑	✔
Duration $_{mean}$ [d]	9.4 ± 0.01	9.4 ± 0.01	7.2 ± 0.01	6.1 ± 0.02	-	4.4 ± 0.01
Labor costs [€]	21,996	26,226	21,384	20,862	-	18,810

The production processes of P_{INA} and P_{EMB} are comparable; however, the number of batches per campaign for P_{EMB} is 2.5 times larger than that for P_{INA}. This makes the P_{EMB} models both more susceptible and more relevant for choosing the optimal shift system.

The combination of six operators working in a TS yielded the lowest duration (P_{INA}: −50%; P_{EMB}: −53%) and the lowest labor costs (P_{INA}: −9%; P_{EMB}: −14%) for both products compared to the initial scenario. This is not initially surprising, as both duration and headcount seem to correlate, as duration is one parameter of labor cost calculations. As shown in Table 3, the durations of OHS with four and six operators differs slightly for P_{INA} due to its different capacity utilizations. There was also no difference in the duration of the TS with four operators. Ultimately, having six operators in a TS decreased the duration by about 30% compared to having four operators in a TS or six operators in an OHS. This decreased labor costs by 4% (OHS, four operators) and 13% (OHS, six operators) for the TS with the six operators. This noteworthy difference in labor costs highlighted P_{INA} in the OHS as the second-best option for P_{INA}.

For P_{EMB}, the TS with six operators was found to be 28% faster and 10% cheaper than the second-best option. This OHS used six operators, as it was significantly different (15%) to the OHS with four operators in the campaign duration. This extra production day produced only minimally higher labor costs (2%).

As previously mentioned, the obtained results indicate that production with six operators in a TS is superior to all types of production with four operators for both products. Here, only four operators were fully qualified. Therefore, the best option under a headcount of four is of great importance. For P_{INA} and P_{EMB}, OHS is the best option due to its faster production (P_{INA}: 25%; P_{EMB}: 23%) and lower labor costs (P_{INA}: 5%; P_{EMB}: 3%) compared to the prevailing OS system. The head of production confirmed the superiority of OHS compared to OS post-hoc based on his own experiences.

4. Discussion

Computer simulations enable testing and evaluating different production scenarios by changing relevant parameters in the according model and running it. In this way, the best possible scenarios were found in this case study. The prevailing conditions of a small headcount and limited resources led to a user-friendly, practice-oriented simulation approach for optimizing two approved pharmaceutical production processes. The majority of computer simulation studies in pharmaceutical supply chain and manufacturing, as reported in the introduction [5,16,19,24] concerned with planning or conducting the complex issue of a whole production process, were performed by experts in modelling and process design. In contrast, this study was conducted by non-computer experts, but experts in pharmaceutical technology and production. We have deliberately modeled an already established process to show that there is still a lot of untapped optimization potential. Our determined potential savings of 50% campaign duration, and of up to 14% labor costs, highlight the significance of this approach. Other less complex optimization attempts, as published by Bähner et al., who evaluated the machine utilization, benefit from the fact, that the time-consuming modeling is not required [30]. However, only the processor utilization is mapped and the operator utilization is disregarded. With modeling, especially having a small amount of historical data, one has a problem with a formal proof of validity, but on the other hand, after implementation, one has more application possibilities. With the still feasible efforts, our simulations intend to close the gap of published industrial case studies.

4.1. Case Study Limitations

While a case study allows to obtain detailed and usually well-protected information on specific production processes, it is also limited to it. The deliberate exclusion of extraordinary events, such as breakdowns or process times with deviations, in the very beginning, defined the simulations to represent standard processes without any incidence. Also, data collection and analysis highlighted two challenges in this case study. Time recording was performed manually minute-wise. This was disadvantageous for short processing times. Setting up of scales takes 8 min on average, which is only 3% of the compaction time. However, the time resolution for both process steps is the same. Additionally, each process step was started, and sometimes stopped, manually. Therefore, the processing times strongly depend on the availability of an operator. This produces high relative standard deviations for short processing times (P_{INA}: 5–115%; P_{EMB}: 6–75%), although the production process is still within the necessary specifications. As an example, the process step of compulsory mixing takes 11–20 min in the historical batch documentation for P_{INA}. This time must be split into the actual mixing time (10 min, fix) and the manual setting up/starting/stopping time (1–10 min, operator dependent). Here, a waiting time of 10 min has a stronger impact on the relative standard deviation than the same waiting time has on a compaction process with a mean duration of 269 min. The impact of these limitations on the significance of this case study are, despite everything, acceptable. Even though FlexSim offers the integration of breakdowns, a substantiated analysis of past quality issues would have been necessary and was beyond the scope of this work. Nevertheless, relevant assumptions can be made. The investigated solid production processes have a linear structure. A total breakdown of one processor stops the production of all subsequent batches. Moreover, the start of some batches will additionally be delayed since intermediate products have limited shelf lives in the validated processes. Such factors prolong the campaign duration but do not influence the processing times of single process steps. The effects of the unprecise time recording and low time resolution on the overall model are also not critical, as short processing times have only a small impact on the duration of the entire campaign. The dependency on available operators also has small impact on the campaign duration, but great impact on the model logic. This factum is negligible since all employees are well trained to prioritize between different steps and since those priorities are implemented in the models as well.

4.2. Case Study Outcomes

In addition to the main aim of this paper (the optimization of existing processes by discrete-event simulations), the implementation of this intuitive process can be evaluated. Initially, the steps of data collection up to model verification were very time-intensive but gave precious insight into the production processes. The as-is states of the actual production processes were scrutinized more intensively than those during daily routines. The obtained results challenged the workflow and the dependencies between different working steps. During data analysis, process steps with significant economic potential were identified; thus, analyses were performed, and possible improvements were developed. Even without actually running any discrete-event simulations, significant knowledge was gained. Therefore, it can be assumed that process owners already profit from such case studies regardless of the simulation outcomes.

Knowledge of the relevant processes and regulations (e.g., GMP, quality management, galenics, and marketing authorization) is important for successful model development in a pharmaceutical production setting. Model implementation is manageable and worth the effort, as the present case study demonstrates. Most elements of model building are performed using simple drag and drop options, apart from the implementation of complex logic. Creating possibilities to track processing times and process flow requires more complex logic and, therefore, requires low-level programming. It is, however, possible to overcome these obstacles, especially after gaining some experience. Consequently, the necessary efforts during model building and verification strongly decrease for any other comparable production process. The next step of model validation is already described as being a crucial point for model-based debottlenecking approaches. Irregular or unsteady production steps are known to complicate or even preclude a successful model validation [30]. Unfortunately, unfavorable circumstances also led to a failed predictive model validation. It is therefore of great importance to choose the best validation strategy. The efforts for a historical data validation or a predictive validation are easily manageable, since the data handling identical to the one during model verification. Establishing optimization scenarios, is less time-intensive and challenging, even though the optimization module in FlexSim is not part of the applied student version. This means that all optimization scenarios had to be developed and implemented from scratch. While the results are clear on the superiority of six compared to four operators, more factors need to be considered. The examined work only covers bulk production, which is only one part of the entire production process and strongly depends on other departments, such as warehousing and quality control. Whenever warehousing and quality control work in an OS system, sufficient cooperation with the bulk production working in the OS or OHS system is granted. Modifications and adjustments must only be made when a TS system is established. The establishment of a TS system in bulk production can also produce a more rigid structure, decrease spontaneity, and increase the headcount. The necessary financial and work inputs needed to qualify two more operators for the production of about 20 products must also be taken into consideration. The impact of these disadvantages could be tested with the pilot run of a TS system. It could also be further explored whether an increased headcount in other departments should be mandatory, and whether the pressure on the involved staff is bearable.

Besides analysis of the effects of different shift systems, further simulations could examine more extensive questions, such as a change of the process layout. The simulated processes are based on a long-established production site. Hence, the layout and consequently the equipment localization depend on the floor plan and the structural conditions, such as media supply systems and electrical equipment. As shown in Figure 4, the current floor plan does not allow a lean production flow; an according spaghetti diagram would reveal inefficient transportation and employee movements. A re-layout including a rebuilding of the manufacturing premises of the site would enable an efficient and continuous workflow with significantly reduced non-value-added time (NVA), such as work in process inventory (WIP) or repeating pathways. Simulations could test the benefits of a re-layout and thereby also substantiate such far-reaching considerations.

5. Conclusions

Due to stiff regulations and a large operational complexity in multi-product, batch-operated sites, pharmaceutical companies sometimes operate in suboptimal conditions. Optimizing such pharmaceutical production processes via computer simulations not only appears to be practical but is also highly recommendable, even for relatively small companies. This paper demonstrates that discrete-event simulations of pharmaceutical productions (i) are feasible with FlexSim, (ii) can be conducted by non-computer experts, and iii) may significantly improve the production performance. Modeling of the entire production process does mean some effort at first. However, by considering the entire process, this model can be applied with more flexibility compared to other bottleneck identification methods focusing only on the technical equipment. In this case study, the entire process simulation was conducted by one person and it was possible to reduce the campaign duration by 50% for both products. If more resources were available, the benefits would likely improve and be determined more quickly. From a user perspective, more easily applicable tools and possibilities to effortlessly standardize modeling are desirable, especially for data acquisition and model verification. However, this process is only compelling for software developers if the demand is high enough. A widespread use of discrete-event simulations in pharmaceutical companies may, therefore, potentiate the present possibilities of such simulations.

Supplementary Materials: The following are available online at https://www.mdpi.com/article/10.3390/pr9010067/s1, Table S1: Listing of all integrated process steps with the best fitting statistical distributions according to ExpertFit; Table S2: Listing of additional work steps that are product independent.

Author Contributions: Conceptualization, S.H., F.S., and C.-M.L.; methodology, S.H., N.S., T.M.B., F.S., and C.-M.L.; software, S.H.; validation, S.H.; formal analysis, S.H.; investigation, S.H.; resources, T.M.B. and T.R.; data curation, S.H.; writing—original draft preparation, S.H.; writing—review and editing, S.H., N.S., T.M.B., B.L., T.R., F.S., and C.-M.L.; visualization, S.H.; supervision, N.S., T.M.B., B.L., F.S., and C.-M.L.; All authors have read and agreed to the published version of the manuscript.

Funding: This research received no external funding.

Institutional Review Board Statement: Not applicable.

Informed Consent Statement: Not applicable.

Data Availability Statement: Patent protection. Restrictions apply to the availability of these data. Data was obtained from protected company information. Data sharing is not applicable to this article.

Acknowledgments: The authors are thankful to Ralph Gruber, Ingenieurbüro für Simulationsdienstleistung, Kirchlengern, Germany for providing support in the programming and handling of FlexSim. We acknowledge support by the Deutsche Forschungsgemeinschaft (DFG, German Research Foundation) and Saarland University within the funding programm Open Access Publishing.

Conflicts of Interest: The authors declare no conflict of interest.

References

1. Kvesic, D.Z. Product lifecycle management: Marketing strategies for the pharmaceutical industry. *J. Med. Mark.* **2008**, *8*, 293–301. [CrossRef]
2. Federsel, H. Process R&D under the magnifying glass: Organization, business model, challenges, and scientific context. *Bioorg. Med. Chem.* **2010**, *18*, 5775–5794. [CrossRef] [PubMed]
3. Brandes, R. Der Entwurf zum neuen Annex 1. *Pharm. Ind.* **2018**, *5*, 671–680.
4. Kleemann, A. Metamorphosis of the Pharmaceutical Industry. *Pharm. Ind.* **2013**, *75*, 562–574.
5. Habibifar, N.; Hamid, M.; Bastan, M.; Azar, A.T. Performance optimisation of a pharmaceutical production line by inte-grated simulation and data envelopment analysis. *Int. J. Simul. Process Model.* **2019**, *14*, 360–376. [CrossRef]
6. Behr, A.; Brehme, V.; Ewers, C.; Grön, H.; Kimmel, T.; Küppers, S.; Symietz, I. New Developments in Chemical Engineering for the Production of Drug Substances. *Eng. Life Sci.* **2004**, *4*, 15–24. [CrossRef]
7. WHO. Tuberculosis. Available online: https://www.who.int/en/news-room/fact-sheets/detail/tuberculosis2020 (accessed on 22 March 2020).
8. WHO. *Guidelines for Treatment of Drug-Sesceptible Tuberculosis and Patient Care*; 2017 update Annex 6; World Health Organization: Geneva, Switzerland, 2017.
9. Banks, J.; Carson, J.S.; Nelson, B.L.; Nicol, D.M. *Discrete-Event System Simulation*, 4th ed.; Prentice Hall: Upper Saddle River, NJ, USA, 2005.
10. Neelamkavil, F. *Computer Simulation and Modelling.*, 1st ed.; Wiley: Chichester, UK, 1987.
11. Oberkampf, W.L.; Roy, C.J. *Verification and Validation in Scientific Computing*; Cambridge University Press (CUP): Cambridge, UK, 2010.
12. Rotstein, G.; Papageorgiou, L.; Shah, N.; Murphy, D.; Mustafa, R. A product portfolio approach in the pharmaceutical industry. *Comput. Chem. Eng.* **1999**, *23*, S883–S886. [CrossRef]
13. Laínez, J.M.; Schaefer, E.; Reklaitis, G.V. Challenges and opportunities in enterprise-wide optimization in the pharma-ceutical industry. *Comput. Chem. Eng.* **2012**, *47*, 19–28. [CrossRef]
14. Blau, G.E.; Pekny, J.F.; Varma, V.A.; Bunch, P.R. Managing a Portfolio of Interdependent New Product Candidates in the Pharmaceutical Industry. *J. Prod. Innov. Manag.* **2004**, *21*, 227–245. [CrossRef]
15. Shah, N. Pharmaceutical supply chains: Key issues and strategies for optimisation. *Comput. Chem. Eng.* **2004**, *28*, 929–941 [CrossRef]
16. Sundaramoorthy, A.; Karimi, I.A. Planning in Pharmaceutical Supply Chains with Outsourcing and New Product In-troductions. *Ind. Eng. Chem. Res.* **2004**, *43*, 8293. [CrossRef]
17. Chen, Y.; Mockus, L.; Orcun, S.; Reklaitis, G.V. Simulation-optimization approach to clinical trial supply chain manage-ment with demand scenario forecast. *Comput. Chem. Eng.* **2012**, *40*, 82–96. [CrossRef]
18. Parker, J.; Lamarche, K.; Chen, W.; Williams, K.; Stamato, H.; Thibault, S. CFD simulations for prediction of scaling effects in pharmaceutical fluidized bed processors at three scales. *Powder Technol.* **2013**, *235*, 115–120. [CrossRef]
19. Sen, M.; Chaudhury, A.; Singh, R.; John, J.; Ramachandran, R. Multi-scale flowsheet simulation of an integrated continu-ous purification-downstream pharmaceutical manufacturing process. *Int. J. Pharm.* **2013**, *445*, 29–38. [CrossRef] [PubMed]
20. Benyahia, B.; Lakerveld, R.; Barton, P.I. A Plant-Wide Dynamic Model of a Continuous Pharmaceutical Process. *Ind. Eng. Chem. Res.* **2012**, *51*, 15393–15412. [CrossRef]
21. Matsunami, K.; Sternal, F.; Yaginuma, K.; Tanabe, S.; Nakagawa, H.; Sugiyama, H. Superstructure-based process synthesis and economic assessment under uncertainty for solid drug product manufacturing. *BMC Chem. Eng.* **2020**, *2*, 1–16. [CrossRef]
22. Makrydaki, F.; Georgakis, C.; Saranteas, K. Dynamic Optimization of a Batch Pharmaceutical Reaction using the Design of Dynamic Experiments (DoDE): The Case of an Asymmetric Catalytic Hydrogenation Reaction. *IFAC Proc. Vol.* **2010**, *43*, 260–265. [CrossRef]
23. Kaylani, H.; Atieh, A.M. Simulation Approach to Enhance Production Scheduling Procedures at a Pharmaceutical Company with Large Product Mix. *Procedia CIRP* **2016**, *41*, 411–416. [CrossRef]
24. Saadouli, H.; Jerbi, B.; Dammak, A.; Masmoudi, L.; Bouaziz, A. A stochastic optimization and simulation approach for scheduling operating rooms and recovery beds in an orthopedic surgery department. *Comput. Ind. Eng.* **2015**, *80*, 72–79. [CrossRef]
25. Lin, J.T.; Chen, C.-M. Simulation optimization approach for hybrid flow shop scheduling problem in semiconductor back-end manufacturing. *Simul. Model. Pract. Theory* **2015**, *51*, 100–114. [CrossRef]
26. Caggiano, A.; Bruno, G.; Teti, R. Integrating Optimisation and Simulation to Solve Manufacturing Scheduling Prob-lems. *Procedia CIRP* **2015**, *28*, 131–136. [CrossRef]
27. Cognition IfHaM. CmapTools. Available online: https://cmap.ihmc.us/cmaptools/2020 (accessed on 22 March 2020).
28. Minitab. MINITAB SOLUTIONS Lean Six Sigma. Available online: https://www.minitab.com/en-us/products/companion/solutions/lean-six-sigma/2020 (accessed on 24 March 2020).
29. Stachowiak, H. *Allgemeine Modelltheorie*; Springer: Wien, Austria; New York, NY, USA, 1973.

30. Bähner, F.D.; Huusom, J.K. A Debottlenecking Study of an Industrial Pharmaceutical Batch Plant. *Ind. Eng. Chem. Res.* **2019**, *58*, 20003–20013. [CrossRef]

processes

Article

Abrasive Water Jet Cutting of Hardox Steels—Quality Investigation

Tibor Krenicky [1,*], Milos Servatka [2], Stefan Gaspar [1] and Jozef Mascenik [1]

[1] Faculty of Manufacturing Technologies with a Seat in Presov, Technical University of Kosice, Sturova 31, 08001 Presov, Slovakia; stefan.gaspar@tuke.sk (S.G.); jozef.mascenik@tuke.sk (J.M.)
[2] IMSLOV, P. Horova 19, 08001 Presov, Slovakia; ms@imslov.sk
* Correspondence: tibor.krenicky@tuke.sk; Tel.: +421-55-602-6337

Received: 9 November 2020; Accepted: 12 December 2020; Published: 14 December 2020

Abstract: The paper aims to study the surface quality dependency on selected parameters of cuts made in Hardox™ by abrasive water jet (AWJ). The regression process was applied on measured data and the equations were prepared for both the *Ra* and *Rz* roughness parameters. One set of regression equations was prepared for the relationship of *Ra* and *Rz* on cutting parameters—pumping pressure, traverse speed, and abrasive mass flow rate. The second set of regression equations describes relationships between the declination angle in kerf as the independent variable and either the *Ra* or the *Rz* parameters as dependent variables. The models can be used to predict cutting variables to predict the surface quality parameters.

Keywords: abrasive water jet; cutting; surface quality; quality prediction

1. Introduction

Cutting of materials by abrasive water jets has been studied for several decades. The pioneer scientists dealing with this topic were Hashish [1,2] and Zeng and Kim [3,4]. Later, some further investigations occurred aimed at the machining process, e.g., by Kovacevic and Yong [5,6]. The current state of research of abrasive water jet technology shows that one of the important problems is the quantification and modeling of the influence of technological parameters on surface quality parameters, particularly on wear-resistant steels. Evaluation of cutting quantity and quality was continuously studied by various groups [7–10]. Sutowska et al. [11] studied the influence of cutting parameters on the kerf quality in detail. Some of the recent experiments were performed on Hardox™ 400, 450, and 500 steel plates by Filip, Vasiloni, and Mihail [12,13].

Evaluation of the cutting quality is related to the quality of the cut walls. The typical characteristics of the walls are roughness and waviness. The most common characteristics used for the evaluation of the surface roughness were measured and analyzed. These characteristics are *Ra*, the mean arithmetic deviation of the profile, and *Rz*, the height of the profile unevenness. These two quantities can be measured by contact profilometers or by non-contact profilometers [14–16]. Nevertheless, the values depend not only on cut material or depth in the kerf but also on abrasive material quality and grain size [17]. Former research works aimed at the problem of abrasive material changes in the mixing process show, however, that the problem is not easy to solve [18,19] because not only the average mean size a_0 plays the decisive role in changes to new one a_n but also the amount and type of the original damage of abrasive grains. The influence of the abrasive material and its granularity is constant for one selected material sort. This is the most common case in all commercial firms. Therefore, the influence of abrasive material can be considered as disturbance quantity identical in all experiments. The surface waviness has much higher values than roughness, generally in the order of millimeters. The quality of this part of the cut walls is incredibly low, therefore, beyond the interest of this paper.

The Hlaváč group has presented another approach to the determination of the cutting wall quality than the use of the *Ra* and *Rz* values, proposing a direct relationship between the declination angle (measured between the tangent to the striation curve in the definite depth h and the impinging jet axis) and respective cutting wall quality. The angle is calculated either for a certain depth in material or some assigned traverse speed from the presented model [17]. Nevertheless, angle values are incredibly low in quality cutting, and thus even relatively small imperfections in measurements bring quite large uncertainty in quality results. Therefore, this method is better for evaluating the part of the cut walls with predominant waviness.

Although there is a constantly growing set of developed solutions to the problem, including methodologies and evaluations of experiments valid for specific measurement conditions, the current solutions still do not cover several variations. The microscopic models describing the mechanism of material cutting were prepared [18] as well as the macroscopic model of cutting front behavior [19,20]. An interesting multi-parametric phenomenological description of the cutting process has also been presented by the Ostrava group [21–23]. The group of TU Kosice researchers entered this research area as a part of systematic studies of the operational states of manufacturing processes using progressive technologies [14,15,24] and influence of the process parameters on the surface quality [25,26] a few years ago.

The recent research is focused on complementing existing models and preparing some new ones that would be simple enough to be applicable in industrial conditions to help predict and control the production quality. The most important results are presented in this paper. They can be used for the preparation of the regression models describing surface quality relationship to the cutting factors, water pressure, traverse speed, and abrasive mass flow rate.

2. Experimental Section

2.1. Characteristics of the Samples

All samples were cut from Hardox^{TM} 500 abrasion resistant plates with a nominal hardness of 500 HBW developed for applications with high demands on abrasion resistance. Material properties were obtained by a combination of quenching and tempering performed by manufacturer SSAB Oxelösund AB, Sweden. Sheet thicknesses of 6, 10, 15, and 40 mm were used for the individual sets of experiments with the following characteristics [27]:

Hardness (Brinell hardness, HBW according to EN ISO 6506-1, on a milled surface 0.5–2 mm below plate surface per heat and 40 tons): 486 (6 mm)–497 (40 mm).

Impact Properties (longitudinal Charpy-V; typical impact energy for 20 mm plate thickness at temperature −40 °C): 30 J.

The chemical composition of the material is presented in Table 1.

Table 1. Chemical composition of the Hardox^{TM} 500 plate samples [27].

Plate Thickness mm	C Max %	Si Max %	Mn Max %	P Max %	S Max %	Cr Max %	Ni Max %	Mo Max %	B Max %
4-13	0.27	0.70	1.60	0.025	0.010	1.00	0.25	0.25	0.004
(13)-32	0.29	0.70	1.60	0.025	0.010	1.00	0.50	0.30	0.004
(32)-40	0.29	0.70	1.60	0.025	0.010	1.00	1.00	0.60	0.004

To study the dependencies of parameters, a 3-level full 3-factor experiment was designed with a total number of combinations of technological parameter values of 27 (Table 2). These combinations were applied to 4 different sample thicknesses (6, 10, 15, and 40 mm). It follows that 9 samples with 3 cut surfaces were cut from each sheet thickness. Therefore, the shape with the plan view of an equilateral triangle was chosen as the most suitable sample shape. Transverse speeds v was used for sample thicknesses of 10 and 15 mm; v^{+} are increased speeds for 6 mm samples because the speeds v for

6 mm sheet metal would leave minimal roughness and at the same time almost identically rough-cut surfaces. Traverse speeds v^- were used for 40 mm sheet metal, as v would not be enough to cut the plate, so decreased speeds were chosen.

Table 2. Combinations of technological parameter values for sets of experiments.

Combination of Technological Values Parameters (Cutting No.)	Technological Parameter				
	m_a	p	v	v^+	v^-
1	170	300	40	60	10
2	170	300	60	90	15
3	170	300	80	120	20
4	170	340	40	60	10
5	170	340	60	90	15
6	170	340	80	120	20
7	170	380	40	60	10
8	170	380	60	90	15
9	170	380	80	120	20
10	220	300	40	60	10
11	220	300	60	90	15
12	220	300	80	120	20
13	220	340	40	60	10
14	220	340	60	90	15
15	220	340	80	120	20
16	220	380	40	60	10
17	220	380	60	90	15
18	220	380	80	120	20
19	270	300	40	60	10
20	270	300	60	90	15
21	270	300	80	120	20
22	270	340	40	60	10
23	270	340	60	90	15
24	270	340	80	120	20
25	270	380	40	60	10
26	270	380	60	90	15
27	270	380	80	120	20

2.2. *Characteristics of the AWJ System and Procedure*

The experiments were performed on the AWJ system comprising technological (cutting) head PaserIII^{TM}, X-Y table WJ1020-1Z-EKO with X-Y Computer Numerical Controlled (CNC) system and pump Flow HSQ 5X (see Figure 1) with combination of the following parameters:

Water orifice diameter d_o	0.25 mm
Stand-off distance L	2 mm
Focusing tube diameter d_a	1.02 mm
Focusing tube length l_a	76 mm
Abrasive material average grain size a_o	0.275 mm (MESH 80)
Abrasive material type	Australian garnet
Angle of impact θ	0 rad
Water jet pressure p	300, 340, 380 MPa
Abrasive mass flow rates m_a	170, 220, 270 g/min
Experimental traverse speeds v	40, 60 80 mm/min for each thickness 6, 10 15 mm
	10, 15, 20 mm/min for thickness 40 mm
	60, 90, 120 mm/min for thickness 6 mm

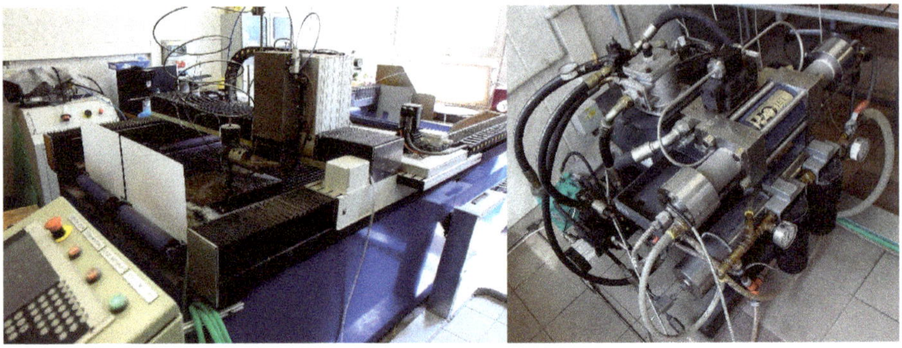

Figure 1. Experimental AWJ system (on the **left**) and pump Flow HSQ 5X (on the **right**).

These combinations represent 135 single cuts. Therefore, 45 triangle-shaped samples were cut, each side being cut with a different traverse speed v (Figure 2). All samples' cut surfaces were chemically treated with a passivation bath immediately after the end of the experiments—a solution of 5 g of sodium nitrite per 1 liter of water. The samples were immersed for 2–3 s in a solution at a temperature of about 60 ± 5 °C. Immediately after application of the solution, drying with hot air and storage in a dry environment followed. Surfaces treated in this way will resist corrosion for sufficient time to perform measurements.

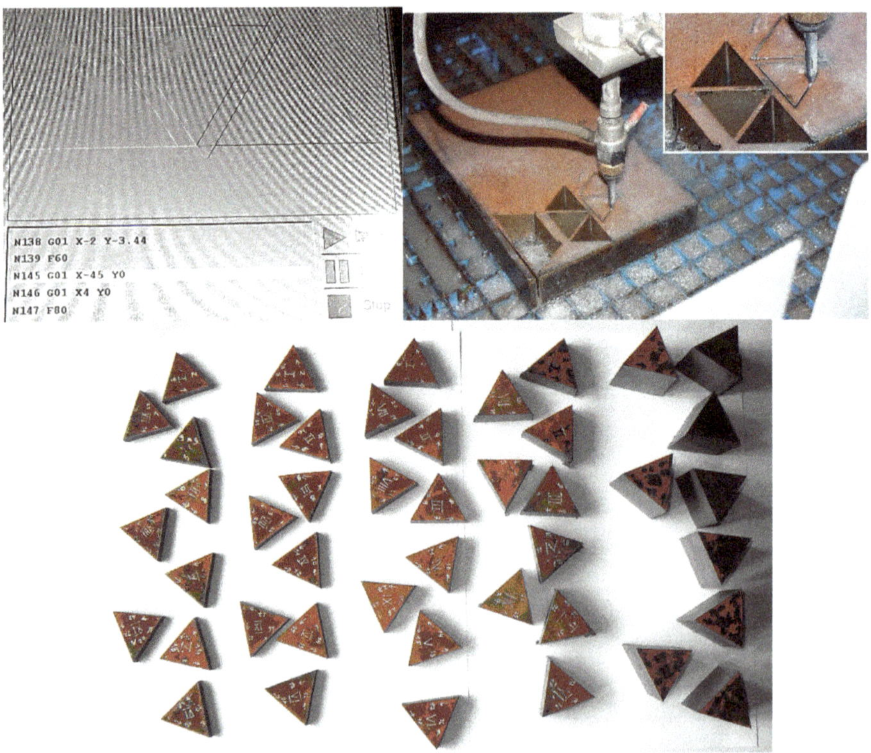

Figure 2. The cutting program's screen and detail of the Hardox^{TM} steel plates cutting (on the top); below samples prepared from (thicknesses from left to right 6, 6, 10, 15 and 40 mm).

The samples with thickness 6 mm were cut two times, ones with the same traverse speed as thicknesses 10 and 15 mm, ones with higher traverse speeds. Thickness 40 mm was not possible to cut using the same traverse speeds as other thicknesses. Therefore, the lower ones were utilized.

2.3. Roughness Measurement of Cut Surfaces

The roughness parameters *Ra* and *Rz* were measured in the middle height of the sample, i.e., at half the cut material's thickness. The roughness parameters Ra_4 and Rz_4 were measured on the cut surfaces of samples of all examined thicknesses (6, 10, 15, 40 mm and 6 mm+) at a distance of 4 mm from the upper cutting edge (from the surface of the sheet where the jet enters the material) using the Mitutoyo Surftest SJ-301 roughness tester. Repeated control measurements were performed for the reliability of all measured sets of values. The control measurements' total errors for the roughness *Ra*, *Rz*, Ra_4, and Rz_4 are in the range <3.06; 5.09> percent.

The Dixon test of extreme values is applied to selected sets of measured values in which some values differ significantly from the other values of the set. Based on the comparison of the calculated value and the tabular critical value of the test criterion, it can be stated with 95% probability that Rz and *Ra*'s assessed values are not extreme values and can therefore remain in the sets of measured values for evaluation.

Results of *Ra* and *Rz* measurements performed on the middle line of the thickness of samples cut from 6, 10, and 15 mm thick plates using the identical jet parameters mentioned above are summarized in Table 3. The additional results were measured on samples prepared from a 40 mm thick plate applying lower traverse speeds. They are presented in Table 4.

Table 3. Values of roughness *Ra* and *Rz* measured on cut surfaces of samples.

Sample Number	Cut Surface Number	m_a g/min	*P* MPa	*v* mm/min	6 mm *Ra* μm	6 mm *Rz* μm	10 mm *Ra* μm	10 mm *Rz* μm	15 mm *Ra* μm	15 mm *Rz* μm
I	1	170	300	40	3.17	21.88	4.44	23.82	4.92	26.76
	2	170	300	60	3.52	22.02	4.36	24.58	6.12	32.26
	3	170	300	80	3.69	22.11	4.98	29.25	7.93	35.97
II	4	170	340	40	3.06	21.73	3.87	23.41	4.62	26.32
	5	170	340	60	3.42	21.90	4.44	23.66	5.96	29.49
	6	170	340	80	3.63	22.02	4.72	24.77	7.11	35.25
III	7	170	380	40	3.04	21.59	3.76	22.80	4.23	24.45
	8	170	380	60	3.18	21.61	3.67	22.87	5.74	28.42
	9	170	380	80	3.56	21.91	4.26	23.78	6.86	34.65
IV	10	220	300	40	2.98	21.45	3.68	22.50	4.18	23.14
	11	220	300	60	3.05	21.50	3.58	23.12	5.36	28.21
	12	220	300	80	3.52	21.77	4.40	24.15	6.24	34.49
V	13	220	340	40	2.87	21.05	3.32	21.18	3.84	22.24
	14	220	340	60	3.02	21.41	3.47	21.51	5.08	26.73
	15	220	340	80	3.28	21.52	3.33	22.00	6.02	33.20
VI	16	220	380	40	2.73	19.39	3.07	20.00	3.22	20.48
	17	220	380	60	2.99	21.12	3.33	20.70	3.78	21.30
	18	220	380	80	3.26	21.08	3.30	21.58	5.62	31.66
VII	19	270	300	40	2.71	18.39	3.05	19.37	3.11	19.48
	20	270	300	60	2.92	21.02	3.40	20.30	3.56	20.25
	21	270	300	80	3.20	20.90	3.28	20.82	5.49	30.80
VIII	22	270	340	40	2.42	17.45	2.79	18.54	2.95	19.02
	23	270	340	60	2.78	19.87	3.14	19.75	3.27	19.68
	24	270	340	80	3.08	20.22	3.25	20.01	5.40	28.00
IX	25	270	380	40	2.27	17.20	2.75	17.81	3.10	18.27
	26	270	380	60	2.44	19.11	2.89	19.05	3.08	19.33
	27	270	380	80	2.80	19.35	3.22	19.62	3.76	24.01

Table 4. Measured values of declination angle θ and roughness characteristics *Ra* and *Rz* on the cut walls at samples with a thickness of 40 mm.

Sample Number	Surface Number	m_a g/min	p MPa	v mm/min	θ deg	*Ra* μm	*Rz* μm
I	1	170	300	10	17.7	3.65	20.84
	2	170	300	15	24.7	4.09	21.90
	3	170	300	20	30.1	6.95	24.96
II	4	170	340	10	15.0	2.90	19.79
	5	170	340	15	18.2	3.92	21.00
	6	170	340	20	26.1	5.87	23.64
III	7	170	380	10	14.5	2.83	19.11
	8	170	380	15	17.1	3.66	20.81
	9	170	380	20	21.2	4.10	22.90
IV	10	220	300	10	14.3	2.75	18.67
	11	220	300	15	16.9	3.46	20.56
	12	220	300	20	20.5	4.07	21.10
V	13	220	340	10	13.9	2.71	17.14
	14	220	340	15	15.8	3.22	20.60
	15	220	340	20	19.8	3.65	20.93
VI	16	220	380	10	13.5	2.60	16.83
	17	220	380	15	15.4	3.02	19.66
	18	220	380	20	18.8	3.46	20.69
VII	19	270	300	10	13.0	2.44	16.52
	20	270	300	15	14.1	3.01	19.30
	21	270	300	20	18.3	3.33	20.10
VIII	22	270	340	10	11.5	2.28	16.25
	23	270	340	15	12.7	2.93	18.29
	24	270	340	20	16.8	3.21	19.77
IX	25	270	380	10	9.6	2.27	16.02
	26	270	380	15	10.5	2.67	16.77
	27	270	380	20	14.6	2.96	18.40

Results measured on samples 6 mm thick, cut at higher traverse speeds than samples presented in Table 3, were used to broaden the confidence interval of a regression derivation of the relationships useful for analyses, simulations, and control of the surface quality.

Results of surface roughness characteristics *Ra* and *Rz* presented in Table 3 indicate supposed relationships—increasing quality for lower traverse speeds, higher pressures in pump, and higher abrasive mass flow rates. Similar conclusions can be drawn from the results presented in Table 4. Nevertheless, the relationship between roughness and declination angle values can be derived from values in Table 4. Subsequently, the results can be compared with the model presented by Hlaváč [17].

Summarizing all combinations of factors, it is possible to obtain functions describing speed-dependent roughness for each doublet m_a and p. However, it is necessary to measure the values in a certain selected identical depth on the cut wall for all samples (to compare the values). The depth equal to 4 mm was selected for presentation in this paper (values are marked Ra_4 and Rz_4). The typical series of roughness values for all used traverse speeds is presented in Table 5. The selected abrasive mass flow rate is typical for applied nozzle diameter, focusing tube characteristics, abrasive material type, grain size, and pump pressure. Traverse speeds were completed from all experimental sets.

Table 5. Typical series of roughness values Ra_4, Rz_4 (abrasive mass flow rate 220 g/min and pressure 380 MPa are typical technological parameters used for cutting).

v mm/min	m_a g/min	p MPa	Ra_4 µm	Rz_4 µm
10	220	380	2.16	16.90
15	220	380	2.44	18.09
20	220	380	2.71	19.53
40	220	380	2.89	20.11
60	220	380	2.98	20.75
80	220	380	3.32	21.58
90	220	380	3.70	23.45
120	220	380	3.85	23.22

The graph of relation between traverse speed and surface roughness parameters is presented in Figure 3 (for values presented in Table 5).

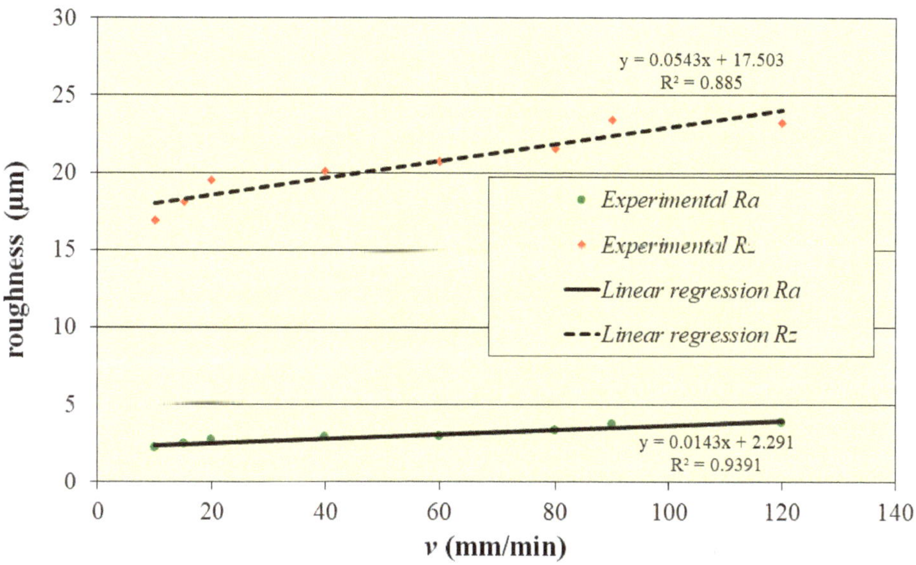

Figure 3. Roughness dependence on traverse speed with linear regression (graph).

The summary regression models for factors x_1 (m_a), x_2 (p) and x_3 (v) can be written as

$$Ra = 7.905 - 0.012x_1 - 0.007x_2 + 0.011x_3 \quad (1)$$

$$Rz = 39.103 - 0.049x_1 - 0.027x_2 + 0.046x_3 \quad (2)$$

Values calculated from these models were compared with further experimental results, and the comparison is presented in Table 6. The modeled surface roughness values for the respective combinations of technological parameter values were subsequently experimentally verified. The result of the verification confirmed the correctness of the verified mathematical models and the subsequently performed calculation. The deviation between the values of *Ra*, *Rz* obtained from the simulation, and the values from the subsequent verification experiment ranges from −5.4 to +5.6% for *Ra* and in the range of −4.9 to +0.3% for *Rz*. From that, it is evident that the model is simple but works effectively.

Table 6. Comparison of calculated and measured roughness values *Ra* and *Rz*.

Technological Parameters			Quality Parameters				Deviation of Calculated Value Regarding the Experimental Value	
Abrasive Mass Flow Rate	Pump Pressure	Traverse Speed	Values Calculated from Model		Measured Experimental Values			
m_a (x_1) g/min	p (x_2) MPa	v (x_3) mm/min	Ra (y) µm	Rz (y) µm	Ra µm	Rz µm	for Ra %	for Rz %
160	270	35	4.48	25.58	4.70	26.91	−4.7	−4.9
180	285	38	4.17	24.34	4.41	25.22	−5.4	−3.5
190	310	45	3.95	23.49	3.84	24.04	2.9	−2.3
190	310	45	3.82	22.99	3.95	22.90	−3.3	0.3
200	320	50	3.70	22.53	3.90	22.85	−5.1	−1.4
210	330	57	3.41	21.37	3.48	22.27	−2.0	−4.0
230	350	65	3.13	20.26	3.11	20.95	0.6	−3.3
250	360	68	3.04	19.92	2.88	20.08	5.6	−0.8
260	370	77	2.75	18.76	2.81	19.55	−2.1	−4.0

2.4. Measurements of the Angle of Declination of the Jet

The declination angle θ of the abrasive water jet for 40 mm thick samples was measured at 5 locations on each cut surface on series of successive measurements distinct approximately 5 mm from the previous measurement in the cutting direction according to Figure 4. The resulting values of the jet deflection on the cut surfaces were obtained by the arithmetic mean of the measured repeated values on the individual cut surfaces (θ_1–θ_5).

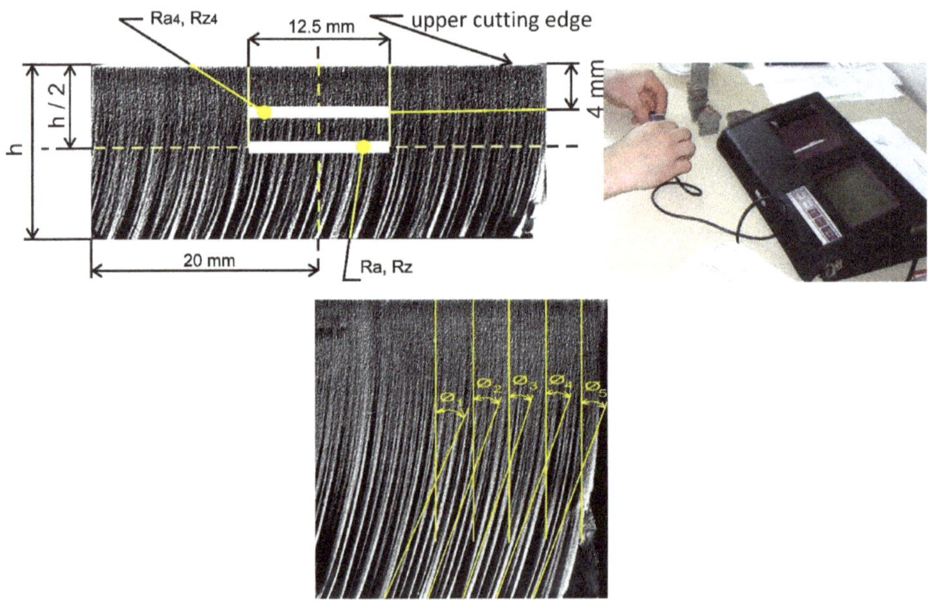

Figure 4. Locations of zones for measuring roughness parameters *Ra, Rz*, and Ra_4, Rz_4 (**upper left**); measurement of roughness *Ra, Rz*, and Ra_4, Rz_4 on cut surfaces of samples (**upper right**); declination angle on the cut surface (**bottom**).

A Vogel-Germany Universal Winkelmesser device with measuring range distribution 4 × 90° and scale resolution 5′ was used to measure jet declination on cut surfaces of 40 mm thick samples. Presented results also make it possible to prepare the regression equations describing the relations between declination angle value and roughness parameters. The equations obtained from regression by the processing of all measured values for both the *Ra* and *Rz* characteristics of roughness for factor x_4 (θ) are as follows

$$Ra = 0.2195x_4 - 0.2239 \quad (3)$$

$$Rz = 0.4442x_4 + 12.25 \quad (4)$$

The obtained model reveals linear behavior within the tested range, as shown in Figure 5. Testing of this model accuracy on other materials is just in the stage of preparation for future work.

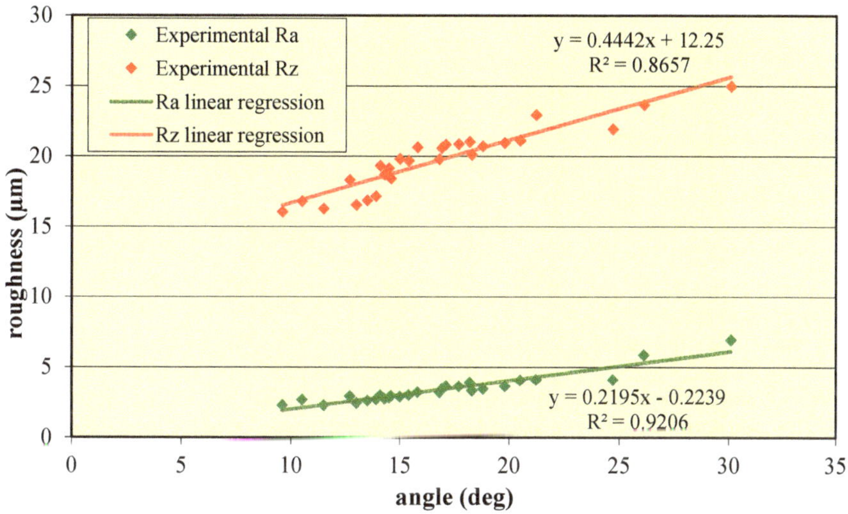

Figure 5. Roughness dependence on traverse speed with linear regression (typical graph).

3. Discussion

The experimental investigation of cuts made in very hard and wear-resistant steel concludes that relationships between the declination angle and roughness parameters are linear. The studied problem is close to the tailback and the taper investigations performed in the past by Hashish [16] or Ma and Deam [28]. Precise experiments studying surface quality on the selected process parameters of the Hardox steel plates cutting were revealing the traverse speed as the most important parameter influencing the accuracy of AWJ cutting were performed previously [11–13].

Calculation of the tilting of the cutting head for compensation of the declination angle effect on cut walls, presented, e.g., by a group of Ostrava researchers [17,29,30], should help to improve the cutting process and to minimize the typical defects caused by the abrasive water jet declination when the cut starts, ends, changes direction in the corners and in the curved parts of trajectories. Because according to the up-to-date results, no simple and direct relationship has been proved between common material properties and cutting quality for all types of cut material, the presented relations seem to be quite important. Their further investigation and confirmation for a broad spectrum of materials are necessary.

Models proposed in the present article may be evaluated from some points of view as simple; lacking physical linking of the parameters; but still in accordance with the main features of the more complex Hlaváč model and with the experimental findings of other teams studying the quality of this type of cut materials and confirmed by experiments reported in Table 6. Thus, it may complement the variety of the materials and range of studied parameters and widen existing models for application in particular conditions.

The relations for these quantities should be the aim of further research because the miniaturization of abrasive water jets needs a strong and stable description to predict and control the production quality. Therefore, studying the surface topography on the cut walls is still important. Experimental and theoretical studies of the interaction problems are important for lifting the abrasive water jet tool to a higher level of operational excellence.

4. Conclusions

The presented research was aimed at the dependence of selected technological parameters of the AWJ system on selected parameters of the quality of the cut surface. The multiple linear regression function describing the cut wall roughness as a function of the mentioned selected cutting variables has been determined for both *Ra* and *Rz*.

The main results can be summarized as follows:

With increasing material thickness from 6 to 10 and 15 mm, the roughness in its central part increases by more than 18% at *Ra* and 5.5% at *Rz* for 6 and 10 mm thicknesses and by one third for *Ra* and 21% for *Rz* for 10 and 15 mm thicknesses.

The values of technological parameters m_a = 170 g/min, p = 300 MPa, v = 80 mm/min represent the combination with which the highest roughness values while the values of technological parameters m_a = 270 g/min, p = 380 MPa, v = 40 mm/min represent the combination with which the smallest roughness values were achieved. By increasing m_a from 170 to 270 g/min at p = 340 MPa, it is possible to twice the speed v with an unchanged roughness value *Ra* of the cut surface.

The largest influence of the monitored technological parameters on the roughness (*Ra*, *Rz*) was found for the abrasive mass flow, a smaller influence was revealed for the cutting speed v.

Derived regression models (Equations (1) and (2)) show linear relationships have been determined between studied independent variables of the cutting process (traverse speed, liquid pressure, and abrasive mass flow rate) and roughness characteristics. Simultaneously, the linear relationships (Equations (3) and (4)) have also been found between declination angle values and roughness parameters *Ra* and *Rz*.

The models can be used both for a prediction of cutting variables and for a calculation of the cutting characteristics, such as traverse speeds, abrasive flow rates, and other influencing cutting walls quality. The achieved results are utilizable for improvement of the control software of the CNC machines used for water jet and abrasive water jet cutting and complement the existing solutions in the scientific field and can be used to reduce operating costs and increase the economic efficiency of production systems with AWJ technology.

The authors plan to include non-contact measurements on the samples cut using the AWJ systems, more complex roughness and waviness parameter analysis, and texture modeling of measured surfaces using their merging for future work.

Author Contributions: Conceptualization, M.S. and T.K.; methodology, M.S.; validation, T.K. and M.S.; formal analysis, M.S.; data curation, T.K.; writing—original draft preparation, M.S.; writing—review and editing, T.K.; visualization, J.M.; supervision, T.K.; project administration, J.M.; funding acquisition, S.G. All authors have read and agreed to the published version of the manuscript.

Funding: This work was supported by the Slovak Ministry of Education within project VEGA No. 1/0116/20 and by the Slovak Research and Development Agency under contract No. APVV-18-0316.

Acknowledgments: Authors would like to acknowledge Stanislav Fabian for valuable comments.

Conflicts of Interest: The authors declare no conflict of interest and the funders had no role in the design of the study; in the collection, analyses, or interpretation of data; in the writing of the manuscript, or in the decision to publish the results.

Nomenclature

θ	angle between impinging jet axis and tangent to the striation curve in the selected depth h ... [°]
a_n	average mean size of abrasive particles formed in the mixing process ... [m]
a_o	average mean size of abrasive particles entering the mixing process ... [m]
d_o	water nozzle diameter ... [mm]
d_a	focusing tube diameter ... [mm]
H	material thickness ... [mm]
l_a	focusing tube length ... [mm]
L	stand-off distance ... [mm]
p	water jet pressure ... [MPa]
m_a	abrasive mass flow rate ... [g/min]
v	traverse speed ... [mm/min]
Ra	arithmetic average roughness ... [μm]
Rz	maximum peak to valley height of the profile ... [μm]

References

1. Hashish, M. Modeling study of metal cutting with abrasive waterjets. *J. Eng. Mater. Technol. Trans. ASME* **1984**, *106*, 88–100. [CrossRef]
2. Hashish, M. A Model for Abrasive—Waterjet (AWJ) Machining. *J. Eng. Mater. Technol. Trans. ASME* **1989**, *111*, 154–162. [CrossRef]
3. Zeng, J.; Kim, T.J. Development of an abrasive waterjet kerf cutting model for brittle materials. In *Jet Cutting Technology*; Lichtarowicz, A., Ed.; Kluwer Academic Publishers: Dordrecht, The Netherlands, 1992; pp. 483–501.
4. Zeng, J.; Kim, T.J. An erosion model of polycrystalline ceramics in abrasive waterjet cutting. *Wear* **1996**, *193*, 207–217. [CrossRef]
5. Kovacevic, R.; Yong, Z. Modelling of 3D abrasive waterjet machining, part 1—Theoretical basis. In *Jetting Technology*; Gee, C., Ed.; Mechanical Engineering Publications Ltd.: Bury St Edmunds, UK; London, UK, 1996; pp. 73–82.
6. Yong, Z.; Kovacevic, R. Modelling of 3D abrasive waterjet machining. Part 2—Simulation of machining. In *Jetting Technology*; Gee, C., Ed.; Mechanical Engineering Publications Ltd.: Bury St Edmunds, UK; London, UK, 1996; pp. 83–89.
7. Hlaváč, L.M. JETCUT—Software for prediction of high-energy waterjet efficiency. In *Jetting Technology*; Louis, H., Ed.; Prof. Eng. Pub. Ltd.: Bury St Edmunds, UK; London, UK, 1998; pp. 25–37.
8. Chen, F.L.; Wang, J.; Lemma, E.; Siores, E. Striation formation mechanism on the jet cutting surface. *J. Mater. Process. Technol.* **2003**, *141*, 213–218. [CrossRef]
9. Henning, A.; Westkämper, E. Analysis of the cutting front in abrasive waterjet cutting. In *Water Jetting*; Longman, P., Ed.; BHR Group: Cranfield, UK; Bedford, UK, 2006; pp. 425–434.
10. Monno, M.; Pellegrini, G.; Ravasio, C. An experimental investigation of the kerf realised by AWJ: The influence of the pressure fluctuations. In *Water Jetting*; Longman, P., Ed.; BHR Group: Cranfield, UK; Bedford, UK, 2006; pp. 309–321.
11. Sutowska, M.; Kapłonek, W.; Pimenov, D.Y.; Gupta, M.K.; Mia, M.; Sharma, S. Influence of variable radius of cutting head trajectory on quality of cutting kerf in the abrasive water jet process for soda–lime glass. *Materials* **2020**, *13*, 4277. [CrossRef] [PubMed]
12. Filip, A.C.; Mihail, L.A.; Vasiloni, M.A. An experimental study on the dimensional accuracy of holes made by abrasive waterjet machining of Hardox steels. *MATEC Web Conf.* **2017**, *137*, 02003. [CrossRef]
13. Filip, A.C.; Vasiloni, M.A.; Mihail, L.A. Experimental research on the machinability of Hardox steel by abrasive waterjet cutting. *MATEC Web Conf.* **2017**, *94*, 03003. [CrossRef]
14. Krenický, T. Non-contact study of surfaces created using the AWJ technology. *Manuf. Technol.* **2015**, *15*, 61–64. [CrossRef]
15. Maščenik, J.; Gašpár, Š. Experimental assessment of roughness changes in the cutting surface and microhardness changes of the material S 355 J2 G3 after being cut by non-conventional technologies. *Adv. Mater. Res.* **2011**, *314–316*, 1944–1947. [CrossRef]

16. Pahuja, R.; Ramulu, M.; Hashish, M. Surface quality and kerf width prediction in abrasive water jet machining of metal-composite stacks. *Compos. Part B Eng.* **2019**, *175*, 107134. [CrossRef]
17. Hlaváč, L.M. Investigation of the abrasive water jet trajectory curvature inside the kerf. *J. Mater. Process. Technol.* **2009**, *209*, 4154–4161. [CrossRef]
18. Deam, R.T.; Lemma, E.; Ahmed, D.H. Modelling of the abrasive water jet cutting process. *Wear* **2004**, *257*, 877–891. [CrossRef]
19. Orbanic, H.; Junkar, M. Analysis of striation formation mechanism in abrasive water jet cutting. *Wear* **2008**, *265*, 821–830. [CrossRef]
20. Rahman, M.A.; Mustafizur, R.; Kumar, A.S. Chip perforation and 'Burnishing–like' finishing of Al alloy in precision machining. *Precis. Eng.* **2017**, *50*, 393–409. [CrossRef]
21. Hlaváč, L.M.; Hlaváčová, I.M.; Arleo, F.; Viganò, F.; Annoni, M.; Geryk, V. Shape distortion reduction method for abrasive water jet (AWJ) cutting. *Precis. Eng.* **2018**, *53*, 194–202. [CrossRef]
22. Hlaváč, L.M.; Štefek, A.; Tyč, M.; Krajcarz, D. Influence of material structure on forces measured during Abrasive Waterjet (AWJ) machining. *Materials* **2020**, *13*, 3878. [CrossRef] [PubMed]
23. Hlaváč, L.M.; Hlaváčová, I.M.; Plančár, Š.; Krenický, T.; Geryk, V. Deformation of products cut on AWJ x-y tables and its suppression. *IOP Conf. Ser. Mater. Sci. Eng.* **2018**, *307*, 012015. [CrossRef]
24. Mascenik, J. Experimental determination of cutting speed influence on cutting surface character in material laser cutting. *MM Sci. J.* **2016**, 3, 960–963. [CrossRef]
25. Servátka, M. Modelling, simulation and optimization of the technological parameters in binding on the demanded quality of products in manufacturing technologies with water jet. Ph.D. Thesis, FMT TUKE, Prešov, Slovak, 2009.
26. Olejárová, Š.; Ružbarský, J.; Krenický, T. Introduction into the issue of water jet machining. In *Vibrations in the Production System. SpringerBriefs in Applied Sciences and Technology*; Springer: Cham, Swizerland, 2019; pp. 1–10. ISBN 978-3-030-01736-1. [CrossRef]
27. Hardox 500 Data Sheet, Version 2005-07-04. Available online: www.ssab.com (accessed on 11 July 2012).
28. Ma, C.; Deam, R.T. A correlation for predicting the kerf profile from abrasive water jet cutting. *Exp. Therm. Fluid Sci.* **2006**, *30*, 337–343. [CrossRef]
29. Štefek, A.; Hlaváč, L.M.; Tyč, M.; Barták, P.; Kozelský, J. Remarks to Abrasive Waterjet (AWJ) Forces Measurements; Advances in Water Jetting. Water Jet 2019. In *Lecture Notes in Mechanical Engineering*; Klichová, D., Sitek, L., Hloch, S., Valentinčič, J., Eds.; Springer: Cham, Switzerland, 2021. [CrossRef]
30. Hlaváč, L.M.; Hlaváčová, I.M.; Gembalová, L.; Jonšta, P. Experimental investigation of depth dependent kerf width in abrasive water jet cutting. In *Water Jetting*; Trieb, F.H., Ed.; BHR Group: Cranfield, UK; Bedford, UK, 2010; pp. 459–467.

processes

MDPI

Article

Integration and Evaluation of Intra-Logistics Processes in Flexible Production Systems Based on OEE Metrics, with the Use of Computer Modelling and Simulation of AGVs

Krzysztof Foit, Grzegorz Gołda and Adrian Kampa *

Department of Engineering Processes Automation and Integrated Manufacturing Systems, Silesian University of Technology, Konarskiego 18A, 44-100 Gliwice, Poland; krzysztof.foit@polsl.pl (K.F.); grzegorz.golda@polsl.pl (G.G.)

* Correspondence: adrian.kampa@polsl.pl; Tel.: +48-32237-1863

Received: 13 November 2020; Accepted: 9 December 2020; Published: 14 December 2020

Abstract: The article presents the problems connected with the performance evaluation of a flexible production system in the context of designing and integrating production and logistics subsystems. The goal of the performed analysis was to determine the parameters that have the most significant influence on the productivity of the whole system. The possibilities of using automated machine tools, automatic transport vehicles, as well as automated storage systems were pointed out. Moreover, the exemplary models are described, and the framework of simulation research related to the conceptual design of new production systems are indicated. In order to evaluate the system's productivity, the use of Overall Equipment Efficiency (OEE) metrics was proposed, which is typically used for stationary resources such as machines. This paper aims to prove the hypothesis that the OEE metric can also be used for transport facilities such as Automated Guided Vehicles (AGVs). The developed models include the parameters regarding availability and failure of AGVs as well as production efficiency and quality, which allows the more accurate mapping of manufacturing processes. As the result, the Overall Factory Efficiency (OFE) and Overall Transport Efficiency (OTE) metrics were obtained. The obtained outcomes can be directly related to similar production systems that belong to World Class Manufacturing (WCM) or World Class Logistics (WCL), leading to the in-depth planning of such systems and their further improvement in the context of the Industry 4.0.

Keywords: AGV—Automated Guided Vehicles; DES—Discrete Event Simulation; FMS—Flexible Manufacturing Systems; Industry 4.0; OEE—Overall Equipment Efficiency; WCLcWorld Class Logistic

1. Introduction

One of the key factors, which determine the level of competitiveness of manufacturing plants, is the ability to achieve flexible production and delivery of goods in accordance with customer requirements. Therefore, logistics and Supply Chain Management play an important role in market competition [1]. Logistic operations are carried out in two areas, internal and external to the organization. Therefore, the term "Intra-logistics" describes the organisation and realisation of internal material flow and logistic technologies as well as the goods transhipment in the industry, using technical components, partial and full systems and services [2].

In connection with the flexible production, there are also increased requirements regarding the flexibility and reliability of internal transport systems, associated with the production of short series of various products and thus requiring more transport operations [3]. In response to these needs, the Automated Guided Vehicles (AGVs) are used a lot more widely, which enables full automation of transport operations and can handle various transport routes using only computer navigation systems [4–6].

In order to evaluate the system's productivity, the use of Overall Equipment Efficiency (OEE) metrics has been proposed that is usually used for stationary resources, like machines. The aim of the paper was to prove the hypothesis that the OEE metric can also be used for transport facilities such as AGVs, and there is a dependence between machine and transport effectiveness.

To investigate the influence of the intra-logistic system over the performance of the manufacturing system we have developed a conceptual model of Flexible Manufacturing System (FMS) with an automated transport system with AGVs. The model of the FMS was built with the use of FlexSim software and the OEE metric was used for system integration and description of the availability, reliability and performance parameters for machines and vehicles. Then the Discrete Event Simulation (DES) method was used for performing the series of experiments and an analysis of results is presented in form of Overall Factory Efficiency (OFE) and Overall Transport Efficiency (OTE). Our previous works [7–9] show that the computer simulation of the detailed model of the production line with machines, operators and robots with reliability parameters allows better representation and understanding of a real production process which is important for early design and enables front-end planning.

The AGV parameters play an important role for FMS performance; therefore, literature review about FMS, AGV and logistics issues, are presented in the next section. The subsequent sections of the paper include description of the problem, modelling and simulation experiments, results analysis and discussion and final conclusions.

2. State-of-the-Art

Many researchers in logistics have examined the influence of high-performance logistics practices on organizational performance [1,10,11]. In an attempt to drive performance improvements, managers often struggle with multiple, seemingly conflicting, objectives [1]. Logistics management is faced with a tough choice: either strive for efficiency; or strive for effectiveness. Some recent logistics research has suggested that these two performance objectives are mutually exclusive [12]. Performance measures are essential for effective management of any organization. Performance measurement provides a needed assessment of service and cost aspects of logistics execution in the supply chain. Specifically, there is little guidance regarding where a specific measure should be used and, more pointedly, where the use of the measure would be less appropriate.

Fugate et al. [13], have presented the model of logistics performance with the concept of simultaneous pursuit of efficiency, effectiveness and differentiation. However, most companies' priorities change over time due to market and competitive dynamics. In light of this business reality, enterprises and managers must be able to identify and select new or different measures consistent with evolving organizational priorities [14].

Muthiah and Huang [15] reviewed and categorized various productivity improvement methods and productivity metrics. For example, Overall Equipment Efficiency (OEE is an established technique in World Class Manufacturing (WCM). It is used as a key performance indicator (KPI) in conjunction with lean manufacturing efforts to provide a quantifiable measurement of success. There are a few examples of the performance evaluation of manufacturing systems with the use of the OEE metric, including [16,17], but without considering the efficiency of the transport subsystem.

Muñoz-Villamizar et al. [18], have used OEE to evaluate the effectiveness of urban freight transportation systems and a framework for Overall Transport Efficiency (OTE) based on OEE factors was proposed by Dalmolen et al. [19]. McCalion [20] ask the question: is OEE relevant to logistics management and Automated Guided Vehicle (AGV) operations?

Hayes [21] suggests that the OEE can be used for eliminating the ripple effect caused by stopped vehicles and along with Six Sigma [22] can be used as a measure for World Class Logistic (WCL).

Comparing WCL with WCM, they have a lot in common. The common area is related to intra-logistic in manufacturing systems. Intra-logistic performance has a great influence over the manufacturing performance, because of inter-operational breaks which have a great impact on materials flow in flexible manufacturing system [23].

The literature review only shows a few publications on the design methodology of AGV systems and most of the them use simple KPIs as metrics [24–29]. At the time of preparing this paper, no publication was found concerning the detailed assessment of the impact of the AGVs system on the manufacturing process effectiveness that includes the OEE metric. Therefore, the studies have been undertaken in order to elaborate this problem in terms of transport and production effectiveness and to strengthen the logistics potential of the organization.

2.1. Issues Related to FMS and AGV

The flexible manufacturing system (FMS) is a fully automated production system that interconnects machines and workstations with the logistics equipment, where the entire manufacturing process is coordinated by the digital control systems such as Computer Numerical Control (CNC) or Programmable Logic Controller (PLC). Such flexible, automated manufacturing systems are intended for tasks of large typological diversity, high complexity, ensuring on-time delivery and minimal manufacturing costs, while production is unpredictable, being organized in small batches, with frequent changes [30].

The FMS has been studied over the last couple of decades and the researchers have found a variety of problems, which can be distributed in three major categories: workshop design, transportation network design and scheduling problems [31,32]. Different methods were used to solve them, including mathematical (linear, constraints, stochastic) programming, combinatorial optimization, Petri nets and scenario analysis, but computer simulation, especially Discrete Event Simulation (DES), is the most universal and widely used one [33], e.g., for the design of manufacturing systems [27], efficiency [9] and stability analysis [34] of production systems and the design of warehouse transportation systems with Automated Guided Vehicles (AGVs) [35–37].

The AGVs are classified as service robots for professional purposes in manufacturing environments and broad review of AGV is presented in [6,28].

Modern AGV vehicles are characterized by precision of operation, speed of movement and high reliability. They can have various equipment to perform numerous transport tasks, such as transporting pallets and containers, pulling trailers with cargo, lifting with a forklift or manipulating details using an integrated robot arm. Examples of AGV carriages are shown in Figure 1.

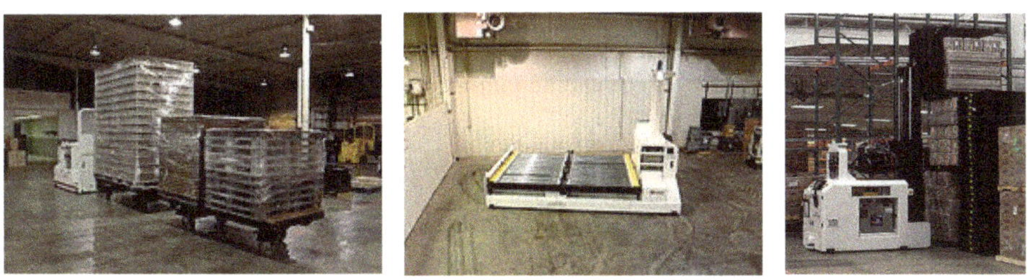

(a) (b) (c)

Figure 1. Examples of Automated Guided Vehicles (AGVs), (**a**) pulling trailers with cargo, (**b**) transporting pallets, (**c**) lifting loads with a forklift [38].

In comparison to the other solutions of transport systems, AGV vehicles show many advantages including [5,6]:

- they do not require an operator's service, which allows reducing the labour costs,

- increased work efficiency—it can work 24 h a day,
- high positioning precision—less material losses during transport,
- high security—the replacement of the operator reduces the number of accidents at work, safety systems reduce the risk of collision,
- flexibility of use—the ability to program the route according to the requirements of the process, easy route change and system expansion.

Typical features of AGV are related to the following parameters, as [5]:

- weight and size,
- load capacity (from a few kgs to several tons),
- driving speed 1–2 m/s,
- drive power,
- navigation method, positioning accuracy,
- time of loading and unloading,
- battery capacity,
- working time, battery charging time.

AGV uses electric drive and efficient batteries, however, it requires periodic recharging. Depending on the battery capacity and load, a typical work cycle includes 8–16 h of work and 4–8 h of battery charging, which takes place completely automatically. Some solutions for recharging the battery during short interruptions in the work of AGV (Opportunity Battery Charging) can be found. There are also solutions based on manual or automatic replacement of a depleted battery with a new fully charged battery (Battery Swap). This action takes about 10 min and allows to take full advantage of AGV's working time but requires a more advanced service system and additional battery packs.

The design of a transport system based on AGVs requires an advanced navigation system and appropriate delineation of transport routes and reloading points. Based on the technique, various navigation systems are used, such as [37]:

- photo-optical—with a passive lead line,
- inductive—with an active lead line,
- without a lead line—autonomous navigation with different location methods: incremental, infrared, ultrasonic, laser, gyroscopic, satellite (GPS).

Various methods can be used to design AGV systems, including mathematical programming methods, heuristic methods, Petri nets and computer simulations. These methods are used in order to improve the transport network in terms of criteria, for minimizing the length of transport routes, maximizing the production flow, scheduling transport tasks, number and location of transhipment points, parking zones and others [28,39].

Transport routes can be one- or two-way. Due to the reduction of the risk of collision, one-way roads in the form of closed loops, which enable cyclical transport operations, are the most commonly used [22]. In this case, it is easier to develop traffic control algorithms than in the case of two-way traffic, which requires additional passing and parking zones. Therefore, during the design of the AGV system, the most frequently used zones are defined including specific segments of the route (Segmented Flow Configuration) and individual transport loops (Single Loops) in a given segment. The advantages of such a solution are related to [40]:

- All AGV vehicles move in the same direction, which practically excludes collisions,
- system control is simplified due to the lack of alternative routes.

In turn, some drawbacks are connected with [35]:

- Small fault tolerance, in the event of failure of one vehicle, the others usually cannot pass it by,
- if the vehicle passes the given transfer point, it cannot turn back, but it must cross the entire loop once again to reach it again,
- vehicles hold each other, which may lead to blockages of the system (deadlock).

When designing the AGV transport system, the most important problem is determining the number of vehicles needed to achieve the required production volume or the minimum number of vehicles required to obtain the optimal production volume [36,40].

There are a lot of situations in which the AGV system may stall because of a deadlock. A variety of deadlock-detecting algorithms are available in literature [41], but these methods work mainly for manufacturing system where the network layout is simple and uses only a small number of AGVs. The paper [42] discusses the development of an efficient strategy for predicting and avoiding the deadlocks in a large scale AGV systems. The integration of the scheduling of production and transport tasks tends to also be problematic because of computational complexity [43,44]. In initial papers, the transportation times between machines have not been considered. Their authors claimed that because transportation times are very small in comparison with the processing times, they are negligible [45]. On the other hand, in recent decades, the more researchers have been attracted by some issues that the transportation times were considerable and ignoring them can have impacts on the solution of scheduling problems.

2.2. Evaluation of FMS and AGV

The performance of the AGV logistics system can have a great influence over the performance of the whole FMS system; therefore, a performance evaluation should be conducted. The key performance indicators (KPI) of the production system include [16,46]:

- Production throughput,
- time of the production process (Manufacturing Lead Time),
- average waiting time for transport,
- length of queues in storage buffers,
- work in progress (WIP),
- downtime of workstations,
- delayed execution of production orders,
- OEE—Overall Equipment Effectiveness,
- OTE—Overall Throughput Effectiveness.

Work efficiency and the use of the means of production can be expressed by using the OEE metric that depends on three factors: availability, performance and quality [16].

$$\text{OEE} = (\text{Availability}) \times (\text{Performance}) \times (\text{Quality}) \tag{1}$$

Availability is the ratio of the time spent on the realization of a task to the scheduled time. Availability is reduced by disruptions at work and machine failures.

$$\text{Availability} = \frac{\text{available work time} - \text{failure time}}{\text{scheduled time}} \tag{2}$$

Machinery failures may cause severe disturbances in production processes and the loss of availability. Inherent availability can be calculated with Formula (3).

$$\text{Availability} = \frac{\text{MTBF}}{\text{MTBF} + \text{MTTR}} \tag{3}$$

where:

- MTBF—Mean Time Between Failures,
- MTTR—Mean Time To Repair.

The OEE metric was developed for single component maintenance. In the case of complex systems including serial or parallel subsystems the availability is changed. For the series system to be available, each subsystem should be available. For the parallel system to be available, whichever subsystem should be available.

Performance is the ratio of the time to complete a task under ideal conditions compared to the realization in real conditions; or the ratio of the products obtained in reality, to the number of possible products to obtain under ideal conditions. Performance is reduced (loss of working speed) by the occurrence of any disturbances, e.g., human errors.

$$\text{Performance} = \frac{\text{ideal cycle time}}{\text{real cycle time}} = \frac{\text{real number of products}}{\text{ideal number of products}} \tag{4}$$

Quality is expressed by the ratio of the number of good products and the total number of products.

$$\text{Quality} = \frac{\text{number of good quality products}}{\text{total number of products}} \tag{5}$$

To compare the influence of the AGV logistic system over the manufacturing system, we will consider different OEE factors. However, according to the lean manufacturing paradigm, the flow of production through bottlenecks is the most important, therefore some equipment should be fully utilized whilst other equipment does not require full utilization. The literature review [16,46] indicates that OEE metrics are lacking at complex manufacturing systems and the factory level. In order to address this gap, an overall throughput effectiveness metric can be used [47]. It measures the factory-level performance and can also be used for performing factory-level diagnostics such as bottleneck detection and identifying hidden capacity. It also accounts for subsystems processing multiple products. Any factory layout can be modelled using a combination of the predefined subsystems (serial, parallel), which allows a determination of the Overall Factory Effectiveness (OFE). Note that the OEE equation can be further simplified as [46,47]:

$$\text{OEE} = \frac{\text{Actual throughput (units) from equipment in total time}}{\text{Theoretical throughput (units) from equipment in total time}} \tag{6}$$

By extending this definition to the factory level, we have Overall Factory Effectiveness (OFE):

$$\text{OFE} = \frac{\text{Actual throughput (units) from factory in total time}}{\text{Theoretical throughput (units) from factory in total time}} \tag{7}$$

Similarly, the Overall Transport Effectiveness (OTE) can be defined:

$$\text{OTE} = \frac{\text{Actual throughput (units) from vehicle in total time}}{\text{Theoretical throughput (units) from vehicle in total time}} \tag{8}$$

3. Description of the Problem—Materials and Methods

Let us consider a production system with eight automated machine tools, such as a CNC machining centre, which performs a two-stage process of machining a family of typical machine parts, like sleeves or discs of different sizes.

The machines are arranged as shown in Figure 2, which allows the series-parallel flow of production.

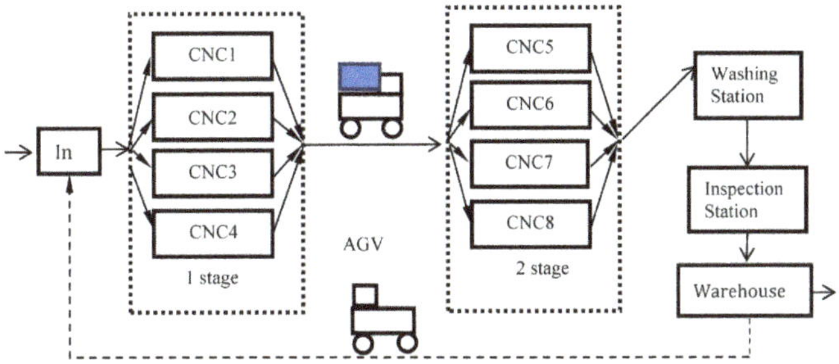

Figure 2. The schema of Flexible Manufacturing System (FMS) with AGV transport.

The system also includes an Automatic Washing Station and Inspection Station as well as a Storage System with an automatic rack stacker. Randomly generated various production orders are delivered to the system, differing in the duration of the operation (from 2.5 to 15.2 min). As a means of product transportation devices, several AGV vehicles are used, which will move along the planned transport routes. We assume that the manufacturing process meets the lean manufacturing, i.e., the flow of a single product and minimal buffers capacity to limit production in progress.

When designing a production system, we strive to achieve maximum production efficiency and, in particular, maximum utilization of machines and devices constituting bottlenecks in the manufacturing process. Other machines and devices will usually be used to a lesser extent, but they are necessary for the production process. By introducing changes to the model, we can analyse the formation of bottlenecks in the production system and their impact on the production volume. This allows, i.e., to determine the required storage capacity and capacity of the transport system. Particularly, the most interested issue is the impact of the number of AGV transport resources on the production efficiency of the entire system. For this purpose, a simulation model was developed in the FlexSim 2018 environment, shown in Figure 3.

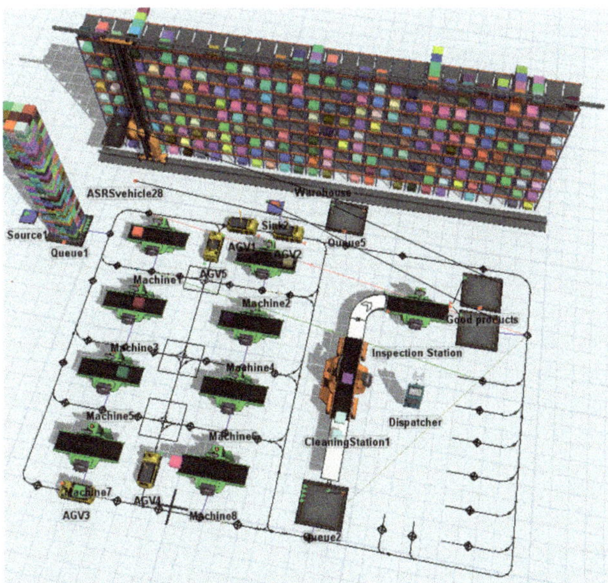

Figure 3. Simulation model of the FMS with AGV transport.

Initially, the reference system consisted only of machines, without taking transport into account. It represents the manufacturing system in ideal conditions. At the next step, transport-related constraints were added to the model. The layout takes into account the dimensions of individual model objects and the distance between them. According to the recommendations given in the literature, the model uses unidirectional transport routes forming three main loops. Several control points have been introduced to designate places of loading and unloading as well as parking spaces. For the most used intersections, control zones were used to reduce the risk of collisions and blockages.

Typical parameters of AGV were assumed, including the speed of 2 m/s and a loading/unloading time of 30 s. A FIFO (First In First Out) control strategy was applied. In addition, the warehouse system was expanded by adding components such as the high storage warehouse with an automated storage retrieval system (ASRS).

4. Results of the Simulation Experiments

The developed model was used to conduct a series of simulation experiments. In subsequent experiments, the number of AGV vehicles from 0 to 8 was changed. The simulation time of 24 h was assumed as the time of automatic maintenance of the entire system. A random generation of production orders was assumed according to the exponential distribution with the expected value of 100 s. As a result of a lack of data and the need for simplification, the retooling of the system, charging of AGV batteries and the failure of machines and vehicles was omitted. As part of each experiment, 30 simulation runs were carried out. The results are presented in Table 1 and Figure 4, respectively. Due to the stochastic parameters of the model, the production value P_{avg} obtained in the experiment is a random variable with a distribution close to normal.

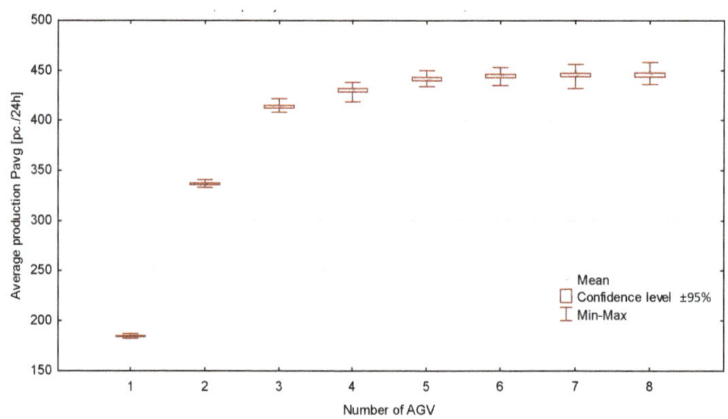

Figure 4. The relationship between average production Pavg and the number of AGV used for 24 h of simulation.

The first row in the Table 1, where the number of AGV amount is 0, represents the reference system consisting only of machines and not taking transport into account (transport time equal to zero).

Figure 4 presents the relationship between average production P_{avg} and the number of AGVs used for transport in the form of a box-and-whisker plot.

The box represents a 95% confidence interval, which means the average production volume is within this range with a probability of 95%, whilst, whiskers represent the maximum and minimum value of production obtained in a given experiment (there is a small spread of results, therefore some of the boxes in the chart are very small).

In the Figure 4, it can be seen that initially the increase of the AGV number results in a rapid increase of obtained production volume. On the other hand, increasing the AGVs number above

5 units results in a slight increase in production, as more vehicles are used to a lesser extent. A similar phenomenon is described in the literature [36] as the effect of the mutual blocking of AGVs.

Table 1. The results of simulation experiments (Average production completed P_{avg}, in [pieces] for 24 h of work, 30 simulation runs in each experiment, 95% confidence level).

Number of AGVs N_{AGV}	Minimum Production P_{min} [Pieces]	Lower Limit of 95% Confidence Interval [Pieces]	Average production P_{avg} [Pieces]	Upper Limit of 95% Confidence Interval [Pieces]	Maximum Production P_{max} [Pieces]	Standard Deviation Σ
0	548	559.0	561	563	573	6.3
1	182	183.66	184.7	184.87	187	1.62
2	333	335.97	336.7	337.43	341	1.95
3	408	412.1	413.4	414.7	422	3.6
4	419	428.6	430.3	432	438	4.5
5	434	439.8	441.5	443.3	450	4.7
6	435	443.0	444.7	446.4	453	4.6
7	432	443.8	445.6	447.4	456	4.9
8	436	443.6	445.7	447.8	458	5.6

To eliminate them, it would be necessary to use multi-lane transport routes or passages. The excessive increase in the number of AGVs is in turn associated with high costs and brings little effects and a relatively small increase in production efficiency [34] with a drop in the effectiveness of AGV being used.

The comparison with the results obtained in the ideal conditions (AGV = 0, P_{avg} = 561 pc.) shows a great difference with the other results. For example, for 5 AGVs system, there was a P_{avg} = 441.5 pc., average machine utilization of about 70% and average AGVs utilization of about 50% (from the range of 28–71% for AGV1 and AGV5). The increase of the AGV number caused very little increase in machine utilization and a considerable decrease in the utilization of the additional AGVs.

That problem requires a more detailed investigation, but the traditional metrics for measuring productivity as throughput or utilization rate are not very helpful for identifying the problems and underlying improvements needed to increase productivity. In this situation, a more rigorously defined productivity metric is needed [44]. Therefore, in this case, OEE metrics can be used, which take into account equipment availability, breakdowns, performance (reduced speed, idling) and quality (good and bad quality products).

Second Experiment

A more detailed model of the FMS system was developed taking into account the quality, availability and reliability of AGVs and battery charging.

AGVs have very advanced design and are considered very reliable, but there are certainly few publications about AGV reliability, compared to publications about machine reliability including [48,49]. With the use of fault tree analysis, a reliability block diagram and a hazard decision tree of AGV components, reliability evaluation of the failure rate λ [1/h] was estimated to be 0.003 [48] and 0.0014 [49]. That can also be described with the Mean Time Between Failures (MTBF) as reciprocal of the failure rate λ. Basing on the λ values, we have the range of MTBF = 300 ÷ 700 h and we have assumed an average of $MTBF_{agv}$ = 500 h for modelling the reliability of AGV. We have also assumed Mean Time To Repair ($MTTR_{agv}$) = 8 h. The reliability of CNC machine tools was omitted because its reliability should be much better than that of AGVs, and we will concentrate on the failure effect of the AGVs. The machines are working parallelly; therefore, the effect of machine failures would be very small. On the other hand, another random factor could hinder the analysis of results.

The AGV can work 24 h per day but sometimes battery loading is required. We have assumed a working schedule for 6 AGVs with a 4 h pause for battery loading. It means that AGVs charge the batteries alternately, and in each moment 5 AGVs are working and one is charging the battery. In the case of malfunction, the AGV is automatically moving to the parking place for maintenance or should be manually removed to prevent blockage.

The scenario includes continuous work of the FMS for 3 shifts per day and 5 days per week. As a result of the long-time effect of AGVs failures, long-time simulations were performed, including work for 24, 120, 500 and 1500 h. The experiment's results without and with reliability parameters of AGVs are presented in Tables 2 and 3, respectively. (The raw data are included in the Supplementary Materials).

Table 2. The results of simulation experiments (Average production completed Pavg, in [pieces] for 6 AGVs with battery charging, without failures, 30 simulation runs in each experiment, 95% confidence level).

Time [Hours]	Minimum Production P_{min} [pc.]	Lower Limit of 95% Confidence Interval [pc.]	Average Production P_{avg} [pc.]	Upper Limit of 95% Confidence Interval [pc.]	Maximum Production P_{max} [pc.]	Standard Deviation σ	Average Throughput [pc./Hour]
24	593	611	614.2	617.5	634	8.7	25.59
120	3094	3131.8	3139.9	3148.1	3188	21.8	26.17
500	13,061	13,116.7	13,128.3	13,139.9	13,188	31.1	26.26
1500	39,288	39,387	39,406	39,425	39,534	51	26.27

Table 3. The results of simulation experiments (Average production completed Pavg, in [pieces] for 6 AGVs with battery charging, with AGVs failures, MTBF = 500 h, MTTR = 8 h, 30 simulation runs in each experiment, 95% confidence level).

Time [Hours]	Minimum Production P_{min} [pc.]	Lower Limit of 95% Confidence Interval [pc.]	Average Production Pavg [pc.]	Upper Limit of 95% Confidence Interval [pc.]	Maximum Production P_{max} [pc.]	Standard Deviation σ	Average Throughput [pc./Hour]
24	549	607.3	613.2	618.3	634	14.8	25.55
120	2810	3091	3116	3141	3174	68	25.97
500	12,327	12,965	13,030	13,094	13,162	173	26.06
1500	38,402	39,044	39,151	39,258	39,463	287	26.10

An analysis of the previous model showed that blockage of the machines sometimes occurs; therefore, small loading/unloading buffers with a capacity of one piece were added to machines in order to improve the production flow. Quality parameters were defined as 99.9% of good products according to the OEE quality factor.

Comparing the results from Tables 2 and 3, a small but significant effect of AGVs failures on production can be seen (a decrease of about 0.7%). For a more detailed analysis, the OFE metrics can be used. Since the model was built based on the OEE components and contains parameters of availability, performance and quality, the production value from the simulation P_{avg} can be directly used to calculate the OFE indicator [25,41].

$$OFE = \frac{P_{avg}}{P_{limit}} \tag{9}$$

The value P_{limit} represents the theoretically available maximal production in ideal conditions. For the average machining time of t_m = 530 s, the limit is equal to 6.79 pc./hour for one machine and 652 pc./24 h for the whole machining system (P_{limit} = 27.17 pc./hour).

The juxtaposition of the OFE indexes is included in Table 4.

The differences between OFE_2 and OFE_3 are related to the warmup of the system in a short time and to the effect of AGV failures in a long time. This result is consistent with assumed reliability parameters and inherent availability (Equation (3)) of AGV and parallel system. As there is a small probability of simultaneous failure of all AGVs, the effect is connected with the loss of performance including loss of speed during loading/unloading, waiting for transport and blocking.

Table 4. The Overall Factory Effectiveness (OFE) metrics for model of 6 AGVs without failures OFE_2 and with failures OFE_3 and OFE_1 from the previous experiment.

Time [Hours]	P_{limit}	P_{avg1}	OFE_1	P_{avg2}	OFE_2	P_{avg3}	OFE_3
24	652	444.7	0.68206	614.2	0.94203	613.2	0.94049
120	3260	2252.3	0.69089	3139.9	0.96316	3116	0.95583
500	13,583.33	9407.6	0.69258	13,128.3	0.96650	13,030	0.95926
1500	40,750	28,230.7	0.69278	39,406	0.96702	39,151	0.96076

5. Discussion

The question is—which KPI should be used for an evaluation of the whole manufacturing system and the transport subsystem?

The key factor is the production flow through the machines, as there is the bottleneck, which is related with the utilization of the machines and transport vehicles.

The relationship between average utilization of machines and AGVs and the number of AGVs used for 24 h of simulation is shown in the Figure 5. With the increasing number of AGVs, the utilization of machines is increasing, and at the same time the utilization of vehicles is decreasing. These two performance goals are mutually exclusive. The maximum average machine utilization was about 95% compared to about 53% for 6 AGVs (from the range 46–62%).

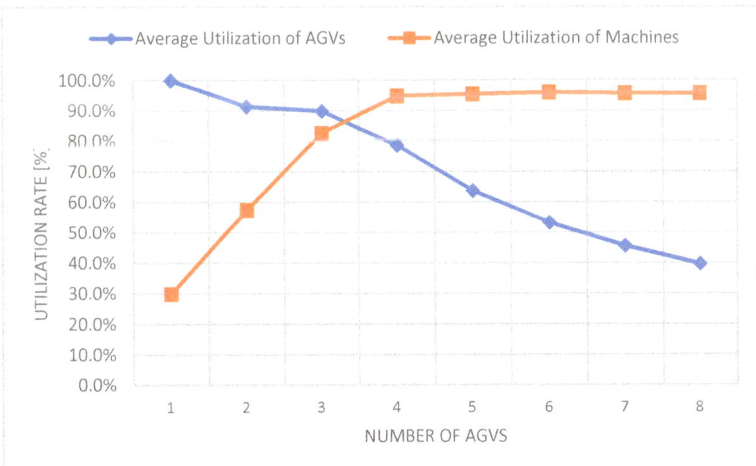

Figure 5. The relationship between average utilization of machines and AGVs and the number of AGV used for 24 h of simulation.

We propose an analysis of the effectiveness of the transport system by the Overall Transport Effectiveness (OTE) metric that can be determined on the basis of the number of transport operations carried out and the theoretical planned limit of transport operations per vehicle. There are 4 transport operations for each product in the production flow, therefore, the production of P_{limit} = 652 products requires AGV_{limit} = 2608 transport operations. One AGV can make about 782 transport operations during 24 h. As the AGVs are working parallelly, then theoretically the 3.3 AGVs should be enough, but if there are more vehicles the blocking can occur more frequently. Therefore, the overall transport effectiveness per vehicle was also calculated and is presented in the Table 5.

Table 5. OTE—Overall Transport Effectiveness and OFE—Overall Factory Effectiveness (24 h of simulation, 30 simulation runs in each experiment, without failures and battery charging).

Nr of AGVs N_{agv}	P_{limit} [Pc.]	AGV_{limit} ($4 \times P_{limit}/N_{agv}$) [Pc.]	P_{avg} [pc.]	Finished Transport Operation [pc.]	Average Transport Oper./AGV [pc.]	OTE	OFE
1	652	2608	186.7	781.9	781.9	0.29981	0.28635
2	652	1304	347.23	1428.2	714.1	0.54762	0.53256
3	652	869.3	516.7	2107.5	702.5	0.80812	0.79249
4	652	652.0	606.0	2461.4	615.4	0.94379	0.92945
5	652	521.6	614.0	2493.0	498.6	0.95590	0.94172
6	652	434.7	615.0	2497.0	416.2	0.95737	0.94325
7	652	372.6	614.0	2494.4	356.3	0.95637	0.94172
8	652	326.0	611.7	2484.5	310.6	0.95265	0.93819

There is a close relationship between the number of achieved products and the number of required transport operations. However, there is a small difference in the values of OFE and OTE, because of work in progress and related transport operations. The value of OTE is depended on the number of AGVs. The maximal value of OTE = 95.737% (±0.25%) was achieved for 6 AGVs. In the case of battery charging, the value of OTE = 95.621% was achieved. The failures have decreased the value by about 0.27% to OTE = 95.353%.

However, for longer simulation time, the system is more stable and a maximal value of OFE = 96.076% and OTE = 96.194% was achieved for the longest simulation time of 1500 h.

It should be noted that there can be different versions of the OTE metric due to the scope of the data taken into account. The main difference in this version is that only planned transport operations are taken into consideration (not all possible working time as in utilization rate).

According to principles of lean manufacturing, an unnecessary movement of people, information or materials wastes time and increases costs. Any unnecessary transport of raw materials in the plant is a waste, and thus should be reduced.

Any non-critical resource such as AGV should be "utilized", such that the bottleneck is never starved for work and all work that is processed by the bottleneck is of high quality. Otherwise, additional activation of these resources just generates excess work-in-process and additional costs. This condition will be met if the OTE is greater than or equal to OEE (OFE).

$$\text{OTE} \geq \text{OEE (OFE)} \tag{10}$$

Therefore, the hypothesis that there is dependence between plant effectiveness and transport effectiveness, which can be expressed by the use of the OEE metric, has been proven.

In the case of other logistics systems (e.g., transport of multiple products, different routes with returns), the difference in value between OFE and OTE may be greater. These problems and industrial implementation of the proposed methodology in the context of the digital twin for Industry 4.0, will be taken into account in further research.

6. Conclusions

Due to the complexity of AGV systems, they cause many decision problems, which are difficult to solve. The paper has presented an example of the Flexible Manufacturing System solution with the AGV transport system and discusses some issues related to the design and simulation of such systems. The stage of initial system design optimization is very important, and computer simulation enables the relatively easy elaboration and testing of various variants of manufacturing and logistics systems. On the other hand, excessive simplifications may be applied at the modelling stage, which will make the simulation not reflect the production system properly. It should be noted that detailed modelling is very labour intensive and requires the involvement of experienced specialists. Therefore, choosing what parameters are used in the modelling process and which metric is used to evaluate the model is very important. In order to make the simulation more accurate and to evaluate the system's productivity, the use of Overall Equipment Effectiveness (OEE) metrics was proposed.

The results obtained from the presented simulations show that the OEE metrics can be used for the modelling and productivity evaluation of manufacturing and logistics systems, with the generalization of Overall Factory Effectiveness (OFE) and Overall Transport Effectiveness (OTE). The use of OEE factors also allows to compare the results obtained from different manufacturing systems. In the real world, most of manufacturing companies have OEE scores closer to 60%, but there are many of them with OEE scores lower than 45%, and a small number of world-class companies that have the OEE value higher than 85% [50]. According to that, the results of simulation can be also used to analyse the costs involved in the implementation of a given project and at the stage of in-depth design of the production system.

Supplementary Materials: The following are available online at http://www.mdpi.com/2227-9717/8/12/1648/s1.

Author Contributions: Conceptualization, A.K. and G.G.; methodology, A.K.; validation, K.F., G.G.; investigation, K.F.; data curation, G.G.; writing—original draft preparation, A.K.; visualization, G.G. All authors have read and agreed to the published version of the manuscript.

Funding: This research received no external funding.

Conflicts of Interest: The authors declare no conflict of interest.

References

1. Rutner, S.M.; Langley, C.J. Logistics Value: Definition, Process and Measurement. *Int. J. Logist. Manag.* **2000**, *11*, 73–82. [CrossRef]
2. Schuhmacher, J.; Baumung, W.; Hummel, V. An Intelligent Bin System for Decentrally Controlled Intralogistic Systems in Context of Industrie 4.0. *Procedia Manuf.* **2017**, *9*, 135–142. [CrossRef]
3. Paszkiewicz, A.; Bolanowski, M.; Budzik, G.; Przeszłowski, Ł.; Oleksy, M. Process of Creating an Integrated Design and Manufacturing Environment as Part of the Structure of Industry 4.0. *Processes* **2020**, *8*, 1019. [CrossRef]
4. Avelar-Sosa, L.; Alcaraz, J.L.G.; Maldonado-Macías, A.A.; Mejía-Muñoz, J. Application of structural equation modelling to analyse the impacts of logistics services on risk perception, agility and customer service level. *Adv. Prod. Eng. Manag.* **2018**, *13*, 179–192. [CrossRef]
5. Berman, S.; Schechtman, E.; Edan, Y. Evaluation of automatic guided vehicle systems. *Robot. Comput. Manuf.* **2009**, *25*, 522–528. [CrossRef]
6. De Ryck, M.; Versteyhe, M.; DeBrouwere, F. Automated guided vehicle systems, state-of-the-art control algorithms and techniques. *J. Manuf. Syst.* **2020**, *54*, 152–173. [CrossRef]
7. Kampa, A.; Golda, G.; Paprocka, I. Discrete Event Simulation Method as a Tool for Improvement of Manufacturing Systems. *Computers* **2017**, *6*, 10. [CrossRef]
8. Kampa, A.; Gołda, G. Modelling and simulation method for production process automation in steel casting foundry. *Arch. Foundry Eng.* **2018**, *18*, 47–52. [CrossRef]
9. Barosz, P.; Golda, G.; Kampa, A. Efficiency Analysis of Manufacturing Line with Industrial Robots and Human Operators. *Appl. Sci.* **2020**, *10*, 2862. [CrossRef]
10. Lambert, D.M.; Burduroglu, R. Measuring and Selling the Value of Logistics. *Int. J. Logist. Manag.* **2000**, *11*, 1–18. [CrossRef]
11. Lynch, D.; Keller, S.; Ozment, J. The effects of logistics capabilities and strategy on firm performance. *J. Bus. Logist.* **2000**, *21*, 47–67.
12. Griffis, S.E.; Cooper, M.; Goldsby, T.J.; Closs, D.J. Performance Measurement: Measure Selection Based upon Firm Goals and Information Reporting Needs. *J. Bus. Logist.* **2004**, *25*, 95–118. [CrossRef]
13. Fugate, B.S.; Mentzer, J.T.; Stank, T.P. Logistics Performance: Efficiency, Effectiveness, and Differentiation. *J. Bus. Logist.* **2010**, *31*, 43–62. [CrossRef]
14. Lambert, D.M.; Pohlen, T.L. Supply Chain Metrics. *Int. J. Logist. Manag.* **2001**, *12*, 1–19. [CrossRef]
15. Muthiah, K.M.; Huang, S.H. A review of literature on manufacturing systems productivity measurement and improvement. *Int. J. Ind. Syst. Eng.* **2006**, *1*, 461. [CrossRef]

16. Roda, I.; Macchi, M. Factory-level performance evaluation of buffered multi-state production systems. *J. Manuf. Syst.* **2019**, *50*, 226–235. [CrossRef]
17. Andersson, C.; Bellgran, M. On the complexity of using performance measures: Enhancing sustained production improvement capability by combining OEE and productivity. *J. Manuf. Syst.* **2015**, *35*, 144–154. [CrossRef]
18. Muñoz-Villamizar, A.; Santos, J.; Montoya-Torres, J.R.; Jaca, C. Using OEE to evaluate the effectiveness of urban freight transportation systems: A case study. *Int. J. Prod. Econ.* **2018**, *197*, 232–242. [CrossRef]
19. Dalmolen, S.; Moonen, H.; Iankoulova, I.; Van Hillegersberg, J.; Simon, D.; Hans, M.; Iliana, I.; Jos, V.H. Transportation Performances Measures and Metrics: Overall Transportation Effectiveness (OTE): A Framework, Prototype and Case Study. *Proc. Annu. Hawaii Int. Conf. Syst. Sci.* **2013**, 4186–4195. [CrossRef]
20. McCalion, R. Is OEE relevant to lift truck fleet & warehouse operations? *Eureka* **2013**. Available online: https://eurekapub.eu/fleet-management/2013/09/14/fleet-management (accessed on 26 September 2020).
21. Hayes, J. AGV IIoT Monitoring: Lean Six Sigma Monitoring. *RFID J.* **2018**. Available online: https://www.rfidjournal.com/agv-iiot-monitoring-lean-six-sigma-monitoring-2 (accessed on 27 September 2020).
22. Palací-López, D.; Borràs-Ferrís, J.; Oliveria, L.T.D.S.D.; Ferrer-Riquelme, A.J. Multivariate Six Sigma: A Case Study in Industry 4.0. *Process* **2020**, *8*, 1119. [CrossRef]
23. Golda, G.; Kampa, A.; Foit, K. Study of inter-operational breaks impact on materials flow in flexible manufacturing system. *IOP Conf. Ser. Mater. Sci. Eng.* **2018**, *400*, 022030. [CrossRef]
24. Alzubi, E.; Atieh, A.M.; Abu Shgair, K.; Damiani, J.; Sunna, S.; Madi, A. Hybrid Integrations of Value Stream Mapping, Theory of Constraints and Simulation: Application to Wooden Furniture Industry. *Process* **2019**, *7*, 816. [CrossRef]
25. Zhang, X.; Li, Y.; Ran, Y.; Zhang, G. Stochastic models for performance analysis of multistate flexible manufacturing cells. *J. Manuf. Syst.* **2020**, *55*, 94–108. [CrossRef]
26. Hoshino, S.; Ota, J.; Shinozaki, A.; Hashimoto, H. Hybrid Design Methodology and Cost-Effectiveness Evaluation of AGV Transportation Systems. *IEEE Trans. Autom. Sci. Eng.* **2007**, *4*, 360–372. [CrossRef]
27. Um, I.; Cheon, H.; Lee, H. The simulation design and analysis of a Flexible Manufacturing System with Automated Guided Vehicle System. *J. Manuf. Syst.* **2009**, *28*, 115–122. [CrossRef]
28. Vis, I.F. Survey of research in the design and control of automated guided vehicle systems. *Eur. J. Oper. Res.* **2006**, *170*, 677–709. [CrossRef]
29. Yan, R.; Dunnett, S.J.; Jackson, L. Novel methodology for optimising the design, operation and maintenance of a multi-AGV system. *Reliab. Eng. Syst. Saf.* **2018**, *178*, 130–139. [CrossRef]
30. Florescu, A.; Barabas, S.A. Modeling and Simulation of a Flexible Manufacturing System—A Basic Component of Industry 4.0. *Appl. Sci.* **2020**, *10*, 8300. [CrossRef]
31. Deroussi, L. *Flexible Manufacturing Systems: Metaheuristics Logist*; Wiley: Hoboken, NJ, USA, 2016; pp. 143–160. [CrossRef]
32. Tolio, T. (Ed.) *Design of Flexible Production Systems: Methodologies and Tools*; Springer: Berlin, Germany, 2009. [CrossRef]
33. Banks, J.; Nelson, B.L.; Carson, J.S.; Nicol, D.M. Discrete-Event System Simulation. *Technometrics* **1984**, *26*, 195. [CrossRef]
34. Burduk, A. Stability Analysis of the Production System Using Simulation Models. In *Process Simulation and Optimization in Sustainable Logistics and Manufacturing*; Pawlewski, P., Greenwood, A., Eds.; Springer Science and Business Media LLC: Cham, Switzerland, 2014; pp. 69–83.
35. Plaia, A.; Lombardo, A.; Nigro, G.L. Robust Design of Automated Guided Vehicles System in an FMS. In *Advanced Manufacturing Systems and Technology*; Kuljanic, E., Ed.; Springer: Vienna, Austria, 1996; pp. 259–266.
36. A Vis, I.F.; De Koster, R.; Roodbergen, K.J.; Peeters, L.W.P. Determination of the number of automated guided vehicles required at a semi-automated container terminal. *J. Oper. Res. Soc.* **2001**, *52*, 409–417. [CrossRef]
37. Qi, M.; Li, X.; Yan, X.; Zhang, C. On the evaluation of AGVS-based warehouse operation performance. *Simul. Model. Pract. Theory* **2018**, *87*, 379–394. [CrossRef]
38. Automated Guided Vehicle. Wikipedia. Available online: https://en.wikipedia.org/wiki/Automated_guided_vehicle (accessed on 30 September 2020).
39. Nieoczym, A.; Tarkowski, S. The modeling of the assembly line with a technological automated guided vehicle (AGV). *LogForum* **2011**, *7*, 35–42.

40. Vivaldini, K.C.T.; Rocha, L.F.; Martarelli, N.J.; Becker, M.; Moreira, A.P. Integrated tasks assignment and routing for the estimation of the optimal number of AGVS. *Int. J. Adv. Manuf. Technol.* **2016**, *82*, 719–736. [CrossRef]
41. Li, Z.; Wu, N.; Zhou, M. Deadlock Control of Automated Manufacturing Systems Based on Petri Nets—A Literature Review. *IEEE Trans. Syst. Man. Cybern. Part C Appl. Rev.* **2011**, *42*, 437–462. [CrossRef]
42. Moorthy, R.L.; Hock-Guan, W.; Wing-Cheong, N.; Chung-Piaw, T. Cyclic deadlock prediction and avoidance for zone-controlled AGV system. *Int. J. Prod. Econ.* **2003**, *83*, 309–324. [CrossRef]
43. Lv, Y.L.; Zhang, G.; Zhang, J.; Dong, Y.J. Integrated Scheduling of the Job and AGV for Flexible Manufacturing System. *Appl. Mech. Mater.* **2011**, *80*, 1335–1339. [CrossRef]
44. Mousavi, M.; Yap, H.J.; Musa, S.N.; Tahriri, F.; Dawal, S.Z.M. Multi-objective AGV scheduling in an FMS using a hybrid of genetic algorithm and particle swarm optimization. *PLoS ONE* **2017**, *12*, e0169817. [CrossRef]
45. Saidi-Mehrabad, M.; Dehnavi-Arani, S.; Evazabadian, F.; Mahmoodian, V. An Ant Colony Algorithm (ACA) for solving the new integrated model of job shop scheduling and conflict-free routing of AGVs. *Comput. Ind. Eng.* **2015**, *86*, 2–13. [CrossRef]
46. Muthiah, K.M.N.; Huang, S.H. Overall throughput effectiveness (OTE) metric for factory-level performance monitoring and bottleneck detection. *Int. J. Prod. Res.* **2007**, *45*, 4753–4769. [CrossRef]
47. Huang, S.H.; Dismukes, J.P.; Shi, J.; Su, Q.; Razzak, M.A.; Bodhale, R.; Robinson, D.E. Manufacturing productivity improvement using effectiveness metrics and simulation analysis. *Int. J. Prod. Res.* **2003**, *41*, 513–527. [CrossRef]
48. Fazlollahtabar, H.; Saidi-Mehrabad, M. *Autonomous Guided Vehicles*; Springer Science and Business Media LLC: Cham, Switzerland, 2015; Volume 20.
49. Yan, R.; Jackson, L.; Dunnett, S.J. Automated guided vehicle mission reliability modelling using a combined fault tree and Petri net approach. *Int. J. Adv. Manuf. Technol.* **2017**, *92*, 1825–1837. [CrossRef]
50. OEE Benchmark Study. Available online: https://sageclarity.com/articles-oee-benchmark-study (accessed on 28 September 2020).

Publisher's Note: MDPI stays neutral with regard to jurisdictional claims in published maps and institutional affiliations.

processes

MDPI

Article

Stability of Optimal Closed-Loop Cleaning Scheduling and Control with Application to Heat Exchanger Networks under Fouling

Federico Lozano Santamaria * and Sandro Macchietto

Department of Chemical Engineering, Imperial College London, South Kensington Campus, London SW7 2AZ, UK; s.macchietto@imperial.ac.uk
* Correspondence: f.lozano-santamaria16@imperial.ac.uk

Received: 27 October 2020; Accepted: 3 December 2020; Published: 9 December 2020

Abstract: Heat exchanger networks subject to fouling are an important example of dynamic systems where performance deteriorates over time. To mitigate fouling and recover performance, cleanings of the exchangers are scheduled and control actions applied. Because of inaccuracy in the models, as well as uncertainty and variability in the operations, both schedule and controls often have to be revised to improve operations or just to ensure feasibility. A closed-loop nonlinear model predictive control (NMPC) approach had been previously developed to simultaneously optimize the cleaning schedule and the flow distribution for refinery preheat trains under fouling, considering their variability. However, the closed-loop scheduling stability of the scheme has not been analyzed. For practical closed-loop (online) scheduling applications, a balance is usually desired between reactivity (ensuring a rapid response to changes in conditions) and stability (avoiding too many large or frequent schedule changes). In this paper, metrics to quantify closed-loop scheduling stability (e.g., changes in task allocation or starting time) are developed and then included in the online optimization procedure. Three alternative formulations to directly include stability considerations in the closed-loop optimization are proposed and applied to two case studies, an illustrative one and an industrial one based on a refinery preheat train. Results demonstrate the applicability of the stability metrics developed and the ability of the closed-loop optimization to exploit trade-offs between stability and performance. For the heat exchanger networks under fouling considered, it is shown that the approach proposed can improve closed-loop schedule stability without significantly compromising the operating cost. The approach presented offers the blueprint for a more general application to closed-loop, model-based optimization of scheduling and control in other processes.

Keywords: closed-loop scheduling; scheduling stability; optimal control and scheduling; fouling; heat exchanger networks

---

1. Introduction

In batch plants, continuous plants, and general manufacturing plants with multiple processing units, multiple products or time-decaying performance, scheduling of production and maintenance is essential to ensure a feasible and economically profitable operation. The aim of scheduling is to define the production sequence, order, allocation, and timing for execution of all production and maintenance tasks. For example, a closed-loop nonlinear model predictive control (NMPC) approach has been developed to simultaneously optimize the cleaning schedule and the flow distribution for refinery preheat trains under fouling [1]. Production scheduling and maintenance scheduling belong to the same kind of problem (i.e., they follow the same principles, assumptions, and modeling approaches) and, in some instances, have been integrated [2–4]. One of the main assumptions used to address these problems is a perfect knowledge of the current and future operating conditions, which includes demand, unit performance, availability, and cost of resources.

However, all processes are by nature dynamic and subject to uncertainty and disturbances. For example, in batch processing, unplanned events such as unit breakdown, new orders, changes in order quantity, performance decay of the unit, and variation in costs and prices affect the performance (technical and economic) and even feasibility of a schedule previously determined [5]. Therefore, a re-evaluation of the scheduling decision is necessary and advantageous. Traditionally, two alternatives schemes have been defined: (i) rescheduling, where the main objective is to recover feasibility of the operation after a (significantly large) disturbance is observed, and (ii) online scheduling, where the schedule is updated at regular intervals [6,7]. Rescheduling can be done via a full re-evaluation of the scheduling problem, via partial modification of the previous scheduling decisions, or by postponing the execution of some actions [8]. Typically, this is done over the same time horizon as the original schedule and with no new decision variables. Most of the approaches for rescheduling are based on heuristics and aim to do minimal, yet significant, modifications to recover feasibility [5]. Some others are based on mathematical programming and solve a nonlinear programming (NLP), mixed integer linear programming (MILP) or mixed integer nonlinear programming (MINLP) problem representing a partial scheduling problem (i.e., with a subset of the decisions fixed based on the solution of the initial schedule) [5,9]. In the above classification, online scheduling uses all available decision variables, and aims to maximize the performance of the system at every evaluation so that it does not just reject disturbances, but also generates improvements when the system dynamics allow so [10]. This alternative relies on the solution of optimization problems in a feedback loop using a receding horizon approach (i.e., the time horizon of each schedule evaluation rolls forward and includes new future decisions). The update interval may be fixed and constant, or conditional to the detection of disturbances to the system.

Online scheduling, also referred to as closed-loop scheduling, aims to automate a production and/or maintenance schedule of a plant despite disturbances and variability. However, it has been noted that such a rolling update of the schedule can produce instability in the operation [10,11]. Schedule instability, also called schedule nervousness, may be loosely defined as changes in scheduling decisions between consecutive updates which are undesired (the opposite defines schedule stability). Such changes often have important consequences for the operation. For example, some tasks may not be included in the scheduling model (or not included in sufficient detail) and a change in schedule requires revising them as well. Some tasks may require manual intervention and some resources may require a long procurement time. If scheduling decisions change too frequently or too suddenly, there may not be sufficient time to implement those tasks or procure those resources. In addition, from the operator perspective, too many and sudden schedule changes may be perceived as "erroneous" and "nonintuitive", leading the operator to manually overwrite some decisions. This in turn will most likely generate delays in execution, introduce further disturbances to the operation that have to be corrected later on, and negatively affect performance [5,8].

In principle, increasing schedule stability within the closed loop would often facilitate the implementation of scheduling decisions, avoid other disturbances occurring in the long term, and improve the closed-loop performance. This will, however, reduce the ability of the system to react to disturbances. Ensuring a rapid schedule response to changes in conditions and schedule stability are, therefore, both desired objectives.

Refining operations are an example of highly dynamic processes with a high energy demand and environmental impact, which are also subject to many uncertainties and variability. They can benefit from an online optimization of their operation to reduce energy consumption, operating cost, and carbon emissions. A key section of a refinery is the preheat train, a large heat exchanger network that recovers around 70% of the energy in the products of the main distillation column [12]. An efficient operation of this section ensures satisfying the production targets, while reducing energy consumption. However, it is subject to a wide range of disturbances such as changes in flow rates, operating temperature, and crude blends processed (which occur on the timescale of hours or days), as well as to efficiency losses, among which the most important is fouling. To maintain an efficient operation of the preheat train in the presence of such disturbances and process variability, the flow distribution through the network and the cleaning schedule of the units have to be optimized.

Usually, the cleaning scheduling and the flow distribution problems of preheat trains have been considered independently, ignoring the inherent variability of the process, and solved using heuristics [13–15]. This leads to suboptimal operations because key elements of the problem are ignored, or to infeasible operations because operating limits (e.g., the firing limit of the furnace, the limit capacity of the pumps) are reached, causing a need for emergency cleaning actions or a reduction in production rates. It has been shown that, for these type of processes, integrating flow control in the network and scheduling of exchanger cleaning is advantageous because of the strong synergies between them [16,17]. Optimizing these two elements in a closed loop is, therefore, important to reject disturbances and improve performance. A closed-loop nonlinear model predictive control (NMPC) approach that does this has been developed [1]. However, to achieve a successful implementation of an online cleaning scheduling and flow control of preheat trains, issues related to schedule stability have to be addressed first. Schedule stability is of particular importance in this application because (i) the time scale involved spans from weeks to years, which requires the integration of short-term and long-term decisions, and (ii) the nature of the scheduling decisions (i.e., cleaning of units) requires planning ahead of the specialized resources necessary (e.g., crews, cleaning equipment, cranes, usually contracted out with long notices). Refinery operators, therefore, invariably demand some stability in the future scheduling decisions. Schedule stability, disturbance rejection, and performance optimality are all desired objectives for the problem at hand.

Several approaches have been proposed to balance this trade-off between schedule stability and closed-loop schedule performance in various applications related to batch or manufacturing processes. However, to our knowledge, they have not been proposed related to maintenance or cleaning scheduling. Dynamic effects and variability have been considered by using heuristic algorithms to modify the starting time of the task online [18], by solving an MILP problem that swaps the order or allocation of the task to minimize wait time [19], and by using constraint programming to repair the schedule [20]. All of these methods relay an incumbent schedule as a reference and ignore the effects on economic performance. Other rescheduling approaches penalize in an objective function the changes with respect to the incumbent schedule and may include penalties for reallocation of tasks [21], penalties for changes in the starting time of tasks [22], or a more detailed discrimination of all rescheduling costs (i.e., starting time deviation cost, unit reallocation cost, resequencing cost) as penalties in the objective function [8]. As noted, most of these approaches are designed to be used reactively to recover feasibility when large disturbances are observed and not online for closed-loop optimization of a schedule. An early system for online scheduling (SuperBatch) dealt with highly complex processing configurations (plant, recipes, orders, etc.) in batch manufacturing. Schedules were updated every minute, adjusting for external and process variations on a rapid basis, using an unpublished heuristic method evolved from [18]. The system was successfully applied industrially to scheduling and design of very complex, large-scale food productions [23,24] (Figure 1).

In online or closed-loop scheduling, variability is considered explicitly on a rolling horizon. In this case, the objective function or the constraints of the scheduling problem can be modified to additionally include closed-loop schedule stability requirements. For instance, this may be done by retaining some allocations from previous evaluations and promoting early task allocations as a penalty in the objective function [25]. Another formulation minimizes the earliness/tardiness in the execution of the tasks and the cost of flexible tasks [26]. More recently, a state-space representation of the scheduling problem was proposed according to the nonlinear model predictive control (NMPC) paradigm, where the scheduling problem is solved online and automatically includes the effect of disturbances [10,11,27]. The objective is economically driven but does not consider schedule stability.

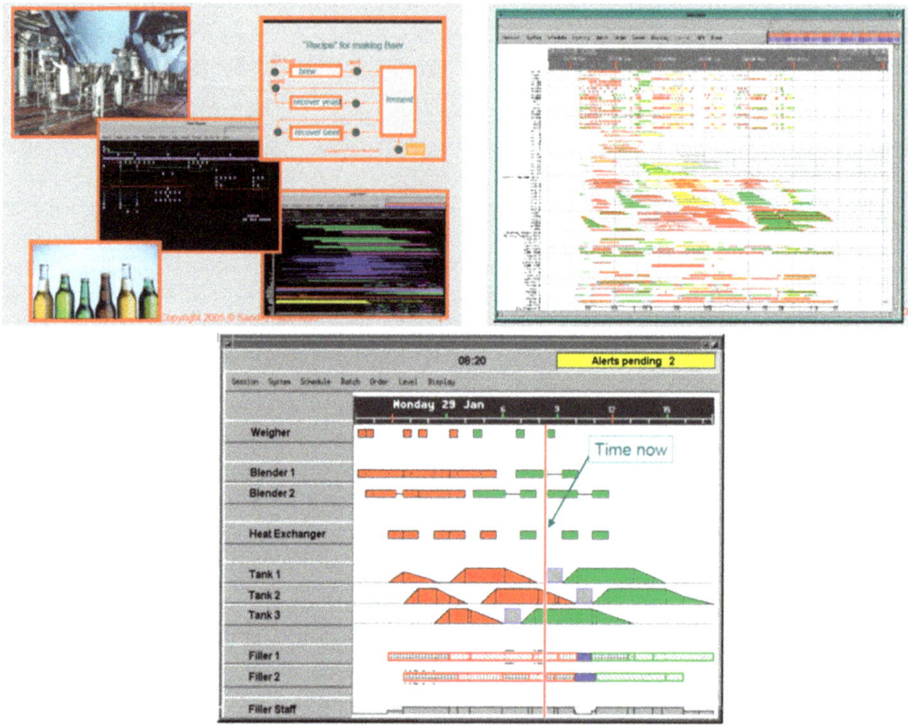

Figure 1. Top left: 14 day schedule of beer production (140 process units, 26 recipe families). Top right: 5 day production schedule of chilled desserts for northern Europe (35 product families, daily changes) as of 4:00 a.m. on a Saturday morning. Bottom: Online rescheduling every minute; the red vertical line separates the historical schedule (as actually happened) from the predicted one (adapted from [23]).

The previous survey indicated that schedule stability for online scheduling is still an open issue, and there is no single, general approach that optimizes the trade-off between closed-loop performance and schedule stability. First, stability is not well defined and quantified, and there are different metrics for various rescheduling actions. Second, most of the rescheduling formulations have focused on just restoring feasibility while ignoring optimality and opportunities arising from the process dynamics and disturbances. Third, only certain sources of schedule instability have been considered, with no clear definition or guidelines for setting the penalty factors. Fourth, most methods so far do not include the possibility of optimizing continuous control decisions at the same time as discrete scheduling decisions. These methodological limitations result in practical barriers to the online optimization of flow distribution and cleaning scheduling in refinery preheat trains, as well as of other dynamic process systems with analogous features.

The aims of this paper are (i) to present a method for the online optimization of operational schedules and continuous controls under high input and disturbance variability, while considering schedule stability explicitly in the closed loop, and (ii) to demonstrate its application and benefits for the online cleaning scheduling and flow distribution control of refinery preheat trains. The remainder of the paper is structured as follows: Section 2 briefly presents the modeling framework used to describe the dynamics of preheat trains under fouling and for online integration and optimization of the flow distribution and cleaning scheduling considering disturbances. In Section 3, some metrics to quantify schedule instability are presented and discussed. Section 4 introduces three alternative ways to include schedule stability objectives within the closed-loop optimization formulation. Section 5 introduces a small case study that is used to demonstrate the use of the instability metrics, and it compares the performance (in terms of stability and total cost) of the various formulations aimed at increasing schedule stability. Section 6 demonstrates the application of the framework to a realistic industrial case study, using historical refinery data and the actual variability observed in the operation of the preheat train. Lastly, the conclusions of the work are drawn in Section 7.

2. Closed-Loop Optimal Cleaning Scheduling and Control of Preheat Trains

The online optimization approach of the cleaning schedule and dynamic flow distribution of preheat trains under fouling is based on an advanced nonlinear model predictive control (NMPC) strategy, presented in detail in a previous study by the authors [1]. It defines two feedback control loops, one for the fast dynamics of the process associated with flow distribution (of the order of hours) and another for the slow dynamics associated with fouling and cleaning (of the order of weeks and months). Figure 2 shows a simplified block diagram of the control loops, their components, and interactions. In this figure, the plant block corresponds to the actual system or a representation of it, the control layer refers to the advanced control and state estimator that defines the control elements of the system for rejection of fast disturbances (its inputs are the set schedule, the current state of the system, and disturbances, and the outputs are the control actions), and the scheduling layer refers to the algorithm defining the online scheduling strategy and its corresponding state estimator (its inputs are the current state of the system and a forecast of the disturbances, and the output is the schedule for the current time). Each control loop has two components: a moving horizon estimator (MHE) to update the model parameters and predict the current state of the system on the basis of the latest plant data and a nonlinear model predictive controller (NMPC) to optimize the future operation of the network. These two elements solve optimization problems using a realistically accurate and representative mechanistic, dynamic model of the plant. In particular, the model describes heat transfer, deposition rates, temperature changes, and hydraulic performance of the heat exchangers, as well as their interactions within the network that constitutes the preheat train. A brief, general description of the modeling components is given next, whereas a more detailed presentation of the model formulation and assumptions can be found in [16].

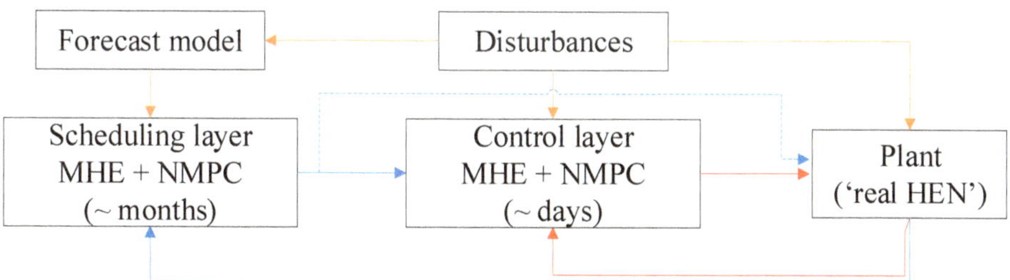

Figure 2. Simplified representation of the online, integrated optimal cleaning scheduling and control of preheat trains subject to fouling and disturbances.

The preheat train model is based on a directed multigraph representation of the heat exchanger network, where each graph corresponds to a stream (e.g., crude oil, naphtha, residue) and the nodes are exchangers, furnace, sources, sinks, mixers, and splitters. At each node, mass and energy balances must be satisfied to ensure network connectivity. The operation of the heat exchangers, all assumed to be of shell and tube type, is represented using an axially lumped, but radially distributed model based on the P-NTU concept [28,29], an explicit description of the heat transfer and temperature profiles in the radial direction through different domains (shell, deposit layers, tube wall, tube), as well as hydraulic relations for the tube side pressure drop. The semiempirical reaction fouling model of Ebert–Panchal, Equation (1) [28], is used to characterize the evolution over time of the thermal resistance of the deposit in a unit. This affects the thermal performance of the unit and is related to the deposit thickness, Equation (2), affecting its hydraulic performance (all variables are defined in the Nomenclature). Experimental or plant data are required to estimate the parameters of the fouling model (α, γ, E_f). It has been demonstrated that this model adequately captures the main effect of the operating variables of the exchangers (e.g., surface temperature, velocity, shear stress) on the fouling rate [30]. In addition to these modeling components, operational limits such as the maximum duty of the furnace, the pressure drop limits in the network, bounds of flow split fractions to parallel branches, and pressure drop equalization constraints over parallel branches are included in the form of inequalities in the problem formulation. The resulting large set of nonlinear equality and inequality constraints is a sufficiently accurate [31] yet compact dynamic model of each exchanger and the network.

$$\frac{dR_{f_i}}{dt} = \alpha_i Re_i^{-0.66} Pr_i^{-0.33} \exp\left(-\frac{E_{f,i}}{RT_{f.i}}\right) - \gamma_i \tau_i, \qquad \forall i \in HEX. \tag{1}$$

$$\delta_i = \frac{D_{in,i}}{2}\left[1 - \exp\left(-\frac{\lambda_d R_{f,i}}{D_{in,i}/2}\right)\right], \qquad \forall i \in HEX. \tag{2}$$

This dynamic model for preheat trains under fouling is used in the MHE and NMPC problems in both the scheduling and the control layers (labeled with subscript s and c, respectively) for parameter estimation, as well as to simultaneously optimize the flow distribution in the network and the cleaning schedule. It has been demonstrated, using actual refinery data, that this model has good predictive capabilities over a wide range of operating conditions and long operating times, with an average absolute prediction error in each exchanger of 0.9 °C for the tube-side exit temperature, 1.3 °C for the shell-side exit temperature, and 0.05 bar for the tube-side pressure drop [1].

Table 1 summarizes the main components, assumptions, and considerations of each feedback loop and their elements. In each layer, the MHE and NMPC formulations use the dynamic model described above to represent the operation of the preheat train and the effects of fouling. In the NMPC formulation of the scheduling layer, which includes binary decision variables, additional inequality constraints are included to represent the changes in operating modes of the exchangers (i.e., "operating" or "being cleaned") and any conditions optionally imposed on the cleaning sequence (e.g., units to be simultaneously cleaned, periods of no cleanings, exclusive cleanings).

Table 1. Summary of the characteristics of each feedback loop for the online optimization of the cleaning scheduling and flow control. NMPC, nonlinear model predictive control; MHE, moving horizon estimator.

	Control Layer (C)		**Scheduling Layer (S)**	
	$NMPC_C$	**MHE_C**	**$NMPC_S$**	**MHE_S**
Objective function	Minimization of operating cost Equation (4)	Minimization of error between data and model Equation (3)	Minimization of operating and maintenance cost Equation (5)	Minimization of error between data and model Equation (3)
Decision variables	Flow rate profiles	Fouling parameters of each exchanger	Flow rate profiles Allocation of cleanings to exchangers Starting time of the cleanings	Fouling parameters of each exchanger
Inputs (parameters)	Cleaning schedule Operating conditions Model parameters	Past actions Past measurements	Operating conditions Model parameters	Past actions Past measurements
Time discretization	Discrete time	Discrete time	Continuous time + orthogonal collocation	Discrete time
Type of problem	NLP	NLP	MINLP	NLP
Sampling interval	From hours to days	From hours to days	From weeks to months	From weeks to months
Future prediction horizon (FPH)	From days to weeks	-	From months to years	-
Past estimation horizon (PEH)	-	From days to weeks	-	From months to years
Disturbances	Forecast as constant current value	Explicit (past)	Forecast as moving average of past month	Explicit (past)
Complicating aspects	Nonlinearities of the model	Nonlinearities of the model Large data set in the PEH	Nonlinearities of the model Number of binary variables Interaction of control and scheduling decisions	Nonlinearities of the model Large data set in the PEH

The control layer deals with the fast dynamics, and its main objective is to reject disturbances and minimize operating cost by manipulating flow split profiles, knowing the short-term cleaning schedule to be executed. The objective of the MHE_C is to determine the model parameters that best explain the observations (i.e., temperature and pressure measurements from the plant) over a past estimation horizon, as represented in Equation (3). The adjustable parameters are the deposition and removal constants in the Ebert–Panchal model and the surface roughness, for each of the exchangers in the preheat train. The resulting formulation is an NLP problem. Once the MHE_C problem is solved, the parameters thus obtained are used in the $NMPC_C$ problem (also formulated as an NLP) to determine the optimal flow distribution over a future prediction horizon that minimizes the operating cost, Equation (4). The latter includes the cost of the fuel consumed in the furnace and associated carbon emissions. The prediction time horizon is discretized using a discrete representation. Although the optimal solution covers a long horizon, only the first action is implemented in the plant; the remainder are discarded, and the problem is solved again in the next sampling interval, in the usual MPC scheme. The sampling (update) intervals are much shorter than the control prediction horizon. In this control layer, a forecast is required of the disturbances (changes in input variables) over the future prediction horizon. Here, each input variable (flowrate, temperature, and pressure of input streams) is forecast to remain constant at its last measured value for the entire horizon. As control updates are frequent, this is deemed to be adequate.

$$\min_{\alpha_i,\gamma_i,\varepsilon_i} \sum_{n\in PEHc} \sum_{i\in HEX} \omega_{T_t}\left(T_{t_{i,n}} - \hat{T_{t_{i,n}}}\right)^2 + \omega_{T_s}\left(T_{s_{i,n}} - \hat{T_{s_{i,n}}}\right)^2 + \omega_P\left(\Delta P_{i,n} - \Delta \hat{P}_{i,n}\right)^2. \tag{3}$$

$$\min_{m_{a,t}} \int_0^{FPHc} \left(P_f Q_f(t) + P_c m_c Q_f(t)\right) dt. \tag{4}$$

The scheduling layer deals with the slow dynamics of the process over long periods of operation. It integrates scheduling and control decisions to minimize the operating cost and to define the future cleaning actions. The MHE_S problem is similar to that of the control layer, and they share the same objective. However, the past estimation horizon of the scheduling layer is longer than in the control layer because more data is necessary to capture the slow dynamics of the system. On the other hand, the $NMPC_S$ problem is significantly different from that of the control layer. First, the future prediction horizon FPH_S is much longer, as it must be able to schedule cleaning actions and quantify their effects and benefits. Second, the objective function includes both operating cost and cleaning cost, Equation (5). Third, the prediction time horizon is here discretized using a continuous rather than a discrete representation, to reduce the number of binary variables of the scheduling problem. Each period of variable length is further discretized using orthogonal collocation on finite elements in order to accurately integrate the differential equations in the model. Fourth, in this scheduling layer, a forecast is also required of the disturbances over the future prediction horizon. Here, each input stream variable (flowrate, temperature, and pressure) is forecast to remain constant for the entire horizon, but fixed at the value of its moving average over the past month, to account for recent variability. Alternative forecasting estimates (e.g., reflecting predicted trends or known planned changes) could be used. Lastly, the optimization problem involves binary decision variables associated with the operating mode of the units at every time point, resulting in an MINLP instead of an NLP formulation. This is a challenging optimization problem because of the large number of binary variables, few constraints on the cleaning sequence, nonlinearities, nonconvexities, and the degeneracy of the objective function (i.e., multiple solutions may have similar values). To solve the MINLP problem that integrates cleaning scheduling and flow control, a reformulation using complementarity constraints is implemented, which allows finding local optimal solutions online in reasonable computational times [32].

$$\min_{y_{i,t},t_s,m_{a,t}} \int_0^{FPHs} \left(P_f Q_f(t) + P_c m_c Q_f(t)\right) dt + \sum_{n\in FPHs} \sum_{i\in HEX} P_{cl} y_{i,n}. \tag{5}$$

The scheduling layer is not updated as frequently as the control layer because of the different time scales involved. However, the two layers interact strongly so as to ensure that scheduling and control decisions are properly integrated and their synergies exploited. The optimal scheduling actions determined at the scheduling layer until the next schedule update are executed in the plant. They are also sent to the control layer, which determines the best flow distribution according to those cleanings and the disturbances observed. Other schedule decisions in the schedule prediction horizon beyond this first interval are discarded.

For the purpose of this paper, the actual plant is simulated using the same predictive model as used in the NMPC/MEH loops. However, its parameters are modified in order to create a controlled degree of (parametric) model mismatch. The plant parameters are unknown to the feedback loops.

3. Closed-Loop Schedule Stability Metrics

Closed-loop schedule instability must be quantified to determine efficient strategies to reduce it, but no single metric is adequate. In production scheduling, it has been quantified as the difference in the overall quantity of a given product produced at a given time between two consecutive evaluations of the schedule [33]. Other attempts have quantified the changes in starting time of the same task between two consecutive solutions [34] or the changes in task allocations among the units [34,35]. Reference [30] considered batch plants; thus, their criterion is not immediately applicable to the problem of interest here, which is a type of maintenance scheduling for a continuous process. An analogous concept will be developed later which is applicable to continuous processes. On the other hand, the differences in the starting time of tasks (the cleanings) and in the task allocations (which exchanger is cleaned and number of cleanings per exchanger) will be used to quantify schedule instability, according to the notation in Figure 3.The figure shows the cleaning schedules for a five exchanger network at two consecutive evaluations (at the top, schedule $k-1$ evaluated at time t_{k-1}; at the bottom schedule k, evaluated at time t_k) and a representation of the main schedule differences, including changes in task allocations (which units are cleaned) and the starting time of the tasks (when cleanings start). The schedule instability is defined taking into account those actions within the overlapping interval, OI, in the future prediction horizons of the two consecutive schedules. With constant schedule update interval and length of the scheduling prediction horizon, FPH_s, the duration of this overlapping interval is also constant and simply equal to their difference, $FPH_s - (t_k - t_{k-1})$. With variable intervals, the same definitions are indexed according to the schedule evaluation index, k, i.e., the overlapping interval at evaluation k, OI_k, has duration $FPH_{s,k} - (t_k - t_{k-1})$.

Four metrics of schedule instability are defined next on the basis of these definitions: (1) task time instability, (2) task allocation instability, (3) overall instability, and (4) overall weighted instability. They are defined for consecutive schedule evaluations assuming a continuous time representation, although they also apply with a discrete time representation. Instability metrics are generated every time a schedule is updated, and, in an online application, their time evolution can be tracked on a rolling horizon at each update.

The metrics defined here can provide useful insights into schedule stability regardless of how the schedule is defined. The only condition for their application is the existence of two consecutive evaluations or predictions of the schedule with a common period. The definition of these metrics is based on the changes occurring within a common period shared by the schedule evaluations. Hence, these metrics can be calculated for two consecutive instances even in cases where their control horizons, scheduling horizons, or update frequencies are different.

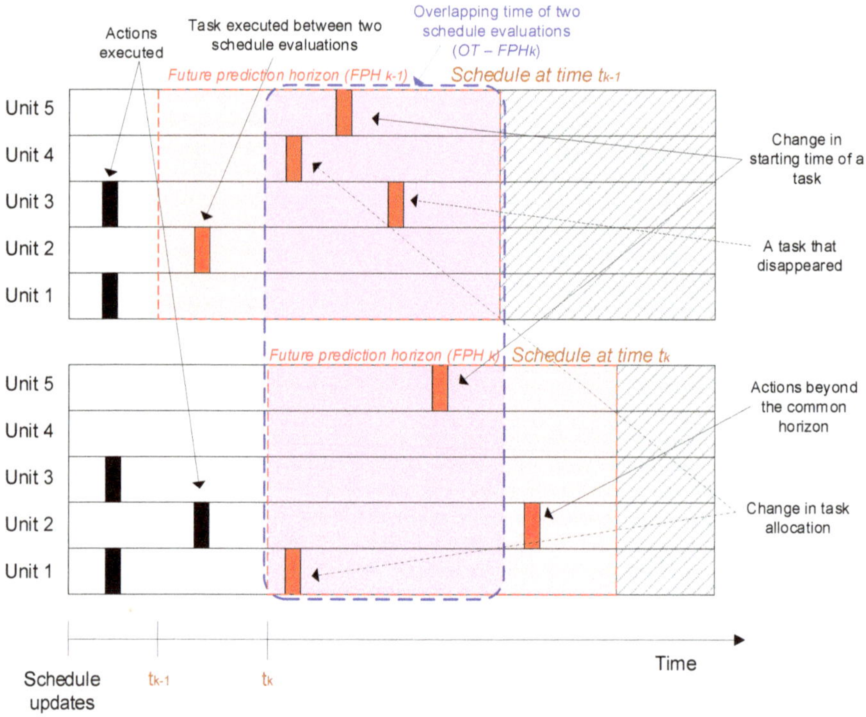

Figure 3. Representation of sources of scheduling instability and the elements used to quantify it.

The following definitions, sets, and indices are used to define the instability metrics for the online scheduling problem:

- $Units = \{1, 2, \ldots, N_U\}$. Set of units.
- $Tasks = \{1, 2, \ldots, N_{TS}\}$. Set of tasks that can be allocated to the units.
- $T = \{1, 2, \ldots, n_T\}$. Set representing time in the *FPH*.
- $SE = \{1, 2, \ldots, n_{SCH}\}$. Set of schedules evaluated over time.
- $y_{I,i,j,t,k} \in \{0, 1\}\ \forall i \in Units, j \in Tasks, t \in T, k \in SE$. Binary variable indicating the allocation of a task j to a unit i starting at a time t in schedule k.
- $t'_k \forall k \in SE$. Time when schedule k is evaluated.
- $T^*_{S,k} = t'_k - t'_{k-1} \forall k \in SE\backslash\{1\}$. Time interval between two consecutive schedule evaluations.
- $\tau'_{i,j,k} = \left\{t'_k + t \middle| y_{I,i,j,t,k} = 1 \wedge 0 \le t \le FPH_{k-1} - T^*_{S,k} \forall t \in T\right\} \forall i \in Units, j \in Tasks, k \in SE\backslash\{1\}$. Set of the starting times of all tasks j allocated to unit i in schedule k and within the time interval OI_k.
- $\tau^0_{i,j,k} = \left\{t'_{k-1} + t \middle| y_{I,i,j,t,k-1} = 1 \wedge T^*_{S,k} \le t \le FPH_{k-1} \forall t \in T\right\} \forall i \in Units, j \in Tasks, k \in SE\backslash\{1\}$. Set of the starting times of all tasks j allocated to unit i in a schedule evaluation $k-1$ and within the operating interval OI_k $OT - FPH_k$.
- $\tau^*_{i,j,k} = argmin\left\{\left|\tau'_{i,j,k}\right|, \left|\tau^0_{i,j,k}\right|\right\} \forall i \in Units, j \in Tasks, k \in SE\backslash\{1\}$. Set assigned to $\tau'_{i,j,k}$ or $\tau^0_{i,j,k}$ based on which one has the minimum number of elements.
- $\tau^{*C}_{i,j,k} = \left\{\tau'_{i,j,k}, \tau^0_{i,j,k}\right\} - \tau^*_{i,j,k} \forall i \in Units, j \in Tasks, k \in SE\backslash\{1\}$. Set defined as the complement of $\tau^*_{i,j,k}$.

Although, in this paper, fixed update intervals are used, the formulation is suitable for both fixed and variable update intervals.

For the application at hand, i.e., cleaning scheduling of preheat trains under fouling, it is assumed that only one type of cleaning is available (i.e., mechanical cleaning) so that the set *Tasks* has a single element, i.e., *Tasks* ={1}. Moreover, the set *Units* is the set of heat exchanges in the network, and the variable y_I defined here has the same role as variable y in the problem formulation detailed in [16], which is associated with the cleaning state of the units over time (i.e., 1 for *being cleaned*, 0 for *operating*). In this formulation, it is possible to assign multiple mechanical cleanings (i.e., multiple instance of the same type of task) to a unit, at different times.

As the (in)stability of a schedule is a relative concept (i.e., it only applies with reference to a previous one), all metrics apply from the second evaluation only ($k \geq 2$) and are undefined (and arbitrarily set to 0) for $k = 1$.

3.1. Task Timing Instability

A *Task timing instability* of schedule k, $I_{ts,k}$, is defined as the difference in the starting times of all tasks j in units i which are common to schedules k and $k-1$ over the overlapping interval, OI_k. Its mathematical representation is presented in Equation (6). Note that this includes only tasks j that are defined in both schedules k and $k-1$. If multiple executions of task j are included over OI_k in both schedules, the difference in their starting times is only relevant for the minimum number of instances of task j predicted in either one. In addition, if in schedule k, or $k-1$, there are no predicted executions of task j in unit i, there is no contribution of this task-unit pair to the overall task timing instability metric.

$$I_{ts,k} = \frac{1}{FPH_k} \sum_{i \in Units} \sum_{j \in Tasks} \left[\sum_{t \in \tau^*_{i,j,k}} \min\left\{ (t - \hat{t})^2, \forall \hat{t} \in \tau^{*C}_{i,j,k} \right\} \right]^{1/2}, \qquad \forall k \in SE \backslash \{1\}. \tag{6}$$

This instability metric is divided by the future prediction horizon of the scheduling problem at update k, FPH_k, to transform it into a dimensionless quantity. The task timing instability takes a value of zero when there is no difference in the predicted starting time of all the common tasks allocated to all the units in two consecutive schedule evaluations, or when no task of the same type is allocated to the same unit in two consecutive schedules (i.e., all the tasks allocated to a unit disappeared from the schedule or were reallocated to another unit). The task timing instability increases when the difference in the starting times of a task allocated to a unit in two successive schedules is large.

3.2. Task Allocation Instability

A *Task allocation instability* of schedule k, $I_{T,k}$, is defined as the change in the total number of executions of tasks j allocated to unit i in schedule k during the OI_k, with respect to the total number of executions of the same task in the same unit in the previous schedule, $k-1$. This is expressed mathematically in Equation (7). This expression assumes that all tasks have the same relative importance for the stability and only considers their total number of executions. In the cleaning scheduling application considered, this refers to the change in the total number of cleanings of each exchanger within OI_k, regardless of their starting time.

$$I_{T,k} = \frac{1}{\sum_{i \in Units} \sum_{j \in Tasks} N^{max}_{i,j}} \sum_{i \in Units} \sum_{j \in Tasks} \left[\sum_{t \in T | t \leq FPH_{k-1} - T^*_{S,k}} \left(y_{I,i,j,t,k} - y_{I,i,j,t+T^*_{S,k},k-1} \right)^2 \right], \qquad \forall k \in SE \backslash \{1\}. \tag{7}$$

This definition of instability is standardized by dividing it by the sum of the maximum number of executions of task j that are allowed in unit i, $N^{max}_{i,j}$.This is a parameter of the scheduling problem, and is specified by the analyst. For example, in the cleaning scheduling problem of preheat trains under fouling, it is the maximum number of cleanings per exchanger that can be executed in the future prediction horizon, which is usually a constraint imposed by the operators.

The task allocation instability becomes zero when there are no changes in the number of tasks of a given type scheduled in each unit regardless of their starting time or when there are no tasks of a given type scheduled in the future prediction horizon. This instability metric increases when one or more instances of a task are added to or deleted from one or multiple units in the current schedule with respect to the previous one.

3.3. Overall Schedule Instability

A metric of *Overall schedule instability* of schedule k, should consider all the changes from the previous schedule $k-1$, such as changes in the starting time of the tasks, changes in task allocation, addition of new tasks, and disappearance of previous tasks. To compute it, the overlapping interval OI_k is discretized using a time step that is lower than or equal to the shortest duration of any of the tasks present in either schedule. In the case of preheat trains, the sampling time of the process is used, which is 1 day, as plant measurements are available as daily averages. With this time discretization, a schedule matrix is defined representing a schedule, with N_U rows, one per each unit, and N_D columns, each representing a snapshot of the tasks scheduled at each time step during the OI_k. Each element of the schedule matrix is referred to as x(i,j,k) where i is an index for the units (rows), j is an index for the time instances in the discretized OI_k (columns), and k is the schedule index. The entries in the matrix are either 0, representing no task allocated, or 1, representing a task allocation. This definition assumes that there is a single task type to be scheduled, as applicable to the single type of cleaning in the scheduling of preheat trains discussed in this paper. However, it can be easily extended to a more general formulation with multiple tasks, by associating different integer values to each task type or different instability metrics for each task.

Figure 4 illustrates such a schedule matrix encoding for a simple example for schedule $k-1$ evaluated at time t_{k-1} (top schedule in Figure 4) and schedule k, evaluated at time t_k (bottom schedule in Figure 4). The corresponding schedule matrices (on the right in Figure 4) have the same dimensions because OI_k, the sampling time, and the number of units do not change between evaluations. With this encoding, it is possible to rapidly calculate the difference between two successive schedules on the basis of the differences in individual elements of the corresponding schedule matrices.

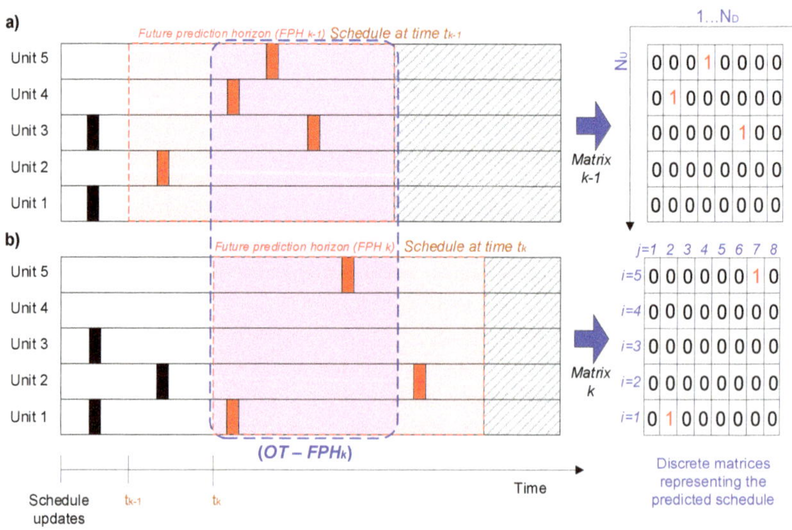

Figure 4. Schedule matrix representation of a schedule for a simple example with a unique task and the overlapping interval discretized in 8 intervals. (**a**) Schedule $k-1$ updated at time t_{k-1}; (**b**) schedule k updated at time t_k.

The *Overall schedule instability* of schedule k, $I_{ov,k}$, is defined in Equation (8), where the quadratic difference between two consecutive schedule matrices, k and $k-1$, is calculated element by element, and all the differences are added up. This instability metric is standardized by dividing it by the size of the schedule matrix ($N_U N_D$). All schedule changes are assumed to have the same effect on the overall schedule instability metric. They affect it by the same magnitude and do not differentiate between schedule differences due to changes in the starting time of the tasks, time delays, or changes in task allocations. Because this metric is standardized, it is bound between zero and one and increases with the number of differences between consecutive schedule evaluations.

$$I_{ov,k} = \frac{1}{N_U N_D} \sum_{j=1}^{N_D} \sum_{i=1}^{N_U} \left(x_{i,j,k} - x_{i,j,k-1} \right)^2, \qquad \forall k \in SE \backslash \{1\}. \tag{8}$$

In the example presented in Figure 4, there are five changes in the schedule matrices between schedules k and $k-1$ (see columns 2, 4, 6, and 7 of the matrices in the figure). Then, applying Equation (8), the overall schedule instability of schedule k is 0.125.

3.4. Time-Weighted Overall Schedule Instability

The above Overall schedule instability definition ignores when the difference in the schedules occurs. For example, the values of the overall schedule instability for two different schedules can be the same when changes in the schedule are observed at the beginning of the FPH_S, which has large implications on the operation because those are the actions to be executed in the current time step, or at the end of the FPH_S, when they may not be very important and are subject to future changes. The metric described below addresses the case when changes in the schedule closer to the current execution time are undesirable.

A *Time-weighted overall schedule instability* metric of schedule k, $I_{ovw,k}$, is defined in Equation (9), where weights are used to represent the relative importance of each difference in the schedules with respect to time. This expression uses the same definitions of overall schedule instability, Equation (8), which are based on a matrix representation of the schedule over OI_k. Here, the weights are selected to decrease linearly from one at the beginning of the overlapping interval OI_k, to zero at is end, according to Equation (10). In terms of the N_D discretizations used in the schedule matrix, we have $w_j = 1$ for $j = 0$ and $w_j = 0$ for $j = N_D$. The differences in the schedules closer to the current time are, thus, given a higher relative importance than those that occurring later in the prediction horizon.

$$I_{ovw,k} = \frac{1}{N_U \sum_{j=1}^{N_D} \left(w_j \right)} \sum_{j=1}^{N_D} \sum_{i=1}^{N_U} w_j \left(x_{i,j,k} - x_{i,j,k-1} \right)^2, \qquad \forall k \in SE \backslash \{1\}. \tag{9}$$

$$w_j = 1 - \frac{j-1}{N_D - 1}, \qquad \forall j \in \{1, 2, \ldots, N_D\}. \tag{10}$$

Using weights to characterize the relative importance changes in the schedule with respect to time was proposed to calculate schedule instability on the basis of the production quantity of different products [33,36], but not for differences in task allocation and timing. Those studies used an exponential decay function to define the weights as a function of time. This could also be used here without adding complexity to the problem. The only difference is that the exponential decay function requires the analysts to set a parameter for the rate of decay, which can be translated as a preference to ignore or not schedule modifications occurring at a future time.

The time-weighted overall schedule instability metric explicitly accounts for the effects of time to indicate that large variability close to the current time is undesirable, whereas that occurring later can be tolerable. However, it does not distinguish whether the source of variability is due to changes in the task allocation or starting time of the task.

Applying the metric defined in Equation (9), the time-weighted overall schedule instability of schedule k in Figure 4 is 0.136, which is higher than its overall schedule instability, 0.125. This happens because most of the differences between the two consecutive schedules occurs close to the current time, t_k, which is reflected in a larger number of differences between the schedule matrices in columns with low indices (columns 2 and 4 of the matrices in Figure 4). This example shows that the time-weighted metric gives more importance to changes in the schedule that occur closer to the current time and that may require an immediate action.

4. Including Stability Considerations in the Online Optimization Problem

There are various alternatives to improve the closed-loop stability of online scheduling, but their actual benefits are not clear, nor is their effect on the overall economics of the process. Three alternatives based on MPC theory and their practical implementation are evaluated here. They are (1) introducing a terminal cost penalty with respect to a steady state in the objective function, (2) freezing or fixing a subset of the scheduling decisions in the FPH_S in consecutive schedules, and (3) penalizing changes in scheduling decisions between consecutive evaluations. These alternatives are described next in the context of online cleaning scheduling of preheat trains under fouling.

4.1. Terminal Cost Penalty

In MPC for continuous systems, the closed-loop stability properties have been widely studied from practical and theoretical perspectives (as noted below, the stability definition for continuous control is different from the schedule stability used in this paper). One alternative to ensure closed-loop stability with a finite prediction horizon is to include a "terminal cost" in the objective function of the optimization problem solved at every sampling time [37] in the form of Equation (11). This represents a general objective function, J_k, where x are continuous variables, y are integer variables, and u are manipulated variables, which are minimized in an MPC scheme at each sampling time. The function V is the "running cost", which, in tracking problems, is defined as the quadratic difference between the states and their reference point. The function l is the "terminal cost", and it is only a function of the variables at the end of the prediction horizon, t_{FPH_S} (for example, the cost of missing a final target). The parameter ρ_l represents a penalty on the terminal cost and indicates its relative importance with respect to the running cost.

$$\min J_k = \rho_l l\big(x(t_{FPH_S}), y(t_{FPH_S})\big) + \int_0^{FPH_S} V(x(t), y(t), u(t))dt, \qquad \forall k \in SE\backslash\{1\}. \tag{11}$$

In our case, the overall integral of the running cost, V, in Equation (11) is just the total operating cost of the preheat train (i.e., the objective function of the online scheduling problem, defined in Equation (5). There are, however, important differences between the general MPC formulation and that for closed-loop scheduling that hinder the applicability of adding a "terminal cost" to improve stability. First, the assumptions to guarantee stability in MPC state that the objective function must decrease with the number instances evaluated; however, in the case of closed-loop scheduling, the objective is economic, and this assumption may be violated. Second, the scheduling stability problem is nonconvex and nonlinear; thus, global optimality cannot be guaranteed. Third, the maintenance/cleaning scheduling problem considered does not have an obvious stable reference point to use in a tracking function. The clean state of the network (ideal) is not achievable without infinite cleanings. A possible reference for the problem on hand is proposed below. Fourth, the scheduling problem includes discrete variables over a long time scale instead of continuous variables over short time scales. Lastly, stability for MPC is defined according to whether the system remains in the same operating point (outputs of the plant) regardless of small disturbances [37], while, for closed-loop scheduling, stability is defined as a function of the intensity of the changes in scheduling variables (inputs to the plant) between consecutive solutions.

Here, according to the rationale to avoid leaving the network unnecessarily clean at the end of the scheduling horizon, the reference point for use in the terminal cost is defined as the operation where each exchanger has reached its asymptotic or maximum fouling level or an operational constraint has been reached. This limit operation is determined by performing a simulation of the system, assuming average operating conditions and no mitigation actions. Alternatively, a stable reference operating point could be defined on the basis of engineering judgement (i.e., a realistic, possible operating state expected or observed in the past). Other scheduling problems may have cyclic solutions that can be used as references for stability [38]. The terminal cost for the online optimization of flow distribution and cleaning schedule is defined in Equation (12), as the sum for all streams in the network of the quadratic difference between the stream temperature predicted at the end of the FPH_S, $T_{a,t=FPH_S}$, and the corresponding one in the limit operation, T_a^{∞}. The set *Arcs* in Equation (12) is the set of all arcs (streams) in the network.

$$l = \rho_y \sum_{a \in Arcs} \left(T_{a,t=FPH_S} - T_a^{\infty} \right)^2 \quad (12)$$

4.2. Freezing Decisions

Fixing or freezing some of the scheduling decisions within the prediction horizon, i.e., retaining those of a previous schedule, explicitly reduces scheduling instability [39,40]. Every time a scheduling problem is solved, a fraction of the decisions from the previous schedule evaluation are frozen, and the remainder are considered free. The fixed actions are defined as equality constraints in the next scheduling optimization problem. The time intervals for online scheduling and the nature of the decisions at each evaluation are schematically shown in Figure 5 for three successive updates. The actions executed between sampling times are a mix of those frozen from the previous solution and those obtained at the current evaluation. The length of the frozen interval and the scheduling decisions included, such as task allocated and starting time of the tasks, give a trade-off between stability and closed-loop performance.

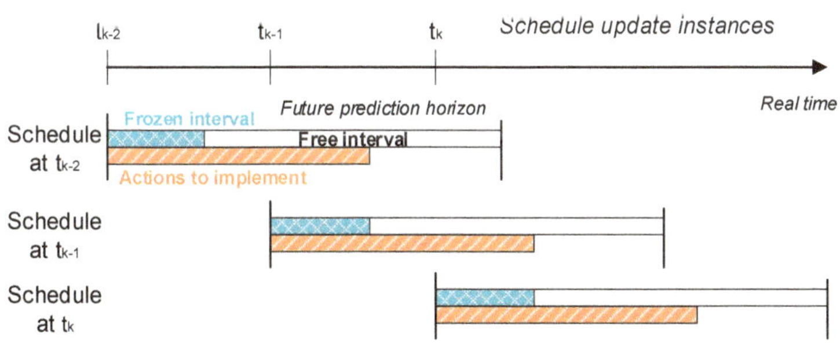

Figure 5. Schematic representation of freezing some scheduling decisions for improving closed-loop schedule instability.

In the online cleaning scheduling and flow distribution problem of preheat trains, there are two kind of decisions that can be kept constant between consecutive schedule updates: the assignment of cleanings to periods and units and the starting time of the cleaning actions. Equality constraints are introduced to fix the selected cleaning actions in the current schedule to the values calculated in the previous one. Equation (13) shows this constraint for the binary decisions, where the asterisk denotes the optimal value in the previous schedule. These equality constraints assign the cleanings to the units and periods; however, because the periods have variable length (with a continuous time representation), the starting time of the cleanings is not fixed. To allow more flexibility, inequality constraints are introduced to restrict the variability of the cleaning starting time with respect to that of the previous optimal schedule. This is shown in Equation (14) which can be transformed into an equality constraint if necessary.

$$y_{i,t(n),k} = y^{*}_{i,\mathrm{T}^{*}_{sch}+t(n),k-1}, \qquad \forall i \in HEX, n \in \{1,\ldots,N_z\}, k \in SE\backslash\{1\}. \tag{13}$$

$$-\Delta\mathrm{T}^{cl} \leq \tau^{cl}_{i,t(n),k} - \tau^{cl*}_{i,\mathrm{T}^{*}_{sch}+t(n),k-1} \leq \Delta\mathrm{T}^{cl}, \ \forall\, i \in HEX, n \in \{1,\ldots,N_z\}, k \in SE\backslash\{1\}. \tag{14}$$

4.3. Penalizing Variability

Penalizing the change in scheduling decisions between two consecutive evaluations is another alternative to improve closed-loop schedule stability. Instead of using constraints to reduce the variability between consecutive schedule evaluations, the variability is penalized in the objective function. The changes between two schedules are only penalized within their overlapping horizon OI_k, as detailed in Figure 3.

The schedule variability penalty is divided into two independent terms: one for the changes in the allocation of tasks, Equation (15), which is related to the task allocation instability metric, and another for the changes in the starting time of the tasks, Equation (16), which is related to the task timing instability metric. Each of these expressions has a penalty parameter ρ that characterizes its importance relative to the other and to the economic objective function of the scheduling problem. The final overall objective function for each schedule update, Equation (17), shows the compromise between stability and process economics. As noted, the integral of the running cost, V, is the total operating cost of the preheat train, which is the objective function defined in Equation (5).

$$l_{y,k} = \rho_y \sum_{i\in Units} \sum_{j\in Tasks} \sum_{t\in T|t\leq FPH_{k-1}-\mathrm{T}^{*}_{S,k}} \left(y_{I,i,j,t,k} - y^{*}_{I,i,j,t+\mathrm{T}^{*}_{sch},k-1}\right)^2, \qquad \forall k \in SE\backslash\{1\}. \tag{15}$$

$$l_{\tau,k} = \rho_\tau \sum_{i\in Units} \sum_{j\in Tasks} \sum_{t\in T|t\leq FPH_{k-1}-\mathrm{T}^{*}_{S,k}} \left[t y_{I,i,j,t,k} - \left(t+\mathrm{T}^{*}_{sch}\right) y^{*}_{I,i,j,t+\mathrm{T}^{*}_{sch},k-1}\right]^2, \ \forall k \in SE\backslash\{1\}. \tag{16}$$

$$\min J_k = l_{y,k} + l_{\tau,k} + \int_0^{FPH_S} V(x(t),y(t),u(t))dt, \qquad \forall k \in SE\backslash\{1\}. \tag{17}$$

5. Comparing Alternatives and Metrics to Improve Closed-Loop Schedule Stability

This section evaluates all the instability mitigation approaches presented in Section 4 that utilse the metrics in Section 3 for a simple, yet realistic, case study.

5.1. Case Study 1—Illustrative Example

The preheat train considered here consist of four heat exchangers, three of which are located on parallel branches. This case study was adapted from [41,42], where all the details of the equipment, costs, and operation can be found, although the most important ones, including the fouling model and cost parameters, are summarized in Tables A1 and A3 (Appendix A). Figure 6 shows the structure of the network, which is commonly found in refining operations with more units in the branches. All exchangers are shell and tube and exhibit significant levels of fouling.

For this case study, a nominal operation was assumed (i.e., with constant inlet streams flow rates and temperatures) with no model–plant mismatch; thus, the predictive models used in the feedback loops perfectly represent the plant. These assumptions lead to a simpler problem than a real application of the online approach, while they allow isolating the analysis of the stability of the online scheduling from other aspects. This case also demonstrates that there can be schedule instability even under these ideal conditions.

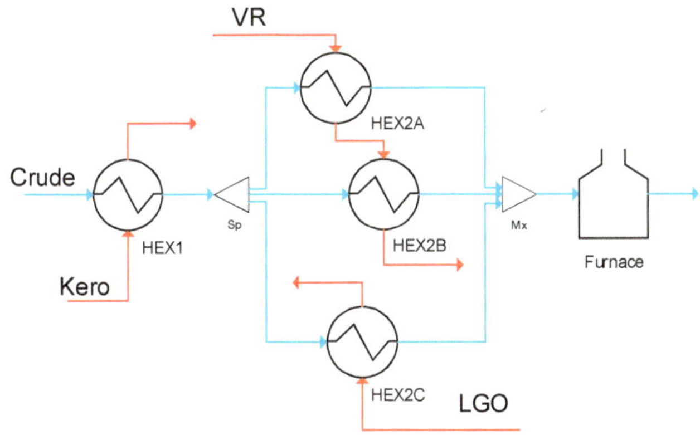

Figure 6. Preheat train structure of the simple case study.

5.2. Results and Discussion

The online optimal control and cleaning scheduling problem was solved with the following settings: for the control layer, a control horizon FPH_C of 10 days and update intervals of one day; for the scheduling layer, a scheduling horizon FPH_S of 120 days, update intervals of 15 days, and 15 periods of variable length in the scheduling horizon. The MHE problems were not solved in the feedback loops because there is no plant–model mismatch. These settings of the closed-loop scheme led to 25 solutions of the optimal cleaning scheduling problem over 1 year of operation. The schedules obtained were used to calculate the schedule instability metrics for 24 consecutive solutions after the first one.

The closed-loop optimization was first performed with the usual economic objective function without including stability (base case). An example of changes in schedule between successive updates is shown in Figure 7 with reference to the predicted fouling state (in terms of fouling resistance) of exchanger HEX2B. At the update on day 91, no cleanings of HEX2B were scheduled over the predicted horizon (red line in Figure 7, top). At the day 106 update, a cleaning was introduced, scheduled for day 120 (red line in Figure 7, bottom). Due to intervening variations in the plant, the cleaning was eventually shifted and executed on day 170 (black line in Figure 7).

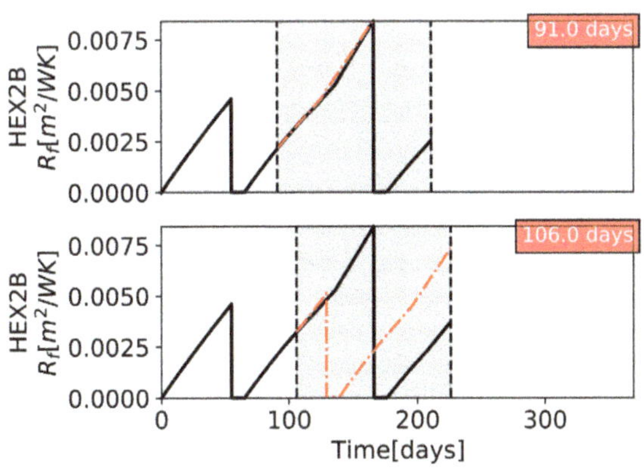

Figure 7. Case study 1, base case—predicted (red) fouling resistance in HEX2B at 91 and 106 days. Black lines show the fouling resistance observed in the final record. Vertical lines correspond to cleanings, and their shifts in time are sources of instability.

Figure 8 presents the cleaning schedule predicted (red) at six successive sampling times each over two time windows, together with the schedules eventually executed over the entire horizon (black). Figure 9 shows the schedule instability metrics calculated at every cleaning schedule update. In all updates, there are some changes in the optimal cleaning schedule with respect to the previous one, and this is reflected in the evolution of the instability metrics. The peaks of task timing instability occur when a cleaning was postponed, and the task allocation instability changes when cleanings are included or removed from the predicted schedule. The overall instability and the time weighted overall instability are good single indicators of instability as their behavior aligned with that of the other metrics.

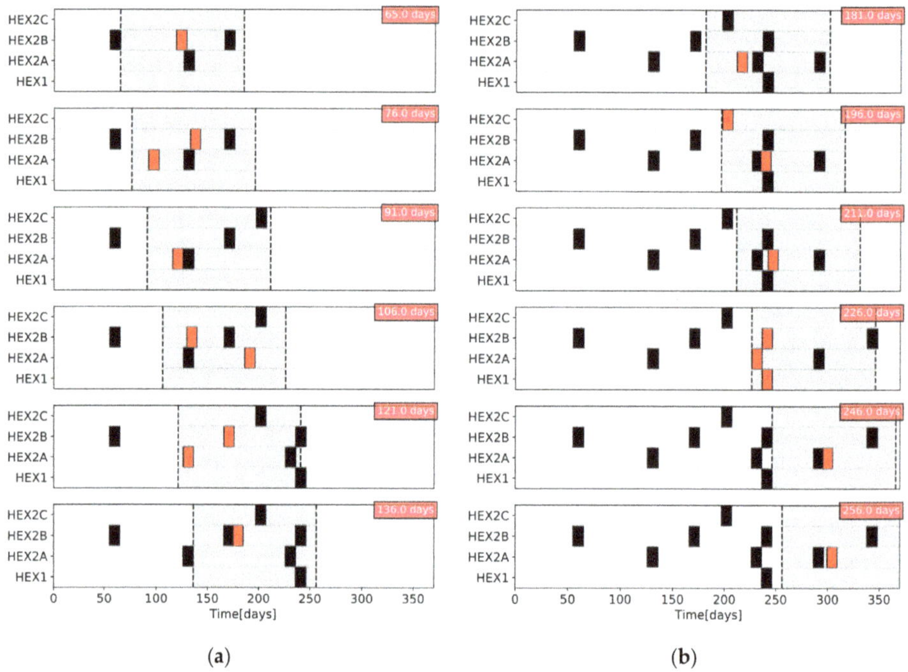

Figure 8. Case study 1, base case—evolution of the cleaning schedule as predicted (red) and executed (black) at various sampling times (marked in the upper right corner): (**a**) from 65 to 136 days; (**b**) from 181 to 256 days.

The variation in scheduling instability metrics observed in Figure 9 can be explained by the evolution of the cleaning schedule. For instance, the maximum value of task timing instability is observed between 90 and 120 days of the operation because the starting time of the cleanings predicted for HEX2A and HEX2B change significantly, and even their precedence order is reversed. As another example, between the schedule evaluations at 65 days and 76 days, there is one additional cleaning introduced, causing an increase in the task allocation instability. The final example relates to the overall instability and the time-weighted overall instability metrics. Consider the consecutive schedule solutions at 211 and 226 days, when two new cleanings are predicted and the starting time of the HEX2A cleaning are shifted closer to the current time. The weighted overall instability metric is higher than the overall instability because all the changes occur closer to the execution time.

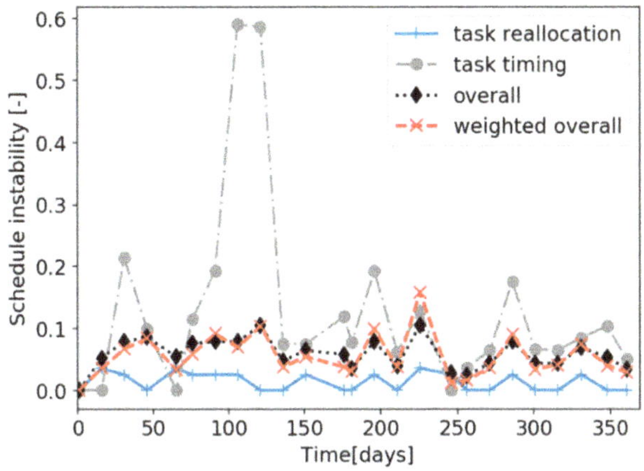

Figure 9. Case study 1, base case—evolution of closed-loop scheduling instability metrics.

Next, the three alternatives proposed in Section 4 to improve closed-loop scheduling stability were implemented, with the parameters varied as follows:

- In the terminal cost penalty (Equation (12)), ρ_l was varied between 1×10^{-1} and 1×10^{-9} on a logarithmic scale.
- In the freezing horizon alternative (Equation (14)), the number of periods in which decisions are kept constant, N_z, was varied between 2 and 10, and the maximum allowed variation in the cleaning starting time, ΔT^{cl}, was varied between 1 day and 100 days.
- In the variability penalty alternative (Equations (15)–(17)), the penalty parameter of the cleaning allocation variability, ρ_y, was evaluated between 1×10^{-3} and 1×10^{2}, while the penalty parameter of the cleaning starting time variability, ρ_τ, was varied between 1×10^{-4} and 1×10^{0}. The different ranges were due to the differences in the order of magnitude of the metrics.

Figure 10 shows the closed-loop economic performance and stability metrics obtained when using the terminal cost alternative to reduce instability, for various values of the penalty parameters. The bars in Figure 10b represent the standard deviation of each metric. Increasing the terminal cost penalty improves the schedule stability but increases the operating cost. At the upper value of the terminal cost penalty, $\rho_l = 1 \times 10^{-1}$, the closed-loop solution has no cleanings scheduled over the entire horizon. This is the most stable solution, but also the most costly. Low penalties reduce the total operating cost as they allow more variability and a higher reactivity in the scheduling actions. For terminal cost penalties lower than 1×10^{-7}, the total operating cost and the average of most instability metrics do not change significantly, but the task timing instability changes. When large variability in the scheduling decisions is allowed (low penalties), the effect of changes in the starting time of the cleanings dominates over that of the assignment of cleanings to units. The results show that proposed metrics may be used as good indicators of the overall closed-loop schedule stability.

Figure 11 illustrates the effect of freezing some scheduling decisions beyond the scheduling sampling time. It shows the total cost and the average of the overall weighted schedule instability as a function of the number of periods frozen and the maximum change allowed in the starting time of cleanings (for clarity, only the average value is shown without indicating its variability). Only the overall weighted instability is used from now on as it is the most comprehensive and illustrative metric among those proposed. The total operating cost increases with the number of periods frozen, while the schedule instability decreases, although no clear trend is observed, as the nonlinearities, nonconvexities, and combinatorial nature of the problem potentially lead to local optimal solutions. A higher number of periods frozen results in fewer degrees of freedom in the scheduling problem, limiting the opportunity to react optimally. When the range of changes in the cleaning starting time is also restricted, it is observed that, for low values of this bound, the closed-loop schedule is more stable than for high values, and its total operating cost is higher. For a cleaning starting time variability bound greater than 10 days, there is no significant change in the schedule stability, but the operating cost varies. In these scenarios, the total number of cleanings and their allocations have a higher impact on the process economics than their starting time.

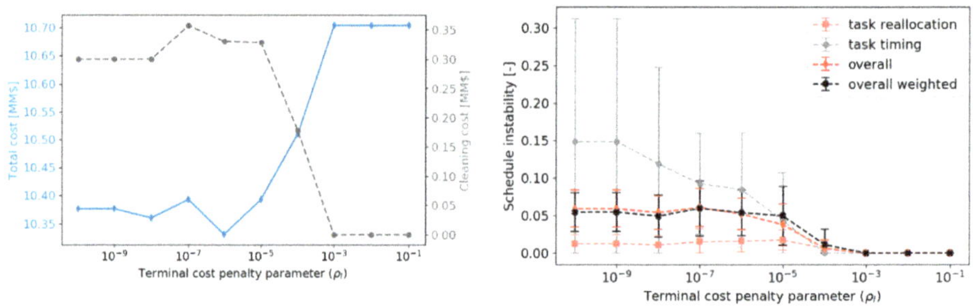

(**a**) (**b**)

Figure 10. Case study 1—effect of the terminal cost penalty (ρ_l) on the closed-loop performance: (**a**) process economics, as total cost (left axis) and cleaning cost (right axis); (**b**) average schedule instability metrics and their standard deviation.

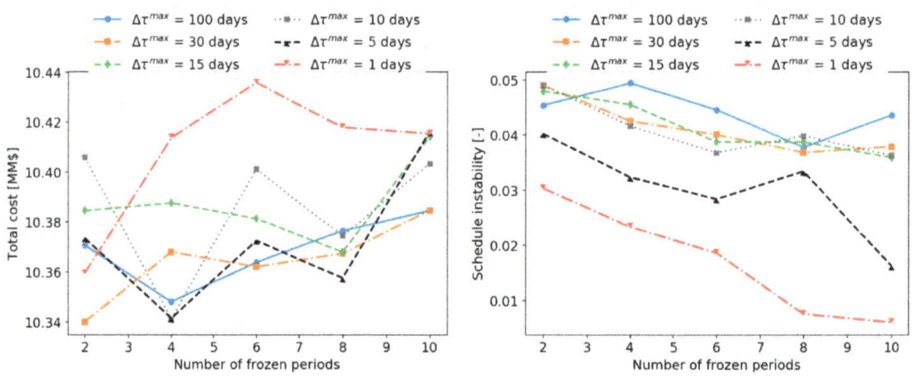

(**a**) (**b**)

Figure 11. Case study 1—effect of the number of frozen periods (N_z) and the maximum allowed variation in the cleaning starting time (ΔT^{cl}) on the closed-loop performance: (**a**) total operating cost; (**b**) average schedule overall weighted instability.

For the variability penalty alternative, which penalizes changes between consecutive schedules, Figure 12 presents the effect of its parameters on the closed-loop performance and schedule stability. Although no clear trend is observed, the total operating cost increases when the penalties on the variability are higher, while the schedule instability decreases. For values of ρ_τ lower than 1×10^{-3} the schedule instability values did not change, but the operating cost could still vary, indicating that the effect of cleaning starting times is not as significant as the cleaning allocation. Furthermore, the two penalties in this alternative are correlated, and there are different combinations leading to similar closed-loop performance of the overall system.

All the alternatives presented to improve closed-loop schedule stability were effective in doing so. In all scenarios considered, improving schedule stability came at the expense of the total operating cost, demonstrating a trade-off between how fast the system reacts to disturbance and the long-term predictability of the schedule. A data envelope analysis (DEA) [43,44] was used to evaluate this trade-off for all the scenarios simultaneously considered in all the alternatives. Each point in the DEA represents a solution of the closed-loop optimal scheduling problem, using any alternative to improve stability and the specifications of its parameters. Hence, there are 71 points in total—one base case, 10 for the terminal cost alternative, 30 for the freezing decisions alterative, and 30 for the penalizing schedule variability alternative. The total operating cost and the average overall weighted instability of each schedule were considered as "inputs" to the standard representation of the DEA, while there were no "outputs", and an "efficiency" was calculated for each point by solving a linear programming problem.

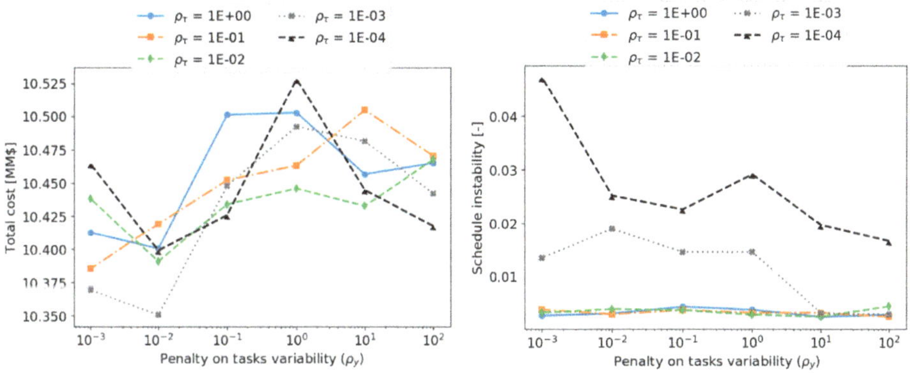

(**a**) (**b**)

Figure 12. Case study 1—effect of the penalty on task variability (ρ_y) and the penalty on cleaning starting time (ρ_τ) on the closed-loop performance: (**a**) total operating cost; (**b**) average overall weighted instability.

The results of the DEA are presented in Figure 13, where the points are classified according to the closed-loop alternative used to improve stability, and the efficiency frontier was constructed from the DEA. The points that lay on the frontier have a 100% efficiency (i.e., represent the best combination of the inputs, and no other data point available can be as good or better) and all other points underperform with respect to those. All the points for the terminal cost alternative lay inside the frontier; thus, they are not as efficient as those defined by the other alternatives or even as the base case, which did not consider stability in the online schedule optimization. This underperformance of the terminal cost alternative is because the reference point used to ensure closed-loop stability, although stable, correspond to the worst conditions to operate the preheat train. The freezing decisions and variability penalty alternatives both improved the closed-loop schedule stability but compromised the operational cost. For these data points, two clusters are observed: one for the freezing decisions alternative that, on average, reduces the schedule instability without a large cost penalty and another for the variability penalty alternative that, on average, achieves a larger improvement in stability but with a larger increase in operating cost. Some of the data points of the two clusters overlap, representing intermediate solutions.

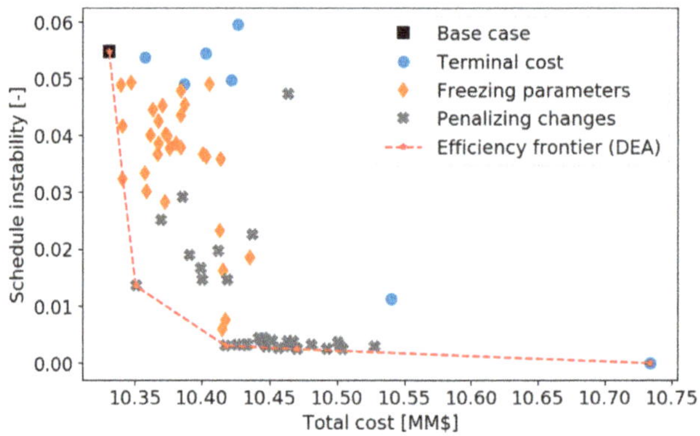

Figure 13. Case study 1—data envelope analysis (DEA) for all the closed-loop solutions of the scheduling and control problem.

For this case study, the DEA suggests that the variability penalty method is the better approach to reduce schedule instability, while still achieving a good economic performance. The terminal cost alternative proved to be the least efficient, whereas freezing decisions in successive schedules increased stability, but could be too restrictive in the presence of disturbances, such that not all the economic benefits of implementing an online fouling mitigation strategy were achieved.

6. Closed-Loop Schedule Stability of an Industrial Preheat Train

This section analyzes the closed-loop performance and stability of the online optimal cleaning scheduling and control of an industrial preheat train under dynamic and variable operation.

6.1. Case Study 2—Definition

This case study involves a network with five heat exchangers, four of which are double shells (modeled as nine exchangers overall) in the hot end of a real refinery preheat train (Figure 14). It was based on the network and operating conditions presented in [45,46]. There is one control degree of freedom, as the flow split of crude oil through the parallel branches is not constrained but bound between 20% and 80%. Plant measurements of flow rates and streams temperature were available as daily averages over 1240 days. Figure 15 shows the flowrates and temperatures of the five inlet streams to the network (crude oil and five recycle streams) which were used to characterize the variability of the inlet streams. The design specifications of the heat exchangers are presented in Table A2 (Appendix A), together with fouling and aging parameters, while cost parameters used are presented in Table A3 (Appendix A). A dynamic model of this preheat train was validated against the plant data in [1] with excellent results and was used here as the "plant" model.

A controlled degree of model plant mismatch was introduced by modifying the fouling deposition constants of each exchanger in the plant model at every sampling time. This aimed to mimic the effect of processing different crudes or crude blends, as they have different fouling propensity. The deposition constants in the plant simulation changed over time, but their actual value was unknown to all predictive models used in the online optimization approach. The variability of the deposition constants in each exchanger was modeled as a pseudo random process around their average values, estimated using the actual measurements (i.e., outlet temperature of the tube side and shell side of each exchanger). Because all exchangers process the same crude at a given time, their deposition constants are not independent, and their correlation was captured when defining their variability. The average deposition constant and its variability were different for each exchanger in the network. The box plots in Figure 16 show their median and ranges for all exchangers.

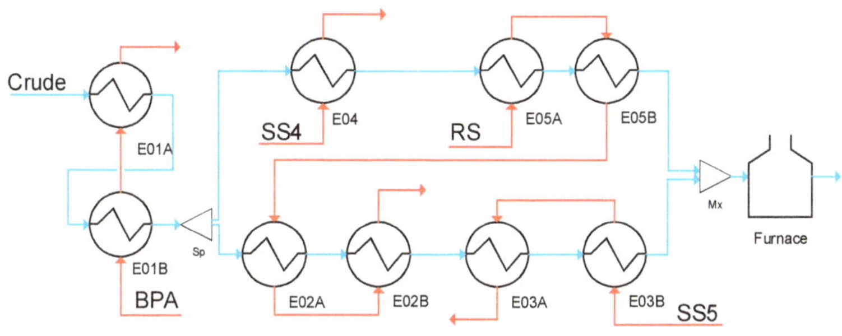

Figure 14. Industrial case study 2—preheat train structure.

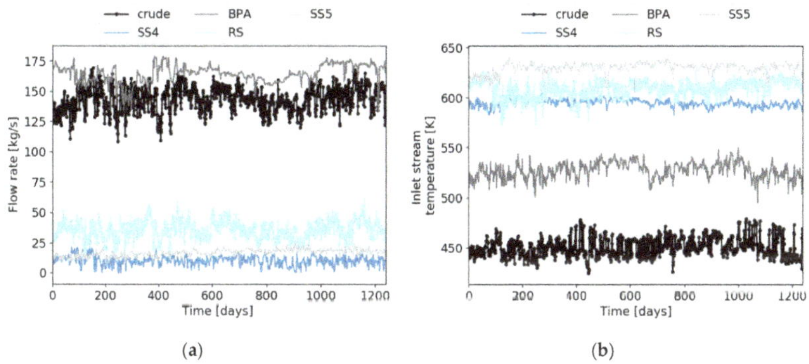

Figure 15. Industrial case study 2—actual measurements of the inlet stream flow rates (**a**) and temperature (**b**).

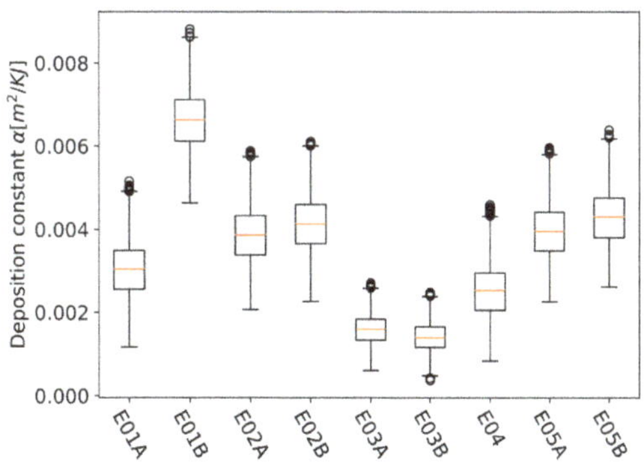

Figure 16. Industrial case study 2—box plot representing the variability in the deposition constant for each exchanger in the preheat train.

A closed-loop simultaneous optimization of the flow distribution and cleaning schedule was carried out over 1240 days, starting with all exchangers in a clean state. The economic performance optimization case, ignoring schedule stability (base case), was detailed in [1]. For this case, Figure 17 presents the cleaning schedule and a key temperature (the crude oil temperature at the entrance of the furnace, CIT), as predicted and executed at three successive sampling times.

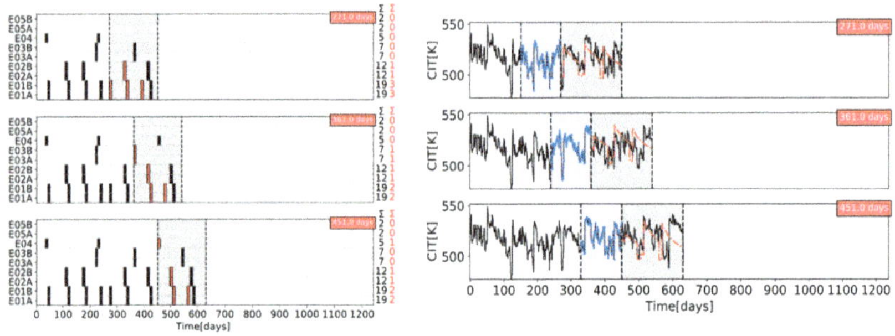

(a) (b)

Figure 17. Industrial case study 2—base case online optimization (no instability reduction). (**a**) Cleaning schedule as predicted in three successive updates (red), and executed (black); (**b**) crude oil temperature at the exit of the last exchanger/inlet to furnace (CIT) as estimated (blue), predicted (red), and eventually observed (black), in three successive updates.

6.2. Results and Discussion

The online optimal cleaning scheduling and flow control problem was solved using the following settings: for the feedback loops, a predictive control horizon FPH_C of 10 days, an update frequency of 1 day, and a PEH_C of 20 days for the control layer; for the scheduling layer, a PEH_S of 120 days, an update frequency of 90 days, and FPH_S of 180 days. The variability penalty alternative to penalize changes between consecutive schedules, described in Section 4.3, was used. The two penalty parameters on task allocation, ρ_y, and on task timing, ρ_τ, were varied, defining different settings for the online optimization. Results are compared against the base case, which did not consider instability, in terms of schedule stability and total operating cost.

The closed-loop performance with the variability penalty formulation is presented in Figure 18 as the schedules and profiles of a key temperature (the crude oil temperature at the entrance of the furnace, CIT) at three successive schedule updates.

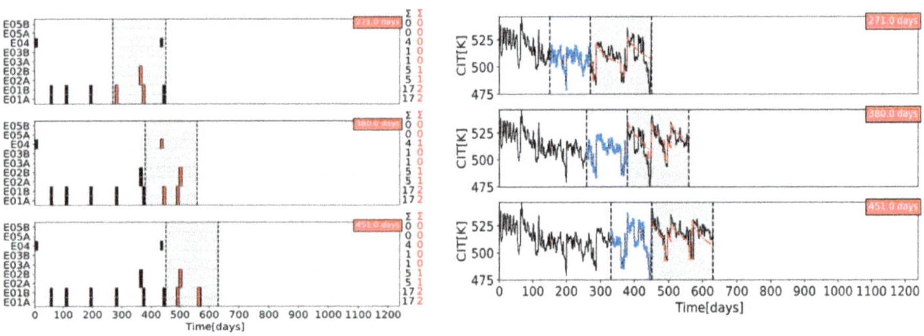

(a) (b)

Figure 18. Industrial case study 2—online optimization: variability penalty with $\rho_\tau = 1 \times 10^{-3}$ and $\rho_y = 1 \times 10^{-1}$. (**a**) Cleaning schedule as predicted in three successive updates (red) and executed (black); (**b**) crude oil temperature at the exit of the last exchanger/inlet to furnace (CIT) as estimated (blue), predicted (red), and eventually observed (black), in three successive updates.

Figure 19a–c show the energy cost (a), the cleaning cost (b), and the total operating cost (c). The closed-loop average overall time-weighted instability is presented in Figure 19d (without the variability bars in the metric for clarity). The scenarios with a task allocation penalty, ρ_y, of 1×10^{-1} have consistently a larger operating cost, up to 1.0 million USD, than the base case. This cost increase is due to larger energy cost and fewer cleanings during the overall online operation. However, these scenarios exhibit the lowest schedule instability. Fewer cleanings are predicted at every schedule update because adding new cleanings to or removing some from a previous schedule are heavily penalized. The predicted cleaning schedules, therefore, have minimal changes between updates. This also inhibits the ability of the scheduling feedback loop to react to disturbances and introduce operational changes to mitigate fouling and minimize the cost of the operation.

With values of the task allocation penalty parameter $\rho_y < 1 \times 10^{-1}$, the closed-loop performance is not very different from the base case, and it could even improve it, reducing the total operating cost by 0.37 million USD in one case. In addition, the corresponding schedules have a lower instability than the base case, meaning that they improved both the closed-loop performance and the closed-loop stability at the same time. All scenarios with $\rho_y < 1 \times 10^{-1}$ have lower cost than the base case, but larger closed-loop instability than those with $\rho_y = 1 \times 10^{-1}$. This observed simultaneous improvement in the two metrics of closed-loop performance contradicts the expectations. A possible explanation is that this case reflects a specific realization of the uncertainty, disturbances, variability in the operation, and forecasting scenarios adopted. The input flow rates and stream temperature in the plant were assumed to change constantly, while the predictive model of the scheduling layer uses at each evaluation only a constant forecast for each input, defined as a time-moving average of recent past values.

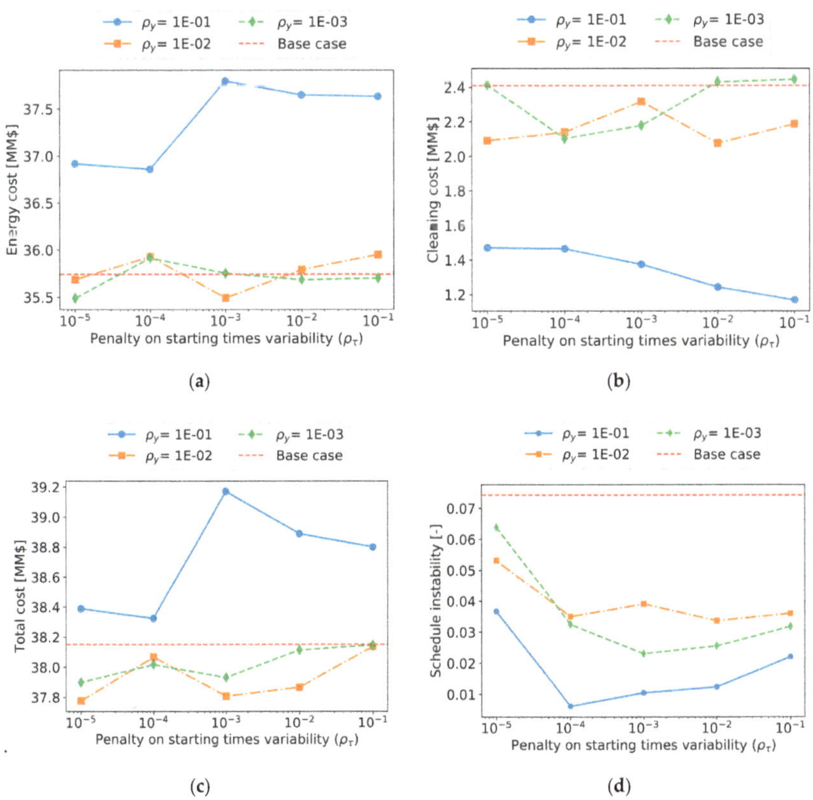

Figure 19. Industrial case study 2—closed-loop performance for different penalty parameters of schedule variability: (**a**) energy cost; (**b**) cleaning cost; (**c**) total cost; (**d**) average overall time-weighted instability.

The effect of the penalty parameter on the task timing instability is not as significant as that of the penalty parameter on the task allocation instability. The operating cost increases only slightly when the task timing penalty increases from 1×10^{-5} to 1×10^{-1}, but this difference is no more than 0.37 million USD for $\rho_y < 1 \times 10^{-1}$, while, for $\rho_y = 1 \times 10^{-1}$, the operating cost ranged from 38.4 million to 39.2 million USD, depending on the task timing penalty, ρ_τ. For the overall closed-loop performance, the starting time of the cleanings is not as important as the allocation of cleanings to heat exchangers. Under variable and uncertain operating conditions, modifying the starting time of an already scheduled cleaning task for a given unit did not have a big potential to reduce the energy cost, but could improve schedule stability. A reduction in schedule instability was observed between $\rho_\tau = 1 \times 10^{-5}$ and $\rho_\tau = 1 \times 10^{-4}$, whereas the changes in instability were minimal when ρ_τ increases further. The lowest penalty, ρ_τ, allowed the largest variability in the cleaning starting times between consecutive schedules. For larger values of ρ_τ the changes in the predicted cleaning time were minimal, and most of the schedule instability came from changes in the allocation of cleanings to the heat exchanger as new cleanings were predicted.

Figure 20 compares the cleaning schedule obtained (as executed by the end of the 1240 day operation) for three scenarios: (i) the base case without stability consideration, (ii) a variability penalty case that achieved better closed-loop stability while increasing the operating cost ($\rho_y = 1 \times 10^{-1}, \rho_\tau = 1 \times 10^{-3}$), and (iii) a variability penalty case that improved both the closed-loop stability and the operating cost with respect to the base case ($\rho_y = 1 \times 10^{-2}$, $\rho_\tau = 1 \times 10^{-3}$). The total number of cleanings, indicated for the exchanger in the columns on the right of Figure 20, changes significantly between the scenarios. The second scenario, with $\rho_y = 1 \times 10^{-1}$, resulted in fewer cleanings and longer times between cleanings of the same exchangers, while the other two scenarios had similar cleaning schedules. Although the final closed-loop schedules of the base case (i) and scenario (iii) are similar, their generation in a receding horizon and their performance were different. In the base case (i), the calculated instability value, 0.075, results mainly from adding to a predicted schedule new cleanings that have to be immediately executed. This is not practical from a planning perspective, if the response to cleaning decisions and supply of resources needed for their execution is not immediate. In scenario (iii), such actions are penalized, thus occur less often, and most of the schedule variability can be attributed to the changes in the starting time of the predicted cleanings. A Pareto plot of all solutions explored is given in Figure 21.

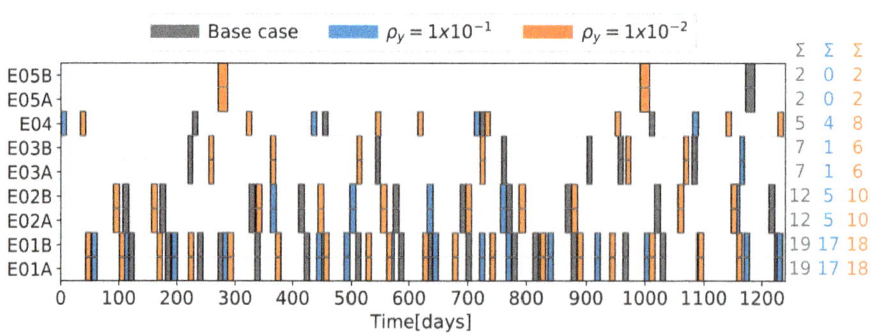

Figure 20. Industrial case study 2—final cleaning schedules as executed online for the base case (no schedule instability mitigation) and two schedule instability mitigation (variability penalty with penalty parameters $\rho_y = 1 \times 10^{-1}$and $\rho_y = 1 \times 10^{-2}$). The columns on the right give the total number of cleanings of each exchanger in each case.

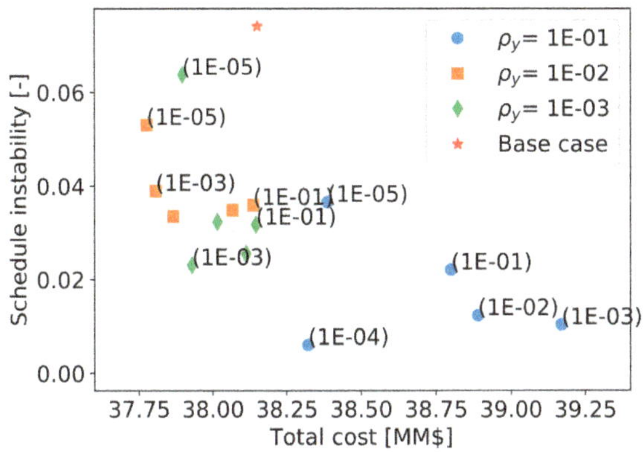

Figure 21. Industrial case study 2—Pareto plot of schedule stability vs. performance, for various penalty parameters (with the value of ρ_τ indicated in parenthesis).

In general, the penalization of schedule changes achieves more stable schedules, but reduces the ability of the system to react to changes. Due to the high variability and uncertainty of the operation (and possibly the existence of multiple local optima to the nonconvex MINLP scheduling problem), the trade-off between stability and performance is less clear.

7. Conclusions and Perspectives

The closed-loop schedule stability problem was addressed in this work with an application to the online cleaning scheduling and control of refinery preheat trains under fouling. The various metrics developed to quantify schedule instability for online scheduling account for distinct aspects, such as changes in task allocation, task sequence, starting time of the task, and the earlier or later occurrence of such changes in the future scheduling horizon. The results show that the metrics are useful to characterize the stability of successive schedules, as well as to identify sources of instability and ways to mitigate it. Further stability metric variations could be easily developed (for example, ways of assigning weights to distinct contributions to a schedule change) on the basis of the methods proposed.

It was demonstrated that such stability considerations can be practically and, in a rather general way, introduced in a closed-loop NMPC formulation of the optimal scheduling and control problem, and solved online over a moving horizon, in terms of penalties in an economic objective or via additional constraints. The result is a formulation which enables to specify both schedule stability and performance requirements, explore the balance between schedule reactivity and disturbance rejections, and establish the optimal trade-off between schedule stability and economic benefits.

The above methods were demonstrated for the online cleaning scheduling and flow control of refinery preheat trains, a challenging application with significant economic, safety, and environmental impact. An illustrative, small but realistic case study was followed by a demanding industrial case study. Results show that, of the three alternatives evaluated, the terminal cost penalty proved to be inefficient in this case. The other two (fixing some decision in the prediction horizon, and penalizing schedule changes between consecutive evaluations) showed improvements in the closed-loop schedule stability, against various degrees of economic penalties. The results highlight the importance of including stability considerations in an economically oriented online scheduling problem, as a way to obtain feasible solutions for operators over long operating horizons without sacrificing the benefits of a reactive system to reject disturbances or take advantage of them.

Nevertheless, there are still open questions related to the definition of the penalties or bounds in the schedule instability mitigation strategies, as well as the definition of acceptable ranges of schedule stability or instability. Extensions of this work include dealing with multiple task types (as already outlined in the paper) and, for longer-term development, the use of global solution methods and formally incorporating uncertainty in models and solutions.

Application of the metrics developed in this manuscript is not restricted to the specific closed-loop NMPC scheduling implementation detailed here. They are useful to assess schedule stability in general regardless of how schedules are calculated, only relying on the existence of two consecutive evaluations or predictions of the schedule with a common period. The two consecutive instances may have different control horizons, scheduling horizons, or update frequency. Lastly, although this work dealt with a specific application (the optimization of refinery heat exchanger networks subject to fouling), the formulations and solution approach demonstrated here should be applicable with small modifications to other cases where closed-loop scheduling and control of dynamic systems is important, such as batch and semi-continuous processes.

Author Contributions: Data curation, writing—original draft preparation, F.L.S.; formal analysis, investigation, software, and validation, F.L.S.; conceptualization, methodology, F.L.S. and S.M.; project administration, resources, supervision, and writing—review and editing, S.M. All authors have read and agreed to the published version of the manuscript.

Funding: One of the authors (FLS) gratefully acknowledges the receipt of a President's Scholarship from Imperial College London in support of his PhD studies.

Conflicts of Interest: The authors declare no conflict of interest

Nomenclature

Subscripts	Description	
c	Related to the control layer in the online optimization approach	
d	Deposit layer	
ov	Overall schedule instability	
ovw	Overall weighted schedule instability	
s	Related to the scheduling layer in the online optimization approach	
ts	Task timing instability	
T	Task allocation instability	
Symbol	**Units**	**Description**
D_{in}	mm	Tube inner diameter
E_f	J/mol	Activation energy of fouling reaction
I	-	Schedule instability
l	-	Terminal cost in the objective function of MPC
m_c	ton CO_2/MWh	Carbon emission factor (0.015)
N_D	-	Discrete points (columns) in a schedule representation
N_U	-	Units (rows) in a schedule representation
P	Pa	Pressure
P_c	\$/ton Co_2	Carbon price (30 USD/ton)
P_{cl}	\$	Cleaning cost per unit
P_f	\$/MW	Energy price (25 USD/MW)
Pr	-	Prandtl number
Q_f	MW	Furnace duty
R	J/molK	Universal gas constant
Re	-	Reynolds number
R_f	m^2K/W	Thermal resistance of the deposit—fouling resistance
t	days	Time
T_f	K	Film temperature
T_s	K	Shell temperature
T_t	K	Tube side temperature
V	-	Running cost in the MPC objective function
w	-	Decreasing sequence of weights for I_{ovw} calculation
x	-	Single entry of a schedule matrix
y	-	Binary variable for cleanings (1, cleaning, 0 operating)

α	m^2K/J	Deposition constant
γ	m^2K/JPa	Removal constant
δ	mm	Deposit thickness
ΔT^{cl}	days	Bounds on the changes of cleaning times
λ	W/mK	Thermal conductivity
ρ	-	Penalty in objective functions
τ	Pa	Shear stress
τ^{cl}	days	Starting time of a cleaning action
T^*_{sch}	day	Update interval of scheduling feedback loop
ω	-	Weights in the MHE objective function
HEN	-	Heat exchanger network
HEX	-	Set of heat exchangers
FPH	days	Future prediction horizon
NLP	-	Nonlinear programming
NMPC	-	Nonlinear model predictive control
MHE	-	Moving horizon estimator
MPC	-	Model predictive control
$OT-FPH$	days	Overlapping time in the future prediction horizon
PEH	days	Past estimation horizon

Appendix A

The model parameters and specifications for the simple case study are presented in Table A1, while those for the industrial case study are presented in Table A2. Other costs and furnace efficiency are given in Table A3.

Table A1. Exchanger, fouling model, and cleaning specifications for the simple case study.

	HEX1	HEX2A	HEX2B	HEX2C
Shell diameter (mm)	1295	1400	1400	1400
Tube inner diameter (mm)	19.86	19.86	19.86	19.86
Tube outer diameter (mm)	25.40	25.40	25.40	25.40
Tube length (m)	6100	5800	5800	6100
Number of tubes	800	600	600	600
Number of passes	2	2	2	4
Baffle cut (%)	25	25	25	25
Tube layout (°)	45	45	45	45
Number of baffles	8	6	6	7
Surface roughness	0.046	0.046	0.046	0.046
Deposition constant (m^2 K/J)	0.0075	0.0085	0.0085	0.0065
Removal constant (m^4 K/NJ)	4.5×10^{-12}	4.0×10^{-12}	4.0×10^{-12}	4.5×10^{-12}
Fouling activation energy (J/mol)	35,000	33,000	33,000	38,000
Ageing frequency factor (1/day)	0.00	0.00	0.00	0.00
Ageing activation energy (J/mol)	50,000	50,000	50,000	50,000
Cleaning time (days)	10	10	10	10
Cleaning cost (USD)	30,000	30,000	30,000	30,000

Table A2. Exchanger, fouling model and cleaning specifications for the industrial case study.

	E01A	E01B	E02A	E02B	E03A	E03B	E04	E05A	E05B
Shell diameter (mm)	1245	1194	1397	1397	990	990	1270	1397	1397
Tube inner diameter (mm)	19.86	19.86	19.86	19.86	13.51	13.51	19.86	19.86	19.86
Tube outer diameter (mm)	25.40	25.40	25.40	25.40	19.05	19.05	25.40	25.40	25.40
Tube length (m)	6090	6090	6090	6090	6090	6090	6090	6090	6090
Number of tubes	764	850	880	880	630	630	888	880	880
Number of passes	2	2	4	4	2	2	4	4	4
Baffle cut (%)	25	22.5	25	25	25	25	17	25	25
Tube layout (°)	45	45	45	45	45	45	45	45	45
Number of baffles	5	7	18	18	20	20	9	18	18
Surface roughness	0.046	0.046	0.046	0.046	0.046	0.046	0.046	0.046	0.046
Deposition constant (m^2 K/J)	0.0045	0.0041	0.0036	0.0036	0.0012	0.0014	0.0032	0.0040	0.0038
Removal constant (m^4 K/NJ)	1.69×10^{-11}	1.53×10^{-11}	1.28×10^{-11}	1.30×10^{-11}	3.77×10^{-11}	4.54×10^{-12}	1.14×10^{-11}	1.46×10^{-11}	1.39×10^{-11}
Fouling activation energy (J/mol)	28,500	28,500	28,500	28,500	28,500	28,500	28,380	28,500	28,500
Ageing frequency factor (1/day)	0.00	0.00	0.00	0.00	0.00	0.00	0.00	0.00	0.00
Ageing activation energy (J/mol)	50,000	50,000	50,000	50,000	50,000	50,000	50,000	50,000	50,000
Cleaning time (days)	9	9	10	10	8	8	9	16	16
Cleaning cost (USD)	27,000	27,000	30,000	30,000	24,000	24,000	27,000	48,000	48,000

Table A3. Costs and furnace specifications.

Fuel cost (USD/MWh)	27
Production cost (USD/kg)	0.23
Carbon cost (USD/t)	30
Carbon emission factor (t CO_2/MWh)	0.015
Furnace efficiency (%)	90

References

1. Santamaria, F.L.; Macchietto, S. Online Integration of Optimal Cleaning Scheduling and Control of Heat Exchanger Networks under Fouling. *Ind. Eng. Chem. Res.* **2019**, *59*, 2471–2490. [CrossRef]
2. Bajestani, M.A.; Banjevic, D.; Beck, J.C. Integrated maintenance planning and production scheduling with Markovian deteriorating machine conditions. *Int. J. Prod. Res.* **2014**, *52*, 7377–7400. [CrossRef]
3. Li, R.; Ma, H. Integrating Preventive Maintenance Planning and Production Scheduling under Reentrant Job Shop. *Math. Probl. Eng.* **2017**, *2017*, 1–9. [CrossRef]
4. Kopanos, G.M.; Puigjaner, L. Integrated operational and maintenance planning of production and utlity systems. In *Solving Large-Scale Production Scheduling and Planning in The Process Industries*; Springer: London, UK, 2019.
5. Vieira, G.E.; Herrmann, J.W.; Lin, E. Rescheduling Manufacturing Systems: A Framework of Strategies, Policies, and Methods. *J. Sched.* **2003**, *6*, 39–62. [CrossRef]
6. Chu, Y.; You, F. Integration of scheduling and control with online closed-loop implementation: Fast computational strategy and large-scale global optimization algorithm. *Comput. Chem. Eng.* **2012**, *47*, 248–268. [CrossRef]
7. Suwa, H.; Sandoh, H. *Online Scheduling in Manufacturing*; Springer: London, UK, 2013.
8. Kopanos, G.M., Capón-García, E.; Espuña, A.; Puigjaner, L. Costs for Rescheduling Actions: A Critical Issue for Reducing the Gap between Scheduling Theory and Practice. *Ind. Eng. Chem. Res.* **2008**, *47*, 8785–8795. [CrossRef]
9. Méndez, C.A.; Cerdá, J.; Grossmann, I.E.; Harjunkoski, I.; Fahl, M. State-of-the-art review of optimization methods for short-term scheduling of batch processes. *Comput. Chem. Eng.* **2006**, *30*, 913–946. [CrossRef]
10. Gupta, D.; Maravelias, C.T. On deterministic online scheduling: Major considerations, paradoxes and remedies. *Comput. Chem. Eng.* **2016**, *94*, 312–330. [CrossRef]
11. Gupta, D.; Maravelias, C.T.; Wassick, J.M. From rescheduling to online scheduling. *Chem. Eng. Res. Des.* **2016**, *116*, 83–97. [CrossRef]
12. Panchal, C.B.; Huangfu, E.-P. Effects of Mitigating Fouling on the Energy Efficiency of Crude-Oil Distillation. *Heat Transf. Eng.* **2000**, *21*, 3–9. [CrossRef]
13. Diaby, A.L.; Luong, L.; Yousef, A.; Addai-Mensah, J. A Review of Optimal Scheduling Cleaning of Refinery Crude Preheat Trains Subject to Fouling and Ageing. *Appl. Mech. Mater.* **2012**, *148*, 643–651. [CrossRef]
14. Nategh, M.; Malayeri, M.R.; Mahdiyar, H. A Review on Crude Oil Fouling and Mitigation Methods in Pre-Heat Trains of Iranian Oil Refineries. *J. Oil Gas Petrochem. Technol.* **2017**, *4*, 1–17.
15. Coletti, F.; Hewitt, G.F. *Crude Oil Fouling. Deposit Characterization, Measurements, and Modelling*, 1st ed.; Gulf Professional Publishing: London, UK, 2015.
16. Santamaria, F.L.; Macchietto, S. Integration of Optimal Cleaning Scheduling and Control of Heat Exchanger Networks Undergoing Fouling: Model and Formulation. *Ind. Eng. Chem. Res.* **2018**, *57*, 12842–12860. [CrossRef]
17. Tian, J.; Wang, Y.; Feng, X. Simultaneous optimization of flow velocity and cleaning schedule for mitigating fouling in refinery heat exchanger networks. *Energy* **2016**, *109*, 1118–1129. [CrossRef]
18. Cott, B.; Macchietto, S. Minimizing the effects of batch process variability using online schedule modification. *Comput. Chem. Eng.* **1989**, *13*, 105–113. [CrossRef]
19. Méndez, C.A.; Cerda, J. Dynamic scheduling in multiproduct batch plants. *Comput. Chem. Eng.* **2003**, *27*, 1247–1259. [CrossRef]
20. Novas, J.M.; Henning, G.P. Reactive scheduling framework based on domain knowledge and constraint programming. *Comput. Chem. Eng.* **2010**, *34*, 2129–2148. [CrossRef]
21. Vin, J.P.; Ierapetritou, M.G. A New Approach for Efficient Rescheduling of Multiproduct Batch Plants. *Ind. Eng. Chem. Res.* **2000**, *39*, 4228–4238. [CrossRef]

22. Ave, G.D.; Alici, M.; Harjunkoski, I.; Engell, S. An Explicit Online Resource-Task Network Scheduling Formulation to Avoid Scheduling Nervousness. *Comput. Aided Chem. Eng.* **2019**, *46*, 61–66.
23. Macchietto, S. Integrated Batch Processing: A model for advanced manufacturing. In Proceedings of the APACT Conference Advances in Process Analytics and Control Technology, Birmingham, UK, 20–22 April 2005.
24. Macchietto, S. Engineering success: What does it take to get PSE technologies used? In *Computer Aided Chemical Engineering*; Kiss, A.A., Zondervan, E., Lakerveld, R., Özkan, L., Eds.; Elsevier: Amsterdam, The Netherlands, 2019; Volume 46, Part A, pp. 85–90.
25. Rodrigues, M.T.M.; Gimeno, L.; Passos, C.A.S.; Campos, M.D. Reactive scheduling approach for multipurpose chemical batch plants. *Comput. Chem. Eng.* **1996**, *20*, S1215–S1220. [CrossRef]
26. Ferrer-Nadal, S.; Méndez, C.A.; Graells, M.; Puigjaner, L. Optimal Reactive Scheduling of Manufacturing Plants with Flexible Batch Recipes. *Ind. Eng. Chem. Res.* **2007**, *46*, 6273–6283. [CrossRef]
27. Subramanian, K.; Rawlings, J.B.; Maravelias, C.T. Economic model predictive control for inventory management in supply chains. *Comput. Chem. Eng.* **2014**, *64*, 71–80. [CrossRef]
28. Hewitt, G.F.; Shires, G.L.; Bott, T.R. *Process Heat Transfer*; CRC Press: Boca Raton, FL, USA, 1994.
29. Thulukkanam, K. Shell and tube heat exchanger design. In *Heat Exchanger Design Handbook*; CRC Press: Boca Raton, FL, USA, 2013; pp. 237–336.
30. Ebert, W.; Panchal, C.B. *Analysis of Exxon Crude-Oil-Slip Stream Coking Data*; Argonne National Lab: Lemont, IL, USA, 1995; retrieved from University of North Texas Digital Library.
31. Santamaría, F.L.; Macchietto, S. Model validation for the optimization of refinery preheat trains under fouling. In Proceedings of the Heat Exchanger Fouling and Cleaning XIII, Warsaw, Poland, 2–7 June 2019; Zettler, H.U., Ed.; Available online: http://www.heatexchanger-fouling.com/proceedings19.htm (accessed on 1 January 2020).
32. Santamaria, F.L.; Macchietto, S. Integration of optimal cleaning scheduling and control of heat exchanger networks under fouling: MPCC solution. *Comput. Chem. Eng.* **2019**, *126*, 128–146. [CrossRef]
33. Sridharan, S.V.; Berry, W.L.; Udayabhanu, V. Measuring Master Production Schedule Stability Under Rolling Planning Horizons. *Decis. Sci.* **1988**, *19*, 147–166. [CrossRef]
34. Pujawan, I.N. Schedule nervousness in a manufacturing system: A case study. *Prod. Plan. Control* **2004**, *15*, 515–524. [CrossRef]
35. Van Donselaar, K.; van den Nieuwenhof, J.; Visschers, J. The impact of material coordination concepts on planning stability in supply chains. *Int. J. Prod. Econ.* **2000**, *68*, 169–176. [CrossRef]
36. Kadipasaoglu, S.N.; Sridharan, S.V. Measurement of instability in multi-level MRP systems. *Int. J. Prod. Res.* **1997**, *35*, 713–737. [CrossRef]
37. Maciejowski, J.M. *Predictive Control with Constraints*; Prentice Hall: Upper Saddle River, NJ, USA, 2002.
38. Risbeck, M.J.; Maravelias, C.T.; Rawlings, J.B. Unification of closed-loop scheduling and control: State-space formulations, terminal constraints, and nominal theoretical properties. *Comput. Chem. Eng.* **2019**, *129*, 106496. [CrossRef]
39. Blackburn, J.D.; Kropp, D.H.; Millen, R.A. A Comparison of Strategies to Dampen Nervousness in MRP Systems. *Manag. Sci.* **1986**, *32*, 413–429. [CrossRef]
40. Jacobs, F.; Berry, W.; Whybark, D.; Vollmann, T. Advanced MRP. In *Manufacturing Planning and Control for Supply Chain Management*; McGraw-Hill: New York, NY, USA, 2005.
41. Coletti, F.; Macchietto, S. Refinery Pre-Heat Train Network Simulation Undergoing Fouling: Assessment of Energy Efficiency and Carbon Emissions. *Heat Transf. Eng.* **2011**, *32*, 228–236. [CrossRef]
42. Macchietto, S.; Macchietto, S. A Dynamic, Distributed Model of Shell-and-Tube Heat Exchangers Undergoing Crude Oil Fouling. *Ind. Eng. Chem. Res.* **2011**, *50*, 4515–4533. [CrossRef]
43. Farrell, M.J. The Measurement of Productive Efficiency. *J. R. Stat. Soc. Ser. A* **1957**, *120*, 253. [CrossRef]
44. Charnes, A.; Cooper, W.W.; Rhodes, E. Measuring the efficiency of decision making units. *Eur. J. Oper. Res.* **1978**, *2*, 429–444. [CrossRef]
45. Coletti, F. *Multi-Scale Modelling of Refinery Pre-Heat Trains Undergoing Fouling for Improved Energy Efficiency*; Imperial College of London: London, UK, 2010.

46. Lanchas-Fuentes, L.; Diaz-Bejarano, E.; Coletti, F.; Macchietto, S. Management of cleaning types and schedules in refinery heat exchangers. In Proceedings of the 12th International Conference on Heat Transfer, Fluid Mechanics and Thermodynamics HEFAT2016, Costa de Sol, Spain, 11–13 July 2016.

processes

MDPI

Article

An Agricultural Products Supply Chain Management to Optimize Resources and Carbon Emission Considering Variable Production Rate: Case of Nonperishable Corps

Mohammed Alkahtani [1,*], Muhammad Omair [2], Qazi Salman Khalid [2], Ghulam Hussain [3] and Biswajit Sarkar [4,*]

1 Department of Industrial Engineering, College of Engineering, King Saud University, P.O. Box 800, Riyadh 11421, Saudi Arabia

2 Department of Industrial Engineering, Jalozai Campus, University of Engineering and Technology, Peshawar 25000, Pakistan; muhamad.omair87@gmail.com (M.O.); qazisalman@uetpeshawar.edu.pk (Q.S.K.)

3 Faculty of mechanical engineering, GIK Institute of Engineering Sciences & Technology, Topi 23460, Pakistan; gh_ghumman@hotmail.com

4 Department of Industrial Engineering, Yonsei University, 50 Yonsei-ro, Sinchon-dong, Seodaemun-gu, Seoul 03722, Korea

* Correspondence: moalkahtani@ksu.edu.sa (M.A.); bsarkar@yonsei.ac.kr or bsbiswajitsarkar@gmail.com (B.S.)

Received: 18 September 2020; Accepted: 11 November 2020; Published: 20 November 2020

Abstract: The management of the man–machine interaction is essential to achieve a competitive advantage among production firms and is more highlighted in the processing of agricultural products. The agricultural industry is underdeveloped and requires a transformation in technology. Advances in processing agricultural products (agri-product) are essential to achieve a smart production rate with good quality and to control waste. This research deals with modelling of a controllable production rate by a combination of the workforce and machines to minimize the total cost of production. The optimization of the carbon emission variable and management of the imperfection in processing makes the model eco-efficient. The perishability factor in the model is ignored due to the selection of a single sugar processing firm in the supply chain with a single vendor for the pragmatic application of the proposed research. A non-linear production model is developed to provide an economic benefit to the firms in terms of the minimum total cost with variable cycle time, workforce, machines, and plant production rate. A numerical experiment is performed by utilizing the data set of the agri-processing firm. A derivative free approach, i.e., algebraic approach, is utilized to find the best solution. The sensitivity analysis is performed to support the managers for the development of agricultural product supply chain management (Agri-SCM).

Keywords: agri-supply chain management; variable production rate; optimal resources; imperfect production; eco-efficient production

1. Introduction

Owing to the escalating awareness of resource depletion, climate change, and increasing population, firms in the agriculture domain need to redesign their current supply chain models by taking economic and environmental impacts into account [1]. The life cycle of products held in inventory and processing produce a major concern of perishability among agri-products throughout the supply chain. Therefore, replenishment strategies, product supply, and processing indicators are crucial to consider in the research models. The global market for perishable goods, such as refrigerated products and prepared meals, is growing due to changing lifestyles and overall decreasing tariffs. Owing to their common fragility and limited lifetime, handling those goods is far more complex and includes much higher risks compared to non-perishable products [2]. However, this work deals with the sugar processing from sugarcane as a raw material in local industry with outsourcing operation as a non-perishable product because of the long life of raw sugarcane. Also, the supply chain considers a small portion of the whole network, i.e., a single sugar processing firm with a single outsourcing vendor.

It is evident that the incidence of the uncertain factors is unforeseeable, which may induce a number of decision-making mistakes through the application of a traditional supply chain, thus incurring a high cost and unclean production environment [3]. Moreover, variable demand alters the former assumptions in which demand follows a discrete known distribution for different agri-products [4]. To cover up the deficit caused by the positive and negative surges in agri-product demand, an intelligent variable production model should be integrated within the supply chain. Moreover, controllable production will result in cleaner production as it will optimize the resources without losses and wastes. Because it is estimated that, by 2050, the overall production of food should increase by approx. 70% in order to feed the increasing global population [5]. Hence, the best utilization of resources in a supply chain is a key factor for cleaner agri-production.

Production technology has played a vital role in the upgradation agriculture supply chain and has been a limelight for governments and agri-business sectors. In the basic production model, the assumption of constant production rate was observed predominantly. Later on, the variable machine production rate was also included by considering optimal production costs in manufacturing systems. Although variable production rate remained the point of interest for researchers many years ago through its effects on machine tool cost with increasing production rate [6,7]. However, Moutaz Khouja (1995) [8] was among the pioneers to extend the basic production model and consider production rate as a decision variable for the volume flexibility of production. The model suggested the volume-flexibility of manufacturing systems for larger sized lots with a lesser production rate. Moreover, an increased production rate decreases the repeatability [9] of a robot and affects the quality, as discussed by [10].

This research contributes to transforming the idea of an intelligent, green supply chain production into a mathematical model. The aim of the model is to represent tangible analysis of human-machine interaction and imperfect production system in order to optimize use of resources and minimize wastes. The inclusion of carbon emission cost as an eco-efficient attribute along with the variable production rate, satisfying agribusiness firms' demand, is also a limelight of this work, which is hardly observed in previous literature. Further, the impact of this work is extended to quantify production loss due to improper human-machine correspondence. The investigation provides a plan of action for agri-product manufacturing managers to invest in favor of optimized production with effective resource utilization, which ultimately leads to less rejection in a production environment.

The article is structured in a well possible way i.e., background and challenges to the agricultural supply chain management (agri-SCM) are discussed in this section. In Section 2, the literature is well represented from author contributions, which are presented in consideration of the research gap. Section 3 covers the detailed mathematical formulation of controllable production rate, labor-machine interaction, inventory management, and eco-friendly agri-SCM. The solution method, i.e., algebraic approach application, is also given in Section 3. Afterwards, Section 4 deals with the numerical experiment, which consists of the required data for performing the experiment using the proposed SCM model. The numerical results are also explained and illustrated significantly in Section 4 along the sensitivity analysis of the SCM model to mathematically check the significant cost parameters with respect to the total cost are also performed. In Section 5, conclusion of the research study is discussed.

2. Research Reviews

In a cleaner production environment, prime attention is given to the reduction of production and associated costs. Fluctuation in agri-products' raw materials, fuel prices and falling sale rates drive the agribusiness firms to incorporate technologies and processes which controls expenses. In order to fulfill the requirements of the future generations, the agri-product supply chain should eliminate existing wastefulness and lay emphasis over green defective free production. Such resource waste elimination requires decision assisted tools that covers intrinsic characteristics of an agri-based product. Furthermore, a green agri-product supply chain needs more than only economic validation objective (profit), thus, it should be also able to handle eco-efficient objective. Hence, the decision assisted tool requires the evaluation of both the economic and environmental aspects simultaneously. For this purpose, mathematical optimization is fairly viable to discover best values from the domain and set a better trade-off for managerial insights [11], and for establishing eco-efficient results-based system [12]. The detailed author contributions are given in Table 1.

In the agricultural supply chain, most of the work is found in the logistic of food supply [13], food safety [14], and imperfect information system. As there is a desirable need to encounter the requirements of lean manufacturing, supervision of scraps and reworks due to the significant concerns for production systems [15–17]. In this aspect, Agri-SCM should be integrated with imperfect and green production. From the perspective of solution methodology, Minjung Kwak (2015) [18] recommended a mixed-integer linear-programming (MILP) model that optimizes the re-manufacturing plan in order to validate both the environmental and economic benefits of products. Some researchers have taken carbon footprint into account for development of cleaner production SCM models. For instance, Xiao et al. (2016) [19] optimized SCM cost via minimizing carbon footprint of both the retailers and manufacturers. In this domain, Chia-Chin Wu and Ni-Bin Chang (2004) [20] presented a grey theory model for uncertain conditions which reflects environmental impact by taking the production planning tax into account. Wang et al. (2011) [21] established a bi-objective model that evaluates the associated costs with environmental plans along a green SCM. Additionally, [22] suggested an electricity monitoring-system that considers a multi-objective linear-programming model taking carbon footprint into account by electricity usage as a supply of energy. As referred earlier, (Banasik (2019) [1]) that work included model development of an uncertain eco-efficient supply chain, however, their model lacks integration of imperfect production.

The effect of the workers' cost on production and inventory is a significant aspect to cleaner production and can be analyzed in numerous models. Most of the researchers, i.e., [23–25] studied an imperfect production environment to assist the managers in dealing with poor quality products. Though, very few studies have analyzed the cause to reduce an imperfect production in the setup. A number of factors that affect the production flow and cause imperfection include reworks, rejections, and scraps etc. The management and planning of imperfect production in the model provide a cleaner production int the system and wastes are managed with modeling of the imperfect production [26]. In another study, Sarkar et al. (2018) [27] developed a global sustainable supply chain model with constant production rate having short-term production period in which synchronize mechanism is used to set the cycle time for each production stage. Tiwari et al. (2018) [28] presented a green production quantity model with random imperfect quality products, service level constraints, and failure in reworking. These are depending on the combined efficiency of the machines and workers. Further, the role of controllable production is effective in dealing with imperfect production.

Moutaz Khouja and Abraham Mehrez (1994) [29] proved the deterioration in quality of product with increase in production rates in an economic production inventory model. He reassessed Rosenblatt [30] work with an assumption of quality function. In another study, Khouja et al. [31] further extended his earlier model by assuming that the production rate had a probability to shift production system from in control to the out-of-control state. Later on, Somkiat Eiamkanchanalai and Avijit Banerjee (1999) [32] advanced the work and established model that determines both the optimal production cycle length and variable production rate for a single item. Giri et al. (2005) [33] introduced a flexible production rate EPQ model that addressed the issue of higher stress level of the human-machine interaction with the increase of production rate. In this EPQ model, the unit production cost was stated as a function of the production rate, under general failure and overhaul time. Moreover, [34] presented an EPQ model where the production cycle consisted of multiple runs at various production rates. The author revealed that the production rates should take values between demand rate and production rate that reduces the production cost. Later on, Shib Sankar Sana (2010) [35] studied unit production cost as a function of product reliability and variable production rate in imperfect production system. Giri et al.'s (2005) [33] model was later extended with stochastic demand by [36], sampling in inspection by [37], and stochastic repair time by [38]. Also, Zanoni (2014) [39] examined the case of energy consumption in two stage production system where production depends on the variable production rate.

Table 1. Author Contribution.

Author(s)	Two-Echelon SCM	Imperfection		Production Rate		Resources Optimization			Eco-Efficient	Agri-SCM	Methodology
		Scrap	Rework	Constant	Variable	Workforce	Machines	Combined			
Bansik et al. [1]	√				√				√	√	Bi-stage stochastic programming
Yasmine et al. [13]	√			√						√	Analytical method
Francesco Zecca [14]	√									√	Analytical method
Pablo Biswas [15]		√	√	√			√				Analytical optimization
Tayyab [16]	√	√		√							Analytical optimization
Wang et al. [21]		√	√	√							Normalized constraint method
S Sarkar [26]	√		√		√		√				Algebraic Approach
Moutaz Khouja [29]					√		√				Analytical method
Martin Linde-Rahr [40]						√				√	Game theory
Sarkar [41]			√	√							Analytical method
Shib Sankar Sana [42]	√		√		√	√	√				Analytical method
Sarkar [43]	√	√	√	√			√				Algebraic approach
Xueli Ma et al. [44]	√			√						√	Algebraic Approach
Proposed research	√	√	√	√	√	√	√	√	√	√	Algebraic approach (SQP)

Production inventory outsourcing policy was studied by [45] for a firm with Markovian in-house production capacity that faced independent stochastic demand operating with outsourcing operation. Also, Pablo Biswas and Bhaba R Sarker (2008) [15] proposed a manufacturing process whereby finished goods are produced along with a proportion of undesirable defective products and scrap. As the system is not always perfect, some scrap is produced during the manufacturing and/or rework processes. According to Wang et al. (2013) [46], when the outsourcing quantity and wholesale price are decision variables, the competitive contract manufacturer sets a wholesale price sufficiently low to allow both parties to coexist in the market, and the original equipment manufacturer outsources its entire production to contract manufacturer. Bettayeb et al. (2014) [47] presented a risk-based approach for quality control of complex discrete manufacturing processes to prevent massive scraps. The advancement targeted from this work is the proposal of a model, aiming at the quality control allocation of the products and an understandable algorithm to prevent the production of excessive amount of scrap.

This research deals with the agri-product supply chain management (Agri-SCM). Abundant work on variable production models on realistic case scenarios exists. Particularly, most of the research on eco-efficient (reducing carbon emission) supply chain assumes deterministic demands and constant production rate, and hardly flexible productivity is taken into account. Further, technology development urges for intelligent models in which man-machine interactions are optimized in order to attain minimal wastes. Such intelligent models can hardly be seen in the literature. Additionally, to assert an efficient green production through the supply chain, imperfect production plays a vital role to cut down the cost and reduce the consumption of extra resources. Moving forward, it is worth mentioning that no traces related to agri-products supply chain with intelligent eco-efficient model is found. This work contributes to the latest literature by: (1) providing a centralized, two-echelon supply chain model with variable production rates, (2) presenting an intelligent model in which human–machine interactions are optimized, (3) carbon emission cost and imperfect production are integrated with the proposed model to assure a cleaner production environment, and (4) Agri-SCM with deteriorated products is introduced.

3. Method and Materials

3.1. Research Modeling

This human–machine interface is more significant in the supply chain management of agricultural products. The research contributes to transform the theoretical idea into a mathematical model with an aim to represent the tangible analysis of the imperfect production system in supply chain management. The model is based on Agri-SCM for deteriorated agri-product by considering controllable production rate from the interaction of the human and workers. The flow diagram of the Agri-SCM is illustrated in Figure 1. This research deals with the two-echelon Agri-SCM and covers the agri-food processing firm, where the vendor is involved into few operations because of capacity limitation. The first stage includes the basic food operations, second stage is dedicated to the vendor operations, and the third stage is the finishing stage. The raw material in manufacturing firms is first processed through the basic cleaning operations. Then, the semi-finished parts are outsourced to the vendor firm. The inspection operations are carried out by the vendor, where the parts are sorted as good and defective. In order to compensate for the rejection to meet the required demand, the same quantity of rejections is ordered to manufacture from the first stage of the processing firm. The good agri-products are further delivered to the final stage for further processing and packaging.

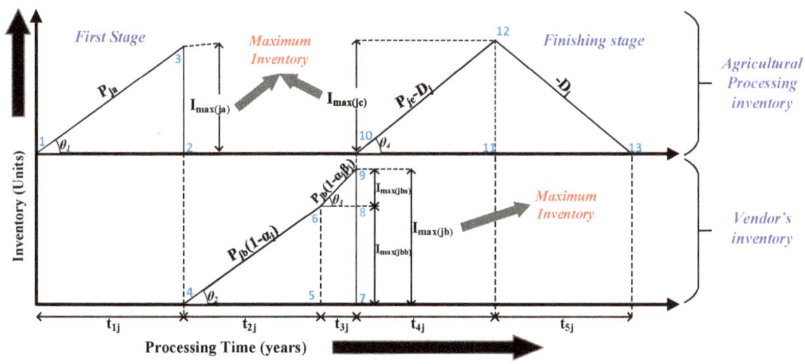

Figure 1. The inventory diagram of the agricultural supply chain management (Agri-SCM).

3.1.1. Assumptions

The following assumptions were used for the proposed model.

1. The mathematical model is based on multiple types of agri-product. The constant and variable production rate with given demand is considered in respectively to avoid shortages [17].
2. The unit production cost is taken as [see for reference: [8] $f(p) = C + \frac{g}{p} + bP$. Here, C is the unit cost of raw material, g represents the per unit cost component that is reduced as the production rate increases, and bP is the unit cost component that increases in the production rate (e.g., tools costs) [48,49].
3. The agri-product processing firm outsources few operations due to limited resources. The imperfect products are produced, for which reworking is done and inspection cost is incurred. The rejected products are disposed and recycled.
4. The model is applicable for non-perishable crop because it considers only a sugar processing firm with outsourcing operation.
5. The processing firm consisting the combination of labor and equipment/machines to process the agri-product.
6. Management of imperfection and carbon emission is considered to make the Agri-SCM a cleaner and eco-effective model.

3.1.2. Notation

The decision variables and the parameters used in the proposed mathematical modeling are denoted by the notations enlisted comprehensively in Appendix A.

3.2. Model Formulation

The inventory diagram of the imperfect agri-food processing firm with vendor/supplier is shown as in Figure 1, where the cycle time of production is given on the x-axis and inventory is given on the y-axis. The upper portion is showing the inventory of the agri-product processing firm, while the lower portion is associated with the vendor inventory. The objective of the research is to minimize the total cost of Agri-SCM, and the formulation of the cycle time of the processing firm is prerequisite to calculate the total cost (TC). Cycle time is taken as a decision variable in the production model, which is dependent upon the production rates of processing firm. The production rate of the processing firm, i.e., P_{ja} and P_{jc} for first and final stage is relying on the production rate of the machines (ε_{ja} and ε_{jc}). In order to meet the customer demand and no shortages in the processing firm, the production rates are considered as a variable (i.e., ε^*_{ja} and ε^*_{jc}) to take an advantage of flexible production, where, $\varepsilon^*_{ja} \in [\varepsilon^*_{ja-min}, \varepsilon^*_{ja-max}]$

and $\varepsilon^*_{ja} \in [\varepsilon^*_{jc-min}, \varepsilon^*_{jc-max}]$. The total cost of production is the sum of Agri-processing firm and vendor cost as given in Equation (1) and their formulations are further represented as following.

$$\text{Total cost of production} = \text{Agri-processing cost} + \text{Vendor cost} \tag{1}$$

3.2.1. Agri-Processing Cost

The total cost associated with the manufacturer include the cost related to the first stage and final stage of the processing firm, where the basic cost includes setup, production, labor, holding, carbon emission, and stress, as expressed in Equation (2).

$$\begin{aligned}\text{Total cost of agri-processing} = \text{Setup cost} + \text{processing cost} + \text{Labor cost}\\ + \text{Holding cost} + \text{Carbon emissions cost}\end{aligned} \tag{2}$$

The breakup of the agri-processing cost is necessary to understand each cost clearly. That is the reason all the costs are described mathematically and theoretically in Equations (3)–(10) as follows.

3.2.2. Setup Cost

The incurred cost is subject to the initial cost required to operate the production setup. Generally, it may contain setup cost, substitution, and tool setting cost. Further, this is fixed nature cost, independent of quantities, though time dependent. The expression is given in Equation (3).

$$SC = \sum_{j=1}^{J} A_j \tag{3}$$

3.2.3. Production Cost

Explicitly, cost incurred in the processing, which is largely dependent on quantity and variable in nature. The cost is accumulation of all costs utilized on resources that are required to manufacture a product. It contains the processing and machining cost, utilities needed for machines, and labor cost. Further, it is linked with the manufacturing processes of initial stage manufacturing. and are the production rates of the initial stage processing plant and machines respectively. Lastly, the production cost is the summation of all the costs linked with the raw materials, production, and tool-die cost [50]. The expression is given as in Equation (4).

$$MC = \sum_{j=1}^{J} (C_{rm} + TD_{maj}P_{ja} + \frac{g_{maj}}{P_{ja}})P_{ja}t_{1j} + (TD_{mcj}P_{jc} + \frac{g_{mcj}}{P_{jc}})Q_{jc} \tag{4}$$

3.2.4. Holding Cost

Holding cost is the basic cost need to understand the basic production model, which is variable and dependent upon the variable inventory at any instant of time. This is the carrying cost of holding inventory. Holding cost depends on production of semi-finished and finished goods used as stock. It is incurred on production and crashing quantities in the proposed model and includes costs such as wedges, warehouse rent, and insurance. It also depends on the holding time of the product in inventory. The holding cost will be applied on the inventory supported by manufacturing firm, outsourcing operations. The average inventory is calculated as the ratio of the sum of inventories in the form of area under the curve to the cycle time of the production, which is expressed as in Equation (7). The cycle time and inventory levels of the processing firm are calculated by a step-by-step procedure gives in Appendix B.

$$TotalInventory = Area_{123} + Area_{10,11,12} + Area_{11,12,13} \tag{5}$$

$$= \frac{Q_j^2}{2P_{ja}} + \frac{(Q_j - u)^2}{2P_{jc}}(1 - \frac{D_j}{P_{jc}}) + \frac{(Q_j + u)^2}{2D_j}(1 - \frac{D_j}{P_{jc}})^2 \quad (6)$$

$$HC_{mj} = \sum_{j=1}^{J} h_{mj}[\frac{Q_j^2}{2P_{ja}} + \frac{(Q_j - u)^2}{2P_{jc}}(1 - \frac{D_j}{P_{jc}}) + \frac{(Q_j + u)^2}{2D_j}(1 - \frac{D_j}{P_{jc}})^2]. \quad (7)$$

3.2.5. Carbon Emission Costs

The model optimized the carbon emission cost as a function of production rate to develop an eco-friendly and eco-effective agri-product SCM. The carbon emission in the agri-processing firm depends upon the source of energy consumed by the equipment/machines. The cost of carbon is incurred by the state government to support the global warming issue. The cost of carbon emission is generated during the life cycle production, which is given as in Equation (9), where A is the emissions function parameter (ton.year/unit), B is the emissions function parameter (ton year/unit), and C is the emissions function parameter (ton/unit) [51].

$$Carbon\ emission\ cost(CEC_j) = Costofcarbonemissioninmanufacturing(CEM) \quad (8)$$

$$= \sum_{j=1}^{J} \gamma_2[(AP_{ja}^2 - BP_{ja} + C)P_{ja}t_{1j} + (AP_{jc}^2 - BP_{jc} + C)Q_{jc}] \quad (9)$$

3.2.6. Labor Cost

The cost is associated with the utilization of the workforce in the agri-SCM. The wages paid to the workers on the basis of the level of the skilled. Here, the cost is incurred to reflect the importance of the unskilled workers to understand the importance of the human factor in the processing firm. The labor cost is calculated on the basis of the machines required in the agri-product processing firm, which is expressed in Equation (10).

$$Laborcost(LC) = \sum_{j=1}^{J} L_j W_j \quad (10)$$

The number of machines required and the amount of labor required in the processing firm are given in Equations (11) and (13), where l_{aj}, and l_{cj} are the labor rate or the number of labors working on each machine.

$$K_j = K_{ja} + K_{jc} \quad (11)$$

$$Numberoflabors = Laborrate \times Numberofmachines \times Productiontime \quad (12)$$

$$L_j = \frac{l_{aj}}{\rho}K_{ja} + \frac{l_{cj}}{\rho}K_{jc} \quad (13)$$

3.2.7. Stress and Workers' Efficiency

The efficiency of the processing firm depends upon the efficiency of the workers to fulfill the customer demand before deadline. The labor-machine coordination in the processing firm is very important for effective and efficient processing firm. The relationship between the average worker's stress and efficiency is expressed as given in Equation (15), where m is scale factor, ρ_0 is the workers' efficiency affected by stress, and ρ_1 is the efficiency due to the effect of another factor. Therefore, the total efficiency of the worker will be ρ [52,53].

$$\rho_0 = e^{-s/m}, \quad (14)$$

$$\rho = \rho_1 \rho_0. \tag{15}$$

3.2.8. Stress Level and Defective Rate

The relationship is developed between the defective rate and averages stress among worker, where the defective rate is considered as a function of stress having a significant impact on the processing firm. The increasing rate of average stress among workers increases the defective rate. That is the reason, the expression for defective is the sum of initial and variable defective rate as expressed in the Equation (16), where variable defective rate is a function of stress among workers, α_0 is the initial defective rate, τ and, ϵ are the scaling factors, and s is average stress among worker [53]. The analysis of workers' stress on the production system has already been analysed by the research work of Omair et al. [53]. Therefore, the formulation has been incorporated in this study. However, the analysis of workplace stress on the production system is not highlighted.

$$\alpha_j = \alpha_0 + \tau(s)^{\epsilon}. \tag{16}$$

Now, the mathematical form of the total cost of the agri-processing firm is expressed as in Equation (17).

$$\begin{aligned} TC_{aj} = \sum_{j=1}^{J} [A_j + (C_{rm} + TD_{maj}P_{ja} + \frac{g_{maj}}{P_{ja}})P_{ja}t_{1j} + (TD_{mcj}P_{jc} + \frac{g_{mcj}}{P_{jc}})Q_{jc} + L_jW_j \\ +h_{mj}[\frac{Q_j^2}{2P_{ja}} + \frac{(Q_j-u)^2}{2P_{jc}}(1-\frac{D_j}{P_{jc}}) + \frac{(Q_j+u)^2}{2D_j}(1-\frac{D_j}{P_{jc}})^2] \\ +\gamma_2[(AP_{ja}^2 - BP_{ja} + C)P_{ja}t_{1j} + (AP_{jc}^2 - BP_{jc} + C)Q_{jc}] + sSC_j]. \end{aligned} \tag{17}$$

The costs related to the manufacturer in the first and final stages of production are calculated, and the costs associated with the vendor are given in the next section.

3.3. Vendor Cost

The semi-finished agri-products are delivered to the vendor for processing few operations. The cost of vendor is the sum of the costs associated with the production, holding, inspection, and recycling of the processes is given in Equation (18). These costs are expressed in the Equations (19)–(24).

$$\begin{aligned} \text{Vendor cost} = \text{Production cost} + \text{Holding cost} + \text{Inspection cost} + \text{Reworking cost} \\ + \text{Scrap recycle cost} + \text{Buffer cost} \end{aligned} \tag{18}$$

3.3.1. Production Cost of Outsourcing

The expression for production cost is utilized from the research work done by [50] except the cost of raw material, because of receiving the semi-finished products from the manufacturer. The expression is given in Equation (19).

$$MCO = \sum_{j=1}^{J} [(TD_{oj}P_{jb}(1-\alpha_j) + \frac{g_{oj}}{P_{jb}(1-\alpha_j)})P_{jb}(1-\alpha_j)t_{2j}]. \tag{19}$$

3.3.2. Holding Cost of Vendor

The holding cost of outsourcing operations is obtained by the sum of inventories calculated from Appendix B. The expression for calculating the total cost of production is expressed in Equation (22).

$$Total\ Inventory = Area456 + Area5678 + Area689 \tag{20}$$

$$= \frac{Q_j^2(1-\alpha_j)}{2P_{jb}} + \frac{\alpha_j(1-\alpha_j)Q_j^2}{P_{jb}} + \frac{\alpha_j^2 Q_j^2(1-\alpha_j\beta_j)}{2P_{jb}}, \tag{21}$$

$$HC = \sum_{j=1}^{J}[h_{oj}[\frac{Q_j^2(1-\alpha_j)}{2P_{jb}} + \frac{\alpha_j(1-\alpha_j)Q_j^2}{P_{jb}} + \frac{\alpha_j^2 Q_j^2(1-\alpha_j\beta_j)}{2P_{jb}}]]. \tag{22}$$

3.3.3. Inspection Cost

Inspection of the agri-product is carried out by the vendor, where the products are checked according to the quality control dimensions. The parts are categorized into good and rejected. The total inspection cost of the production is the sum of the fixed and variable inspection cost in the processing, as expressed in Equation (23).

$$IC_j = \sum_{j=1}^{J}[\theta_j + \psi_{ja}P_{ja}t_{1j} + \psi_{jb}P_{jb}t_{2j}] \tag{23}$$

3.3.4. Recycling Cost/Disposal Cost

Recycling cost is considered in the agri-processing firm. It is not concerned with the recycling of the agri-product after deteriorated but it is the cost incurred on rejected/defective or imperfect agri-product. The customer/user is not the part of the proposed agricultural supply chain management (Agri-SCM) that is the reason, the perishability factor is not considered. Here, recycling cost is the cost incurred on disposing the imperfect/defective not due to deterioration or perishability factor. These products are bio-waste and further utilized into other byproducts, i.e., fertilizers, bio-fuel, feeds, etc., in the processing expressed as in the equation given below.

$$RC_j = \gamma_1 P_{jb}t_{2j}\alpha_j\beta_j$$

The mathematical expression to sum all the costs equations is represented as given in Equation (24).

$$\begin{aligned} TC_{vj} = \sum_{j=1}^{J} [(TD_{oj}P_{jb}(1-\alpha_j) + \tfrac{g_{oj}}{P_{jb}(1-\alpha_j)})P_{jb}(1-\alpha_j)t_{2j} + \tfrac{Q_j^2(1-\alpha_j)}{2P_{jb}} + \tfrac{\alpha_j(1-\alpha_j)Q_j^2}{P_{jb}} \\ + \tfrac{\alpha_j^2 Q_j^2(1-\alpha_j\beta_j)}{2P_{jb}} + \theta_j + \psi_{ja}P_{ja}t_{1j} + \psi_{jb}P_{jb}t_{2j} + \gamma_1 P_{jb}t_{2j}\alpha_j\beta_j] \end{aligned} \tag{24}$$

The production system of agri-product processing firm is analyzed by the formulation of mathematical model. The mathematically model is based on the cycle time of production. The objective of the proposed model is to minimize the total cost (TC_{sj}) of processing firm. The total cost per cycle is given in Equation (26).

$$Totalcost = Agri - processingcost + MR(Vendorcost) \tag{25}$$

$$\begin{aligned} TC_j = \sum_{j=1}^{J} \frac{1}{T_j}[A_j + (C_{rm} + TD_{maj}P_{ja} + \frac{g_{maj}}{P_{ja}})P_{ja}t_{1j} + (TD_{mcj}P_{jc} + \frac{g_{mcj}}{P_{jc}})Q_{jc} + L_jW_j \\ + h_{mj}[\frac{Q_j^2}{2P_{ja}} + \frac{(D_jT_j)^2}{2P_{jc}}(1-\frac{D_j}{P_{jc}}) + \frac{(Q_j+u)^2}{2D_j}(1-\frac{D_j}{P_{jc}})^2] + \gamma_2[(AP_{ja}^2 - BP_{ja} + C)P_{ja}t_{1j} \\ + (AP_{jc}^2 - BP_{jc} + C)Q_{jc}] + s.SC_j + MR[(TD_{oj}P_{jb}(1-\alpha_j) + \frac{g_{oj}}{P_{jb}(1-\alpha_j)})P_{jb}(1-\alpha_j)t_{2j} \\ + \frac{Q_j^2(1-\alpha_j)}{2P_{jb}} + \frac{\alpha_j(1-\alpha_j)Q_j^2}{P_{jb}} + \frac{\alpha_j^2 Q_j^2(1-\alpha_j\beta_j)}{2P_{jb}} + \theta_j + \psi_{ja}P_{ja}t_{1j} + \psi_{jb}P_{jb}t_{2j} \\ + R_j\alpha_jQ_j(1-\alpha_j\beta_j) + \gamma_1 P_{jb}t_{2j}\alpha_j\beta_j]] \end{aligned} \tag{26}$$

where,

$$L_j = L_{ja} + L_{jc}$$

$$Q_j = \frac{T_j D_j(\Omega)}{1-\alpha_j^2\beta_j}$$

$$t_{1j} = \frac{T_j D_j}{K_{ja}\varepsilon_{ja}(1-\alpha_j^2\beta_j)}$$

$$t_{2j} = \frac{T_j D_j}{P_{jb}(1-\alpha_j^2\beta_j)}$$

$$t_{4j} = \frac{T_j D_j}{K_{jc}\varepsilon_{jc}}$$

$$t_{5j} = (\frac{T_j D_j}{1-\alpha_j^2\beta_j} - u)(\frac{K_{jc}\varepsilon_{jc} - D_j}{D_j K_{jc}\varepsilon_{jc}})$$

The SCM mathematical model is non-linear by minimizing total cost of SCM, where the decision variables are $(T_j, L_{ja}, L_{jc}, K_{ja}, K_{jc}, P_{ja}, \text{and } P_{jc})$.

3.4. Solution Algorithm

The variability in the proposed Agri-SCM model make the model non-linear in nature. The decision variables considered are relying on the decisions of the production planning. Analytically, the proposed model is optimized with the help of improved methodology called algebraic function, which is based on quadratic equation. There are four decision variables, i.e., cycle time (T_j), machines (K_{ja}, K_{jc}), production rate (ε_{ja}, ε_{jc}), and labors (L_{ja}, L_{jc}) to optimize the non-linear imperfect production model. The algebraic method consists of a positive expression type, and can be rewritten as:

$$f(x) = a_1x + a_2/x + a_3 = \frac{a_1}{x}(x^2 + a_2/a_1 - 2x\sqrt{a_2/a_1} + 2x\sqrt{a_2/a_1}) + a_3 \tag{27}$$

$$f(x) = \frac{a_1}{x}(x^2 + a_2/a_1 - 2x\sqrt{a_2/a_1}) + 2a_1\sqrt{a_2/a_1} + a_3 \tag{28}$$

$$f(x) = \frac{a_1}{x}(x - \sqrt{a_2/a_1})^2 + 2\sqrt{a_2a_1} + a_3 \tag{29}$$

Since the quadratic expression is non-negative and a_1 is positive, it is always minimized for $x = \sqrt{a_2/a_1}$, which reaches the minimum at $f(x) = 2\sqrt{a_2/a_1} + a_3$.

In the first step of solution algorithm, by using algebraic function methodology, the form of decision variable T_j can be written as given in Equation (30).

$$\begin{aligned} TC_{sj}(T_j, K_{ja}, K_{jc}, L_{ja}, L_{jc}, \varepsilon_{ja}, \varepsilon_{jc}) = \frac{1}{T_j}[A_j + (C_{rm} + TD_{maj}P_{ja} + \frac{g_{maj}}{P_{ja}})P_{ja}t_{1j} \\ +(TD_{mcj}P_{jc} + \frac{g_{mcj}}{P_{jc}})Q_{jc} + L_jW_j + h_{mj}[\frac{Q_j^2}{2P_{ja}} + \frac{(D_jT_j)^2}{2P_{jc}}(1-\frac{D_j}{P_{jc}}) + \frac{(Q_j+u)^2}{2D_j}(1-\frac{D_j}{P_{jc}})^2] \\ +\gamma_2[(AP_{ja}^2 - BP_{ja} + C)P_{ja}t_{1j} + (AP_{jc}^2 - BP_{jc} + C)Q_{jc}] + s.SC_j + MR[(TD_{oj}P_{jb}(1-\alpha_j) \\ +\frac{g_{oj}}{P_{jb}(1-\alpha_j)})P_{jb}(1-\alpha_j)t_{2j} + \frac{Q_j^2(1-\alpha_j)}{2P_{jb}} + \frac{\alpha_j(1-\alpha_j)Q_j^2}{P_{jb}} + \frac{\alpha_j^2Q_j^2(1-\alpha_j\beta_j)}{2P_{jb}} + \theta_j \\ +\psi_{ja}P_{ja}t_{1j} + \psi_{jb}P_{jb}t_{2j} + R_j\alpha_jQ_j(1-\alpha_j\beta_j) + \gamma_1P_{jb}t_{2j}\alpha_j\beta_j]] \end{aligned} \tag{30}$$

The Equation (30) can be written as in Equation (31)

$$= \frac{1}{T_j}[A_j + s.SC_j + \theta_j + MR \times TD_{oj}P_{jb}(1-\alpha_j)] + T_j[h_{mj}[\frac{D_j^2}{2P_{ja}(1-\alpha_j\beta_j)^2} + \frac{1}{2P_{jc}}(\frac{D_j}{1-\alpha_j^2\beta_j} - \frac{\alpha_j\beta_j D_j}{1-\alpha_j^2\beta_j})^2(1-\frac{D_j}{P_{jc}}) + (\frac{D_j}{1-\alpha_j^2\beta_j} - \frac{\alpha_j\beta_j D_j}{1-\alpha_j^2\beta_j})^2(1-\frac{D_j}{P_{jc}})^2] + MR[(\frac{D_j}{1-\alpha_j^2\beta_j})^2\frac{1-\alpha_j}{2P_{jb}} + \frac{\alpha_j(1-\alpha_j)}{P_{jb}}(\frac{D_j}{1-\alpha_j^2\beta_j})^2 + \frac{\alpha_j^2(1-\alpha_j\beta_j)}{2P_{jb}}(\frac{D_j}{1-\alpha_j^2\beta_j})^2]] \quad (31)$$

Our assumptions are given as in Equations (32) and (33).

$$A_1 = \sum_{j=1}^{J}[A_j + s.SC_j + \theta_j + MR \times TD_{oj}P_{jb}(1-\alpha_j)] \quad (32)$$

$$A_2 = \sum_{j=1}^{J}[h_{mj}[\frac{D_j^2}{2P_{ja}(1-\alpha_j\beta_j)^2} + \frac{1}{2P_{jc}}(\frac{D_j}{1-\alpha_j^2\beta_j} - \frac{\alpha_j\beta_j D_j}{1-\alpha_j^2\beta_j})^2(1-\frac{D_j}{P_{jc}}) + (\frac{D_j}{1-\alpha_j^2\beta_j} - \frac{\alpha_j\beta_j D_j}{1-\alpha_j^2\beta_j})^2(1-\frac{D_j}{P_{jc}})^2] + MR[(\frac{D_j}{1-\alpha_j^2\beta_j})^2\frac{1-\alpha_j}{2P_{jb}} + \frac{\alpha_j(1-\alpha_j)}{P_{jb}}(\frac{D_j}{1-\alpha_j^2\beta_j})^2 + \frac{\alpha_j^2(1-\alpha_j\beta_j)}{2P_{jb}}(\frac{D_j}{1-\alpha_j^2\beta_j})^2]], \quad (33)$$

In the algebraic function approach, the constant term becomes neglected and equal to zero. Therefore, the cost function can be given as in Equation (34).

$$TC_{sj}(T_j^*, K_{ja}, K_{jc}, L_{ja}, L_{jc}, \varepsilon_{ja}, \varepsilon_{jc}) = \frac{A_1}{T_j} + A_2 T_j \quad (34)$$

$$= (\sqrt{\frac{A_1}{T_j}})^2 + (\sqrt{A_2 T_j})^2, \quad (35)$$

$$= (\sqrt{\frac{A_1}{T_j}} - \sqrt{A_2 T_j})^2 + \sqrt{2A_1 A_2}. \quad (36)$$

By the algebraic approach, in Equation (36), having the square term as a maximum value, the square will be zero, i.e.,

$$(\sqrt{\frac{A_1}{T_j}} - \sqrt{A_2 T_j})^2 = 0, \quad (37)$$

$$\sqrt{\frac{A_1}{T_j}} - \sqrt{A_2 T_j} = 0, \quad (38)$$

$$\sqrt{\frac{A_1}{T_j}} = \sqrt{A_2 T_j}, \quad (39)$$

$$T_j^* = \sqrt{\frac{A_1}{A_2}} \quad (40)$$

By putting the value of in Equation (34). The Equation (41) is obtained as.

$$TC_{sj}(T_j^*, K_{ja}, K_{jc}, L_{ja}, L_{jc}, \varepsilon_{ja}, \varepsilon_{jc}) = \sum_{j=1}^{J}\frac{A_1}{T_j^*} + A_2 T_j^*. \quad (41)$$

In the second step, the production rate of each machine in the first stage (ε_{ja}) and final stage (ε_{jc}) of the production system are also calculated by using algebraic approach. First of all, to find optimal (ε_{jc}), the TC_{sj} can be converted into the form given as in the following equation:

$$\begin{aligned} TC_{sj}(T_j^*, K_{ja}, K_{jc}, L_{ja}, L_{jc}, \varepsilon_{ja}, \varepsilon_{jc}) &= \varepsilon_{jc}[TD_{mcj}K_{jc}D_j - BK_{jc}\gamma_2 D_j] \\ &+\frac{1}{\varepsilon_{jc}}[\frac{g_{mcj}D_j}{K_{jc}} + \frac{l_{cj}D_jW_j}{\rho} + \frac{h_{mj}(Q_j-u)^2}{2K_{jc}T_j^*} - \frac{(Q_j+u)^2h_{mj}}{K_{jc}T_j^*}] \\ &+[-\frac{h_{mj}}{T_j^*}\left\{\frac{(Q_j-u)^2}{2K_{jc}}\frac{D_j}{(\varepsilon_{jc})^2} + \frac{(Q_j+u)^2}{2D_j}\frac{D_j}{\rho K_{jc}\varepsilon_{jc}}\right\} + \gamma_2 T_j^* D_T A(K_{jc}\varepsilon_{jc})^2] \\ &+[\gamma_2 D_j\varepsilon_{ja} + \frac{(Q_j+u)^2}{2D_jT_j^*}h_{mj} + \frac{l_{aj}D_jW_j}{\rho\varepsilon_{ja}(1-\alpha_j^2\beta j)}] \end{aligned} \tag{42}$$

$$R_1 = \varepsilon_{jc}[TD_{mcj}K_{jc}D_j - BK_{jc}\gamma_2 D_j] \tag{43}$$

$$R_2 = \frac{1}{\varepsilon_{jc}}[\frac{g_{mcj}D_j}{K_{jc}} + \frac{l_{cj}D_jW_j}{\rho} + \frac{h_{mj}(Q_j-u)^2}{2K_{jc}T_j^*} - \frac{(Q_j+u)^2h_{mj}}{K_{jc}T_j^*}] \tag{44}$$

$$R_3 = [-\frac{h_{mj}}{T_j^*}\left\{\frac{(Q_j-u)^2}{2K_{jc}}\frac{D_j}{(\varepsilon_{jc})^2} + \frac{(Q_j+u)^2}{2D_j}\frac{D_j}{\rho K_{jc}\varepsilon_{jc}}\right\} + \gamma_2 T_j^* D_j A(K_{jc}\varepsilon_{jc})^2] \tag{45}$$

$$R_4 = [\gamma_2 D_j\varepsilon_{ja} + \frac{(Q_j+u)^2}{2D_jT_j^*}h_{mj} + \frac{l_{aj}D_jW_j}{\rho\varepsilon_{ja}(1-\alpha_j^2\beta j)}] \tag{46}$$

$$TC_{sj}(T_j^*, K_{ja}, K_{jc}, L_{ja}, L_{jc}, \varepsilon_{ja}, \varepsilon_{jc}) = R_1\varepsilon_{jc} + \frac{R_2}{\varepsilon_{jc}} + R_3 + R_4 \tag{47}$$

R_3 consists of squared ε_{jc} and also in the denominator, whereas R_4 is constant term. Therefore, both are neglected and considered as zero.

$$TC_{sj}(T_j^*, K_{ja}, K_{jc}, L_{ja}, L_{jc}, \varepsilon_{ja}, \varepsilon_{jc}) = (\sqrt{R_1\varepsilon_{jc}})^2 + (\sqrt{\frac{R_2}{\varepsilon_{jc}}})^2 \tag{48}$$

$$= (\sqrt{R_1\varepsilon_{jc}} - \sqrt{\frac{R_2}{\varepsilon_{jc}}})^2 + 2\sqrt{R_1R_2}\varepsilon_{jc} \tag{49}$$

By the algebraic approach, Equation (49), having the square term as a maximum value, the squared expression will be zero, i.e.,

$$(\sqrt{R_1\varepsilon_{jc}} - \sqrt{\frac{R_2}{\varepsilon_{jc}}}) = 0 \tag{50}$$

$$\varepsilon_{jc}^* = \sqrt{\frac{R_2}{R_1}} \tag{51}$$

In step 2, the optimal ε_{ja} is obtained by using algebraic approach, i.e., the TC_{sj} can be given as in Equation (52).

$$\begin{aligned} TC_{sj}(T_j^*, K_{ja}, K_{jc}, L_{ja}, L_{jc}, \varepsilon_{ja}, \varepsilon_{jc}^*) &= \varepsilon_{ja}[TD_{maj}K_{ja}D_j - \gamma_2 D_j BK_{ja}h_{mj}] \\ &+\frac{1}{\varepsilon_{ja}}[\frac{g_{maj}D_j}{K_{ja}} + \frac{l_{aj}D_jW_j}{\rho(1-(\alpha_j)^2\beta_j)} + \frac{h_{mj}(Q_j)^2}{2T_j^*K_{ja}}] \\ &+[\frac{l_{cj}D_jW_j}{\rho\varepsilon_{jc}^*} + \gamma_2 h_{mj}A(K_{ja}\varepsilon_{ja})^2 D_j + \gamma_2\varepsilon_{jc}^* D_j h_{mj}] \end{aligned} \tag{52}$$

where we can assume that

$$R_5 = \varepsilon_{ja}[TD_{maj}K_{ja}D_j - \gamma_2 D_j B K_{ja} h_{mj}] \tag{53}$$

$$R_6 = \frac{1}{\varepsilon_{ja}}[\frac{g_{maj}D_j}{K_{ja}} + \frac{l_{aj}D_jW_j}{\rho(1-(\alpha_j)^2\beta_j)} + \frac{h_{mj}(Q_j)^2}{2T_j^*K_{ja}}] \tag{54}$$

$$R_7 = [\frac{l_{cj}D_jW_j}{\rho\varepsilon^*_{jc}} + \gamma_2 h_{mj}A(K_{ja}\varepsilon_{ja})^2D_j + \gamma_2\varepsilon^*_{jc}D_jh_{mj}] \tag{55}$$

Therefore, Equation (56) can be written as

$$TC_{sj}(T_j^*, K_{ja}, K_{jc}, L_{ja}, L_{jc}, \varepsilon_{ja}, \varepsilon^*_{jc}) = R_5\varepsilon_{ja} + \frac{R_6}{\varepsilon_{ja}} + R_7 \tag{56}$$

$$= (\sqrt{R_5\varepsilon_{ja}})^2 + (\sqrt{\frac{R_6}{\varepsilon_{ja}}})^2 + R_7 \tag{57}$$

where R_7 is squared ε_{ja}, which can be neglected and considered as zero.

$$TC_{sj}(T_j^*, K_{ja}, K_{jc}, L_{ja}, L_{jc}, \varepsilon_{ja}, \varepsilon^*_{jc}) = (\sqrt{R_5\varepsilon_{ja}} - \sqrt{\frac{R_6}{\varepsilon_{ja}}})^2 + 2\sqrt{R_5R_6} \tag{58}$$

$$= (\sqrt{R_5\varepsilon_{ja}} - \sqrt{\frac{R_6}{\varepsilon_{ja}}}) + 2\sqrt{R_5R_6} \tag{59}$$

The square term has a maximum value, if the squared expression will be zero, i.e.,

$$\varepsilon^*_{ja} = \sqrt{\frac{R_6}{R_5}} \tag{60}$$

In the third step, the other decision variables of the production system i.e., K_{jc}, L_{ja}, and L_{jc}, which are discrete and can be calculated indirectly. Therefore, the total cost from Equation (30) will be given as in Equation (61):

$$\begin{aligned}
TC_{sj}(T_j^*, Kja^*, Kjc^*, Lja^*, Ljc^*, \varepsilon^*_{ja}, \varepsilon^*_{jc}) &= \frac{1}{T_j^*}[A_j + (C_{rm} + TD_{maj}(K^*_{ja}\varepsilon^*_{ja}) \\
&+ \frac{g_{maj}}{(K^*_{ja}\varepsilon^*_{ja})})(K^*_{ja}\varepsilon^*_{ja})(\frac{T_j^*D_j}{K^*_{ja}\varepsilon^*_{ja}(1-\alpha_j^2\beta_j)}) + (TD_{mcj}(K^*_{jc}\varepsilon^*_{jc}) + \frac{g_{mcj}}{(K^*_{jc}\varepsilon^*_{jc})})(K^*_{jc}\varepsilon^*_{jc})(\frac{T_j^*D_j}{K^*_{jc}\varepsilon^*_{jc}}) + L_j^*W_j \\
&+ h_{mj}[\frac{(\frac{T_j^*D_j}{1-\alpha_j^2\beta_j})^2}{(2K^*_{ja}\varepsilon^*_{ja})} + \frac{(D_jT_j^*)^2}{(2K^*_{jc}\varepsilon^*_{jc})}(1-\frac{D_j}{(K^*_{jc}\varepsilon^*_{jc})}) + \frac{(D_jT_j^*)^2}{2D_j}(1-\frac{D_j}{(K^*_{jc}\varepsilon^*_{jc})})^2] \\
&+ \gamma_2[(A(K^*_{ja}\varepsilon^*_{ja})^2 - B(K^*_{ja}\varepsilon^*_{ja}) + C)(K^*_{ja}\varepsilon^*_{ja})(\frac{T_j^*D_j}{K^*_{ja}\varepsilon^*_{ja}(1-\alpha_j^2\beta_j)}) + (A(K^*_{jc}\varepsilon^*_{jc})^2 - B(K^*_{jc}\varepsilon^*_{jc}) \\
&+ C)(K^*_{jc}\varepsilon^*_{jc})(\frac{T_j^*D_j}{K^*_{jc}\varepsilon^*_{jc}})] + s.SC_j + MR[(TD_{oj}P_{jb}(1-\alpha_j) + \frac{g_{oj}}{P_{jb}(1-\alpha_j)})P_{jb}(1-\alpha_j)(\frac{T_j^*D_j}{P_{jb}(1-\alpha_j^2\beta_j)}) \\
&+ \frac{(\frac{T_j^*D_j}{1-\alpha_j^2\beta_j})^2(1-\alpha_j)}{2P_{jb}} + \frac{\alpha_j(1-\alpha_j)(\frac{T_j^*D_j}{1-\alpha_j^2\beta_j})^2}{P_{jb}} + \frac{\alpha_j^2(\frac{T_j^*D_j}{1-\alpha_j^2\beta_j})^2(1-\alpha_j\beta_j)}{2P_{jb}} + \theta_j \\
&+ \psi_{ja}(K^*_{ja}\varepsilon^*_{ja})(\frac{T_j^*D_j}{K^*_{ja}\varepsilon^*_{ja}(1-\alpha_j^2\beta_j)}) + \psi_{jb}P_{jb}(\frac{T_j^*D_j}{P_{jb}(1-\alpha_j^2\beta_j)}) + R_j\alpha_j(\frac{T_j^*D_j}{1-\alpha_j^2\beta_j})(1-\alpha_j\beta_j) \\
&+ \gamma_1 P_{jb}(\frac{T_j^*D_j}{P_{jb}(1-\alpha_j^2\beta_j)})\alpha_j\beta_j]]
\end{aligned} \tag{61}$$

where,

$$L_j = L^*_{ja} + L^*_{jc}, \tag{62}$$

$$L^*_{ja} = \frac{l_{aj} K^*_j}{\rho}, \tag{63}$$

3.5. Numerical Experiment

The pragmatic application of the proposed Agri-SCM n model is performed by considering a sugar processing firm with vendor. The local industry is processing sugarcane as a raw material and converting it into sugar. There are various categories of the sugar obtained from the raw material depending on the quality and grades. In our case, we considered three grades of sugar, i.e., A, B, and C, respectively. The constraints of budget and resources compelled the sugar processing firm to outsource few processes of the sugar to vendors for a successful supply chain management. To avoid shortages, the managers are required to keep the production rate as a controllable to fix as per demand. The production rate of the sugar processing firm is linked with the integrated production rate of man and machine. The vendor operations are influencing the production rate of the system, which is kept constant. Therefore, it is limited to set the production rate of the operations before delivering the products to the vendor, and this should be greater than the rate of outsourcing operation. Furthermore, the rate of production operations performed after the outsourcing operation must be greater than the outsourcing rate. The numerical experiment of the research study is based on argir-SCM including sugar processing firm and vendor. The data utilized to perform the experiment is taken from the local industry of sugar processing SCM. The processing-based data for each agri-product are given in Table 2, which consists of tool-die, production, holding, and production rate, which is taken from the research work of [50]. On the basis of the capacity of the machines inside processing firm, the variable production rate, $[\varepsilon_{aj-min}, \varepsilon_{aj-max}]$ is considered as [(120, 130, 140), (150, 160, 170)] for the machine to process each agri-product at the first stage where $[\varepsilon_{cj-min}, \varepsilon_{cj-max}]$ is considered as [(110, 115, 125), (130, 140, 150)] to process each sugar grade processing in finishing stage of the processing.

Table 2. Agri-processing firm data for processing various sugar grades from sugarcane.

Product Type	Tool-Die Cost 1st Stage ($/Machine)	Tool-Die Cost Finishing ($/Machine)	Fixed Production Cost ($/ton)	Fixed Production Cost ($/ton)	Holding ($/ton/Year)
Sugar (A)	0.012	0.09	650	550	1.1
Sugar (B)	0.012	0.085	660	560	1.21
Sugar (C)	0.013	0.095	665	560	1.25
Product Type	**Setup ($/Year)**	**Raw Material Cost ($/ton)**	**Production Rate (tons/Machine)**	**Labor ($/Labor-Year)**	**Demand (tons/Year)**
Sugar (A)	8	20	150	1000	900
Sugar (B)	8.8	23	160	1000	800
Sugar (C)	9.3	25	170	1000	800

All the data related to the imperfect production are given in Table 3, which cover inspection and recycling. Since the imperfect production is the part of normal production, this model considered few vendor operations. The inspection station is located after outsourcing to check the quality of each agri-product, and sorted the checked parts into good and rejected parts. The inspection cost is categorized as fixed including initial investment and variable cost depending upon the production quantity. The recycling cost includes the operations to recycle the rejected agri-products into other useful product. These costs have a significant impact on the total cost of processing.

Table 3. Vendor data to process various grades of sugar.

Product Type	Tool-Die Cost ($/Machine)	Fixed Production Cost ($/tons)	Holding ($/tons/Year)
Sugar (A)	0.011	670	1.12
Sugar (B)	0.012	670	1.23
Sugar (C)	0.013	675	1.28
Product Type	**Fixed Inspection ($/Year)**	**Variable Inspection ($/Unit)**	**Disposal ($/Unit/Year)**
Sugar (A)	5	0.12	0.83
Sugar (B)	5.5	0.22	0.83
Sugar (C)	6	0.28	0.83

4. Results

4.1. Numerical Results

The objective is to make the production plane, where the number of workstations, workers, and production time cycle are required. The production rate of the system is depending upon the production rate of machines, which is kept in such a way that there are no shortages in the system. The systems of equations generated from the proposed model consisting non-linear equations. There are numerous techniques used to find the optimal solution of non-linear models e.g., interior point optimization (IPO), particle swarm optimization (PSO), pattern search (PS), genetic algorithm (GA), min-max optimization (MMO) etc. Analytically, the methodology is proceeded to search the optimal and global solutions. The proposed model is solved with global optimal result and solution as given in Table 4. The total minimum cost of production is obtained as $478,491, which is optimal and better as compared to PSO, PS, and GA as evidence. The possible optimal production plan for the manufacturing of parts A, B, and C as a solution consider the production cycle time in days (5.4, 5.76, 5.76), machine (9, 8, 7) at the first stage, and (7, 6, 6) at the final stage of sugar processing, respectively. The number of labors as an indirect decision variable are calculated as (32, 29, 25) and (17, 15, 15) for first and final stage of production.

A special case is considered to solve the proposed research model by taking constant production rate of the sugar processing firm. This analysis is important to understand the importance of the controllable production rate by comparing the results. The main objective is to minimize the total cost of production. The proposed research model is solved by using the solution algorithm developed by considering constant production rates of the machines located at first stage and final stage of the production system, i.e., $\varepsilon_{ja} = (140, 150, 160)$ and $\varepsilon_{jc} = (120, 130, 140)$. The comparative results on the basis of the TC_{sj} is represented in Table 5 and Figure 2 to show the evaluation of the controllable production rate. It is found that the total cost of production in case of variable production rate is optimal as compared to the special case by taking constant production rate. The total cost of processing is minimum in case of variable production rate i.e., $478,491 as compared to special case. In addition, the machines required for the production of products at constant production rate required more machines to fulfill the demand. This result provides an important justification for the transformation of the proposed model into a traditional system, considering a constant production rate for the production system.

Theoretically, the study is a proactive approach for the decision-makers to take advantage of the controllable production rate to avoid excess production of agri-product against fluctuating demand with the minimum optimal cost of Agri-SCM. The solution of the research is provided by incorporating a controllable production rate for flexible manufacturing, inventory level control, optimal carbon emission, and best resource utilization to cope with the fluctuating demand. The research is effective for agricultural businesses to understand the role of controllable production rate for cleaner production.

Table 4. The optimal result of the production model with solution for the processing of sugar.

Decision Variable	Algebraic Approach (Constant Production Rate)	Algebraic Approach (Variable Production Rate)	Particle Swarm	Pattern Search	Genetic Algorithm
$T_1(year)$	0.015	0.015	0.016	0.016	0.0163
$T_2(year)$	0.016	0.016	0.016	0.04	0.0080
$T_3(year)$	0.017	0.016	0.016	0.016	0.0166
$K_{ja1}(machines)$	10	9	9	9	9
$K_{ja2}(machines)$	8	8	8	8	8
$K_{ja3}(machines)$	8	7	7	7	7
$K_{jc1}(machines)$	8	7	7	7	7
$K_{jc2}(machines)$	7	6	6	6	74
$K_{jc3}(machines)$	6	6	6	6	6
$TC_{js}(\$)$	513,890	478,491	479,045.1	479,500.2	607,138.3

Table 5. Sensitivity analysis of the SCM with respect to key parameters.

Parameters	Original Value	% Change in Value	New Value	% Effect on TC_{js}
Td_{mj}	0.01	−50	0.005	−14.11
		−25	0.01	−7.16
		+25	0.01	7.16
		+50	0.02	14.52
g_j	610	−50	305.00	−0.71
		−25	457.50	−0.4057
		+25	762.50	0.40
		+50	915.00	0.71
W_j	1000	−50	500.00	−21.2
		−25	750.00	−10.6
		+25	1250.00	10.68
		+50	1500.00	21.2
C_{js}	50	−50	25.00	−3.14
		−25	37.50	−1.46
		+25	62.50	1.312
		+50	75.00	2.513
C_{rm}	19.5	−50	9.75	−9.67
		−25	14.63	−4.83
		+25	24.38	4.83
		+50	29.25	9.67
H_j	0.5	−50	0.25	−3.8
		−25	0.38	−1.92
		+25	0.63	1.7
		+50	0.75	3.9
A_j	11	−50	5.50	−0.5
		−25	8.25	−0.265
		+25	13.75	0.27
		+50	16.50	0.52
O_j	1.2	−50	0.60	−4.9
		−25	0.90	−2.84
		+25	1.50	1.663
		+50	1.80	4.19

4.2. Sensitivity Analysis

The sensitivity analysis of the proposed model is necessary to check the limitation and variation of the system by changing the important factors and parameters. The objective of the proposed model as total cost of agri-processing over short run along is showing a good variation depending upon the change in labor cost, reworking cost, carbon cost, and inspection cost. A sensitivity analysis is necessary to show the effects of varying data on the final objective of the model, i.e., the total profit. Different experiments are also required to test the proposed production model in different situations. There are eight parameters considered as performance indicators for the sensitivity of the proposed model, and the results of the analysis for the SQP technique are given as in Table 5 and also presented graphically in Figure 2.

The sensitivity provides a detailed analysis of the largest effects of the parameters (changes of −50%, −25%, +25% and 50%) on the objective function. This analysis is conducted as follows.

1. By changing the tool-die cost Td_{mj} from −50% to +50%, the results exhibit a direct relation, showing respective increases of +14.11% and −14.52% in the total cost of production system. The original value is at equilibrium, showing symmetric positive and negative effects on the total cost.
2. Similarly, a direct relation is found the production cost (g_j), but the difference is that it had a smaller impact of ±0.71% on the total profit.
3. From the sensitivity analysis, a large impact of almost ±21% on the total cost is observed by the variation of labor cost at ±50%.
4. The impact of the manufacturer's raw material cost is also significant on total cost of production i.e., ±10 at extreme values.
5. The setup cost A_j of the manufacturer shows only small effects of −0.05% to +0.50% on the cost by changing the rate from −50% to +50%, respectively.
6. The outsourcing cost is found to have a less impact as compared to the production costs resulting in −4.9% and +4.9% at the extreme points.
7. The holding costs is carrying a nominal impact on TC_{js}, i.e., −3.8% and 3.9% at extreme values. These parameters lie within range of the equilibrium position and are observed to be directly proportional to the objective.

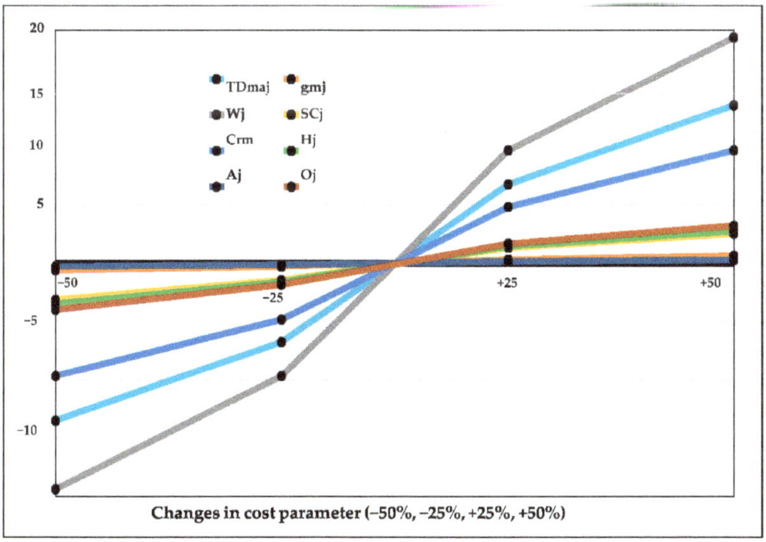

Figure 2. The sensitivity of total cost of Agri-SCM with respect to the cost parameters.

5. Conclusions

Agri-production is intrinsically connected with several uncertain sources, mainly the demand patterns and production yields. From the viewpoint of sugar-cane supply chain planning, few contributions take on tactical and strategic decisions. This paper proposes an optimization model for sugarcane supply chain planning, integrating several agricultural decisions from a strategic-tactical planning perspective. Uncertainties implicate surges in demand which need to be tackled by a variable production supply chain. Indeed, technology development has enforced the agri-business stakeholders to rethink strategies and adopt a better, eco-friendly production environment with optimized costs. Despite high tech procedures, intelligent systems are an hour of need, and this work has proposed an efficient human-machine interactive model embedded, thus going a step ahead for greener production. Further, scrap and reworked products are also dealt with in the model. The right decision at the right time will lead the agricultural industry towards intelligent and smart systems. The study is a form of strategic management approach for the traders, logistics, retailers, manufacturers in agri-SCM to manage resources and to control carbon emissions for cleaner production.

In this work, a non-derivative technique is designed to integrate an algebraic approach in the agri-product based supply chain to optimize the resources and coup with variable demands through a controllable production rate. The analysis is providing a platform for manufacturing managers to invest in favor of advanced technology in agri-SCM, which ultimately leads to a less rejection production environment for clean manufacturing. The solution methodology of the proposed model included manufacturing limitations in the integration of the objective formulations with the developed system. Results findings and sensitivity analysis. The focus of these analyses is to evaluate sensitivity for an optimal solution to the value of uncertain parameters, providing confidence in the solution of the model. Managerial insights are largely beneficial to agri-SCM for the agri-food processing industry and to the people with cleaner production and carbon emission prioritized policies.

The research work can be extended into a three-echelon agri-SCM model by considering the farming industry and agri-retailer. The uncertain factors in the form of costs, prices, inflation, and time value can be dealt with using the fuzzy set theorems. The deterministic model can be converted into probabilistic or stochastic theorems for the implication of real scenarios. Overall, the agri-product supply chain needs be developed globally to make food more secure and accessible.

Author Contributions: Conceptualization, M.A. and M.O.; methodology, M.A and Q.S.K.; software, Q.S.K.; validation, M.A. and Q.S.K.; formal analysis, M.A and G.H.; investigation, G.H and B.S.; data curation, M.O and G.H.; writing—original draft preparation, M.A.; writing—review and editing, Q.S.K and B.S.; supervision, B.S.; funding acquisition, M.A. All authors have read and agreed to the published version of the manuscript.

Funding: This work was supported by Researchers Supporting Project Number (RSP-2020/274), King Saud University, Riyadh, Saudi Arabia.

Acknowledgments: The work was supported by Researchers Supporting Project Number (RSP-2020/274), King Saud University, Riyadh, Saudi Arabia. The authors are also thankful to University of Engineering and Technology, Peshawar and GIK Institute for providing necessary technical assistance.

Conflicts of Interest: The authors declare no conflict of interest.

Appendix A. Model Notation

The list of notation for agri-processing firm and vendor are given in the form of indices, decision variables, manufacturer, and vendor parameters.

Indices

J	the index used to indicate number of agri-product, j = 1, 2, ...n
a	to indicate the parameters for first stage of agri-processing firm
b	used with the parameters of vendor
c	to represent the second stage of agri-processing firm
m	used for agri-processing firm
o	used for vendor

Decision Variables

Tj	cycle time to process jth agri-product (units)
Lja	labors utilized at 1st stage to process j th agri-product (workers)
Ljc	labors utilized to process jth agri-product at 1st during final stage (workers)
Kja	number of machine units utilized at 1st stage to process jth agri-product (workstations)
Kjc	number of machine units utilized to process jth agri-product during finishing stage (workstations)
Pja	plant production rate of jth product at 1st stage (units/year)
Pjc	plant production rate of jth agri-product by the 2nd stage of agri-processing firm (units/year)

Agri-Processing Parameters

Crm	raw material cost of jth agri-product ($/unit)
TDmj	manufacturer total tool-die cost of jth agri-product ($/unit)
TDmaj	tool-die cost of jth agri-product in 1st stage of manufacturer ($/unit)
TDmcj	tool-die cost of jth agri-product in final stage($/unit)
gj	total indirect production cost of jth agri-product ($/unit)
gmaj	indirect cost of jth agri-product in first stage of manufacturer($/unit)
gmcj	indirect production cost of jth agri-product in final stage of manufacturer($/unit)
Dj (Ω)	variable demand depending emergency level due to pandemic ($/unit)
Qj	production quantity (units)
lja	average labor utilized per machine in first stage of manufacturer(labor/machine)
ljc	average labor utilized per machine in final stage of manufacturer(labor/machine)
Wj	average wedge of labor to process jth agri-product ($/labor)
Aj	setup cost for jth agri-product ($/year)
hmj	manufacturer holding cost of each agri-product per cycle ($/unit/year)
Gj	reworking cost of jth agri-product ($/unit)
εja	production rate of each machine unit to process jth agri-product at 1st stage(units/machine)
εjc	production rate of each machine unit at final stage to process jth product (units/machine)
TCaj	total cost of processing agri-product ($/cycle)

Vendor Parameters

TDoj	vendor tool-die cost of jth agri-product during ($/unit)
goj	indirect part production cost of jth agri-product with vendor($/unit)
θj	fixed inspection cost of agri-product jth ($/year)
αj	proportion of rejection produced in defective jth agri-product (%)
βj	defective rate for jth product (%)
ψj	variable inspection cost of product jth ($/unit)
γ1	recycling cost ($/unit)
ρ	efficiency of the labor (%)
Pjb	production rate of jth agri-product processed by the vendor (units/year)
TCvj	total cost of vendor ($/cycle)
MR	marginal rate of vendor ($/cycle)
TCj	total cost of agri-product supply chain management ($/cycle)

Appendix B. Production Cycle Time and Inventory Level Calculations

Appendix B.1. Cycle Time Calculation

The objective of the research is to minimize the total cost of production, and the formulation of the cycle time of the production system is prerequisite to calculate the total cost (TC). Cycle time is taken as a decision variable in the production model, which is dependent upon the production rates of manufacturing system. The production rate of the manufacturing firm and vendor are relying on the production rate of the machines (capital units). In order to meet the customer demand and no shortages in the production system, the production rates of the manufacturing system is different in the manufacturing before outsourcing, during outsourcing and after outsourcing operations are denoted as P_{ja}, P_{jb}, and P_{jc} respectively. The total production cycle time of the item is the sum of the time utilized in production, outsourcing, reworking and delivery as expressed as in Equation (A1).

$$T_j = t_{1j} + t_{2j} + t_{3j} + t_{4j} + t_{5j} \tag{A1}$$

where, the individual time fractions of the total cycle time are mathematically formulated as given from Equations (A2) to (A6).

$$t_{1j} = \frac{Q_j}{P_{ja}} \tag{A2}$$

$$t_{2j} = \frac{Q_j}{P_{jb}} \tag{A3}$$

$$t_{3j} = \frac{Q_j \alpha_j}{P_{jb}} \tag{A4}$$

$$t_{4j} = \frac{Q_j}{P_{jc}} \tag{A5}$$

$$t_{5j} = \frac{I_{max(jc)}}{D_j} \tag{A6}$$

To find the value of $I_{max(jc)}$ from the inventory diagram, the formula of slope in the $Area_{10,11,12}$ can be expressed as in Equation (A7).

$$tan\theta_4 = P_{jc} - D_j \Rightarrow \frac{I_{max(jc)}}{t_{4j}} \tag{A7}$$

$$I_{max(jc)} = (P_{jc} - D_j)\frac{Q_j}{P_{jc}} = Q_j(1 - \frac{D_j}{P_{jc}}) \tag{A8}$$

By putting the value of $I_{max(jc)}$, the simplified form of Equation (A6) can be expressed as in Equation (A10).

$$t_{5j} = \frac{Q_j(1 - \frac{D_j}{P_{jc}})}{D_j} \tag{A9}$$

$$t_{5j} = Q_j(\frac{1}{D_j} - \frac{1}{P_{jc}}) \tag{A10}$$

After calculating all the time fractions (t_{1j} to t_{5j}) for the production of automobile parts, the mathematical form of the total cycle time can be given as in Equation (A11).

$$T_j = \frac{Q_j}{P_{ja}} + \frac{Q_j}{P_{jb}} + \frac{Q_j \alpha_j}{P_{jb}} + \frac{Q_j}{P_{jc}} + Q_j(\frac{1}{D_j} - \frac{1}{P_{jc}}) \tag{A11}$$

Appendix B.2. Average Inventory Costs

The inventory is obtained by calculating the area under the curve of the production, storage, and buffer quantity in complete manufacturing process. As followed the inventory diagram given as in Figure 2, the inventories can be divided into the areas i.e., $Area_{123}$, $Area_{456}$, $Area_{5678}$, $Area_{10,11,12}$, and $Area_{11,12,13}$. These areas are formulated separately from Equation (A12) to Equation (A26) respectively.

$$Area_{123} = \frac{1}{2} t_{1j} I_{max(ja)} \tag{A12}$$

$$tan\theta_1 = \frac{I_{max(ja)}}{t_{1j}} P_{ja} \times t_{1j} = I_{max(ja)} I_{max(ja)} = Q_j \tag{A13}$$

By putting $I_{max(ja)}$, Equation (A12) can be written as in Equation (A14).

$$Area_{123} = \frac{Q_j^2}{2P_{ja}} \tag{A14}$$

Similarly, the area of triangle *456* can be given as in Equation (A15)

$$Area_{456} = \frac{1}{2} I_{max(jbb)} t_{2j} \tag{A15}$$

The maximum inventory of the vendor $I_{max(jbb)}$ is calculated by the slope formula of $Area_{456}$ (Equation (A16)), which is then simplified to express in the form of Equation (A17)

$$tan\theta_2 = \frac{I_{max(jbb)}}{t_{2j}} \tag{A16}$$

$$P_{jb}(1-\alpha_j) \times \frac{Q_j}{P_{jb}} = I_{max(jbb)} I_{max(jbb)} = Q_j(1-\alpha_j) \tag{A17}$$

After substituting, the $Area_{456}$ can be written as in Equation (A18)

$$Area_{456} = \frac{1}{2} Q_j(1-\alpha_j) \frac{Q_j}{P_{jb}} Area_{456} = \frac{Q_j^2(1-\alpha_j)}{2P_{jb}} \tag{A18}$$

The simplified form of area under the curve of the rectangular region *5678* is formulated and expressed as in Equation (A19)

$$Area_{5678} = I_{max(jbb)} t_{3j} = Q_j(1-\alpha_j) \frac{Q_j \alpha_j}{P_{jb}} = \frac{\alpha_j(1-\alpha_j) Q_j^2}{P_{jb}} \tag{A19}$$

The triangular area of the inventory storage $Area_{689}$ by reworking operations during outsourcing operation is expressed as in Equation (A20).

$$Area_{689} = \frac{1}{2} I_{max(jba)} t_{3j} \tag{A20}$$

where, to find the maximum inventory $I_{max(jba)}$ of the reworking operation in the production system, the slope formula is given as in Equation (A21),

$$tan\theta_3 = \frac{I_{max(jba)}}{t_{3j}} P_{jb}(1-\alpha_j\beta_j) \frac{Q_j \alpha_j}{P_{jb}} = I_{max(jba)} I_{max(jba)} = \alpha_j Q_j (1-\alpha_j\beta_j) \tag{A21}$$

By substitution, Equation (A20) can be written in the form of Equation (A22).

$$Area_{689} = \frac{1}{2} \frac{Q_j \alpha_j}{P_{jb}} \alpha_j Q_j (1-\alpha_j\beta_j) = \frac{\alpha_j^2 Q_j^2 (1-\alpha_j\beta_j)}{2P_{jb}} \tag{A22}$$

Likewise, the $Area_{10,11,12}$ and $Area_{11,12,13}$ can be formulated and expressed in simplified form as in Equations (A25) and (A26) respectively, where the maximum inventory is $I_{max(jc)}$.

$$Area_{10,11,12} = \frac{1}{2} I_{max(jc)} t_{4j} \tag{A23}$$

$$tan\theta_4 = P_{jc} - D_j \Rightarrow \frac{I_{max(jc)}}{t_{4j}} I_{max(jc)} = (P_{jc} - D_j) \frac{Q_j}{P_{jc}} \tag{A24}$$

$$Area_{10,11,12} = \frac{1}{2}Q_j(1-\frac{D_j}{P_{jc}})\frac{Q_j}{P_{jc}} \tag{A25}$$

$$Area_{11,12,13} = \frac{1}{2}I_{max(jc)}t_{5j}=\frac{1}{2}\frac{I^2_{max(jc)}}{D_j}=\frac{Q_j^2}{2D_j}(1-\frac{D_j}{P_{jc}})^2 \tag{A26}$$

References

1. Banasik, A.; Kanellopoulos, A.; Bloemhof-Ruwaard, J.M.; Claassen, G.D.H. Accounting for uncertainty in eco-efficient agri-food supply chains: A case study for mushroom production planning. *J. Clean. Prod.* **2019**, *216*, 249–256. [CrossRef]
2. Thron, T.; Nagy, G.; Wassan, N. Evaluating alternative supply chain structures for perishable products. *Int. J. Logist. Manag.* **2007**, *18*, 364–384. [CrossRef]
3. Chen, W.; Li, J.; Jin, X. The replenishment policy of agri-products with stochastic demand in integrated agricultural supply chains. *Expert Syst. Appl.* **2016**, *48*, 55–66. [CrossRef]
4. Cobb, B.R.; Rumí, R.; SalmeróN, A. Inventory management with log-normal demand per unit time. *Comput. Oper. Res.* **2013**, *40*, 1842–1851. [CrossRef]
5. Alexandratos, N.; Bruinsma, J. World Agriculture towards 2030/2050: The 2012 Revision. 2012. Available online: http://www.fao.org/fileadmin/templates/esa/Global_persepctives/ (accessed on 19 April 2020).
6. Drozda, T.J. *Tool and Manufacturing Engineers Handbook: Machining*; Society of Manufacturing Engineers: Dearborn, Michigan, 1983; Volume 1.
7. Conrad, C.; McClamroch, N. The drilling problem: A stochastic modeling and control example in manufacturing. *IEEE Trans. Autom. Control* **1987**, *32*, 947–958. [CrossRef]
8. Khouja, M. The economic production lot size model under volume flexibility. *Comput. Oper. Res.* **1995**, *22*, 515–523. [CrossRef]
9. Offodile, O.F.; Ugwu, K. Evaluating the effect of speed and payload on robot repeatability. *Robot. Comput. Integr. Manuf.* **1991**, *8*, 27–33. [CrossRef]
10. Mehrez, A.; Offodile, O.F.; Ahm, B.-H. A decision analysis view of the effect of robot repeatability on profit. *IIE Trans.* **1995**, *27*, 60–71. [CrossRef]
11. Chaabane, A.; Ramudhin, A.; Paquet, M. Designing supply chains with sustainability considerations. *Prod. Plan. Control* **2011**, *22*, 727–741. [CrossRef]
12. Neto, J.Q.F.; Walther, G.; Bloemhof, J.; van Nunen, J.A.E.E.; Spengler, T. A methodology for assessing eco-efficiency in logistics networks. *Eur. J. Oper. Res.* **2009**, *193*, 670–682. [CrossRef]
13. El Yasmine, A.S.L.; Ghani, B.A.; Trentesaux, D.; Bouziane, B. Supply chain management using multi-agent systems in the agri-food industry. In *Service Orientation in Holonic and Multi-Agent Manufacturing and Robotics*; Springer: Berlin/Heidelberg, Germany, 2014; pp. 145–155.
14. Zecca, F.; Rastorgueva, N. Supply chain management and sustainability in agri-food system: Italian evidence. *J. Nutr. Ecol. Food Res.* **2014**, *2*, 20–28. [CrossRef]
15. Biswas, P.; Sarker, B.R. Optimal batch quantity models for a lean production system with in-cycle rework and scrap. *Int. J. Prod. Res.* **2008**, *46*, 6585–6610. [CrossRef]
16. Tayyab, M.; Sarkar, B. Optimal batch quantity in a cleaner multi-stage lean production system with random defective rate. *J. Clean. Prod.* **2016**, *139*, 922–934. [CrossRef]
17. Omair, M.; Sarkar, B.; Cárdenas-Barrón, L.E. Minimum quantity lubrication and carbon footprint: A step towards sustainability. *Sustainability* **2017**, *9*, 714. [CrossRef]
18. Kwak, M. Planning demand-and legislation-driven remanufacturing for a product family. *Ind. Eng. Manag. Syst.* **2015**, *14*, 159–174.
19. Xiao, Y.; Yang, S.; Zhang, L.; Kuo, Y.-H. Supply chain cooperation with pricesensitive demand and environmental impacts. *Sustainability* **2016**, *8*, 716. [CrossRef]
20. Wu, C.-C.; Chang, N.-B. Corporate optimal production planning with varying environmental costs: A grey compromise programming approach. *Eur. J. Oper. Res.* **2004**, *155*, 68–95. [CrossRef]
21. Wang, F.; Lai, X.; Shi, N. A multi-objective optimization for green supply chain network design. *Decis. Support Syst.* **2011**, *51*, 262–269. [CrossRef]

22. Wang, E.-J.; Lin, C.-Y.; Su, T.-S. Electricity monitoring system with fuzzy multiobjective linear programming integrated in carbon footprint labeling system for manufacturing decision making. *J. Clean. Prod.* **2016**, *112*, 3935–3951. [CrossRef]
23. Goyal, S.K.; Cárdenas-Barrón, L.E. Note on: Economic production quantity model for items with imperfect quality—A practical approach. *Int. J. Prod. Econ.* **2002**, *77*, 85–87. [CrossRef]
24. Wee, H.M.; Yu, J.; Chen, M.C. Optimal inventory model for items with imperfect quality and shortage backordering. *Omega* **2007**, *35*, 7–11. [CrossRef]
25. Hsu, W.-K.K.; Yu, H.-F. EOQ model for imperfective items under a one-time-only discount. *Omega* **2009**, *37*, 1018–1026.
26. Sarkar, S.; Giri, B.C. Stochastic supply chain model with imperfect production and controllable defective rate. *Int. J. Syst. Sci. Oper. Logist.* **2020**, *7*, 133–146. [CrossRef]
27. Sarkar, B.; Ahmed, W.; Kim, N. Joint effects of variable carbon emission cost and multi-delay-in-payments under single-setup-multiple-delivery policy in a global sustainable supply chain. *J. Clean. Prod.* **2018**, *185*, 421–445. [CrossRef]
28. Tiwari, S.; Ahmed, W.; Sarkar, B. Multi-item sustainable green production system under trade-credit and partial backordering. *J. Clean. Prod.* **2018**, *204*, 82–95. [CrossRef]
29. Khouja, M.; Mehrez, A. Economic production lot size model with variable production rate and imperfect quality. *J. Oper. Res. Soc.* **1994**, *45*, 1405–1417. [CrossRef]
30. Rosenblatt, M.J.; Lee, H.L. Economic production cycles with imperfect production processes. *IIE Trans.* **1986**, *18*, 48–55. [CrossRef]
31. Khouja, M. A note on 'deliberately slowing down output in a family production context'. *Int. J. Product. Res.* **1999**, *37*, 4067–4077. [CrossRef]
32. Eiamkanchanalai, S.; Banerjee, A. Production lot sizing with variable production rate and explicit idle capacity cost. *Int. J. Prod. Econ.* **1999**, *59*, 251–259. [CrossRef]
33. Giri, B.C.; Yun, W.Y.; Dohi, T. Optimal design of unreliable production—Inventory systems with variable production rate. *Eur. J. Oper. Res.* **2005**, *162*, 372–386. [CrossRef]
34. Larsen, C. The economic production lot size model extended to include more than one production rate. *Int. Trans. Oper. Res.* **2005**, *12*, 339–353. [CrossRef]
35. Sana, S.S. A production-inventory model in an imperfect production process. *Eur. J. Oper. Res.* **2010**, *200*, 451–464. [CrossRef]
36. Ayed, S.; Sofiene, D.; Nidhal, R. Joint optimisation of maintenance and production policies considering random demand and variable production rate. *Int. J. Prod. Res.* **2012**, *50*, 6870–6885. [CrossRef]
37. Bouslah, B.; Gharbi, A.; Pellerin, R. Joint optimal lot sizing and production control policy in an unreliable and imperfect manufacturing system. *Int. J. Prod. Econ.* **2013**, *144*, 143–156. [CrossRef]
38. Singh, S.; Prasher, L. A production inventory model with flexible manufacturing, random machine breakdown and stochastic repair time. *Int. J. Ind. Eng. Comput.* **2014**, *5*, 575–588. [CrossRef]
39. Zanoni, S.; Bettoni, L.; Glock, C.H. Energy implications in a two-stage production system with controllable production rates. *Int. J. Prod. Econ.* **2014**, *149*, 164–171. [CrossRef]
40. Linde-Rahr, M. Rural Shadow Wages, Labour Supply and Agricultural Production under Imperfect Markets: Empirical Evidence from Vietnam. *Labor Hum. Cap.* **2001**, *31*. Available online: http://ageconsearch.umn.edu/record/20487 (accessed on 26 April 2020). [CrossRef]
41. Sarkar, B.; Moon, I. An EPQ model with inflation in an imperfect production system. *Appl. Math. Comput.* **2011**, *217*, 6159–6167. [CrossRef]
42. Sana, S.S. A production-inventory model of imperfect quality products in a threelayer supply chain. *Decis. Support Syst.* **2011**, *50*, 539–547. Available online: http://www.sciencedirect.com/science/article/pii/S0167923610001934 (accessed on 2 May 2020). [CrossRef]
43. Sarkar, B.; Majumder, A.; Sarkar, M.; Dey, B.K.; Roy, G. Two-echelon supply chain model with manufacturing quality improvement and setup cost reduction. *J. Ind. Manag. Optim.* **2017**, *13*, 1085. [CrossRef]
44. Ma, X.; Wang, S.; Islam, S.M.N.; Liu, X. Coordinating a threeechelon fresh agricultural products supply chain considering freshness-keeping effort with asymmetric information. *Appl. Math. Model.* **2019**, *67*, 337–356. Available online: http://www.sciencedirect.com/science/article/pii/S0307904X18305201 (accessed on 30 April 2020). [CrossRef]

45. Yang, J.; Qi, X.; Xia, Y. A production-inventory system with markovian capacity and outsourcing option. *Oper. Res.* **2005**, *53*, 328–349. [CrossRef]
46. Wang, Y.; Niu, B.; Guo, P. On the advantage of quantity leadership when outsourcing production to a competitive contract manufacturer. *Prod. Oper. Manag.* **2013**, *22*, 104–119. [CrossRef]
47. Bettayeb, B.; Bassetto, S.J.; Sahnoun, M. Quality control planning to prevent excessive scrap production. *J. Manuf. Syst.* **2014**, *33*, 400–411. [CrossRef]
48. AlDurgam, M.; Adegbola, K.; Glock, C.H. A single-vendor single-manufacturer integrated inventory model with stochastic demand and variable production rate. *Int. J. Prod. Econ.* **2017**, *191*, 335–350. [CrossRef]
49. Sarkar, B.; Majumder, A.; Sarkar, M.; Kim, N.; Ullah, M. Effects of variable production rate on quality of products in a single-vendor multi-buyer supply chain management. *Int. J. Adv. Manuf. Technol.* **2018**, *99*, 567–581. [CrossRef]
50. Sarkar, B. An inventory model with reliability in an imperfect production process. *Appl. Math. Comput.* **2012**, *218*, 4881–4891. Available online: http://www.sciencedirect.com/science/article/pii/S0096300311012999 (accessed on 15 April 2020). [CrossRef]
51. Bazan, E.; Jaber, M.Y.; Zanoni, S. Supply chain models with greenhouse gases emissions, energy usage and different coordination decisions. *Appl. Math. Model.* **2015**, *39*, 5131–5151. [CrossRef]
52. Mansour, M. Quantifying the intangible costs related to non-ergonomic work conditions and work injuries based on the stress level among employees. *Saf. Sci.* **2016**, *82*, 283–288. [CrossRef]
53. Omair, M.; Ullah, M.; Ganguly, B.; Noor, S.; Maqsood, S.; Sarkar, B. The quantitative analysis of workers' stress due to working environment in the production system of the automobile part manufacturing industry. *Mathematics* **2019**, *7*, 627. [CrossRef]

processes

MDPI

Article

Material Requirements Planning Using Variable-Sized Bin-Packing Problem Formulation with Due Date and Grouping Constraints

Dejan Gradišar * and Miha Glavan

Department of Systems and Control, Jožef Stefan Institute, 1000 Ljubljana, Slovenia; miha.glavan@ijs.si
* Correspondence: dejan.gradisar@ijs.si

Received: 17 August 2020; Accepted: 28 September 2020; Published: 2 October 2020

Abstract: Correct planning is crucial for efficient production and best quality of products. The planning processes are commonly supported with computer solutions; however manual interactions are commonly needed, as sometimes the problems do not fit the general-purpose planning systems. The manual planning approach is time consuming and prone to errors. Solutions to automatize structured problems are needed. In this paper, we deal with material requirements planning for a specific problem, where a group of work orders for one product must be produced from the same batch of material. The presented problem is motivated by the steel-processing industry, where raw materials defined in a purchase order must be cut in order to satisfy the needs of the planned work order while also minimizing waste (leftover) and tardiness, if applicable. The specific requirements of the problem (i.e., restrictions of which work orders can be produced from a particular group of raw materials) does not fit the regular planning system used by the production company, therefore a case-specific solution was developed that can be generalized also to other similar cases. To solve this problem, we propose using the generalized bin-packing problem formulation which is described as an integer programming problem. An extension of the bin-packing problem formulation was developed based on: (i) variable bin sizes, (ii) consideration of time constraints and (iii) grouping of items/bins. The method presented in the article can be applied for small- to medium-sized problems as first verified by several examples of increasing complexity and later by an industrial case study.

Keywords: mixed-integer linear programming; bin-packing problem; material requirements planning

1. Introduction

Presently, production and logistics processes are commonly supported with production planning and control systems. These are used to support a range of operational activities from production planning to the detailed scheduling and execution of every operation [1–3]. However, it frequently happens that the implemented systems do not adequately cover the problem in a holistic manner. Classical hierarchical organization of the business and production management [4] limits more detailed and tight coupling across the production levels. In practice, different activities of production are supported with separate information systems. Limited data and functional integration among them and additional specifics of the production processes require development and integration of tailor-made solutions.

A holistic view of the production problems become even more important when the environmental footprint of the production is in question. Plant-wide production supervision through data integration and analytics lead to processes that are more effective and consequently the resource demand is reduced. This can even be scaled up by data-sharing and active collaboration with the supply chain, in order to reduce erroneous delivery and consequent undesirable events at the manufacturer's premises such as stoppage of assembly line [5]. On the other hand, evolution from linear business models to circular economy principles will impose novel problems and dilemmas, such as, for example, strategic planning on supply chain level [6] or remanufacturing pricing strategies [7]. All this requires manufacturers to efficiently schedule, manage and optimize resource consumption of their production processes.

The main motivation of the presented work is the problem faced by the steel-processing manufacturer, where specific requirements for the material requirements planning (MRP) could not be implemented by standard solutions, since the purchasing and planning activities are not supported with the same information system. Consequently, manual interactions are needed which are problematic due to the occurrence of human errors, which are inevitable in the case of more complex situations. In addition, in this way it is almost impossible to produce optimal solutions even for simple problems.

The problem and its specifications originate from steel-processing production, where many products with complex MRP structures are being produced. In principle, raw materials must be cut to satisfy the needs for the input material (intermediates) as defined with MRP. The main requirement of the addressed problem is to plan and optimize material requirements while ensuring that a group of work orders for one final product must be produced from the same batch of raw material. Additionally, it is desirable to limit the tardiness and to reduce the material leftover (small material leftovers lead to a scrap). A custom solution is needed that allows for holistic considerations of purchase and work orders. In practice, optimization systems that holistically address the problem and integrate planning and scheduling are needed, where more adopted solutions are based on mathematical models (e.g., [8]).

The addressed problem of material planning and optimization can be translated into a general bin-packing optimization problem, which has been studied since the early 1970s [9,10]. In the bin-packing problem (BPP), we have m items and n identical bins. The goal is to assign each item to one bin so that the total size of the items in each bin does not exceed the capacity of the bin and the number of bins used is minimized. Different variants of the problem continue to attract researchers' attention.

The bin-packing problem can be described as an integer optimization problem. Solving the problem is NP-hard and is usually done with custom heuristics. A detailed review of mathematical models and exact algorithms for bin-packing and cutting stock problems are given in [11]. Coffman et al. [12] present an overview of BPP approximation algorithms. However, advances in computer technology and available solver capabilities enables us to solve mid-sized practical problems [13].

The classical problem has been extended to various forms to mitigate many practical application problems. For a general survey see Wäscher et al. [14]. In manufacturing industries, such as in cutting steel, paper or textiles, a cutting stock problem formulation is used. Here, stock material must be cut into shorter lengths to meet demand while minimizing waste. The problems have an identical structure in common [14]. In this context, Carvalho [15] reviewed several linear programming (LP) formulations for the one-dimensional cutting stock and bin-packing problems.

A natural generalization of the classical bin-packing problem is to consider the problem with two or more dimensions. Lodi et al. [16] discussed mathematical models and survey classical approximation algorithms, heuristic and metaheuristic methods, and exact enumerative approaches for 2-D problems. Epstein et al. [17] studied the oriented multi-dimensional dynamic bin-packing problem for two, three, and multiple dimensions. A more general and realistic model considers bins of differing capacities [18]. Bin-packing problem and its variations are still widely studied [19,20]. Recently, Jansen and Kraft [21] presented a review of a variable bin-sized bin-packing problem and an improved approximation scheme to solve the problem.

A bin-packing formulation is also often used to formally represent problems where time must be considered to be well (e.g., scheduling). A simple case is where a task's durations are represented by items and a job's due dates with bin sizes. Advanced solutions use the due date as an additional constraint factor. Another solution is to soften the due date constraint, which can be violated but penalized proportionally with the delay. In this context, Reinertsen and Vossen [22] have proposed optimization models and solution procedures that solve the cutting stock problem when orders have due dates. The objective in their model is to minimize scrap and tardiness. Recently, Arbib and Marinelli [23] considered a problem, where each item representing a job must be assigned to a minimum number of bins (resources). Additionally, each item is due by a particular date, so minimal tardiness is also being optimized.

The problem we are targeting arises from the steel-processing production process, where specific constraints must be considered. In this kind of industry even small improvements in the production operation can result in large monetary savings [24]. In our case raw materials are ordered and delivered in batches. Depending on the raw material delivery time, there are some minor variations in the visual properties of the raw material. This may result in a final product that is produced of materials with different properties causing unwanted, additional, repair costs. Therefore, a solution is needed that supports the production planner with decisions on how to cut the available raw materials and how to plan new purchase orders in order to obtain production material as to realize work orders, i.e., final products. The plan must consider various constraints such as the quantity, due dates, single material batch constraints, etc. To support the production planner in dealing with this problem, various generalizations of the classical BPP formulation are proposed. The problem can be designated as an one-dimensional multiple bin-size bin-packing problem (MBSBPP) using the topology of Wäscher et al. [14]. A simple bin-packing problem (BPP) formulation using variable bin sizes and strict time constraints is used when the raw material is available on time. For the case when some of the purchase orders are late, we propose using an upgraded Reinertsen's optimization model [22]. Soft constraints are introduced into the linear integer problem as proposed in a work of Srikumar [25]. Main contribution of the article is the formulation of specific optimization problem for the problem where group of work orders for one product have to be produced from the same batch of raw material as one batch of raw material consists of several material stocks, which have in common several immeasurable characteristics. Moreover, the problem had to consider the availability of the materials, order deadlines and must minimize the scrap. This is done through introduction of a custom formulation of the bin-packaging problem with constraints that groups the items and bins in a way that assures a consistent final product, while respecting other constraints as well.

In the next section, the general problem formulation is developed step-by-step. Whole or part of the formulation can be used depending on the problem at hand. A model description of the classical bin-packing problem is extended with variable bin sizes, due date constraints, and grouping of bins and items. In Section 3, two case studies are presented that illustrate the practical implementation of the proposed method. Finally, developed methodology and results are discussed in Section 4.

2. Materials and Methods

The problem of material requirements planning can be formalized as a bin-packing problem, which can be solved using a mixed-integer linear program (MILP). In this section, we describe how a basic BPP can be extended to include constraints such as variable bin and item sizes, time limitations, and the fact that only a group of bins can be used to produce one group of items.

2.1. Mathematical Notations

For a consistent discussion, we first define symbolic notation that is consistently used across different mathematical models:

- y_j—decision variable for bin i (1 used, 0 not used)
- x_{ij}—decision variable—item j assignment to bin i (1 assigned, 0 not assigned)
- g_{ij}—soft constraint criteria—tardiness of bin i depending on item j
- I—set of m items (input material defined with work orders)
- B—set of n bins (raw material defined with purchase orders)

- $| F^I |$—number of item groups
- $| F^B |$—number of bin groups
- F^I—set of $| F^I |$ items
- F^B—set of $| F^B |$ bins
- s_i—item (input material) size
- c_j—bin (raw material) size
- d_i—item due date
- a_j—bin delivery time
- w_i—optimization weighting parameters
- F^I_k—k-th item group
- F^B_l—l-th bin group
- $| F^I_k |$—number of items in k-th item group
- $| F^B_l |$—number of bins in l-th bin group

2.2. Problem Description

In the classical BPP, we are dealing with a set I of m items, each with its size s_i that has to be assigned to a set B of n identical bins with capacity c. The number of used bins must be minimized, where the sum of item sizes in one bin does not exceed the bin capacity. The problem can be described as an integer optimization problem, as shown in Equation (1).

$$\begin{aligned} z_1: \quad & \min \sum_{j=1}^{n} y_j \\ \text{s.t.} \quad & \sum_{i=1}^{m} s_i \cdot x_{ij} \leq c \cdot y_j, j = 1 \ldots n \\ & \sum_{i=1}^{m} x_{ij} = 1, j = 1 \ldots n \\ & y_j \in \{0,1\} \\ & x_{ij} \in \{0,1\}, \end{aligned} \tag{1}$$

where y_j and x_{ij} are binary decision variables

$$y_j = \begin{cases} 1 & \text{if bin } j \text{ is used} \\ 0 & \text{otherwise,} \end{cases}$$

$$x_{ij} = \begin{cases} 1 & \text{if item } i \text{ is assigned to bin } j \\ 0 & \text{otherwise.} \end{cases}$$

Decision variable x_{ij} can be represented as a matrix, depicted in Figure 1.

The first constraint from Equation (1) guarantees that the sum of item sizes assigned to bin B_j does not exceed its capacity, c. The second constraint assures that one item is assigned only to one bin.

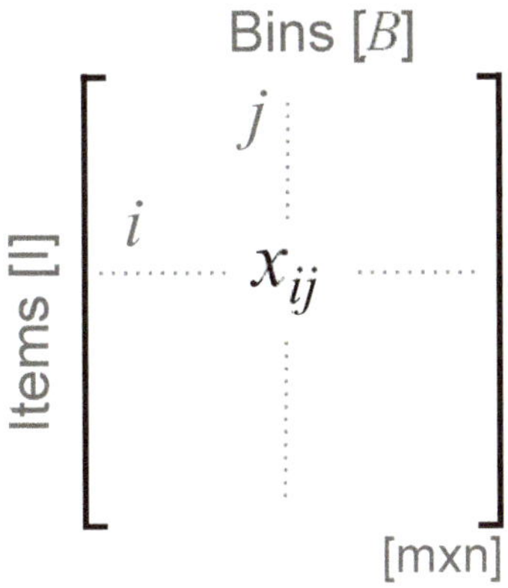

Figure 1. Decision variable x_{ij} in basic BPP.

2.3. Variable Bin Sizes

There are situations where bin sizes are not identical. Variable bin sizes are denoted as c_j in our work. A set of items can be assigned to bin B_j if the total volume of these items does not exceed the capacity of a bin c_j. For this reason, the first constraint in Equation (1) should be rewritten as given in Equation (2).

$$\sum_{i=1}^{m} s_i \cdot x_{ij} \leq c_j \cdot y_j, j = 1 \ldots n \tag{2}$$

The best way to use a bin is to fill it fully, i.e., there is no leftover space. In case of unique bin capacities, the criterion described in Equation (1) minimizes the number of used bins together with the non-assigned capacity of used bin(s)—leftovers. However, this is usually not the case when we have bins of different capacities. In that case, the optimization criteria must be defined as given with Equation (3), where the sum of leftovers of all used bins is being minimized.

$$z_2: \quad \min \sum_{j=1}^{n} (c_j \cdot y_j - \sum_{i=1}^{m} s_i \cdot x_{ij}) \tag{3}$$

2.4. Time Extension

To take into account the time limitations, the problem needs to be additionally extended. The optimization problem needs to consider the situations where bin B_j has to be available before the due date of an item I_i, which is to be scheduled in B_j. The item I_i is therefore defined with a tuple of two elements (s_i, d_i), where d_i is its due date. Similarly, bin B_j is defined as (c_j, a_j), where a_j is its delivery time. We suggest two solutions: (i) considering hard (strict) time constraints, where B_j must be available before the due date of I_i ($d_i \geqslant a_j$) and (ii) considering soft time constraints, where delivery dates are allowed to exceed due dates, but this is penalized within the objective function.

The solution where due date violations are not allowed is achieved with the addition of the following constraint to our optimization problem:

$$a_j \cdot x_{ij} \leq d_i \quad i = 1 \ldots m, j = 1 \ldots n. \tag{4}$$

In practice, the due date constraint violation is commonly allowed, but penalized. In this case, we must consider a slightly different problem formulation. The due date constraint must be considered to be a soft constraint. For this reason, we need to first introduce additional tardiness criteria (g_{ij}), which is used to soften the constraint from Equation (4):

$$g_{ij} = \begin{cases} a_j - d_i & a_j > d_i, \\ 0 & \text{otherwise} \end{cases}. \quad (5)$$

If the hard constraint is violated, g_{ij} evaluates the time for which bin B_j is late for item I_i. When there is no delivery time violation, g_{ij} should be 0. Using this, we can extend our criterion function, where we allow that bin B_j can be late. In this way, tardiness is being minimized in addition to leftovers.

$$\begin{aligned} z_3 :\min \quad & w_1 \cdot \sum_{j=1}^{n} (c_j \cdot y_j - \sum_{i=1}^{m} s_i \cdot x_{ij}) + \\ & + w_2 \cdot \sum_{j=1}^{n} \sum_{i=1}^{m} g_{ij} \cdot x_{ij} \end{aligned} \quad (6)$$

As we are dealing here with a multi-objective problem, the weighting of parameters w_1 and w_2 is introduced, which is used to adjust the importance of leftovers and tardiness criteria. Parameter settings are application-dependent.

2.5. Grouping of Items/Bins

Furthermore, we are targeting the problem of when a group of items can be assigned to only one single group of bins. This implies additional constraints to our optimization problem.

We assume that we have a set F^I of $|F^I|$ item groups and set F^B of $|F^B|$ bin groups. Item I_i is therefore extended to triple (s_i, d_i, F_k^I), where $F_k^I \in F^I$ denotes the group to which item I_i belongs and bin B_j is defined with (c_j, a_j, F_l^B), where $F_l^B \in F^B$ denotes the bin group to which bin B_j belongs. The k-th item group consists of $|F_k^I|$ items and l-th bin group consists of $|F_l^B|$ bins.

Figure 2 illustrates the example where we have three item groups and two bin groups. Items from one group must be assigned to the bins that are grouped within one bin group. One solution to the problem from the observed example is illustrated in Figure 3. Here items from F_1^I and F_2^I are assigned to the bins from bin group F_1^B, and items from F_3^I to a bin from bin group F_2^B. White rectangles indicate the leftovers.

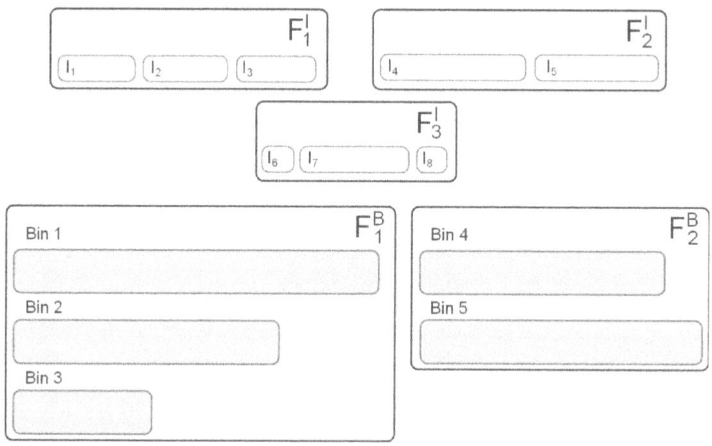

Figure 2. Item and bin groups.

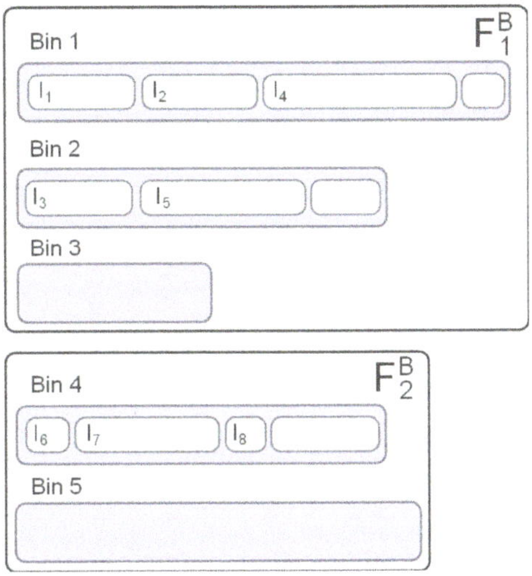

Figure 3. Item and bin groups. Acceptable solution: $F_1^I, F_2^I \in F_1^B$ and $F_3^I \in F_2^B$.

Without loss of the generality, we can assume that the decision variable x_{ij} is organized in a matrix, where items and bins from the same group are given next to each other, as illustrated in Figure 4. Sub-matrices, marked in the figure, indicate all possible group assignments, i.e., combinations of item and bin groups. The items that form k-th item group can be assigned to any of the bins from l-th bin group only if the sum of x_{ij} is equal to the size of item group F_k^I. Here i designates all items from k-th item group and j all bins from l-th bin group. This can be formally represented by adding the equality constraints (Equation (7)) to our optimization problem from Equation (1).

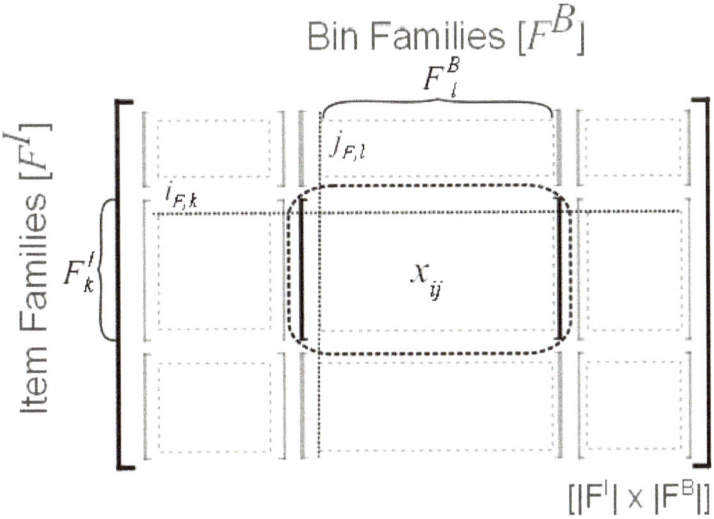

Figure 4. Decision variable x_{ij} - item and bin groups.

$$\sum_{i=i_{F,k}}^{i_{F,k}+|F_k^I|-1} \sum_{j=j_{F,l}}^{j_{F,l}+|F_l^B|-1} x_{ij} = |F_k^I| \cdot \zeta_{l,k}, \tag{7}$$

where $k = 1 \dots |F^I|$ and $l = 1 \dots |F^B|$.

Here $i_{F,k}$ determines the sequence number of the first item in the k-th group of items and is evaluated as:

$$i_{F,k} = \begin{cases} 1 & \text{if } k = 1 \\ 1 + \sum\limits_{p=1}^{F_k^I - 1} |F_p^I| & \text{if } k = 2, \dots |F^I| \end{cases}. \tag{8}$$

$j_{F,l}$ is the sequence number of the first bin in the l-th group of bins, evaluated as:

$$j_{F,l} = \begin{cases} 1 & \text{if } l = 1 \\ 1 + \sum\limits_{q=1}^{F_l^B - 1} |F_q^B| & \text{if } l = 2, \dots |F^B| \end{cases}. \tag{9}$$

Indicator $\zeta_{l,k}$ gives information if the combination of bin and item group is used, i.e., it takes a value of 0 or 1. It is defined as a sum of values from submatrix x_{ij}, where the submatrix is defined for all bins from l-th group and only the first item from k-th group. The indicator is implemented as given with Equation (10).

$$\zeta_{l,k} = \sum_{j=j_{F,l}}^{j_{F,l}+|F_l^B|-1} x_{i_{F,k}j} \tag{10}$$

As equality constraints reduce the solution space, the consideration of item and bin grouping constraints actually reduces the complexity of the problem.

3. Results

In this section, the presented BPP problem formulation will be demonstrated in several case studies of material requirements planning with data sets acquired from an industrial environment. All experimental data are available at [26]. The problem concerns raw materials which must be cut to satisfy the needs of work orders for input material. More specifically, raw materials must be available in desired quantities by a specific time, while the input (intermediate) material for one product has to be acquired from the same raw material batch. Purchase orders and work orders are not linked in the current configuration, and material requirements planning is done manually.

Raw materials are provided as defined by purchase orders. In the presented BPP formulation, purchase orders are described with bins, where c_j determine the package size of raw material and a_j the time when it will be available. a_j is not necessarily the same for all raw materials of a particular purchase order. Every purchase order also has a label F_l^B, which designates a group of raw materials with the same properties. Work orders determine which materials (intermediates) are needed to produce one final product. Every intermediate is presented with an item in our problem formulation, where s_i specifies how much of a material is needed and d_i the work order's due date. As raw materials can slightly deviate in some parameters, we must deal with additional constraints. We must ensure that a group of intermediates used to make one final product are produced from the same raw material. Label F_k^I is used to determine items that belong to one work order, i.e., one final product.

In this way, the problem of material requirements planning can be translated into a generalized bin-packing problem. Depending on the situation encountered by the operator, different criterion with different constraints can be applied. We have analyzed two general situations that can occur in practical implementations. If the time constraints are feasible (Equation (4)), then we should apply the optimization problem presented with Equation (3), where the leftover is being minimizing. If this is not possible, softer constraints have to be chosen (Equation (5)), in which we optimize the problem from Equation (6). In this case, we are solving a multi-criterion problem, where leftover and tardiness are being minimized.

3.1. Implementation

The presented case studies were formally described with a mathematical description as defined in Section 2 and implemented with the linear programming modelling environment PuLP [27]. PuLP is an open source high-level modelling library that allows mathematical programs to be described in the Python computer programming language. PuLP provides an interface to many mixed-integer linear programming solvers such as CPLEX, Gurobi, CBC, GLPK. Gurobi solver [28] was used in our case. The solver can offer more algorithms to solve continuous or mixed-integer models. As we are dealing with an integer linear programming problem, we used the dual simplex method.

The problem was solved using a PC with 3.4 GHz and 8 GB RAM.

3.2. Elementary Case Studies

Let us first overview the simplified example introduced in Section 2.5 that illustrates the realistic environment. We must make three products from five available raw materials, which are supplied with two purchase orders. In the generalized BPP formulation, this means that we have two bin groups with five bins:

- $F_1^B \in \{B_1, B_2, B_3\}$,
- $F_2^B \in \{B_4, B_5\}$.

Every product is composed of more intermediates. In the generalized BPP formulation, the products are represented with item groups and intermediates with items:

- $F_1^I \in \{I_1, I_2, I_3\}$
- $F_2^I \in \{I_4, I_5\}$
- $F_3^I \in \{I_6, I_7, I_8\}$.

Details about available purchase (bins) and work (items) orders are summarized in Tables 1 and 2, respectively. We assume here that the processing time needed to cut the raw material is taken into account within a_j.

Table 1. Purchase orders—Bins.

Bin	c_j	a_j	F^B
B_1	100	10	1
B_2	60	10	1
B_3	20	20	1
B_4	60	50	2
B_5	70	50	2

Table 2. Work orders—Items.

Item	s_i	d_i	F^I
I_1	20	10	1
I_2	23	20	1
I_3	20	40	1
I_4	45	10	2
I_5	38	20	2
I_6	9	50	3
I_7	35	55	3
I_8	8	55	3

The problem was formulated as an integer optimization problem as described in Section 2 and can be solved without delays, as a solution exists that provides intermediate materials before the time is due. Therefore, we are minimizing the leftover only (Equation (3)). The solver finds an optimal solution, which is graphically presented in Figure 5. The figure shows which intermediates (items) are allocated to a specific purchase order (bin). Next to the bin/item label, time information a_j/d_i are given in round parentheses. Bins that belongs to one bin group are framed together. Items from one group are designated with the same color and we can see that the items from item groups F_1^I and F_2^I are assigned to bins from group F_1^B, while items from F_3^I are assigned to bin B_4 from F_2^B. Here white rectangles represent the leftovers. Bins B_1, B_2 and B_4 still have 12, 2 and 8 units of unused space, i.e., the cumulative leftover is $z_2 = 22$, while bins B_3 and B_5 are not used in a resulting plan.

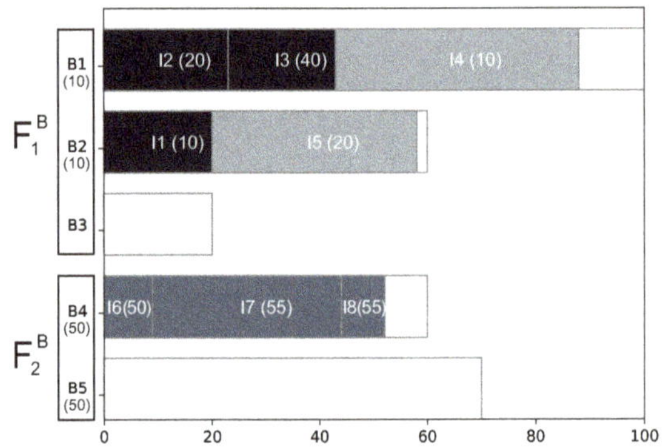

Figure 5. Solution to a simple problem.

In practice, it often happens that one or more raw material deliveries are delayed, or the product would need to be finished earlier. That type of situation could be illustrated when considering a problem where raw material modelled with bin B_2 is late for 30 time units ($a_2 = 40$; all other conditions remain the same). In this case, we cannot achieve a feasible solution using Equation (3). Therefore, an optimization function where delays are allowed but penalized must be used. The multiple criterion function with soft constraints is used to optimize leftover and due date violation (Equation (6)). As for the given problem where tardiness and leftover are equally important, we set the weighting factors as $w_1 = 1$ and $w_2 = 1$. Figure 6 shows the resulting schedule where three raw materials (bins) are used and each of them produces some leftover ($B_1 = 12$, $B_2 = 2$ and $B_4 = 8$). We can see that item I_5 is planned to be made of bin B_2, which is late for 20 time units (see red designation in Figure 6). The resulting objective function is $z_3 = 42$. Cumulative leftover is the same as in the previous example; however items I_1 and I_3 are scheduled vice versa in order to minimize total tardiness.

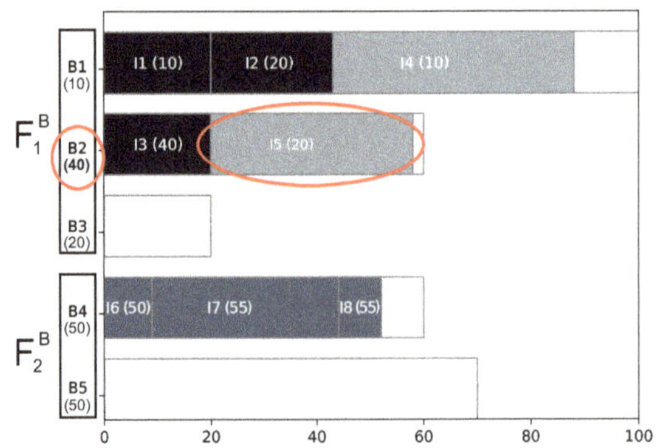

Figure 6. Solution to a simple problem where B_2 is late.

To ensure that the optimization problem works well for large size instances the mathematical formulation of the problem was validated on larger use case problems. Synthetical datasets of four different complexities were generated (UC1–UC4), where in each use cases all the bins are available before item's due date and for each use case at least one optimal solution, where no leftover is produced. Items and bins for each use case were generated based on optimal solution, where approximately 3/4 bin groups were used, where in each bin group one bin is left empty. The complexities of the use cases are as follows:

- UC1: 36 items (9 item groups) with 14 bins (5 bin groups). Optimal solution uses 7 bins. Also, the calculated solution uses 7 bins and no leftover is produced.
- UC2: 78 items (9 item groups) with 28 bins (7 bin groups). Optimal solution uses 17 bins. The calculated solution achieved in 44 min uses 16 bins with leftover of 2 units.
- UC3: 125 items (20 item groups) with 52 bins (9 bin groups). Optimal solution uses 29 bins. The calculated solution achieved in 6 h uses 30 bins with leftover of 5 units.
- UC4: 284 items (42 item groups) with 83 bins (16 bin groups). Optimal solution uses 57 bins. The calculated solution achieved in 7 h uses 55 bins with leftover of 60 units.

Figure 7 illustrates how the optimizer converges to optimal solution. For ease of comparison, the leftover is given in percentage of total items capacity. We can see that we achieve solutions with less than 5% of leftovers in a few seconds even for more complex problems that are present in the considered industrial environment. Such a result is incomparably better than can be achieved with the current planning practice. Let us add that such a plan is flawless (no mistakes and no overlapping activities), with fewer leftover and can be continuously upgraded in case of unexpected changes (e.g., new orders).

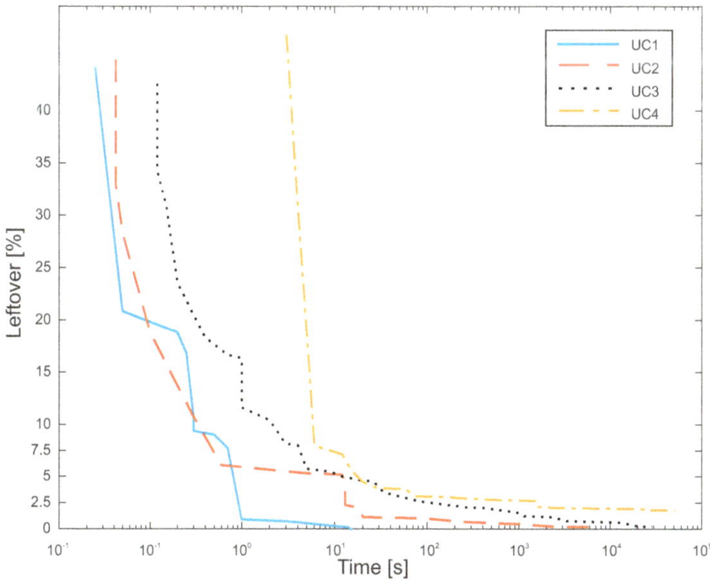

Figure 7. Validation on synthetical datasets (UC1–UC4).

3.3. Case Study of a Steel-Processing Production Process

The method is demonstrated on a data set from a real case study and illustrates a common situation in the observed production company. Typically, the production planner must deal with up to 150 intermediates that have to be produced from up to 50 raw materials. In this study, a problem where

150 intermediates (defined with 12 work orders, i.e., item groups) have to be produced from 100 raw materials grouped in 7 bin groups is used to validate our modelling and problem solving capabilities.

The problem was modelled and solved with the same procedure as described in the simple case study (criterion function from Equation (3)). As the problem is NP-hard, it is not possible to achieve the optimal solution in a short time. In our case, it took 593 s to achieve a close to optimal solution. The graph on the left of Figure 8 illustrates how the optimizer converges to an optimal solution. We do not achieve a better solution even if the optimization is run for a much longer time (24 h). The final solution produces 4923 of leftover when 22 bins are used. The solution and resulting plan are presented in Figure 9.

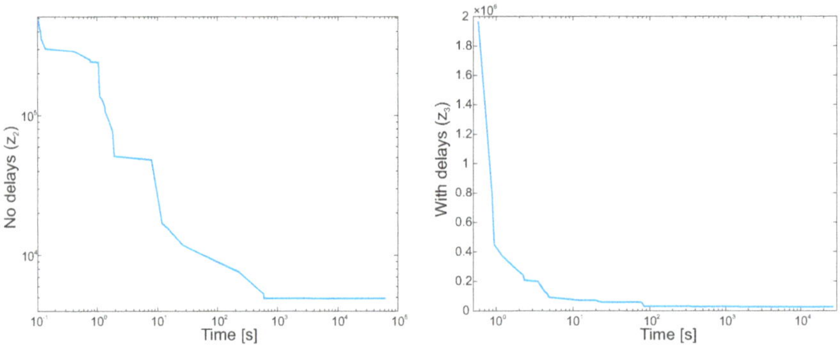

Figure 8. Convergence of criterion functions for the case study with no delays (**left**) and with delays (**right**).

We also analyzed a problem in which two purchase orders are late. In this case, we need to use the criterion function with soft constraints (Equation (6)) to optimize leftover and due date delays simultaneously. The following weighting factor values are considered in this experiment: $w_1 = 1$ and $w_2 = 1/36$. With the given settings, we get a solution ($z_3 = 24,777.45$) in which 25 bins are used and the cumulative leftover is 4238. In the resulting solution, more items are late. Their cumulative tardiness is 665,012 s, i.e., approximately 7.5 days. In this case, the optimization process took more time (26,518 s). The graph on the right in Figure 8 illustrates how the solution converges in this case. However, a close to optimal solution can be achieved in approx. 100 s.

In the case that minimization of the leftover is more important, we would choose other weighting factors ($w_1 = 1$ and $w_2 = 1$). In this case, 26 bins are used with a higher leftover of 11,200, but with a lower total tardiness (551,237 s), which means more than one day shorter than in the previous case. In this case, optimization took approximately 157 s.

Figure 10 illustrates the lateness of the work orders for all three studied cases, i.e., positive and negative difference between the completion time and the due date. We can see that none of the work orders are late in the case when purchase orders are available on time (upper data). In the case when two purchase orders are late, we can see that this will cause some work orders to also be late. However, there will be less delays in the case when the tardiness weighting factor is set to $w_2 = 1$.

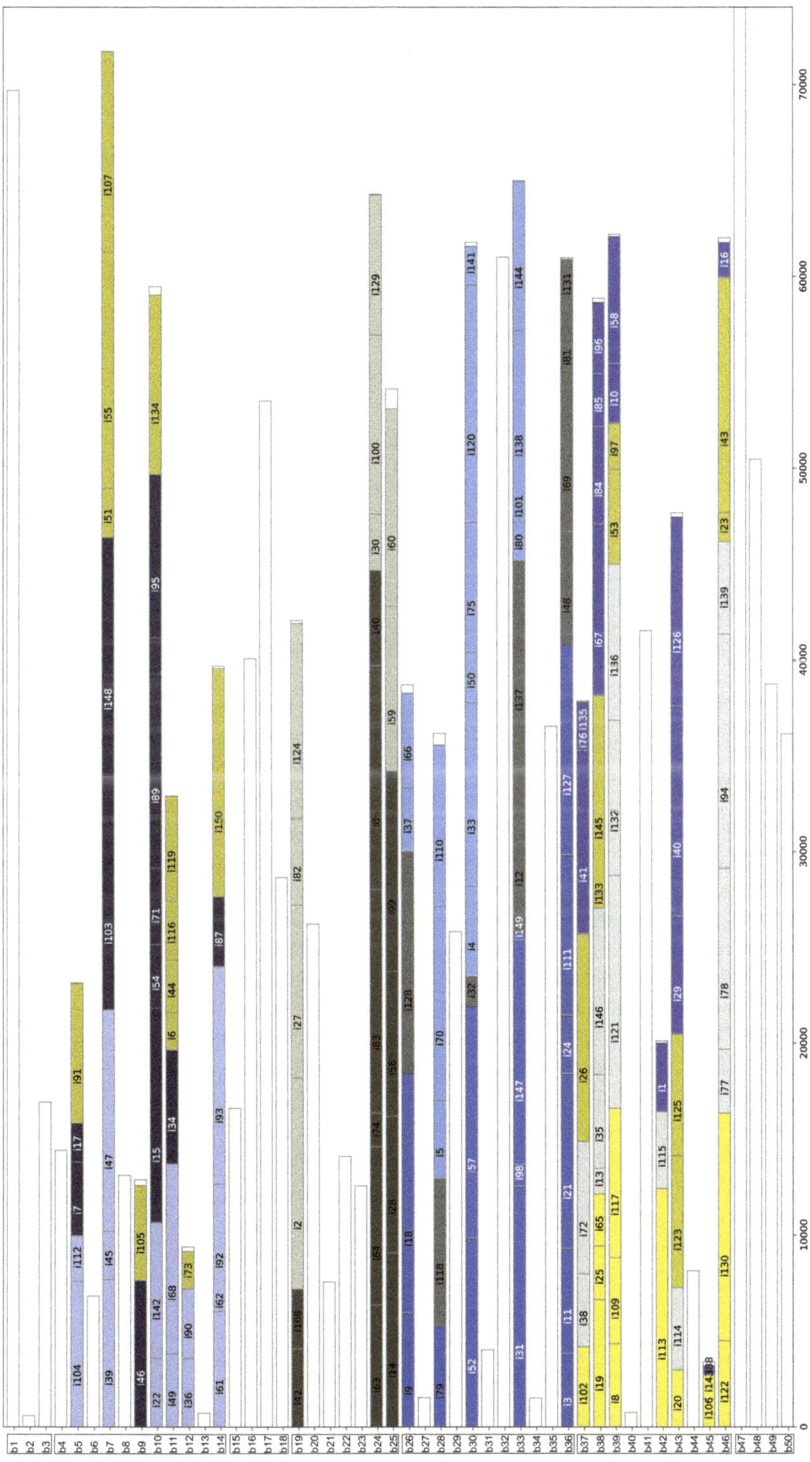

Figure 9. Resulting MRP plan for the case study with no delays (z_2).

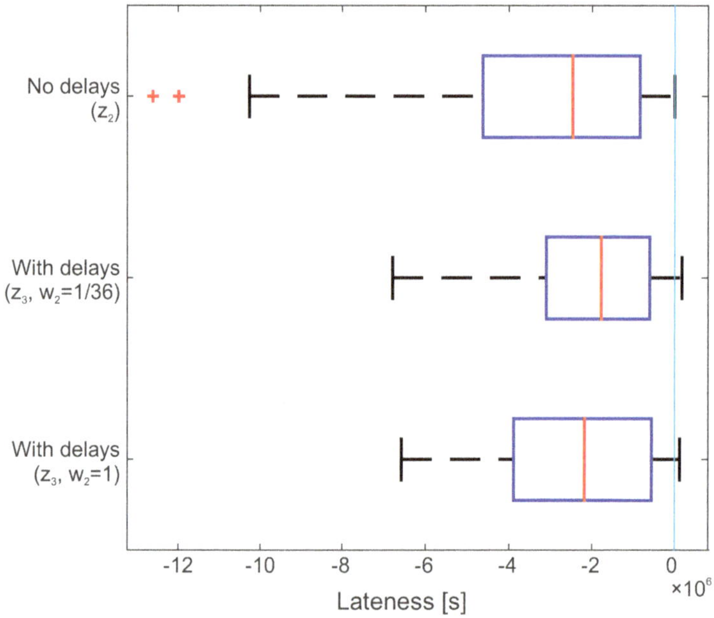

Figure 10. Lateness of input materials.

4. Discussion

General-purpose IT solutions can solve several problems at hand. But often, production specifics require development of special tailor-made solutions. In this paper, we considered a problem from manufacturing in which stock material must be cut to provide a reliable material requirements plan. This plan must satisfy the needs of capacity and must be available by the work order's due date. Besides this, the plan must also consider the fact that a group of work orders must be produced from the same batch of raw material. In this way, the manufacturer can systematically compensate for some undesirable variations in raw material quality. In the current day to day practice, the plan is managed manually, and it is very difficult to maintain the plan up-to-date, even in smaller dimensions. The decisions of the operator are time consuming and prone to errors. This results in situations in which the operator must constantly make plan corrections.

In this paper, we propose using an extended bin-packing problem formulation to systematically solve the material planning. A basic BPP formulation is extended to include constraints such as variable bin and item sizes, time limitations, and the fact that only a group of bins can be used to produce one group of items.

The suggested solution offers a tool for supporting the production planner with his/her decisions. With it, he/she can determine how to efficiently cut the raw material to satisfy the planned work orders. Depending on the situation, the planner can choose between various model formulations. He/she can optimize the leftover, tardiness, or both.

We demonstrated that the proposed solution can solve a problem of realistic dimensions quickly enough to be used in an industrial application. However, in real applications, case-specific requirements would first need to be analyzed to most appropriately set the importance of leftovers and/or tardiness. In this way, the time for producing a feasible material requirements plan is reduced. The operator is relieved and can be focused on other tasks. Also, the need for corrections is decisively reduced. To summarize, variations in raw material quality no longer cause a problem when implementing our solution. We propose to use a rigorous and accessible MILP solution approach, i.e., Gurobi optimization solver, which satisfies the requirements of the problem considered in this paper. To extend the use of the proposed method to more complex problems approximation algorithms should be applied (e.g., [3,12,21]).

To summarize, variations in raw material properties no longer cause a problem when implementing our solution, since end products are made from the same material and, consequently, the products' quality is improved. Moreover, raw material batches are used more rationally, which leads to decrease the needed of storage place and unused batch leftovers.

Author Contributions: Conceptualization, methodology, software and validation, D.G. and M.G.; writing–original draft preparation, D.G.; writing–review and editing, M.G. All authors have read and agreed to the published version of the manuscript.

Funding: The work was supported by the INEVITABLE project which has received funding from the European Union's Horizon 2020 research and innovation program under grant agreement No 869815. The support of the Slovenian Research Agency (P2-0001) is also acknowledged.

Conflicts of Interest: The authors declare no conflict of interest. The funders had no role in the design of the study; in the collection, analyses, or interpretation of data; in the writing of the manuscript, or in the decision to publish the results.

Abbreviations

The following abbreviations are used in this manuscript:

MRP	Material requirements planning
BPP	Bin-packing problem
MBSBPP	Multiple bin-size bin-packing problem
LP	Linear programming
MILP	Mixed-integer linear program

References

1. Onwubolu, G. *Emerging Optimization Techniques in Production Planning and Control*; Imperial College Press: London, UK, 2002.
2. Plenert, G. Focusing material requirements planning (MRP) towards performance. *Eur. J. Oper. Res.* **1999**, *119*, 91–99. [CrossRef]
3. De, A.; Choudhary, A.; Turkay, M.; Tiwari, M.K. Bunkering policies for a fuel bunker management problem for liner shipping networks. *Eur. J. Oper. Res.* **2019**, in press. [CrossRef]
4. ISO/IEC. *IEC 62264-3: Enterprise-Control System Integration—Part 3: Activity Models of Manufacturing Operations Management*; ISO/IEC: Geneva, Switzerland, 2016.
5. Goswami, M.; De, A.; Habibi, M.K.K.; Daultani, Y. Examining freight performance of third-party logistics providers within the automotive industry in India: An environmental sustainability perspective. *Int. J. Prod. Res.* **2020**, 1–28. [CrossRef]
6. Suzanne, E.; Absi, N.; Borodin, V. Towards circular economy in production planning: Challenges and opportunities. *Eur. J. Oper. Res.* **2020**, *287*, 168–190. [CrossRef]
7. Ray, A.; De, A.; Mondal, S.; Wang, J. Selection of best buyback strategy for original equipment manufacturer and independent remanufacturer—game theoretic approach. *Int. J. Prod. Res.* **2020**, 1–30. [CrossRef]
8. Garcia-Sabater, J.P.; Maheut, J.; Garcia-Sabater, J.J. A two-stage sequential planning scheme for integrated operations planning and scheduling system using MILP: The case of an engine assembler. *Flex. Serv. Manuf. J.* **2012**, *24*, 171–209. [CrossRef]
9. Gilmore, P.C.; Gomory, R.E. A Linear Programming Approach to the Cutting Stock Problem—Part II. *Oper. Res.* **1963**, *11*, 863–888. [CrossRef]
10. Martello, S.; Toth, P. *Knapsack Problems: Algorithms and Computer Implementations*; John Wiley & Sons, Inc.: New York, NY, USA, 1990.
11. Delorme, M.; Iori, M.; Martello, S. Bin packing and cutting stock problems: Mathematical models and exact algorithms. *Eur. J. Oper. Res.* **2016**, *255*, 1–20. [CrossRef]
12. Coffman, E.; Csirik, J.; Galambos, G.; Martello, S.; Vigo, D. *Bin Packing Approximation Algorithms: Survey and Classification*; Springer: New York, NY, USA, 2013; Volumes 1–5, pp. 455–531.

13. Kim, T. Production Planning to Reduce Production Cost and Formaldehyde Emission in Furniture Production Process Using Medium-Density Fiberboard. *Processes* **2019**, *7*, 529. [CrossRef]
14. Wäscher, G.; Haußner, H.; Schumann, H. An improved typology of cutting and packing problems. *Eur. J. Oper. Res.* **2007**, *183*, 1109–1130. [CrossRef]
15. De Carvalho, J.M.V. LP models for bin packing and cutting stock problems. *Eur. J. Oper. Res.* **2002**, *141*, 253–273. [CrossRef]
16. Lodi, A.; Martello, S.; Monaci, M. Two-dimensional packing problems: A survey. *Eur. J. Oper. Res.* **2002**, *141*, 241–252. [CrossRef]
17. Epstein, L.; Levy, M. Dynamic multi-dimensional bin packing. *J. Discret. Algorithms* **2010**, *8*, 356–372. [CrossRef]
18. Friesen, D.K.; Langston, M.A. Variable sized bin packing. *SIAM J. Comput.* **1986**, *15*, 222–230. [CrossRef]
19. Hemmelmayr, V.; Schmid, V.; Blum, C. Variable Neighbourhood Search for the Variable Sized Bin Packing Problem. *Comput. Oper. Res.* **2012**, *39*, 1097–1108. [CrossRef]
20. Bang-Jensen, J.; Larsen, R. Efficient algorithms for real-life instances of the variable size bin packing problem. *Comput. Oper. Res.* **2012**, *39*, 2848–2857. [CrossRef]
21. Jansen, K.; Kraft, S. An Improved Approximation Scheme for Variable-Sized Bin Packing. *Theory Comput. Syst.* **2016**, *59*, 262–322. [CrossRef]
22. Reinertsen, H.; Vossen, T.W. Discrete optimization. *Eur. J. Oper. Res.* **2010**, *201*, 701–711. [CrossRef]
23. Arbib, C.; Marinelli, F. Maximum lateness minimization in one-dimensional bin packing. *Omega* **2017**, *68*, 76–84. [CrossRef]
24. Maddaloni, A.; Colla, V.; Nastasi, G.; Seppia, M.D.; Iannino, V. A Bin Packing Algorithm for Steel Production. In Proceedings of the 2016 European Modelling Symposium (EMS), Pisa, Italy, 28–30 November 2016; pp. 19–24.
25. Srikumar, V. Soft Constraints in Integer Linear Programs. Unpublished Note. 2013. Available online: https://svivek.com/research/srikumar2013soft-constraints.html (accessed on 1 October 2020).
26. Gradišar, D.; Glavan, M. Data—MRP Using BPP. Available online: https://doi.org/10.13140/RG.2.2.27414.98885/1 (accessed on 1 October 2020).
27. Mitchell, S.; Kean, A.; Mason, A.; O'Sullivan, M.; Phillips, A.; Peschiera, F. Optimization with PuLP. Available online: https://coin-or.github.io/pulp/ (accessed on 1 September 2020).
28. Gurobi. Gurobi Optimizer. Available online: http://www.gurobi.com (accessed on 1 September 2020).

processes

MDPI

Article

Process of Creating an Integrated Design and Manufacturing Environment as Part of the Structure of Industry 4.0

Andrzej Paszkiewicz [1,*], Marek Bolanowski [1], Grzegorz Budzik [2], Łukasz Przeszłowski [2] and Mariusz Oleksy [3]

[1] Department of Complex Systems, The Faculty of Electrical and Computer Engineering, Rzeszow University of Technology, Al. Powstańców Warszawy 12, 35-959 Rzeszów, Poland; marekb@prz.edu.pl
[2] Department of Machine Design, The Faculty of Mechanical Engineering and Aeronautics, Rzeszow University of Technology, Al. Powstańców Warszawy 12, 35-959 Rzeszów, Poland; gbudzik@prz.edu.pl (G.B.); lprzeszl@prz.edu.pl (Ł.P.)
[3] Department of Polymer Composites, Faculty of Chemistry, Rzeszow University of Technology, Al. Powstańców Warszawy 6, 35-959 Rzeszów, Poland; molek@prz.edu.pl
* Correspondence: andrzejp@prz.edu.pl

Received: 21 July 2020; Accepted: 18 August 2020; Published: 20 August 2020

Abstract: This paper presents the process for creating an integrated design and manufacturing environment supporting 3D printing as part of the structure of Industry 4.0. This process is based on a developed framework for the design of modern automated and computerized infrastructure. The task of the described system is to combine all the steps included in the operating range of incremental systems based on an IT platform by integrating data from individual areas, such as IT systems supporting remote 3D printing. The proposed framework for incremental processes is a universal solution that can be defined in detail by a single organizational unit running 3D printing, as well as by a cluster of entities related to 3D printing. In the initial phase, the framework design includes a set of guidelines for IT (Information Technology) systems that facilitate the construction of individual elements and the creation of communication interfaces. In subsequent stages, the framework may already implement elements of the access and communication program interface, as well as guidelines for the industrial components to be included. The proposed framework for additive technologies is based on modern IT tools that enable the creation of geographically and functionally possible prototyping systems that can be integrated into the structure of Industry 4.0. To create optimal processes and economic systems, the principles of the construction and integration of individual services and equipment were developed. This new comprehensive approach is proposed in the present paper as a coherent framework. Moreover, the proposed solution has great potential for use in the design and production processes of various industries, such as chemicals, materials and construction.

Keywords: manufacturing process; additive manufacturing; IoT; computer systems and networks; 3D printing

1. Introduction

Incremental production is now widely used in production [1,2], scientific [3,4], didactic [5,6] and medical [7] environments. The present examples relate to the use of 3D printing both in SMEs and in large enterprises. Of course, the methods and technologies used in these separate environments may differ, but they share common goals, such as increasing the efficiency, competitiveness and flexibility in the design and production of various components. The above-mentioned papers also refer to the possibility of combining additive and sub-additive technologies within common business and production processes. Additive technologies are fundamentally different from those of traditional production. Since these technologies constitute a relatively new approach, on the one hand, they require considerable research (e.g., in the field of materials, technological processes, etc.); on the other hand, their features and properties enable their use by scientists to develop new innovative components and products, including for medicine. This applies to both organ models and dedicated replicas. The didactics area mentioned previously holds great promise for the use of additive technologies through the possibility of integrating knowledge for creative and demanding projects. At present, 3D printing technologies are an indispensable element of many design and manufacturing processes in modern industry, including Industry 4.0. The world of manufacturing, which focuses on the conceptual, design, test and laboratory environments, as well as the production of all components, has undergone sudden and dynamic changes. In the current era of globalization, individual stages and their individual components can be implemented in geographically remote locations. The same applies to human resources, such as a team of specialists carrying out specific research, development work, or learning the basics of rapid prototyping. A person's location (i.e., their place of work, study or life) is playing an increasingly less important role in his or her professional work or participation in courses. From this point of view, the concept of remote work seems to be crucial [8–11]. The work in [8] deals with issues related to the efficiency of remote work, which is particularly important in the current pandemic situation, and is related to the dissemination of this form of work within industry. The results described in [9] indicate that remote work is positively perceived by employees in the context of productivity and job satisfaction. The work in [10] compares the use of communication channels in remote work and in work based around a central office. At the same time, the authors indicate the threats that companies whose employees work remotely have to face. The authors in [11] propose a new approach for improvement in the area of designing remote work systems to improve their efficiency. Thus, employers and employees are ready for remote work and are aware of the risks associated with it. Companies can, therefore, benefit from the expertise of workers without requiring them to be physically present in a given geographical location. This mechanism can shorten the time needed for designing and implementing new products on the market and reduce the logistics costs related to a company's operation. Taking into consideration the achievements of modern technology in the field of IT tools, devices and means of communication, the remote implementation of research is at our fingertips. Of course, this approach has its own advantages and disadvantages.

The challenges posed by the modern labor market, IT technologies and industrial conditions have forced the introduction of a new approach for the prototyping process. Presently, this process is increasingly implemented in remote rapid prototyping laboratories with distributed structures using asynchronous work models with batch processing elements. Two trends are visible in this area:

- The establishment of teaching laboratories to which students or employees can obtain remote access [12]. In this case, the functionality of such laboratories is aimed at the implementation of readily available (often closed) test scenarios. Often, remote laboratories are built around issues that are relatively easy to implement, e.g., programming, the operation of electronic circuits, and access to CAD (Computer Aided Design) software.
- The second trend is related to the construction of remote research and production laboratories [13]. The analysis of available solutions clearly shows that such solutions are designed for specific applications, built each time from scratch while considering only the needs of a specific customer. Of course, the solutions proposed in these scenarios are optimally suited to specific needs, but the costs of producing and modifying such solutions are disproportionately high. It is often assumed that the construction of this class of solutions will utilize dedicated and expensive elements of IT architecture that will overcome problems with distance and transmission delays. An example of such a solution is the Tactile Internet (TI) network [14–16]. Often, laboratories of this type are considered in terms of their teleoperative applications and are used, for example, to conduct remote surgical procedures or remotely control machines. TI class systems are based on

communication systems characterized by ultra-low latency, strong reliability, and high availability. However, such an approach is characterized by very high construction and operating costs. Often the costs of such a solution exceed the costs related to the physical relocation of employees to carry out project work.

The classic approach to the design and construction of remote testing and manufacturing laboratories does not fit the modern approach to project creation and management. The designed nature of tasks, their geographic dispersion, and the multidisciplinarity of created projects and prototypes enforce flexibility at every stage of prototyping. For the implementation of specific projects, a team is formed and the necessary resources are obtained. This approach fits the agile project management methodologies used in Industry 4.0, but primarily agrees with the trends defined by Industry 5.0 [17], which will increase the quality and efficiency of production thanks to artificial intelligence. However, this development is associated with a high security risk. Therefore, as part of Industry 5.0, it will be necessary to closely integrate people and machines in some decision-making areas. Differences in the approaches to prototyping are presented in Figure 1.

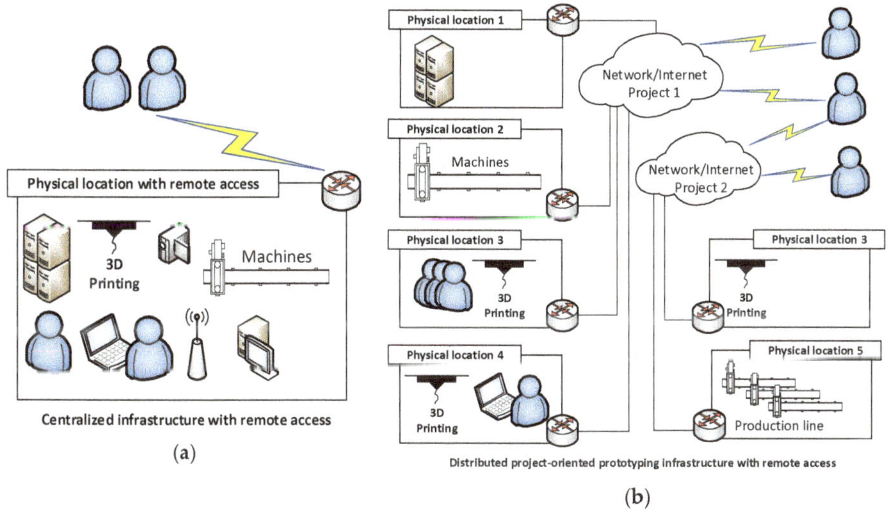

Figure 1. Types of prototyping system architecture: (**a**) Centralized infrastructure with remote access; (**b**) Distributed project-oriented prototyping infrastructure with remote access.

Both architectures shown in Figure 1 may be applicable. In both cases, the following problems can be identified in their implementation: difficulties with the standardization of solutions in the area of IT systems, different approaches to security policies, and different approaches to resource sharing. To further develop Remote Distributed Rapid Prototyping (RDRP) [18] or Remote Rapid Prototyping (RRP), it is necessary to propose a functional, uniform and layered reference model. It is particularly important that RDRP or RRP systems at this stage be perceived as IT systems that must be constructed and designed. Thus, we propose a new design pattern, hereafter referred to as the framework.

The idea of remotely manufacturing components based on 3D printing is not a new concept. An approach based on the Tele-Manufacturing Facility (TMF) concept [19] was proposed at the beginning of the 21st century. This approach enabled semi-automated remote design and 3D printing. Its basic assumption was to avoid errors by verifying and correcting STL (Stereolithography) files and then printing ready elements. A similar approach already applied on a global scale is the implementation of 3D printing services available for Windows 10 operating systems [20]. Thanks to these services, even a complete amateur can become a designer of components that will later be printed on a local 3D printer (using LAN or WiFi links), or by ordering a print service from a remote location. In the first case, this service constitutes an improvement, but is limited from the perspective of the system's scalability, as well as its lack of integration with other elements of the 3D printing process, e.g., a knowledge base or a system supporting the decision-making process of rapid prototyping. On the other hand, the approach in the second case is focused on the individual recipient and cannot be used today to provide professional or industrial solutions for distributed infrastructure. Combinations of both concepts include internet services such as i-materialise [21], 3D Hubs [22] and Igus [23], which enable one to design and print 3D elements. Some of these services also include CNC (Computerized Numerical Control) machining and access to ready-made models. However, even these solutions provide only a limited range of services and also provide a closed environment without the possibility of integration with other distributed design and manufacturing resources. In this area, IT solutions supporting design processes could also be more widely considered. These issues have been described in various publications, which discuss, e.g., issues related to the combination of design processes, analyses using CAx (various Computer Aided systems) systems and elements of information algorithms used in incremental manufacturing processes [24], and topological design and spatial analysis in the PLM (Product Lifecycle Management) area with selected applications of incremental generation systems [25]. Several existing systems also support decision making in the process of rapid prototyping. Initially, these systems only included databases designed to facilitate the selection of an appropriate solution based on a catalog of available products [19]. In the next period, knowledge databases appeared along with advanced mechanisms for choosing the right technology, e.g., 3D printing for current customer needs [26]. However, all the solutions presented above do not introduce a comprehensive approach for the entire rapid prototyping process, taking into account both the design and manufacturing phases and combining ICT (Information and Communication Technologies), analytical, application, and control and measurement infrastructure with distributed software and hardware environments belonging to different entities. Notably, the problem of combining and standardizing the tools available on the market for the rapid prototyping (RP) process was already identified in 2002 [19]. To date, research results and IT tools have been made available to enable the creation of geographically and functionally extensive prototyping systems.

To create systems that are optimal from a process and economic perspective, it is necessary to develop specific principles for the construction and integration of individual services and equipment. A new comprehensive approach for this process is proposed in the present paper as a coherent framework. In the literature, one can find examples of methods and other frameworks that are used in distributed manufacturing environments. However, these systems represent only a limited response to the aforementioned issue in the field of operational management, or refer to the implementation of a selected technology or tools supporting selected areas related to design and manufacturing. The framework proposed in [27] only handles the desire to increase operational efficiency in an organization. For the framework described in [28], the authors proposed a solution that aims to increase the efficiency of designing and reconfiguring multi-level and multi-dimensional manufacturing systems. The framework proposed in this paper cannot be compared to the Cloud-based manufacturing (CBM) architecture [29], which facilitates decentralization, thereby increasing the use of design and manufacturing resources and reducing costs. The CMB model is an extension of web-based [30] and agent-based [31,32] models. The former is based only on the classic client–server communication used on the internet, while the latter allows for more effective integration of independent resources in the form of agents. All three models can enable the remote design or manufacturing of components. However, they do not constitute a coherent environment that enables the integration of various design and manufacturing elements that were initially independent from each other. Moreover, the framework proposed in this paper can be implemented using cloud solutions. Nevertheless, this framework is not identical and depends on a solution to cloud computing.

Generally, the need to integrate production resources can be understood in terms of the ubiquitous manufacturing paradigm described in [33,34]. The assumptions of these models are based on the use of highly unified communication interfaces and rigid process rules, which must all be implemented at the same time so that a given entity can integrate with the production elements belonging to other entities. The use of this class of systems is possible for integrated producer groups, such as production consortia planning to create distributed production lines. The direct and widespread use of the above models in production environments is difficult and sometimes even impossible due to the high costs of their implementation. In contrast to these models, the present framework allows for the evolutionary implementation of successive levels of integration, while maintaining a high level of universalism and openness to the proprietary solutions of companies seeking integration with other entities.

This paper consists of several sections. The second section includes a description of the framework for the design of modern automated and computerized infrastructure. This section presents the main concept and the cross-layer structure of the layers. In section three, the level of IT maturity of the implemented framework is presented, as these levels facilitate the planning and implementation of the framework in terms of its organization and technology. The versatility of this concept allows its use in both homogeneous and heterogeneous environments in terms of ownership and location. Section 4 covers the system architecture of the proposed framework along with the functionality of the individual modules, while Section 5 illustrates the possibility of adopting the proposed architecture in a remote rapid prototyping environment. Section 6 presents an example of the actual implementation of the framework proposed by the authors in cooperation with industry. Section 7 compares the proposed approach with the classic approaches in terms of the actual production costs using three selected Fused Deposition Modeling (FDM) technologies. At the end, the results of the work are summarized, along with the plans for future work and the development of the proposed approach.

2. Proposed Framework

The large variety of technologies, application solutions, communication standards and architectures requires the development of an efficient structure that facilitates their management. The proposed structure divides the entire design area into layers, each of which is responsible for a different functional range (Figure 2), while not introducing technological limitations. Previous models focused on individual elements of the design and manufacturing process [27–30,33,34] Some of them [33,34] covered only the integration of the communication layer on the basis of selected mechanisms, standards and data transmission protocols. Others [27] analyzed and implemented effective process management in the actual analysis/business layer. Other solutions involve architecture models related to the physical structure of the production system [28]. However, a separate group of models consists of solutions related to processing mechanisms that are actually located in the service layer [29,30]. However, the proposed approach can consider several solutions while arranging their functional scope and implementing cross-layer mechanisms [35–37]. Our experiences related to the construction of RP systems have shown that the process of building a laboratory for RDRP can be understood as building a specific IT system in which selected elements will be implemented using the batch processing paradigm. Such an approach entails the possibility of using a project approach based on the use of a framework that reduces the costs of system implementation and promotes the solution [18]. The proposed framework (Figure 2) has the characteristics of a universal solution, i.e., it can be defined in detail by both a single organization as well as by a cluster of organizations. In the initial phase, the framework project provides a set of guidelines for information systems that facilitate the construction of individual elements and the creation of communication interfaces. In the next stages, the framework may have already implemented elements of the software access and communication interface, as well as guidelines for the attached industrial elements.

The layers have been designed in such a way as to combine elements and actions similar to each other.

Hardware layer (w1): This layer includes the physical elements that make up the chain necessary for designing, modeling, manufacturing and control in the rapid prototyping process. Therefore, within this layer, one can distinguish the different production devices, specialized computing units, high resolution digital cameras and data storage necessary to collect the data used at particular stages of the prototyping process.

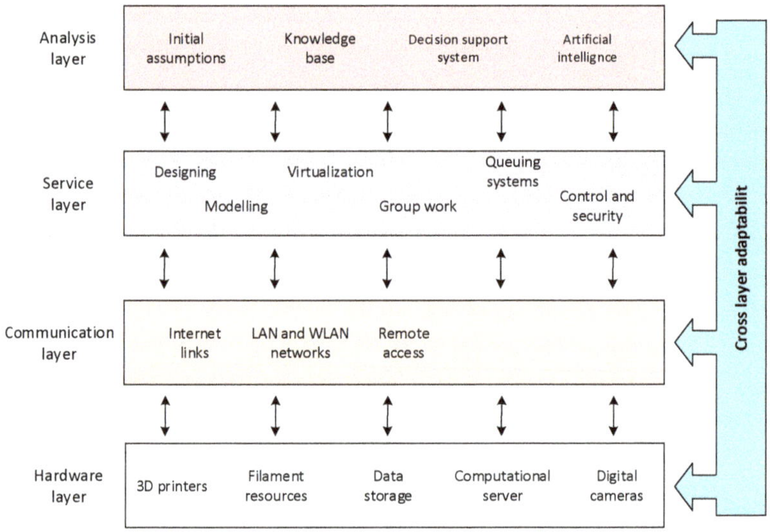

Figure 2. Proposed cross-layer framework.

Communication layer (w2): This layer is responsible for providing a high-throughput communication environment that enables effective data exchange between all elements and persons involved in the rapid prototyping process. Thus, within this layer, the elements of the network infrastructure are separated, including the mechanisms for remote access.

Service layer (w3): This layer includes all activities and corresponding tools (primarily IT) used in most of the stages of rapid prototyping. On the one hand, these tools create a specific digital–virtual environment that enables the optimal use of available hardware platforms; on the other hand, these tools guarantee interoperability, group work, the ability to separate, and the integration of individual works.

Analysis layer (w4): This layer is the most important from the perspective of Industry 4.0 and the optimization of the rapid layer prototyping process. It mainly supports the functioning of the service layer, although it works with the hardware layer, as well as the communication layer (this will be described later in the paper as the cross-layer). This layer implements highly specialized tools based on artificial intelligence or knowledge, as well as systems supporting decisions based on fuzzy models. The proper use of this layer may increase the effectiveness of the manufacturing process, e.g., providing better management and selection of filament, better management of access to particular resources (manufacturing equipment, design applications and component modeling), and increase the quality of manufactured components.

The proposed framework has not only a sequential character (i.e., data from one layer are not only forwarded to the next one) but also a cross-layer feature. This property allows the exchange of information between layers (cross-interactions) that are not directly related to each other, so as to increase the efficiency of the entire infrastructure. Below are examples of cross-layer functioning in relation to the proposed framework:

- Communication of the analysis layer with the hardware layer (with the participation of the communication layer) but omitting the service layer. The communication layer has a dual nature: it can be perceived as a passive computer network or as an active element shaping the RDRP system architecture. Whenever this layer is not explicitly mentioned, it should be treated as a functionally passive communication element. The direct cross-interaction of the analysis layer with the hardware layer is obvious. The hardware layer provides computing resources directly for programs at the analysis layer. Elements of manufacturing programs (e.g., for numerical machine tools) can be directly tested on machine controllers and images from cameras can be transferred directly to image recognition systems placed in the analysis layer. It thus becomes possible to carry out remote research, such as remote analysis of the distribution of machine operators' focus areas, while working to optimize manufacturing procedures. Examples of such research carried out to increase the efficiency and safety of aircraft pilots are described in [38]. However, similar studies could be carried out for operators of production machines, especially in an Industry 4.0 environment. Examples of interactions can be multiplied. An important feature of using cross-interactions is the possibility of partially implementing the framework in a given organization, or the possibility of system interactions within frameworks implemented in different clusters or organizations.

- Communication of the analysis layer with the communication layer. This approach enables the collection of data based on physical and logical communication in the environment of programmable computer networks (e.g., Software-Defined Networking SDN) [18,39,40] based on the data collected and the system's knowledge of the communication environment, as well as the algorithms created for the optimal selection of parameters and communication paths. The system combines (automatically or with the administrator) the transmission environment with performance and logistics aspects.
- Service layer communication with the hardware layer. In Internet of Things (IoT) devices, application modules collect and process telemetric data locally, e.g., the temperature, pressure, tool wear, etc. An example of such a service is hardware and manufacturing process diagnostics. The interactions of these layers are natural in the final stage of each iteration of the RDRP process during the production of the prototype.

3. Levels of IT Framework Maturity

The adopted model has a wider application than just the rapid prototyping process, as it is part of the whole idea of Industry 4.0. A separate issue is how to implement the framework in a given institution, cluster or company. The initial results of the research on and implementation of the test framework in real systems clearly show that full implementation of the entire model (even for newly-built systems) is difficult, and, in some cases, even impossible. This is mainly due to the time constraints of the project and the lack of a sufficient number of qualified specialists on the border of IT and industry. Therefore, the implementation of the proposed framework should adopt an evolutionary character. Due to the dominant nature of IT issues related to the construction of this class system, this criterion was selected for further analysis. To assess the level of implementation, the concept of the level of IT maturity of the implemented framework was introduced.

Definition 1. *The level of IT information maturity of the framework is described by the variable* $p = \{p_1, p_2, \ldots, p_n\}$*. We describe the level of implementation of particular system features* $c = \{c_1, c_2, \ldots, c_m\}$ *via the following matrix D of size* $n \times m$*:*

$$d_{ij} = \begin{cases} 0 & \text{if for a given level } p_i, c_j \text{ feature does not have to be implemented} \\ 1 & \text{if for a given level } p_i, c_j \text{ feature has to be implemented} \end{cases}$$

This definition of levels of maturity allows a person to use his or her own set of key features for the planned implementation. Let us consider an example whose system features are defined as follows (Table 1):

Table 1. Definition of the features of the example system.

Feature	Description
c_1	The system has precisely defined requirements for layers w_1 to w_2
c_2	The system has precisely defined requirements for layers w_3 to w_4
c_3	The system has implemented hardware services in the w_3 layer
c_4	The system has implemented hardware services in the w_4 layer
c_5	The system has implemented software services in the w_2 layer
c_6	The system has implemented adaptive control in the w_3 layer
c_7	The system has implemented remote access for layers w_3 to w_4
...	...
c_m	The system has implemented advanced management algorithms based on artificial intelligence in layer w_1

At the implementation stage, the system designers, in consultation with the business division, industrial partners, employees of the prototyping laboratory and customers, define a matrix with levels D according to Definition 1. For the example considered, this matrix can take the following form:

$$D = \begin{array}{c|cccccccc} & c_1 & c_2 & c_3 & c_4 & c_5 & c_6 & \dots & c_m \\ p_1 & 1 & 1 & 0 & 0 & 0 & 0 & \dots & 0 \\ p_2 & 1 & 1 & 1 & 1 & 0 & 0 & \dots & 0 \\ p_3 & 1 & 1 & 1 & 1 & 1 & 0 & \dots & 0 \\ \vdots & \vdots & \vdots & \vdots & \vdots & \vdots & \vdots & \vdots & \vdots \\ p_n & 1 & 1 & 1 & 1 & 1 & 1 & \dots & 1 \end{array}$$

During the development of information systems supporting prototyping, the proposed framework can determine the features of the information system architecture required at each level of maturity. This allows one to plan the development of the system, but also determines the possibility of its integration with other systems. The levels of IT system maturity in relation to the development of a given system are presented in Figure 3.

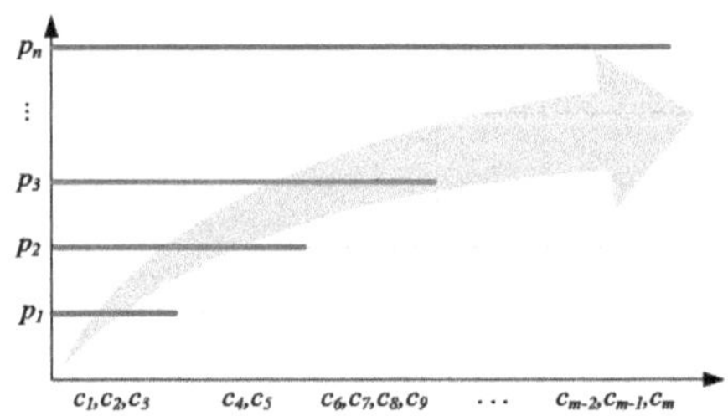

Figure 3. Levels of IT system implementation maturity.

Of course, the proposed framework gives one the opportunity to create custom levels of maturity that correspond to the requirements of a given company, laboratory, or production cluster. To better illustrate the assumptions of the model, we proposed a general set of levels of maturity as the starting point for further work.

For the system characteristics given in Table 2, the IT maturity level matrix will be as follows:

$$D = \begin{array}{c|cccccccc} & c_1 & c_2 & c_3 & c_4 & c_5 & c_6 & c_7 & c_8 \\ p_1 & 1 & 1 & 0 & 0 & 0 & 0 & 0 & 0 \\ p_2 & 1 & 1 & 1 & 1 & 0 & 0 & 0 & 0 \\ p_3 & 1 & 1 & 1 & 1 & 1 & 0 & 0 & 0 \\ p_4 & 1 & 1 & 1 & 1 & 1 & 1 & 0 & 0 \\ p_5 & 1 & 1 & 1 & 1 & 1 & 1 & 1 & 1 \end{array}$$

Table 2. Definition of the general features of the framework model.

Feature	Description
c_1	Analysis of the needs and capabilities of the company and technology
c_2	The system has precisely defined requirements for layers w_1 to w_4
c_3	Reconciliation of network communication standards for layer w_4
c_4	Definition of communication standards and protocols for w_1 and w_2 and remote access technologies
c_5	Deployment of services in layer w_3
c_6	Implementation of mechanisms and analytical tools in the w_4 layer (including specialized knowledge, search and data mapping algorithms, and artificial intelligence, reporting systems)
c_7	Full inter-layer integration, including the implementation of cross-layer mechanisms
c_8	Transition from reactive to active management and a resource sharing model

From the perspective of the matrix analysis, we should determine the correlation properties of the features that are appropriate for subsequent levels $p_1, p_2, \ldots, p_n$. It is natural to include the following form, $p_i = p_{i-1} + c^{p_i}$, where c^{p_i} is a set of features specific to a given level p_i, where $i = 1 \ldots n$. For example, from Figure 3, $p_3 = p_2 + \{c_6, c_7, c_8, c_9\}$. However, one should consider a case where the level does not contain all the features of level p_{i-1}. In other words, if level p_i contains a set of features $p_{i-1} = \left\{c_1^{p_{i-1}}, c_2^{p_{i-1}}, \ldots c_u^{p_{i-1}}\right\}$, is it possible that the following situation may occur:

$$p_i = p_{i-1} + c^{p_i} - \left\{c_1^{p_{i-1}}, c_2^{p_{i-1}}\right\} \tag{1}$$

Admitting this situation will lead to the creation of a matrix with the following form:

$$D = \begin{array}{c|cccccccc} & c_1 & c_2 & c_3 & c_4 & c_5 & c_6 & \cdots & c_m \\ p_1 & 1 & 1 & 0 & 0 & 0 & 0 & \cdots & 0 \\ p_2 & 1 & 1 & 1 & 1 & 0 & 0 & \cdots & 0 \\ p_3 & 0 & 0 & 1 & 1 & 1 & 0 & \cdots & 0 \\ \vdots & \vdots & \vdots & \vdots & \vdots & \vdots & \vdots & \vdots & \vdots \\ p_n & 1 & 1 & 0 & 1 & 0 & 1 & \cdots & 1 \end{array}$$

With this approach, further levels of maturity will become less dependent on each other. Such a situation may apply to the implementation of some IT systems supporting the process of rapid prototyping and Industry 4.0. For example, consider a situation where one of the requirements of level p_{i-1} is that an administrator allocates resources for given prototyping clusters (let us mark this attribute as c_i). Let us assume that one of the features of level p_i is the introduction of automatic resource management systems using the SDN architecture. In this case, feature c_i is unnecessary for level p_i. For the example given, it is appropriate to use the inclusion described by Expression (1).

If the prototyping system or production system is built by one organization, ensuring the compliance of individual elements within one framework using a given matrix D is a relatively simple task. However, if we consider the architecture presented in Figure 1b, then it is necessary to ensure the compatibility of the maturity level in individual organizations for a given project being carried out.

4. System Architecture

Considering the needs of the rapid prototyping laboratory, the following functional modules were defined to form an integrated system supporting the RP process:

3D design support module: A set of CAD tools limited to licenses and the performance of hardware infrastructure. To increase the flexibility of resource management, it is possible to use virtualization techniques and collective licenses.

Filament selection module: This module is based on the programmed mechanisms for selecting the correct production components according to the needs defined for the final product. The system includes a database of available filaments with certain technical parameters; a set of available 3D printers with their technical parameters linked to a set of available filaments that can be used on individual printers; a system for defining the technical parameters of the final element; and a previously established mechanism for selecting a filament for specific production tasks.

Queue management module: An IT system that manages planned tasks in the area of available resources (including filament for 3D printers). This system analyzes the available 3D printers on an ongoing basis in terms of their load, informing the person making orders about the planned start and end dates of the printout. This system considers the priorities of individual tasks. An extension of this module can also manage access to other resources, including CAD modeling tools.

SDN network control module: This module can optimize the use of ICT infrastructure resources, by implementing adaptive mechanisms for controlling access to network resources and flow control through the dynamic creation and modification of communication paths with appropriate Quality of Service (QoS) parameters. This issue is described in detail in [18].

Access and control module (including remote control): A module responsible for access control to individual architectural resources of the rapid prototyping system. Two user groups can be distinguished here: administrators and operating users who are assigned the appropriate pre-defined access rights.

Print control and verification module: An IT system based on high resolution cameras. Data are then passed to the control system, which is placed in the analysis layer. There, the image from the camera is compared to the standard model, and then the level of the match is evaluated. When exceeding the pre-set limit parameters, the system may report the need to stop the manufacturing process.

These individual modules apparently belong to single layers of the framework presented earlier. However, in reality, these modules can (but do not have to) use the functionality of all four layers (Figure 4). For example, the 3D design support module uses available hardware resources (including computing power) and access to this module can be done remotely through the communication infrastructure, thereby implementing the design and modeling service for the end-products and, at the same time, working under the initial assumptions and knowledge (the example model reference) from the analysis layer. Similar relations apply to the other modules.

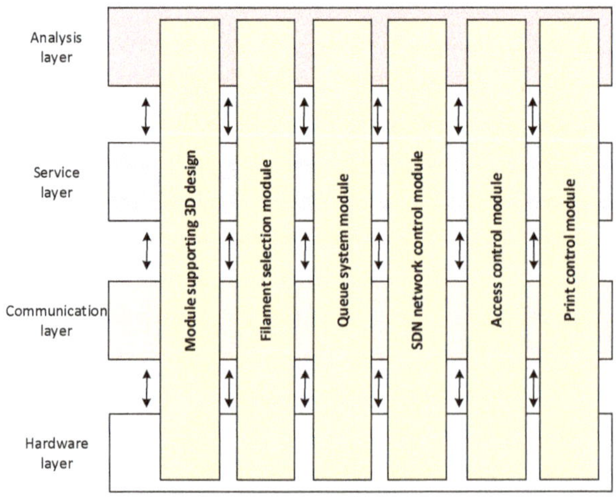

Figure 4. System architecture in the context of the proposed framework.

5. Adopted Architecture for the Remote Rapid Prototyping Environment

The implementation of many research, development and didactic works brings together a team of specialists dispersed across various institutes, academic centers, startups, industrial laboratories, etc. Modern broadband networks and available IT solutions allow one to create a "virtual laboratory", where scientists, engineers and students can carry out the entire rapid prototyping process from a place far from the real laboratory. The architecture of such a virtual–remote rapid prototyping laboratory (as part of the previously presented framework) is presented in Figure 5.

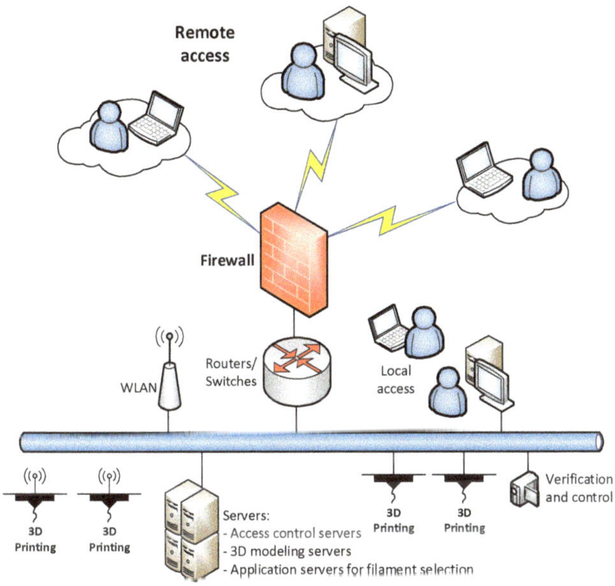

Figure 5. Architecture of a remote rapid prototyping laboratory.

The proposed solution is characterized by the following features:

- Scalability: Assumptions regarding the architecture of the remote laboratory system do not introduce limitations in the number of users, 3D modeling platforms, available 3D printers, type of filament used, etc. Consequently, the system can be freely extended with new additional components.
- Availability: Modern requirements for manufacturing systems, such as in a laboratory, can be available 24/7. Thanks to this, efficiency is significantly increased. This also represents a significant improvement for people who work from different time zones. Access is possible using both stationary and mobile devices.
- Flexibility: The topology in which solutions of this class are created can combine any elements with each other. As an element of infrastructure, industrial resources, IT infrastructure and human resources are considered. Available IT tools allow the creation of many research scenarios without structural time or spatial constraints. The basic restrictions here are the cost of the task and the time of its implementation. During the creation of several projects implemented by industry partners in cooperation with the Rzeszów University of Technology, the key parameter affecting the time of implementation for a given laboratory was shown to be the standardization of communication interfaces and the use of generally known protocols and standards available on the market. This approach significantly increases the flexibility of the planned environment, and reduces the costs of its construction and operation.

- Security: The system allows the application of modern methods for user verification and access control to individual elements of the entire system. This applies to both remote and local access. The main element is the RADIUS (Remote Authentication Dial In User Service) authentication server.
- Universality: The adopted assumptions allow the interoperability of various solutions in the field of printing methods, applied object modeling techniques, authentication mechanisms, IT resource virtualization, etc.
- Openness: The system allows the use of open solutions, including OpenSource and production solutions. This feature of the system was partially described in the previous section.
- Verification/inspection: The verification and inspection of each prototyping process are crucial in order to verify the design prototype. In stages related to CAD design, object properties can be verified by numerical simulations, expert systems, etc. At the production stage, verification is carried out by a camera system and a machine operator whose working time can be shared among many projects. Operator and industrial work can also be supported by a dedicated IT system, e.g., related to anomaly detection or machining optimization.

6. Implementation of the Framework in a Real Environment

The framework proposed in this paper was used for the creation and operation of a rapid prototyping environment designed at Rzeszow University of Technology. An example of such a solution is the iS Rapid at InfoSoftware Polska Ltd., Rzeszów, Poland. This system includes all four layers of the framework with the implementation of selected functionalities (Figure 6). The system was implemented in the production process and offered remote access to the prototyping infrastructure. The entire system was built from scratch according to previously developed assumptions. Four levels of maturity were defined for the developed framework. At present, the system has reached the third level of maturity. Works related to the achievement of the fourth level of maturity are currently being finished. At the analysis layer, a knowledge base used by the engine to select the print parameters was implemented. The selection and configuration of the manufacturing process carried out in the analysis layer directly translates into the configuration of the virtual work environment created by the service layer. Of course, the user/client authentication process itself and remote access to appropriate resources must be carried out beforehand. Subsequently, the users have the ability to design their own components, using the available design and modeling tools, and can also import a unique ready-made object. All operations and processes are available through a so-called "thin client", i.e., through a web browser. The user has the impression that he or she performs all activities locally on his or her computer/mobile device. One of the steps is the selection of the device and filament. This process is based on the predefined parameters of the previously designed prototype. On this basis, the system selects a set of recommended manufacturing devices and types of materials from which the given component will be finally made. Until final approval of the printout, the user can still go back and make any modifications. In one of the last steps, the user can determine the validity of his or her task, i.e., his or her priorities. This information will be used as input for the mechanisms that reserve IT and manufacturing resources.

Resources (CAD system, operating system, network resources, disk resources, etc.) are allocated in a virtual environment at the time of a given user's work and deleted after a session. The effects of the work are saved in the indicated place by the user, and can be used to implement the next stages of creating a prototype or develop another virtual work environment. The creation and the sharing of resources are carried out in the environment via automatically generated network connections implemented at the network layer (Figure 7). At the network layer, there is also a remote access control system supported by NAC (Network Access Control) and Firewall systems, enabling access to resources through a dedicated application.

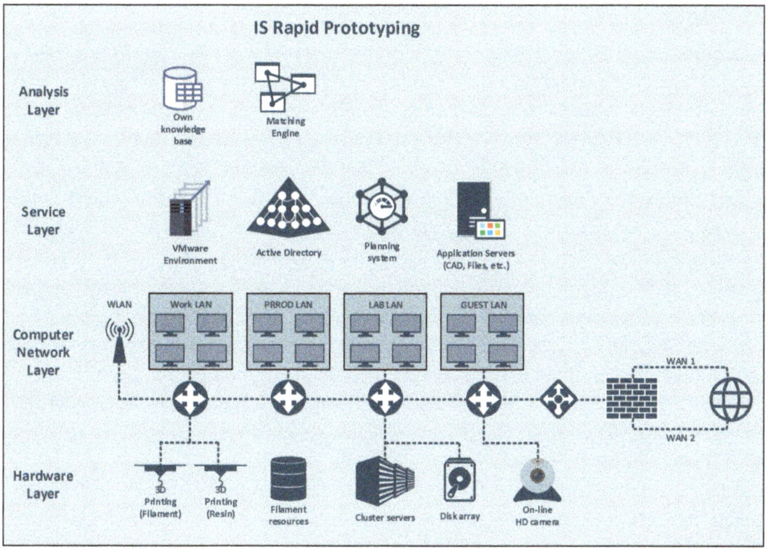

Figure 6. The architecture of the remote prototyping laboratory implemented at InfoSoftware Poland Ltd.

The connection system also provides access to hardware resources deployed in the hardware layer. The entire production process is controlled by the original queuing system, which is responsible for allocating resources from individual layers to specific users at a given time. The system is also responsible for the tariffication and can estimate the costs of physical prototype production as early as the stage of project creation.

At present, research work is in progress that will enable the transformation of the system from a centralized architecture to a distributed architecture. Cooperation has already been established with companies and external laboratories that offer other hardware components not available in the iS Rapid system. In the next stage, these components will be made available as a part of the iS Rapid system. The only condition for the implementation of such a structure is for the relevant partners to achieve the second level of maturity of the present framework. This will entail modifications in the area of the network and proper preparation of devices in the hardware layer.

Figure 7. Example elements of the IT and manufacturing infrastructure.

7. Comparison between the Classic and Proposed Approaches

The introduction of a new approach, or rather a methodology, for rapid prototyping processes requires a comparison with extant or "classic" solutions. From the perspective of modern manufacturing, in addition to flexibility, reliability and availability, real production costs are also a very important issue. For classic manufacturing, most design and manufacturing infrastructure must be available to a given company or person. This approach involves the high purchasing costs of efficient workstations necessary for the modern creation of 3D models, specialized software, a suitable machine park (in the form of, e.g., 3D printers), having appropriate resources (e.g., various types of filaments), and a team of specialists to guarantee the high level of performance of the required components. For the purposes of this paper, we analyzed the infrastructures of three representative technologies, FDM (Stratasys), PJM—Polyjet Modeling (Stratasys) and SLS—Selective Laser Sintering (Sintratec), for which comparative analyses were also performed. As part of the study, the costs for achieving the organization's readiness to implement the rapid prototyping process using the classic AM process were estimated using the resources at the Rzeszow University of Technology laboratory. A list of these costs is presented in Table 3. These costs were selected based on a market analysis in Poland at the beginning of 2019.

Analysis of the data from Table 3 allows us to estimate the costs of the so-called entry threshold: the costs primarily for the purchasing of software and materials as well as one employee (monthly salary). The employee is the operator of three 3D printers and software. For the rapid prototyping process laboratory, these costs are estimated at EUR 188,600. The above table presents the annual costs resulting from the depreciation of hardware and software, monthly costs and hourly averages, including costs related to the space and utilities used. Hourly costs can also be related to the operation of individual devices. The valuation of the implementation of a specific model must consider the specificity of the incremental technology used, the time of production using this technology, and fixed costs. The presented costs can be reduced by renting some of the elements on the market (including software), but in such a model, the integration of the entire manufacturing process is done on the client side.

Table 3. Cost estimations of the basic elements needed for the rapid prototyping process.

Position	Purchase Cost	Annual Costs (Depreciation)	Monthly Costs (Depreciation)	Total Cost for Working Hours of Laboratory Work
Software	4000	800	67	1
Computer	2000	400	33	1
PJM	130,000	26,000	2170	20
SLS	26,000	5200	433	10
FDM	24,000	4800	400	9
Materials	1000	12,000	1000	2
Personal	1600	19,200	1600	10
SUM	188,600	68,400	5703	53

In this process, considering the heterogeneity in the ownership of solutions used in single RP processes, there is a high probability that responsibility for prototyping failure will transfer between entities. For example, Company *A* producing a physical element may explain its faulty performance as a consequence of incorrectly selected parameters by the designer of company *B*, or improper processing by company *C*. In the proposed framework, transfers between companies *A*, *B* and *C* are implemented by strictly defined IT processes that ensure adequate quality, normalization, control compliance, and a number of other relevant parameters for a given case.

Next, the costs for the prototyping and production of two different (overall and structural) elements were analyzed: a rotor 200 m in diameter and 127 mm height (Figure 8a), as a thin-walled element with a complex geometric structure, and a bracket (Figure 8b), as a small element 50 mm long, 25 mm wide and 13 mm high, designed as a solid prototype.

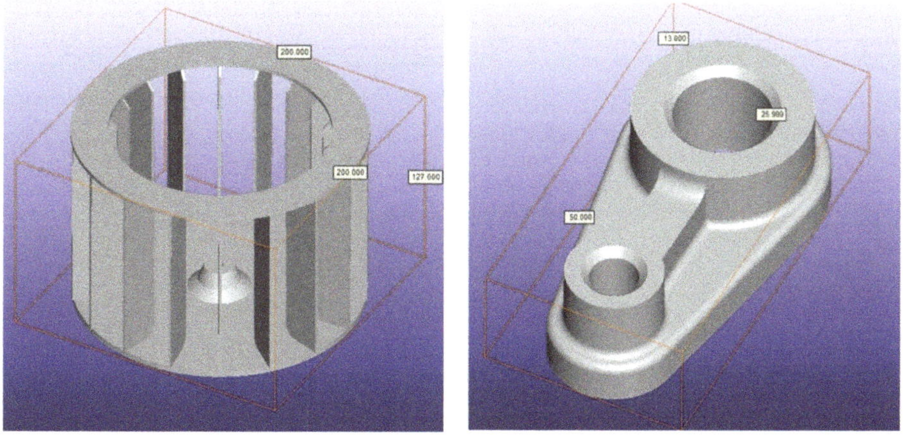

(**a**) (**b**)

Figure 8. 3D-STL models of the test elements: (**a**) rotor; (**b**) bracket.

In the cost assessment process, the printout of elements in three types of processes was analyzed:

1. Process 1 (M1): The company has its own infrastructure, production elements and qualified staff involved in the production of components. In this process, the costs related to the purchase and operation of equipment were also estimated. Estimating the manufacturing of an element in this model is the most difficult because it is necessary to consider the purchase and operation of equipment, and break down those values into the production costs of individual elements. Moreover, having our own machine park allows for the production of one type of element (with given parameters), and significantly reduces the prototyping process, which may require the use of various types of machines.
2. Process 2 (M2): In this model, with the exception of the designer, all elements used in the prototyping process are leased, including the CAD software. The entire operation is coordinated by the client and supported by a prototyping engineer.
3. Process 3 (M3): All prototyping is done remotely (along with the selection of the print parameters) using the system implemented by the company InfoSoftware using the proposed framework, which automatically coordinated and merged the production stages.

Determining the costs for the model was carried out in the first stage based on the resources and operating costs of the rapid prototyping laboratory of the Rzeszów University of Technology. A cost analysis was carried out for the production of prototypes using three industrial technologies: PJM, SLS and FDM. In addition, for process 1 and process 2, the costs were determined on the basis of quotes, cost estimates, or actual project and manufacturing works. The assumptions and controls for dimensional and shape accuracy were not considered because they are a separate issue. The list of costs is presented in Figures 9–14. The cost of producing prototypes for a larger number of pieces was also analyzed.

The results of the simulations carried out for the two different prototypes and three process models, as well as three manufacturing technologies, show significant differences in all cases. The cost of making a prototype of one rotor is the highest using the PJM method and process 1 (M1), which thus involves the most expensive device, an important element of which is machine depreciation. In this case, the price is constant regardless of the number of pieces produced. Using processes 2 (M2) and 3 (M3), the price decreases when the number of pieces is increased to 10, and then stabilizes. The execution of the rotor model with FDM technology under process 1 also assumes a constant value. However, the cost is noticeably lower than that when using processes 2 and 3, for which the costs are also decreased by increasing the number of ordered items (manufacturing 20 pieces stabilizes the price). For SLS technology, the costs also stabilize around 20 pieces. The cost of making the rotor is the lowest for processes 2 and 3. This may be due to the applied technology, SLS, for which support structures are not required. For the prototype of the rotor with a complex spatial structure, the support structures for PJM and FDM technologies entail a significant cost connected with the use of supporting materials and the time necessary to develop supporting structures. For the majority of 3D printers of a medium size, it is possible to make a rotor with these dimensions as one piece in the machine's working space. Hence, each rotor is made in a separate manufacturing process.

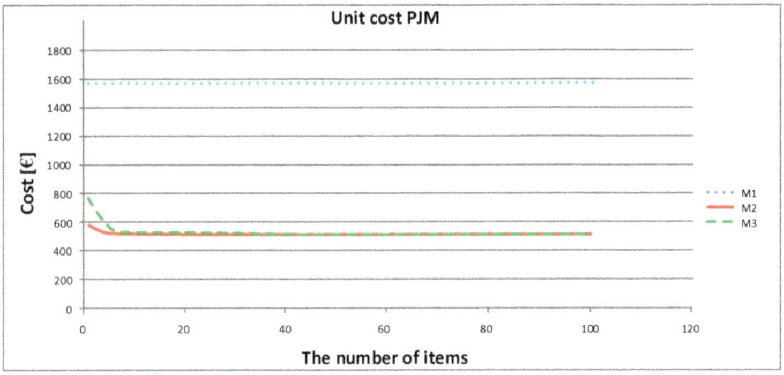

Figure 9. Dependence of the rotor production costs on the number of prototypes for PJM technology.

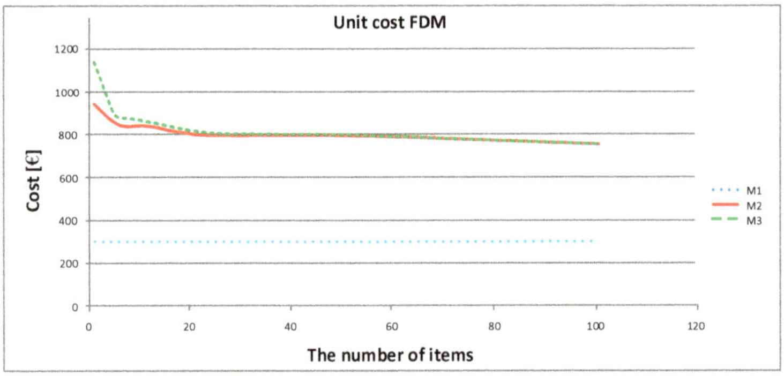

Figure 10. Dependence of the rotor production costs on the number of prototypes for FDM technology.

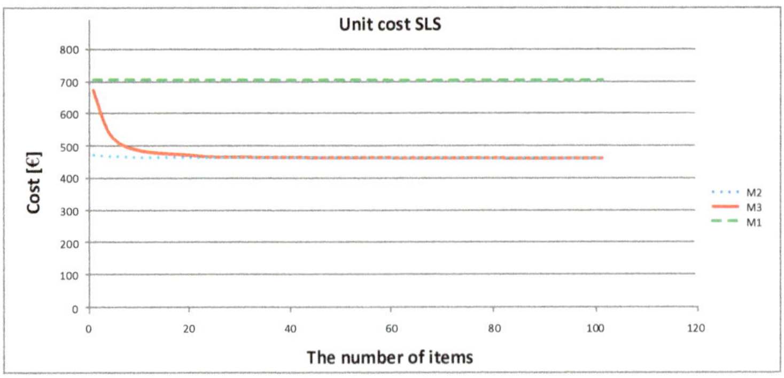

Figure 11. Dependence of the rotor production costs on the number of prototypes for SLS technology.

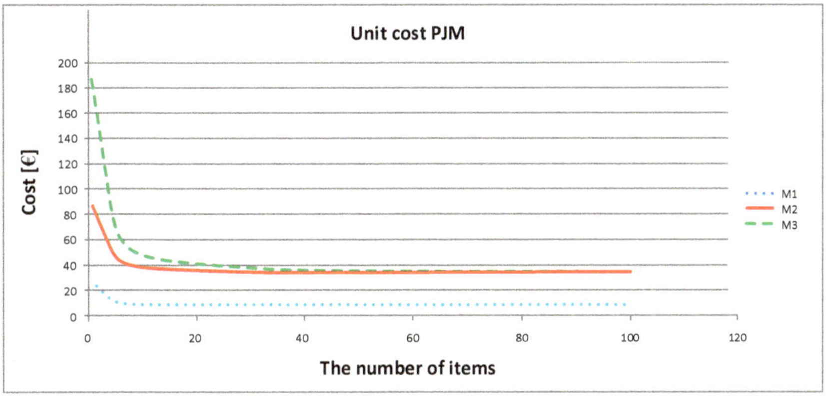

Figure 12. Dependence of the costs of bracket production on the number of prototypes for PJM technology.

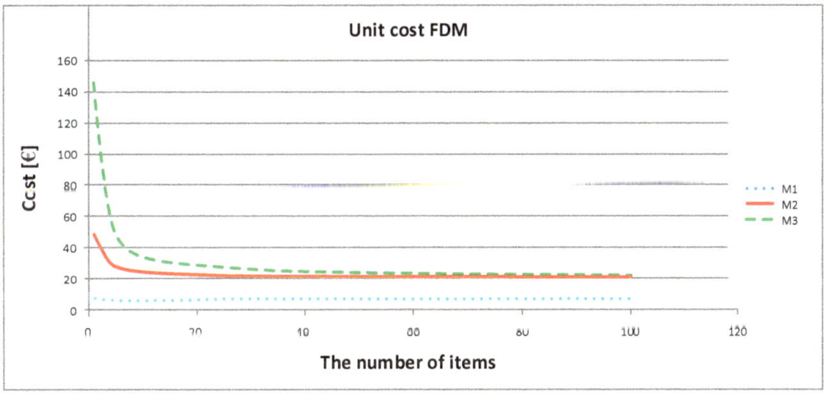

Figure 13. Dependence of the costs of bracket production on the number of prototypes for FDM technology.

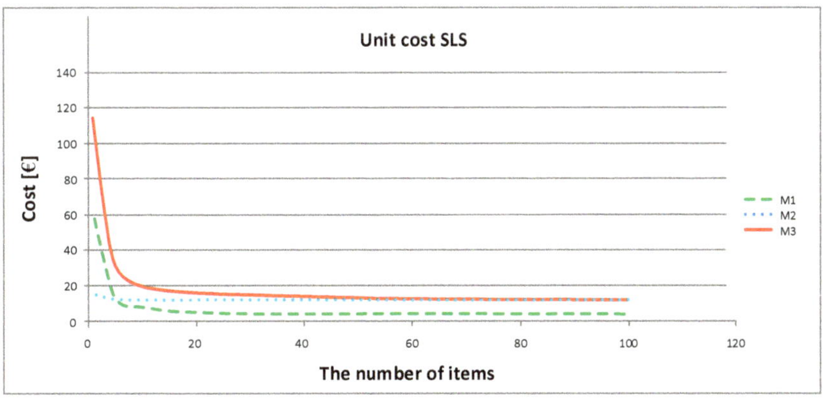

Figure 14. Dependence of the bracket production costs on the number of prototypes for SLS technology.

The costs for bracket manufacturing are different because brackets are a compact element with much smaller dimensions than a rotor. Due to the small size of a bracket, it is important to be able to make a few to a dozen pieces in one process. Thus, the most expensive context is the production of a single piece, which is especially notable for processes 2 and 3, for which cost stabilization occurs at 20 pieces of ordered product. For the model of process 1, the cost of execution is the smallest, and the course of the curve is gentle. For the prototype of the bracket, the highest cost was also observed for PJM technology with process 3. However, the difference between the production costs for individual technologies is not very significant in terms of the value itself, especially for over 20 pieces, compared to the rotor production price.

Analysis of the obtained results shows that the model based on the proposed framework M3 is extremely flexible, and is characterized by the low production costs of its prototype elements when the number of manufactured elements is greater than 15, which confirms the legitimacy of this model's use in the production of around a dozen pieces. This model has also great advantages associated with its ability to select print parameters individually for each sample in a given series. Process 1, despite having the lowest production costs, is often characterized by the highest entry threshold cost in each of the technologies studied, which is associated with the purchase of equipment and the need to hire an employee to operate that equipment (Table 3). However, for a large manufacturing company that intends to implement many prototypes incrementally for its own needs or for external orders, the long-term use of M1 is profitable.

8. Conclusions

This paper proposes a novel approach for the integration of the distributed additive manufacturing process enabling remote designing, the selection of appropriate manufacturing means, and the implementation of a physical production process and control at all stages. This approach was possible thanks to the development of an unprecedented framework, through which we were able to integrate distributed and functionally different elements (IT and manufacturing), forming a coherent design and manufacturing system. Importantly, this framework ensures not only an increase in production efficiency but also shortens production time, reduces costs, and increases the flexibility of and accessibility to the latest methods and design and manufacturing tools. In addition, we presented a mechanism that facilitates the integration of independent manufacturing environments by considering and implementing appropriate levels of maturity in the system. The validity of the presented solution was also confirmed by its implementation in a real production environment, i.e., at Infosoftware Poland. At present, work is in progress to integrate the rapid prototyping laboratory of the Rzeszów University of Technology with the infrastructure of the company to expand the available functionality, i.e., by providing a larger machine park, a wider range of design and modeling tools, and facilitating the development of analytical tools supporting the decision-making process. The presented platform can be widely used in the automotive and aerospace industries, and will facilitate cooperation between industrial clusters and academic centers to a higher degree, as well as encourage cooperation between small enterprises and startups. From the perspective of management, the technical implementation of the presented framework allows one to adapt to the needs of globalization, and facilitates the integration of distributed resources. Thus, this framework affects business, logistics and technological processes. One of the implications of implementing such a framework is the need to develop or adapt existing workflows to the new heterogeneous and distributed work environment. This is an interesting issue and may constitute the background of a new article. Considering the technological aspects, it remains a large challenge to achieve higher levels of maturity in a diverse ownership environment. This process requires cooperation and adoption at the management levels regarding the common assumptions behind the direction of development and implemented investments. In a homogeneous ownership environment, the related processes and decisions are much simpler. One of the significant limitations, that may affect the speed and scope of the integration of distributed heterogeneous design and production environments, is the lack of protocols and standards enabling the use of plug-and-play techniques known from ICT environments to enable the automated integration of physical devices with design applications and the preparation of the final manufacturing process, as well as with job queuing and decision support systems. The current approach based on the individual integration of individual infrastructure components requires a great deal of time and effort to achieve a higher level of maturity. The potential development of technology in this area could contribute to the full use of cross-layer optimization, as well as full automation of the attachment and disconnection of individual production devices. Of course, manufacturers of production devices use selected communication standards with their own dedicated applications, but there is currently no uniform mechanism comparable to the automatic installation of drivers in computer devices. Development in this direction would not only concern technological development, but would also influence the evolution of management models. The proposed solution, especially in terms of achieving the highest level of IT maturity, may enable the further evolution of productions methods toward Industry 5.0. Such progress will entail the implementation of fully autonomous production areas, where, after ensuring integration at the ICT level (i.e., IoT), it will be possible to implement a concept based on self-adaptive IoE (Internet of Everything) systems. Under this approach, IT systems based on artificial intelligence, machine learning, Big Data, and modern and safe communication systems will not only support decision-making processes at various stages of design and manufacturing, but will gradually replace them by creating Integrated Adaptive Generation Systems. In future works, we plan to develop solutions supporting decision-making in the field of design and production based on the experience of engineers, available technological databases and expert systems in the form of an iterative adaptive decision support system improved via the loopback model for Cyber–Human systems. Such a system would shorten the time needed for the design and production process, reduce the costs related to corrections and waste, and thus increase the quality of the services provided.

Author Contributions: A.P., M.B., G.B., Ł.P., and M.O. researched the literature; A.P., M.B., and G.B. formulated the problems and constructed the research framework, A.P., M.B., G.B., Ł.P., and M.O. conceived the methodologies and designed the experiments; A.P., M.B., G.B., Ł.P., and M.O. performed the simulations; A.P., M.B., G.B., Ł.P., and M.O. investigated and developed the application of the results in a real production environment; A.P., M.B., G.B., Ł.P., and M.O. contributed to the writing of this paper. All authors have read and agreed to the published version of the manuscript.

Funding: This project is financed by the Minister of Science and Higher Education of the Republic of Poland within the "Regional Initiative of Excellence" program for years 2019–2022. Project number 027/RID/2018/19, amount granted 11 999 900 PLN.

Conflicts of Interest: The authors declare no conflict of interest.

References

1. González-Varona, J.M.; Poza, D.; Acebes, F.; Villafáñez, F.; Pajares, J.; López-Paredes, A. New Business Models for Sustainable Spare Parts Logistics: A Case Study. *Sustainability* **2020**, *12*, 3071. [CrossRef]
2. Newman, S.T.; Zhu, Z.; Dhokia, V.; Shokrani, A. Process planning for additive and subtractive manufacturing technologies. *CIRP Ann.* **2015**, *64*, 467–470. [CrossRef]
3. Tofail, S.A.M.; Koumoulos, E.P.; Bandyopadhyay, A.; Bose, S.; O'Donoghue, L.; Charitidis, C. Additive manufacturing: Scientific and technological challenges, market uptake and opportunities. *Mat. Today* **2018**, *21*, 22–37. [CrossRef]
4. Melchels, F.P.W.; Domingos, M.A.N.; Klein, T.J.; Malda, J.; Bartolo, P.J.; Hutmacher, D.W. Additive manufacturing of tissues and organs. *Prog. Polym. Sci.* **2012**, *37*, 1079–1104. [CrossRef]
5. Go, J.; Hart, A.J. A framework for teaching the fundamentals of additive manufacturing and enabling rapid innovation. *Addit. Manuf.* **2016**, *10*, 76–87. [CrossRef]
6. Schelly, C.; Anzalone, G.; Wijnen, B.; Pearce, J.M. Open-source 3-D printing technologies for education: Bringing additive manufacturing to the classroom. *J. Vis. Lang. Comput.* **2015**, *28*, 226–237. [CrossRef]
7. Velázquez, J.S.; Cavas, F.; Bolarín, J.M.; Alió, J.L. 3D Printed Personalized Corneal Models as a Tool for Improving Patient's Knowledge of an Asymmetric Disease. *Symmetry* **2020**, *12*, 151. [CrossRef]
8. Thulin, E.; Vilhelmson, B.; Johansson, M. New Telework, Time Pressure, and Time Use Control in Everyday Life. *Sustainability* **2019**, *11*, 3067. [CrossRef]

9. Vega, R.P.; Anderson, A.J.; Kaplan, S.A. A Within-Person Examination of the Effects of Telework. *J. Bus. Psychol.* **2015**, *30*, 313–323. [CrossRef]
10. Lee, H.G.; Shin, B.; Higa, K. Telework vs. central work: A comparative view of knowledge accessibility. *Decis. Support Syst.* **2007**, *43*, 687–700. [CrossRef]
11. Karia, N.; Hasmi, M.; Asaar, A.H. Innovation capability: The impact of teleworking on sustainable competitive advantage. *Int. J. Technol. Policy Manag.* **2016**, *16*, 181–194. [CrossRef]
12. Bolanowski, M.; Paszkiewicz, A. Metody i środki zapewnienia dostępu do specjalizowanych zasobów laboratoryjnych. *Eduk. Tech. Inform.* **2014**, *5*, 334–341.
13. Torrisi, N.M.; Oliveira, J.F.G. Remote control of CNC machines using the CyberOPC communication system over public networks. *Int. J. Adv. Manuf. Technol.* **2008**, *39*, 570–577. [CrossRef]
14. Wazid, M.; Das, A.K.; Lee, J.-H. User authentication in a tactile internet based remote surgery environment: Security issues, challenges, and future research directions. *Pervasive Mob. Comput.* **2019**, *54*, 71–85. [CrossRef]
15. Ren, J.; Lin, C.; Liu, Q.; Obaidat, M.S.; Wu, G.; Tan, G. Broadcast tree construction framework in tactile internet via dynamic algorithm. *J. Syst. Softw.* **2018**, *136*, 59–73. [CrossRef]
16. Vora, J.; Kaneriya, S.; Tanwar, S.; Tyagi, S.; Kumar, N.; Obaidat, M.S. TILAA: Tactile Internet-based Ambient Assistant Living in fog environment. *Future Gener. Comput. Syst.* **2019**, *98*, 635–649. [CrossRef]
17. Özdemir, V.; Hekim, N. Birth of Industry 5.0: Making Sense of Big Data with Artificial Intelligence, "The Internet of Things" and Next-Generation Technology Policy. *OMICS J. Integr. Biol.* **2018**, *22*, 65–76. [CrossRef]
18. Mazur, D.; Paszkiewicz, A.; Bolanowski, M.; Budzik, G.; Oleksy, M. Analysis of possible SDN use in the rapid prototyping process as part of the Industry 4.0. *Bull. Pol. Acad. Sci. Tech. Sci.* **2019**, *67*, 21–30. [CrossRef]
19. Gibson, I. (Ed.) *Software Solutions for Rapid Prototyping*; Wiley: London, UK, 2002.
20. Microsoft. Available online: https://support.microsoft.com/pl-pl/help/4028122/windows-how-to-use-3d-builder-and-3d-scan-for-windows (accessed on 10 April 2020).
21. i.materialise. Available online: https://i.materialise.com/ (accessed on 10 April 2020).
22. 3D HUBS. Available online: https://www.3dhubs.com/ (accessed on 10 April 2020).
23. IGUS. Available online: https://www.igus.pl/ (accessed on 10 April 2020).
24. Kamrani, A.K.; Nasr, E.A. *Engineering Design and Rapid Prototyping*; Springer: Boston, MA, USA, 2010. [CrossRef]
25. Bordegoni, M.; Rizzi, C. (Eds.) *Innovation in Product Design from CAD to Virtual Prototyping*; Springer: London, UK, 2011. [CrossRef]
26. Mellor, S.; Hao, L.; Zhang, D. Additive manufacturing: A framework for implementation. *Int. J. Prod. Econ.* **2014**, *149*, 194–201. [CrossRef]
27. Jaeger, A.; Matyas, K.; Sihn, W. Development of an Assessment Framework for Operations Excellence (OsE), based on the Paradigm Change in Operational Excellence (OE). *Procedia CIRP Var. Manag. Manuf.* **2014**, *17*, 487–492. [CrossRef]
28. Benkamoun, N.; ElMaraghy, W.; Huyet, A.-L.; Kouiss, K. Architecture Framework for Manufacturing System Design. *Procedia CIRP Var. Manag. Manuf.* **2014**, *17*, 88–93. [CrossRef]
29. Lehmhus, D.; Wuest, T.; Wellsandt, S.; Bosse, S.; Kaihara, T.; Thoben, K.-D.; Busse, M. Cloud-Based Automated Design and Additive Manufacturing: A Usage Data-Enabled Paradigm Shift. *Sensors* **2015**, *15*, 32079–32122. [CrossRef] [PubMed]
30. Wang, L.; Shen, W.; Lang, S. Wise-ShopFloor: A Web-Based and Sensor-Driven e-Shop Floor, Transactions of the ASME. *J. Comput. Inf. Sci. Eng.* **2004**, *4*, 56–60. [CrossRef]
31. Shen, W.; Hao, Q.; Yoon, H.J.; Norrie, D.H. Applications of agent-based systems in intelligent manufacturing: An updated review. *Adv. Eng. Inform.* **2006**, *20*, 415–431. [CrossRef]
32. Monostori, L.; Váncza, J.; Kumara, S.R. Agent-based systems for manufacturing. *CIRP Ann. Manuf. Technol.* **2006**, *55*, 697–720. [CrossRef]
33. Suh, S.-H.; Shin, S.-J.; Yoon, J.-S.; Um, J.-M. UbiDM: A new paradigm for product design and manufacturing via ubiquitous computing technology. *Int. J. Comput. Integr. Manuf.* **2008**, *21*, 540–549. [CrossRef]
34. Zuehlke, D. SmartFactory–Towards a factory-of-things. *Annu. Rev. Control* **2010**, *34*, 129–138. [CrossRef]
35. Srivastava, V.; Motani, M. Cross-layer design: A survey and the road ahead. *IEEE Commun. Mag.* **2005**, *43*, 112–119. [CrossRef]

36. Yang, X.; Wang, L.; Xie, J.; Zhang, Z. Cross-layer model design in wireless ad hoc networks for the Internet of Things. *PLoS ONE* **2018**, *13*, e0196818. [CrossRef]
37. Resner, D.; de Araujo, G.M.; Fröhlich, A.A. Design and implementation of a cross-layer IoT protocol. *Sci. Comput. Program.* **2018**, *165*, 24–37. [CrossRef]
38. Gomolka, Z.; Twarog, B.; Zeslawska, E. Cognitive Investigation on Pilot Attention during Take-Offs and Landings Using Flight Simulator. In *International Conference on Artificial Intelligence and Soft Computing*; Rutkowski, L., Korytkowski, M., Scherer, R., Tadeusiewicz, R., Zadeh, L., Zurada, J., Eds.; Springer: Cham, Switzerland, 2017; pp. 432–443.
39. Bolanowski, M.; Twaróg, B.; Mlicki, R. Anomalies detection in computer networks with the use of SDN. *Meas. Autom. Monit.* **2015**, *9*, 443–445.
40. Woojin, S.; Keecheon, K. A Study on Communication Optimization in Multi-SDN Controller. In Proceedings of the International Conference on Information Networking (ICOIN), Kuala Lumpur, Malaysia, 9–11 January 2019; pp. 461–464. [CrossRef]

processes

Review

Digital Twins in Pharmaceutical and Biopharmaceutical Manufacturing: A Literature Review

Yingjie Chen [1], Ou Yang [2], Chaitanya Sampat [2], Pooja Bhalode [2], Rohit Ramachandran [2] and Marianthi Ierapetritou [1,*]

1 Department of Chemical and Biomolecular Engineering, University of Delaware, Newark, DE 19716, USA; cyj@udel.edu
2 Department of Chemical and Biochemical Engineering, Rutgers—The State University of New Jersey, Piscataway, NJ 08854, USA; oy21@scarletmail.rutgers.edu (O.Y.); crs234@soe.rutgers.edu (C.S.); prb63@scarletmail.rutgers.edu (P.B.); rohitrr@soe.rutgers.edu (R.R.)
* Correspondence: mgi@udel.edu; Tel.: +1-302-831-6641

Received: 6 August 2020; Accepted: 31 August 2020; Published: 2 September 2020

Abstract: The development and application of emerging technologies of Industry 4.0 enable the realization of digital twins (DT), which facilitates the transformation of the manufacturing sector to a more agile and intelligent one. DTs are virtual constructs of physical systems that mirror the behavior and dynamics of such physical systems. A fully developed DT consists of physical components, virtual components, and information communications between the two. Integrated DTs are being applied in various processes and product industries. Although the pharmaceutical industry has evolved recently to adopt Quality-by-Design (QbD) initiatives and is undergoing a paradigm shift of digitalization to embrace Industry 4.0, there has not been a full DT application in pharmaceutical manufacturing. Therefore, there is a critical need to examine the progress of the pharmaceutical industry towards implementing DT solutions. The aim of this narrative literature review is to give an overview of the current status of DT development and its application in pharmaceutical and biopharmaceutical manufacturing. State-of-the art Process Analytical Technology (PAT) developments, process modeling approaches, and data integration studies are reviewed. Challenges and opportunities for future research in this field are also discussed.

Keywords: digital twin; Industry 4.0; pharmaceutical manufacturing; biopharmaceutical manufacturing; process modeling

1. Introduction

Competitive markets today demand the use of new digital technologies to promote innovation, improve productivity, and increase profitability [1]. The growing interests in digital technologies and the promotion of them in various aspects of economic activities [2] have led to a wave of applications of the technologies in manufacturing sectors. Over the years, the advancements of digital technologies have initiated different levels of changes in manufacturing sectors, including but not limited to the replacement of paper processing with computers, the nurturing and promotion of Internet and digital communication [1], the use of programmable logical controller (PLC) and information technology (IT) for automated production [3], as well as the current movement towards a fully digitalized manufacturing cycle [4]. The digitalization waves have enabled a broad range of applications from upstream supply chain management, shop floor control and management, to post-manufacturing product tracing and tracking.

Among the new digital advancements, the development of artificial intelligence (AI) [5], Internet of Things (IoT) devices [3,5] and digital twins (DTs) have received attention from governments, agencies, academic institutions, and industries [6]. The idea of Industry 4.0 has been put forward by the community of practice to achieve a higher level of automation for increased operational efficiency and productivity. Smart technologies under the umbrella of Industry 4.0, such as the development of the IoT, big data analytics (BDA), cyber-physical systems (CPS), and cloud computing (CC) are playing critical roles in stimulating the transformation of current manufacturing to smart manufacturing [7–10]. With the development of these Industry 4.0 technologies to assist data flow, a number of manufacturing activities such as remote sensing [11,12], real-time data acquisition and monitoring [13–15], process visualization (data, augmented reality, and virtual reality) [16,17], and control of all devices across a manufacturing network [18,19] is becoming more feasible. The implementation of Industry 4.0 standards by institutions and companies encourages them to implement a more robust, integrated data framework to connect the physical components to the virtual environment [1], enabling a more accurate representation of the physical parts in digitized space, leading to the realization and application of DTs.

The concept of creating a "twin" of a process or a product can be traced back to the late 1960s when NASA ensembled two identical space vehicles for its Apollo project [20–22]. One of the two was used as a "twin" to mirror all the parts and conditions of the one that was sent to the space. In this case, the "twin" was used to simulate the real-time behavior of the counterpart.

The first definition of a "digital twin" appeared in 2002 by Michael Grieves in the context of an industry presentation concerning product lifecycle management (PLM) at the University of Michigan [23–25]. As described by Grieves, the DT is a digital informational construct of a physical system, created as an entity on its own and linked with the physical system [24].

Since the first definition of DT, interpretations from different perspectives have been proposed, with the most popular one given by Glaessegen and Stargel, noting that a DT is an integrated multiphysics, multiscale, probabilistic simulation of a complex product and uses the best available data, sensors, and models to mirror the life of its corresponding twin [26]. It is generally accepted that a complete DT consists of a physical component, a virtual component, and automated data communications between the physical and virtual components [2]. Ideally, the digital component should include all information of the system that could be potentially obtained from its physical counterpart. This ideal representation of the real physical system should be an ultimate goal of a DT, but for practical usage, simplified or partial DTs are the dominant ones in industry currently, including the employment of a digital model where the digital representation of a physical system exists without automated data communications in both ways, and a digital shadow where model exists with one-way data transfer from physical to virtual component [2].

Together with the US Food and Drug Administration (FDA)'s vision to develop a maximally efficient, agile, flexible pharmaceutical manufacturing sector that reliably produces high quality drugs without extensive regulatory oversight [27], the pharmaceutical industry is embracing the general digitalization trend. Industries, with the help of academic institutions and regulatory agencies, are starting to adopt Industry 4.0 and DT concepts and apply them to research and development, supply chain management, as well as manufacturing practice [9,28–31]. The digitalization move that combines Industry 4.0 with International Council for Harmonisation (ICH) guidelines to develop an integrated manufacturing control strategy and operating model is referred to as the Pharma 4.0 [32].

However, according to the recent survey conducted by Reinhardt et al. [33], the preparedness of the industry for this digitalization move is still unsatisfactory. It is noted that most pharmaceutical and biopharmaceutical processes currently rely on quality control checks, laboratory testing, in-process control checks, and standard batch records to assure product quality, whereas the process data and models are of lower impact. Within pharmaceutical companies, there are gaps in knowledge and familiarization with the new digitalization move, resulting in a roadblock in strategic and shop floor implementation of such technologies.

With the rapid development of DT and its building blocks, state-of-the-art review studies concerning pharmaceutical and biopharmaceutical manufacturing are limited. This paper aims to provide a literature review and a discerning summary of the current status of DT development and its application in the pharmaceutical industry, focusing on small and large molecule drug product manufacturing for the purpose of identifying current and future research directions in this area. The remainder of the paper is structured as follows. A description of the general DT framework is provided in Section 2, followed by a detailed review of DT in pharmaceutical and biopharmaceutical manufacturing in Sections 3 and 4, respectively. More specifically, we intend to provide readers with a summary of the critical components of an effective DT and the progress of implementing these components in pharmaceutical and biopharmaceutical manufacturing. After discussing the current status, we discuss the challenges associated with the development and application of DT in each section, with conclusions at the end.

2. Digital Twin Framework

As mentioned in Section 1, a DT has a physical component, a virtual component, and automated data communication in between, which is realized through an integrated data management system. This synergy between the physical, virtual space, and the integrated data management platform is demonstrated in Figure 1. The physical component consists of all manufacturing sources for data, including different sensors and network equipment (e.g., routers, workstations) [34]. The virtual component needs to be a comprehensive digital representation of the physical component in all aspects [8]. The models are built on prior knowledge, historical data, and the data collected in real-time from the physical components to improve its predictions continuously, thus capturing the fidelity of the physical space. The data management platform includes databases, data transmission protocols, operation data, and model data. The platform should also support data visualization tools in addition to process prediction, dynamic data analysis, and optimization [34]. Sections 2.1–2.3 discuss each component in more detail.

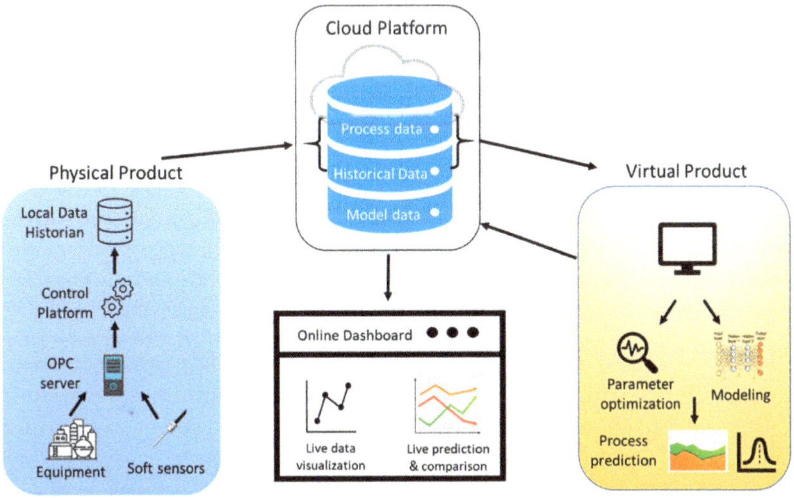

Figure 1. Physical component, virtual component, and data management platform of a general digital twin (DT) framework.

2.1. Physical Component

Sourcing data from the physical process and component is one of the most essential elements in the development of a DT. The critical process parameters (CPPs) for equipment can be obtained either manually from the human–machine interface (HMI) generally provided by the equipment manufacturer or automated using several machine–machine interfaces (MMI). There are several standard MMIs such as Open Platform Communications (OPC), OPC Data Access (OPC DA), OPC Unified Architecture

(OPC UA), and Modbus [35] for automating the data transfer between equipment software to a control or historian software. OPC UA is considered to be the current standard as it has added features such as multiple tags along with their properties [36]. Data can also be transmitted over the network using message queue telemetry transport (MQTT), Hypertext Transfer Protocol (HTTP), Transmission Control Protocol/Internet Protocol (TCP/IP), etc. The critical quality attributes (CQAs) for the product are determined using soft sensors, and they usually employ network protocols for data transmission [37]. Soft sensors are a combination of hardware sensors with their propriety software-enabled models that help obtain information about the process [38]. Soft sensors have been implemented in several process industries for process monitoring and control. These sensors have been used to measure cake resistance in freeze-drying applications [39], measuring temperature from pyrometers [40], estimating product quality during crude distillation [41], and have also found several other industrial applications [42–45]. Continuous acquisition of large amounts of data requires a systematic framework such as a data historian to store the historical data. Several studies have employed local data historians [46,47] to create an information infrastructure enabling the synchronous collection of process and sensor data. Zidek [48] demonstrated the Industry 4.0 concept for small–medium size enterprises (SMEs) where the quality of the product was assessed by a DT, and the communication between the OPC server and PLC system was achieved using OPC-UA. A combination of network and OPC communication protocols was used by Kabugo [35] to develop the cloud-based analytics platform for a waste-to-energy plant. Several other studies focusing on smart factories according to Industry 4.0 standard have utilized similar communication protocols [49–51].

2.2. Virtual Component

The virtual component consists of a collection of models to simulate the physical process and to analyze the current and future state of the system. With appropriate models, the virtual components can be used to perform real-time process simulation and system analyses, including but not limited to sensitivity studies that identify the set of most influential factors [52], design space studies that yield feasible operating conditions [53], and system optimization [54]. Results from real-time process simulation can be sent to the data management platform to visualize the process, and the results of system analyses, together with the preprogrammed expert knowledge, can be used to deliver control commands to the physical counterpart to ensure process and component conformity.

Different model types exist for use in DT, namely mechanistic models, data-driven models, and hybrid models. Mechanistic models strongly rely on process knowledge and understanding, as the development is based on fundamental principles and process mechanisms [55]. The resulting models are highly generalizable with physically interpretable variables and parameters, with a relatively low requirement from process data. Often, however, this comes with high development and computation costs [54,56]. In contrast, data-driven models depend only on process data, and no prior knowledge is needed [55]. The advantages include more straightforward implementation, relatively low development and computational expenses, and convenient online usage and maintenance. However, the poor interpretability, poor generalizability, and the need for large amounts of data present limitations of this modeling method [55,57,58]. A hybrid modeling strategy is then introduced to balance the advantages and disadvantages of the other two model types [57,59–61]. With different hybrid structures, the hybrid modeling method offers improved predictability and flexibility in process modeling [58,61,62].

In addition to the development of models, the computational cost is also a main concern in the virtual component of DT. Since a fully developed DT aims to represent the physical counterpart and perform system analyses, it would require extensive computational power. For a large system, local desktops and consumer-grade Central Processing Units (CPUs) cannot meet the demand. Many computationally intensive models can run in parallel using high-performance computing (HPC) to enhance the computational speed to achieve real-time or near-real-time simulations [63–65].

To develop models, perform simulations, and conduct system analyses for the virtual component of the DT framework, appropriate modeling platforms are needed. Various commercial modeling platforms and software packages have been developed and have become available. Among all the available ones, MATLAB and Simulink (MathWorks) [66], COMSOL Multiphysics (COMSOL) [67], gPROMS FormulatedProducts (Process Systems Enterprise/Siemens) [68], aspenONE products (AspenTech) [69], and STAR-CCM+ (Siemens) [70] are commonly seen in process industries. These platforms offer a large collection of models and/or tools that enable users to create or incorporate unit operations and flowsheet models based on the actual process. Some of these companies have also been developing local and cloud platforms (e.g., gPROMS Digital Applications Platform [71] from Process System Enterprise/Siemens, Siemens Mindsphere [72]) for hosting and computing models, for integrating physical component, and for providing data management functions, providing end-to-end DT solutions. Others have focused on improving compatibility with common data management and Internet of Things (IoT) integration platforms, which are described next in Section 2.3.

2.3. Data Management

In addition to model management and simulation platforms, several commercial IoT Platforms as a Service (PaaS), such as Predix (General Electric) [73], Mindsphere (Siemens) [72], SEEQ [74], TrendMiner [75], TIBCO Cloud [76], etc. have been developed. These platforms offer a large collection of tools that enable users to develop, visualize, analyze, and manage data on cloud servers. Some cloud service companies, such as Amazon Web Services (AWS) [77], Microsoft Azure [78], Google Cloud [79], IBM Watson [80], offer multipurpose platforms which are more versatile [81]. These platforms also offer distributed computing, data analysis tools, interaction protocols, and data and device management tools. Several of the interface protocols mentioned in Section 2.1 are also applicable to data transfer in the cloud. These platforms also provide large data storage capacities at affordable prices. Industrial grade IoT platforms are developed with a higher emphasis on secure device connectivity and cyber-security [82].

Seamless data integration in most cases is mainly hindered by a large amount of heterogeneity between manufacturers and services based on the software used and data formats supported [83]. Some cloud services provide their solutions as optional application program interfaces to integrate with other software, but several are left out due to the large number of software present. Thus, there is a need for a standard file format that needs to be employed to encourage cross-platform integration. The World Wide Web Consortium (W3C) has proposed Extensible Markup Language (XML), Resource Description Framework (RDF), among other markup languages to model information explicitly [84]. XML [85] provides the user with the freedom to define tags and data structures which are both readable by machines and humans. This syntax is further developed to incorporate the graph structure of the information within the RDF framework. The W3C also proposed Web Ontology Language (OWL) for information modeling. OWL is a vocabulary extension of RDF and is currently in use with XML and RDF. Unfortunately, these files become cumbersome when large databases need to be stored [86]; thus, new standard language Structured Query Language (SQL) for relational databases was recommended by the American National Standards Institute (ANSI) [87]. SQL databases are commonly found on cloud servers; however, their difficulty in horizontal scalability has led to the development of Non-SQL (NoSQL) databases, which are easily scalable vertically and horizontally [88] and can be hosted on cloud servers. Cloud servers are not limited to storage, but they offer large and scalable compute capabilities that can be leveraged for quick data analysis and simulations. A web service can also be hosted on a cloud server to create an online dashboard to visualize both the real-time physical data and the data from the simulation/data analysis.

2.4. Applications of Digital Twin

DT frameworks, as presented in Sections 2.1–2.3, are implemented across various industries [2,4,89] for simulation, real-time monitoring, control, and optimization to handle "what-if" or risk-prone [89] scenarios for improving process efficiency, safety analysis, maintenance, and decision-making [24]. This section provides a brief overview of such applications [4] within various industries such as aerospace, energy, manufacturing, automobile, chemical, healthcare, semiconductor, and city planning, as shown in Table 1.

Table 1. Applications of digital twin in various industries.

Areas of Application	Specific Application	Purpose	Component of DT Framework with Software	References
Energy production	Steam turbines	Integrates historical data with real-time process to forecast process wear/tear and suggest modifications	Virtual component using Predix	[4,90–92]
	Wind farm	Integrates historical data to enhance process efficiency and predict maintenance strategies	Virtual component based on General Electric (GE) fleet using Predix	[92]
Smart product manufacturing	Factory smart floor map	Redesign manufacturing platforms	Virtual replica of manufacturing floor to optimize location of machinery and sensors	[2,18,93–96]
	Digitization of manufacturing of packaging machines	Redesigning product to improve production efficiency and digitize overall process design	Virtual model using Siemens mechatronics concept designer	[96,97]
Aviation industry	DT of next-generation aircrafts	Aircraft structural health management and assessment of potential damage analysis	Virtual replica of airplanes using GE's Predix software platform	[98,99]
	Airframe DT simulator (ADT)	Training and engineering solutions	Virtual simulator using GE's Predix software	[100]
Aerospace industry	DT of outer-space vehicles	Replication of health maintenance problems and monitoring for safety and reliability	Virtual replica of the vehicle's on-board integrated system	[26,101]
Automotive transportation	DT of cars	Prediction and assessment of maintenance issues for improvement of durability of automobile parts	Virtual replica of automobiles	[102]
	Automated transport vehicles	Vehicle simulations for safe, automated long-distance transportations	Dassault systems using digital control systems	[102]
Healthcare industry	Virtual replica of patients	Surgical operation training and health monitoring using sensors	Virtual component developed using a simulated environment	[4]
	Living Heart project	3D model of human heart for analysis of blood circulation and pharmacokinetic/pharmacodynamic (PKPD) testing of medicines	Virtual model using finite element-based modeling environment	[4]
Infrastructure planning	City planning	Construction of smart, sustainable city infrastructure	Virtual digital replica using information communication technology	[103]

A commercial application of fully integrated DT was first demonstrated by General Electric (GE) at the Minds + Machines event in 2017 for the GE90 engine [104], with 300 engines integrated together to supply historical and real-time process information for predicting process failure, mitigating risks, and optimizing maintenance costs. Similar applications in the aviation industry include DT of airplanes used for training simulations [100] and aircraft health management [98,99,105,106] for damage assessment and rectification. The aerospace industry focuses on DT applications for the development of next-generation outer-space vehicles, following a successful demonstration of Apollo 13 by NASA [26,101] rectifying maintenance problems. DT applications in the energy sector include GE's wind farm [92] and steam turbines [4,90–92]. These DTs are capable of integrating historical data in terms of process, fuel costs, electricity, process wear and tear, and weather forecasts to suggest possible real-time modifications for reducing operating costs. Smart manufacturing is another sector benefitting from DT applications through digitization of product manufacturing [96,97] and development of digital shop floor (DTS) [2,18,93–96], incorporating real-time information of manufacturing plant, state of production machinery, environmental conditions, and its effects on manufactured products. DT applications in the area of automobile and transportation focus on automation of vehicles [107] and long-distance transportations [102] along with analysis of maintenance [22] and risk-prone issues [108]. The healthcare industry includes applications such as virtual replica of patients used for surgical operation training [4], sensors for health monitoring [109], the study of health of a country's population [110], and the "The Living Heart" [111] project developed for the analysis of blood circulations. Furthermore, city planning is another domain where virtual replica of cities, known as "smart cities" [103] are used for urban city planning and optimal resource allocation [112]. Such efforts promote the construction of smart, sustainable cities [113] while providing a holistic view of cross-vertical optimization of overall city infrastructure [114].

From the applications reviewed, it is clear that the concept of DT is rapidly being employed across various domains, given its advantages. However, it is important to identify the challenges associated with the development and application of integrated frameworks for the systematic utilization of DTs.

2.5. Challenges

Many research and review articles have discussed challenges in the implementation of DTs, and the issues can be categorized as time-, safety-, and mission-critical [115–120]. In this section, issues that are more relevant to the manufacturing sector and modeling community are presented, including data communication, model development and maintenance, cyber-physical security, and real-time capability.

One of the challenges in achieving a DT framework is to establish a stable two-way connection between the physical and virtual components to support real-time integration. Heterogeneity in equipment manufacturers and their software [116] is a major hurdle that needs to be addressed using a common interface or file format that could make interactions between several software easier. Several prominent manufacturers are already making strides by supporting commonly used OPC UA/DA interfaces. The creation of a database system that is not only vertically and horizontally scalable but also structured would also be important in such a framework. Thus, migrating to a NoSQL database would be recommended, but in this case, the manufacturing industry lags since several software currently only save data in SQL databases. Additionally, the resolution of sensor data, latency within the data communication channel, increased volume and variety of data, and the requirement of fast storage and retrieval are all challenges within this context.

The development of virtual models is often costly and challenging due to the lack of a complete understanding of the physical process [93]. This deficiency sometimes leads to inconsistences between models and the physical system. These inconsistencies need to be appropriately identified and handled, which can impose challenges to the modeling and operation teams. To resolve the issue, systematic model development approaches, along with appropriate model maintenance strategies are needed. Moreover, since the models need to perform simulation and system analyses in real-time, efficient and accurate algorithms that can utilize available information in real-time and continuously are crucial, presenting a challenge to both the modelers and allocation of computing resources.

In addition to the modeling aspects, cyber-physical security is another area of concern to ensure the normal operation of physical and virtual components against malicious attacks [121]. In a fully integrated DT, large data sets with important and potentially confidential information are exchanged, which require secure communication and processing among all systems [122].

3. Digital Twin in Pharmaceutical Manufacturing

In pharmaceutical manufacturing, the potential of using DTs to facilitate smart manufacturing can be seen in different phases of process development and production. In the process design stage, the use of a DT can significantly accelerate the selection process of a manufacturing route and its unit operations as it is able to represent physical parts with various models. The understanding of process variations can be obtained from DT simulations, which allows for the prediction of product quality, productivity, and process attributes, reducing the time and costs for physical experiments [123]. In the operation phase, real-time process performance can be monitored and visualized at any time, and the DT can analyze the system in a continuous manner to provide control and optimization insights of the process [123]. The DT can also be used as a training platform for operators and engineers, as the real-time scenario simulation and on-the-job feedback can be realized through DT. With regards to pre- and post-manufacturing tasks, the DT platform can assist with tasks including but not limited to material tracking, serialization, and quality assurance.

Some key requirements for achieving smart manufacturing with DT include real-time system monitoring and control using Process Analytical Technology (PAT), continuous data acquisition from equipment, intermediate and final products, and a continuous global modeling and data analysis platform [29]. The pharmaceutical industry has taken several steps towards this by using techniques such as Quality-by-Design (QbD) [124], Continuous Manufacturing (CM) [124], flowsheet modeling [125], and PAT implementations [126]. Some of the tools have been investigated extensively, but the overall integration and development of DTs are still under infancy.

This section reviews the progress of current research and industry applications towards DTs in pharmaceutical manufacturing from aspects of PAT sensing, model building, and data integration, which corresponds to the physical component, virtual component, and data management parts in the general DT framework. Challenges and opportunities are discussed at the end of this section.

3.1. PAT Methods

A key component in the development of a DT is data collection. In addition to readings from equipment, (critical) quality attributes also need to be collected from physical plants in a timely manner for use in the virtual component. The models and analyses are reliant on good data. Several traditional technologies exist to determine CQAs such as sieve analysis and High-Performance Liquid Chromatography (HPLC), but these cannot provide real-time data and are performed away from the production line rather than in-line or at-line. Thus, PAT tools have been explored and developed to address these issues [127].

PAT tools in the pharmaceutical industry have a wide range of applications, including measuring particle size of crystals [128], blend uniformity [129], testing tablet content uniformity [130], etc. Spectroscopy tools (Nuclear Magnetic Resonance (NMR), Ultraviolet (UV), Raman, near-infrared, mid-infrared, online mass spectrometry) constitute one of the major techniques used to measure the CQAs of pharmaceutical processes. Raman and Near-Infrared Spectroscopy (NIRS) are commonly used in the industry. Raman Spectroscopy has been employed for the on-line monitoring of powder blending processes [131]. Since acquisition times for Raman can be higher, NIRS is preferred for real-time measurements. NIRS has been used for real-time monitoring of powder density [15] and blend uniformity of processes [129]. NIRS has also been integrated with control platforms for process monitoring and control [132]. Baranwal et al. [133] employed NIRS to replace HPLC methods to predict API concentration in bi-layer tablets. PAT tools have also been used by the pharmaceutical industry to determine the particle size distribution of the product [134]. Several available optical tools such as Focused Beam Reflectance Measurement (FBRM) [135], a high-resolution camera system [136] have also been employed in the industry for particle size analysis. Some studies have utilized a network of PAT tools to achieve a monitoring system to help monitor and control a unit process [127,137].

The US FDA has also taken steps in promoting the use of PAT tools in pharmaceutical manufacturing with the goal of ensuring final product quality [138]. The pharmaceutical industry has adopted PAT in various applications throughout the drug-substance manufacturing process [139]. Although this has certainly led to an increase in the usage of PAT tools, their applications still remain focused on research and development rather than in full-scale manufacturing [126]. In the limited number of cases where they were employed in manufacturing, they have been successful in reducing manufacturing costs and improving the monitoring of product quality [140]. The development of different PAT methods, with their compelling application as an integral part of a monitoring and control strategy [141], has established a building block in gathering essential data from the physical component, enabling the further development of process model and DT.

3.2. Process Modeling

DTs highly depend on the use of data and models, and in the pharmaceutical industry, there is a growing interest in the development and application of methods and tools that facilitate that [142]. Different types of models have been developed for batch and continuous process simulations, material property identification and prediction, system analyses, and advanced control. Papadakis et al. recently proposed a framework for selecting efficient reaction pathways for pharmaceutical manufacturing [143], which includes a series of modeling workflows for reaction pathway identification, reaction and separation analysis, process simulation, evaluation, optimization, and operation [142]. The overall framework would yield an optimized reaction process with identified design space and process analytical technology information. The models developed under this framework can all be used as the virtual component within a DT framework to provide further process understanding and control of the manufacturing plant.

As mentioned in Section 2.2, the modeling approaches can be classified as mechanistic modeling, data-driven modeling, and hybrid modeling. For mechanistic modeling approaches in pharmaceutical manufacturing, the discrete-element method (DEM), finite-element method (FEM), and computational fluid dynamics (CFD) are often used [144]. To simulate the particle-level or bulk behavior of the material flow in different pharmaceutical unit operations, DEM is a powerful tool and has been applied widely [145–147], though its high computational cost limits its practical use when running locally. With HPC and cloud computing, it is possible to integrate DEM simulations with the overall process, resulting in a near-real-time model. For model fluid flow in pharmaceutical processes, including API drying and fluidized beds, CFD and FEM are popularly implemented [144]. These two methods are also heavily utilized in biopharmaceutical manufacturing (see Section 4.2).

Data-driven modeling methods involve the collection and usage of a large amount of experimental data to generate models, and the resulting models are based on the provided datasets only. Commonly implemented approaches in pharmaceutical manufacturing include the artificial neural network (ANN) [148,149], multivariate statistical analysis, Monte Carlo [150], etc. These methods are less computationally intensive, but due to the lack of underlying physical understanding in the trained models, the prediction outside of the space of the dataset is often unsatisfactory.

There is also a recent trend in developing various types of hybrid modeling techniques to model complex pharmaceutical manufacturing processes, while lowering the demand of computational cost and data availability. Population balance modeling (PBM), with a comparatively lower computational cost, has been extensively used to model blending and granulation processes [64,151], and a PBM–DEM hybrid model has also been used to improve model accuracy while maintaining reasonable computational costs [152]. Other semi-empirical hybrid models, such as the ones that incorporate material properties into process models [153], and to investigate the effect of material properties in residence time distribution (RTD) and process parameters [146,154–157], have also been developed for different powder processing unit operations [52,158]. These models, when incorporated with a full DT framework, will facilitate the overall product and process design and development, accelerating the drug-to-market timeline.

Table 2 provides a feature-based comparison of various models used in pharmaceutical manufacturing applications. The characterization of computational complexity is based on the typical computational cost for a single unit operation. The feature of real-time capability emphasizes the ability for a model to produce simulation or prediction results in real-time and optimally, in-sync with the equipment. This feature highly depends on computational complexity. Even though mathematical and semi-empirical modeling approaches have this capability, they are mostly trained and implemented offline. Real-time applications are rarely seen in the context of pharmaceutical manufacturing. For adaptive modeling capability, the modeling approaches that are able to incorporate data are advantageous as new data can be used to update the models. The online usage of these models in adaptive mode can hardly be found.

Table 2. Feature-based comparison of various models.

Features	Discrete-Element Method (DEM)/ Computational Fluid Dynamics (CFD)/ Finite-Element Method (FEM)	Population Balance Modeling (PBM)	Mechanistic/ Mathematical	Semi-Empirical/ Hybrid	Data-Driven	Advanced Process Control
Computational complexity	High	Medium	Medium	Low	Low	Low
Real-time capability	No	No	Yes	Yes	Yes	Yes
Adaptive modeling	No	No	No	Yes	Yes	Yes

In addition to developing models for single pharmaceutical unit operations, a flowsheet model integrating the entire manufacturing process can be used to predict the process dynamics affected by material properties and operating conditions of different unit operations. More importantly, systematic process analysis of the flowsheet model, such as sensitivity analysis, design space identification, and optimization, can all be performed with the flowsheet model. This provides insight into the characteristics and bottlenecks of the process and thus facilitates the development of control strategies [125]. Throughout the years of development, many researchers and pharmaceutical companies have developed mature approaches in conducting these analyses offline during the process design phase [52,56,125,159,160]. Flowsheet models are needed for the development of DTs. However, flowsheet models are stand-alone, so they cannot automatically update adapting to the physical plant. In current research, there is limited communication between the flowsheet model and the plant, which is a challenge in the development of a DT.

3.3. Data Integration

The implementation of IoT devices in pharmaceutical manufacturing lines leads to the acquisition of vast amounts of data. This collection of process data and CQAs needs to be transmitted to the virtual component in real-time and in an efficient manner. In addition to these, several pharmaceutical process models also require material properties for accurate prediction. Thus, a central database location is required for access to all datasets for the virtual component [46]. All data transfer protocols discussed in Section 2.3 are applicable here as well. In addition to these, the applications and databases should also be compliant with 21 CFR Part 11 data integrity requirements in accordance with US FDA's guidance [161]. The database not only serves as a warehouse for real product data but can also be used to store results from simulations performed in the virtual component and optimized process parameters. It would also serve the purpose of relaying back these optimized process parameters to the real product.

Several studies have attempted to achieve an integrated data framework in downstream pharmaceutical manufacturing [46,84,132,162–165]. Some of these studies were focused on implementing a control system for the direct compression line [132,157,165]. Cao et al. [46] presented an ISA-88 compliant manufacturing execution system (MES) where the batch data were stored on a cloud database as well as on a local data historian. The communications between the equipment and the control platform were performed in a similar manner for all the studies. The process control system (PCS) created a database based on the input recipe, and the database was replicated directly into the local data historian. The communication between the historian and PCS can be achieved using TCP/IP and OPC since each software is hosted on different computer systems on the same network. The historian database can in turn be duplicated onto the cloud using network protocols such as MQTT, HTTPS, etc. Some authors have also presented ontologies for efficient data flow for laboratory experiments performed during pharmaceutical manufacturing [166–168]. Cao et al. [46] also addressed the collection of laboratory data in an ISA-88 applicable recipe-based electronic laboratory notebook—many of the presented studies focused primarily on integrating one component of a completely integrated data management system. Figure 2 illustrates a sample data integration framework, where data collected from the manufacturing plant as well as laboratory experiments are uploaded to a cloud database using the mentioned protocols. The data can then be used in the virtual component for simulations, and corrective actions can be sent back to the control platform.

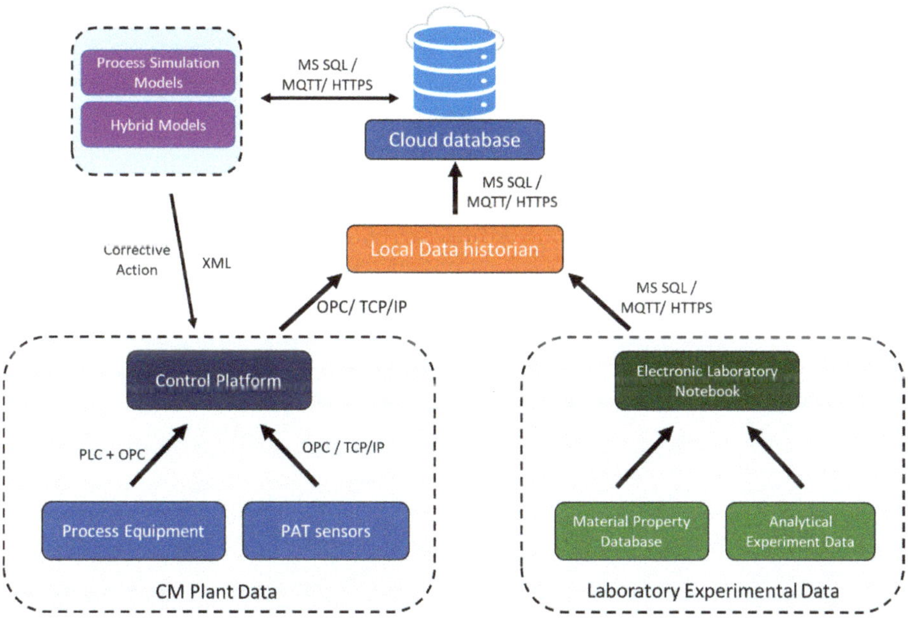

Figure 2. Framework for dataflow in a continuous direct compaction tablet line. The text over the arrow indicates options for data transfer protocols.

3.4. Challenges and Opportunities

Integrating all building blocks mentioned in Sections 3.1–3.3, the authors are visioning a fully integrated, model-centric DT framework for pharmaceutical manufacturing, as shown in Figure 3. The physical plant continuously sends process data to the virtual end, establishing a data inflow to achieve continuous process monitoring and data storage. Once the real-time data are received, process visualization and evaluation can be performed in real-time using visualization tools and process models. Automatic control based on evaluation results can then be executed to modify process operations if it is needed. The overall data and information flow become a continuous, real-time, integrated loop. Models can be updated based on plant measurements and changes by implementing hybrid or adaptive modeling techniques, and real-time model evaluation results that support the identification of critical process parameter boundaries, process optimizations, and material/process characterization can guide the operational updates of the plant. Our review has showcased that the pharmaceutical industry is on the move towards adopting a full DT. Currently, continuous monitoring of processes, storage of operation data, process visualization, and model-predictive control have been implemented in pharmaceutical applications. Building blocks are in place for all three components, but there still exist some key challenges and gaps.

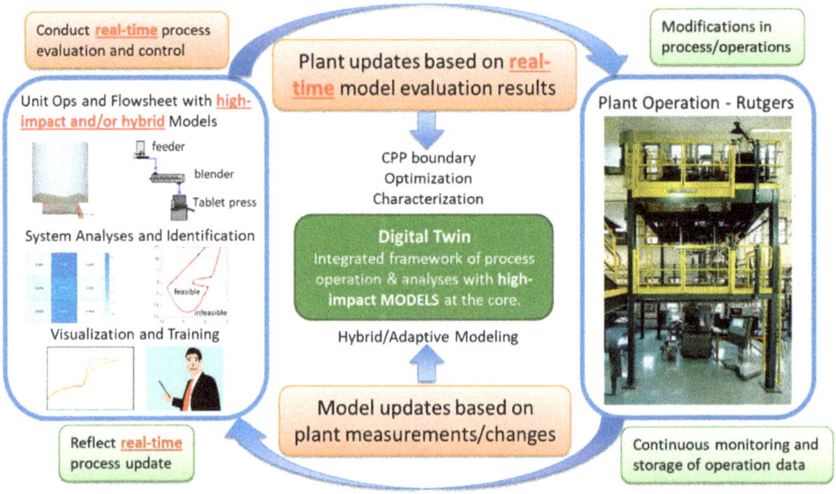

Figure 3. Fully integrated DT framework for continuous pharmaceutical manufacturing.

In terms of process monitoring and the use of PAT, though the use of spectroscopy to estimate product compositions has become a routine, the accuracy of measurements in low-dose drug products, the consideration and handling of outside interferences, and the maintenance of calibration models (i.e., the robustness of calibration) are all common problems. For low-dose drug measurements, though there are new tools such as NIRS and in-line UV spectroscopy, the accuracy can be improved by increasing sampling frequency and spectra analysis. The outside interference issue may be resolved by implementing various iterative optimization technologies, as recent studies have demonstrated the capability of such an approach [169,170]. With regard to the calibration model maintenance, different offline, adaptive methodologies have been well presented by Kadlec et al. [171], but the online, continuous update with streaming data may be an option moving forward.

At the virtual end, recent research and technology development have shaped the general framework and applications. Libraries of models and system analysis tools exist to develop a fully connected virtual model. However, as mentioned in Section 3.2, the computational cost for many complex and integrated models is high, requiring the use of cloud and/or high-performance computing. The high computational requirement also hinders the use of models in real-time, which is a key component of the DT framework [4]. To resolve this issue, efficient computational algorithms and reduced order modeling approaches need to be implemented, as well as the efficient distribution of computational resources. Another relevant issue is that most models developed for the pharmaceutical industry are static, meaning that they only reflect the system at the time that the models are developed. The models do not update themselves as new data become available. Model maintenance is, therefore, required [172], and the goal is that this can be performed automatically by the virtual component [171,173,174]. These model maintenance problems can also be viewed as issues caused by a number of drifts (i.e., concept drift, model drift, data drift, sensor drift). Methodologies in handling drifts have been extensively studied in many electrical and computer engineering papers [175–178], but case studies in pharmaceutical manufacturing have not yet been reported.

One of the most prominent issues includes the information communication between the two components. Table 3 illustrates a comparison between previous data integration frameworks that have been developed for pharmaceutical manufacturing. The limitations of each of these studies highlight the inability of current software tools and solutions to build a complete DT. Though the integration capability has been improving, it is noted that most of the current applications in the pharmaceutical industry only transfer data from the physical plant to the virtual component. The reverse is rarely seen. To have a fully integrated and automated DT, the information flow from the virtual component to the physical plant also needs to be established. The virtual plant should be able to change system settings and control the physical plant to help achieve an optimized process within the design space.

Table 3. A comparison of data integration studies presented for pharmaceutical manufacturing.

Reference	Integration Achieved	Tools Used	Limitation
Hailemariam et al. 2010 [166,167]	Presented a data collection ontology to for laboratory data	Extensive Markup Language (XML), Resource Description Framework (RDF),	A limited number of software and processes were connected to the ontology
Singh et al. 2014 [132,165]	Physical plant level up to control platform to implement model predictive control (MPC) using sensor data	MATLAB, Process Pulse, DeltaV, SynTQ	Data integration was only achieved till the control platform
Cao et al. 2018 [46]	Presented a cloud-based data collection strategy for collecting data from a continuous pharmaceutical manufacturing pilot plant as well as collecting data from analytical equipment	XML, AWS, DeltaV, OSI-PI	A complete integration was presented for data collection, but it lacked its integration with any software for live data prediction
Barenji et al. 2019 [29]	Presented a cyber-physical framework for Process Analytical Technology (PAT) tools for pharmaceutical manufacturing	N/A	Data integration was only performed for PAT tools without any integration of analytics

In addition, integrating data inside the physical manufacturing plant faces issues with homogeneity of the data format used by manufacturers [116]. A full manufacturing cycle requires the collection of online and offline data from different departments and software. Though an increasing number of companies are adopting standard data formats and transfer protocols, the coordination among all different data, software, and platforms is still a challenge. Currently, this coordination is more of a business and engineering decision within the companies using these systems. Poor integration and coordination often lead to the burden of using and maintaining multiple platforms and software. Because of this, many companies now prefer to purchase equipment and systems from a sole vendor, which is both a challenge and an opportunity for equipment and system providers.

The use of cloud databases and cloud-based data management systems, data availability, stability of service, storage volume, and information security are all critical issues to be addressed [118]. As data are stored on the cloud, these data should be available when needed, which demands a high stability service and a rigorous business continuation plan. Many cloud platforms are using distributed technologies and cloud backups to resolve this issue, but the validity and reliability of the solutions need to be carefully studied before implementing them [179]. Moreover, with the implementation of IoT devices and various types of sensors, the volume of data collected from the manufacturing cycle can be extremely large. Even though many cloud platforms claim that they can coordinate the demanded storage capacity, it would result in an increasing burden to the company if the storage cost is high. With regard to information security, the issue is not new to the field of cloud storage, but it is particularly relevant to the pharmaceutical industry since the majority of the information is highly confidential, and cases have shown that a vulnerable cyber system in pharmaceutical companies can cost millions or even billions of dollars. This challenge gives rise to opportunities in research and employment of cyber-physical security systems to ensure the safety and confidentiality of the information being transferred. This field has been a hot topic, especially in electrical and computer engineering disciplines. Methodologies used in securing smart grids, statistical-based authentication systems, physical and virtual cyber barriers, etc. can be implemented in pharmaceutical manufacturing to develop a secure DT.

Finally, regulatory perspective is an important consideration in developing and applying DT in pharmaceutical manufacturing. The US FDA has developed modeling capability and has granted funding to academic institutions to explore the appropriate application of process models and DTs in the field. Various guidelines, reports, and presentations have all demonstrated that the regulatory experience and exposure to the DT concept is currently evolving [27,180]. Though DT development is not required for regulatory approval, its components can definitely offer pharmaceutical companies and regulatory bodies more insight into the process and product.

4. Digital Twin in Biopharmaceutical Manufacturing

Biopharmaceutical manufacturing focuses on the production of large molecule-based products in heterogeneous mixtures, which can be used to treat cancer, inflammatory, and microbiological diseases [181,182]. To fulfill the FDA regulations and obtain safe products, biopharmaceutical operations should be strictly controlled and operate under a sterilized process environment.

In recent years, there is an increasing demand for biologic-based drugs that drives the need for manufacturing efficiency and effectiveness [183]. Thus, many companies are transitioning from batch to continuous operation mode and employing smart manufacturing systems [182]. DT integrates the physical plant, data collection, data analysis, and system control [4], which can assist biopharmaceutical manufacturing in product development, process prediction, decision making, and risk analysis, as shown in Figure 4. Monoclonal Antibody production is selected as an example to represent the physical plant, which includes cell inoculation, seed cultivation, production bioreactor, recovery, primary capture, virus inactivation, polishing, and final formulation. These operations produce and purify protein products. Quality (majorly protein structure and composition) and impurities need to be monitored and transported to a virtual plant for analysis and virtual plant updates. Virtual plant includes plant simulation, analysis, and optimization, which guide the physical plant diagnosis and update with the help of the process control system. Integrated mAb production flowsheet modeling, bioreactor analysis and design space and biomass optimization are selected as examples shown in the three sections in the figure. However, the capabilities of virtual plant are not limited to the examples list above. To understand the progress of DT development in biopharmaceutical manufacturing, this section reviews the process monitoring, modeling and data integration (virtual plant, physical plant communication) in the existed industry and analyzed possibilities and gaps to achieve integrated biopharma-DT manufacturing.

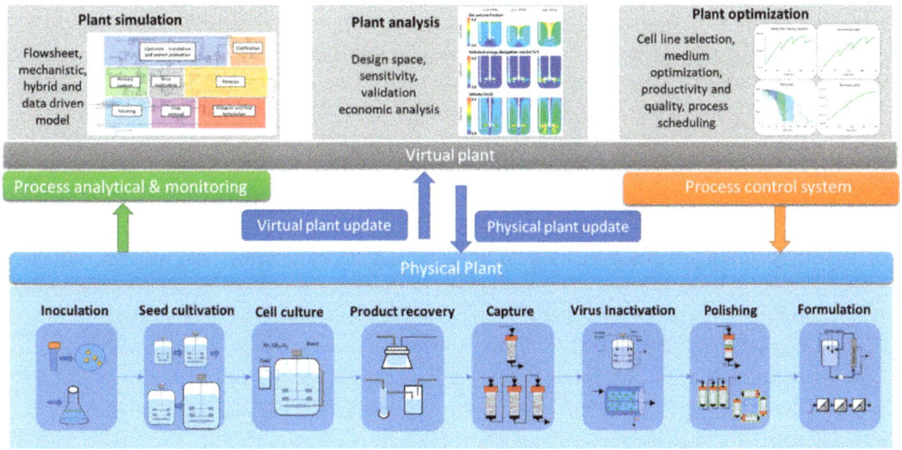

Figure 4. Biopharma process, benefits, and DT connections.

4.1. PAT Methods

Biological products are highly sensitive to cell-line and operating conditions, while the fractions and structures of the product molecules are closely related to drug efficacy [184]. Thus, having a real-time process diagnostic and control system is essential to maintain consistent product quality. However, process contamination needs to be strictly controlled in the biopharmaceutical manufacturing; thus, the monitoring system should not be affected by fouling nor interfere with media to maintain monitoring accuracy, sensitivity, stability, and reproducibility [185]. In general, among different unit operations, process parameters and quality attributes need to be captured.

Biechele et al. [185] presented a review of sensing applied in bioprocess monitoring. In general, process monitoring includes physical, chemical, and biological variables. In the gas phase, the commonly used sensing system consists of semiconducting, electrochemical, and paramagnetic sensors, which can be applied to oxygen and carbon dioxide measurements [185,186]. In the liquid phase, dissolved oxygen, carbon dioxide, and pH values have been monitored by an in-line electrochemical sensor. However, media composition, protein production, and qualities such as glycan fractions are mostly measured by online or at-line HPLC or GC/MS [186,187]. The specific product quality monitoring methods are reviewed by Guerra et al. [188] and Pais et al. [189].

Recently, spectroscopy methods have been developed for accurate and real-time monitoring for both upstream and downstream operations. The industrial spectroscopy applications mainly focus on cell growth monitoring and culture fluid components quantifications [190]. UV/Vis and multiwavelength UV spectroscopy have been used for in-line real-time protein quantification [190]. NIR has been used for off-line raw material and final product testing [190]. Raman spectroscopy has been used for viable cell density, metabolites, and antibody concentration measurements [191,192]. In addition, spectroscopy methods can also be used for process CQA monitoring, such as host cell protein and protein post-translational modifications [187,193]. Research shows that in-line Raman spectroscopy and Mid-IR have capabilities to monitor protein concentration, aggregation, host cell proteins (HCPs), and charge variants [194,195]. The spectroscopy methods are usually supported with chemometrics, which require data pretreatments such as background correction, spectral smoothing, and multivariant analysis for quantitative and qualitative analysis of the attributes. Many different applications of spectroscopic sensing are reviewed in the literature [187,188,190,193].

4.2. Process Modeling

The application of DT in biopharmaceutical manufacturing requires a complete virtual description of physical plant within a simulation platform [4]. This means that the simulation should capture the important process dynamics in each unit operation within an integrated model. Previous reviews have focused on the process modeling methods for both upstream and downstream operations [183,196–200].

For upstream bioreactor, extracellular fluid dynamics [201–203], system heterogeneities, and intracellular biochemical pathways [204–215] can be captured. Process modeling supports early-stage cell-line development, obtains optimal media formulations, and enables prediction of the overall bioreactor performance, including cell activities, metabolites' concentrations, productivity, and product quality under different process parameters [216,217]. The influence from various parameters such as temperature, pH, dissolved oxygen, feeding strategies, and amino acid concentrations can be captured and further used to optimize process operations [218–222].

For downstream operation, modeling strategies have focused on selecting design parameters, adjusting operating conditions, and buffer usage to achieve high protein productivity and purities efficiently. The different operating conditions include (1) flowrate, buffer pH, or salt concentration effects for chromatography operation [223–226]; (2) residence time, buffer concentration, and pH used for virus inactivation; (3) feed protein concentration, flux, retentate pressure operated for filtration [227]. Thus, the product concentration and various types of impurities can be predicted for each unit operation. The detailed modeling methods have been reviewed in the literature [228].

In recent years, biopharmaceutical companies are shifting from batch to continuous operations. It remains an unanswered question if it is feasible to start up a new, fully continuous process plant or replace specific unit operations with continuous units. Integrated process modeling provides a virtual platform to test various operating strategies such as batch, continuous, and hybrid operating modes [229]. These different operating modes can be compared based on life cycle analysis and economic analysis for different target products under various operation scales [229–233].

For flowsheet modeling, there are two approaches available in the literature, which include mechanistic and data-driven models. Due to the high computational cost, mechanistic modeling mostly focuses on the integration of a limited number of units, such as the combination of multiple chromatography operations [234]. Data-driven/empirical models are generally used to integrate all the unit operations in a computationally efficient way. Mechanistic models for a single unit can be integrated with other units that are built by the data-driven model to optimize a specific unit in the integrated process [235]. Mass flow and RTD models [236] can be included in the model to examine different scenarios of adding and replacing new unit operations and adjusting process parameters. Coupling with the control system, flowsheet modeling will be able to achieve real-time decision making and optimize the overall process operation automatically [237].

The data-driven models can be further integrated with Monte Carlo analysis or linear/nonlinear programming for risk assessment and process scheduling. Zahel et al. [238] applied Monte Carlo simulation in the end-to-end data-driven model, which can be used to estimated process capabilities and provide risk-based decision making following a change in the manufacturing operations.

Table 4 shows examples of capabilities and methods for process modeling, that can be potentially used in DT virtual plant model building. However, it needs to note that although process modeling has capabilities to capture all the above operating conditions and critical quality attributes, none of the modeling work incorporates all the process information within a single model. In recent years, hybrid models (for example, ANN + mechanistic model) have become more prevalent in both upstream and downstream model building because they improve the computational speed as well as the broad applications and model robustness.

Table 4. Capabilities and methods for process modeling in biopharmaceutical manufacturing. Note that many studies have used these methods, and the studies cannot be listed one by one. The papers selected in the table are used to represent the capabilities of the specific methods.

Categories	Methods	Platforms	Comments
	Upstream Process		
Bioreactor fluid dynamics, system heterogeneity	CFD simulation [201] CFD + PBM simulation [239] CFD + kinetics model [202]	Ansys Fluent, COMSOL Multiphysics	Support to understand operations such as agitation, aeration, nutrients feeding. Guide process scale-up. Computationally expensive. Can reduce the computational time by using a compartment model, hard to be validated.
Cell growth, nutrients, and metabolism. Product quality (protein glycosylation)	Kinetic model [204,240,241]	MATLAB, gPROMS, Visual Basic for Applications	Capture and predict the dynamic profile of the cell culture. Correlate critical process parameters (CPPs) and critical quality attributes (CQAs). Require a large amount of data for parameter estimations.
	Stoichiometric methods [242]	MATLAB, OptFlux etc.	Deal with a large amount of mechanistic reaction, genome-scale simulation. Need to integrate with the kinetic model to capture the dynamic profiles
	Multivariate tools [243]	MATLAB	Require a large amount of data. Represent input–output correlations. Do not capture the mechanistic correlations.
Media formulation	Multivariate analysis MFA [211,222]	MATLAB	Identify nutrient correlations, improve productivity and cell viability
Product impurities	Regression model and Multivariate analysis [244]	MATLAB	Capture predict titer, aggregation, low molecular weight components, and glycan groups
	Downstream Process		
Bind-elute/flow-through chromatography	Mechanistic: Plate model, mass balance model, general rate model with their simplifications models [245,246]	MATLAB, CADET, ChromX	Capture moving and stationary phases, obtain breakthrough curves, gradient elution curves. Predict the product concentration and impurities (charge variant, aggregates, host cell proteins)
	Transport dispersive model—ANN model [225]	MATLAB	
Filtration/ultrafiltration	Mechanistic: Film theory, Osmotic Pressure Model, boundary layer, mass transfer coefficient) [227]	Aspen Custom Modeler	Capture volumetric flow, flux, and pressure across the filtration membrane. Can be used for model predictive control.
	Hybrid model (ANN-mechanistic film theory) [247]	MATLAB	
Downstream integration (precipitation)	Empirical model (quantitative structure-activity relationship) + Mechanistic model [248]	NA	Physico-chemical process model supported by design of experiment (DoE). Capture CPP and CQA
Downstream integration and optimization	Mechanistic model—Artificial Neural Network-Optimization algorithm [249]	MATLAB	Optimize overall process yield and solvent use by adjusting operation parameters such as duration. However, only high molecular weight contamination was considered.

Table 4. *Cont.*

Categories	Methods	Platforms	Comments
		Integrated Process	
Residence time distribution	Probability distribution function for each unit operation [236]	Python	Correlate input material operating conditions, design parameters with outlet profile. Easy to update.
Activity tracking and decision making	Discrete Event Simulation [250]	Extend Sim, Simul8	Discrete/dynamic system, track activity, scheduling, and resource utilization
Material tracking and decision making	Mechanistic/Empirical model [229,251]	SuperPro Designer, Biosolve	Track material balance and optimize cost-effectiveness. Process debottlenecking, capacity planning
Process risk assessment	Implement process model with Monte Carlo analysis [238]	MATLAB	Evaluate parameter sensitivity, impurity purification, and product quality. Hard to apply to computationally expensive model
Overall process optimization	Integrate flowsheet model with optimization solvers [252]	SuperPro Designer-VB-Matlab	Optimize environment impact and cost-effectiveness by adjusting 4 operating parameters

4.3. Data Integration

Data obtained in the biopharmaceutical monitoring system are usually heterogeneous in data types and time scales. They can be collected from different sensors, production lines (laboratory or manufacturing), and at different time intervals. With the development of real-time PAT sensors, a large amount of data is obtained during biopharmaceutical manufacturing. Thus, data preprocessing is essential to handle missing data, perform data visualization, and reduce dimension [253]. Casola et al. [254] presented data mining-based algorithms to stem, classify, filter, and cluster historical real-time data in batch biopharmaceutical manufacturing. Lee et al. [255] applied data fusion to combine multiple spectroscopic techniques and predict the composition of raw materials. These preprocessing algorithms remove noise from the dataset and allow the data to be used in a virtual component directly.

In DTs, virtual components and physical components should communicate frequently. Thus, the virtual platforms need to have the flexibility to adjust their model-structure for different products and operating conditions. Herold and King [256] presented an algorithm that used biological phenomena to identify fed-batch bioreactor process model structure automatically. Luna and Martinez [257] used experimental data to train the imperfect mathematical model and corrected model prediction errors. Although there are no such applications for the integrated process, these works show the possibilities to achieve physical and virtual component communication.

In biopharmaceutical manufacturing, the integrated database can guide process-wide automatic monitoring and control [258]. Fahey et al. applied six sigma and CRISP-DM methods and integrated data collection, data mining, and model predictions for upstream bioreactor operations. Although the process optimization and control have not been considered in this work, it still shows the capabilities to handle large amounts of data for predictive process modeling [259]. Feidl et al. [258] used a supervisory control and data acquisition (SCADA) system to collect and store data from different unit operations at each sample time and developed a monitoring and control system in MATLAB. The work shows the integration of supervisory control with a data acquisition system in a fully end-to-end biopharmaceutical plant. However, process modeling has not been considered during the process operations, which cannot support process prediction and analysis.

4.4. Challenges and Opportunities

In terms of process monitoring in the physical plant, the application of real-time CQA monitoring methods has not been adapted to industrial applications. The use of NIR or Raman spectroscopy shows potential in real-time multicomponent measurements, although most applications have not yet been applied to industrial practice. To obtain accurate predicting/measurement results, raw material calibration and chemometric methods need to be applied, which increases the complexity of the application of spectroscopy. In addition, the data obtained from biopharmaceutical manufacturing are high dimensional and heterogeneous, which require advanced data integration and synchronization. An automated data aggregation, mining, storage, and visualization system is required to achieve DT automation. The data storage system should have large enough capability, easy accessibility, and high security as described in Section 3.4 to ensure manufacturing data security, patient data privacy, and the communication between the physical and virtual plant successfully.

To build a simulation of the physical plant, although different modeling methods have been developed for both upstream and downstream unit operations, there is no robust model that captures CPPs and CQAs for all the unit operations in the integrated process. As listed in Table 4, upstream CFD, stoichiometric and kinetic models can achieve the bioreactor modeling on different scales (from genome scale to manufacturing scales); however, not all these methods can be implemented within a DT framework because of the high computational cost. Similarly, downstream processes composed of different unit operations that integrate and optimize all the mechanistic models altogether are not realistic. Thus, these can explain the reason why the current integrated process models focus on mass balance and activity plans based on empirical models or simulators. To deal with this problem, one possible way is to apply pre-analysis to the system to reduce the dimension and parameters by evaluating the CPPs and CQAs to ensure productivity and efficacy. Based on the analysis, the system will select models and use the limited number of parameters to analyze or optimize the process. In this case, all different modeling methods need to be built on the same platform or have good model–model communications. An alternative way is to apply hybrid models to reduce the computational burden in the integrated process. In addition, to capture the major unit operations, the auxiliary equipment such as buffer preparation, Cleaning-In-Place (CIP), and Sterilization-In-Place (SIP) also need to be integrated into the process modeling. These operations do affect decision-making, including manufacturing scheduling and cost analysis. However, there is no such model that captures all the auxiliary equipment. Moreover, in the risk analysis in biopharmaceutical manufacturing, process contamination will directly cause batch failure. Lot to lot variations also exist in the bioreactor culture and purification process. Developing a model-based control system that can diagnose the contamination and process variabilities at an early stage is essential to improve the process efficiency. It is known that pharmaceutical or biopharmaceutical industries follow more stringent regulatory pathways; thus, the progress of accepting new technologies usually takes a longer time than other industries. It must be noticed that current technologies such as AI DTs do not conform to the QbD regulatory guidelines. The good news is that regulatory agencies are also seeking the adoption of innovative technologies. If DT can be developed for process operations and control at the same time, this method might be promising to be accepted by regulatory [260]. However, the DT approach is closely related to real-time optimization and operation supports, which are based on already built manufacturing platforms. In this situation, it might be hard to obtain regulatory approval [235].

The integration of virtual plant and physical plant in biopharmaceutical manufacturing is still in its infancy. It is promising to show that the application of data–model–control integration can be achieved for a single unit operation. Additionally, a data acquisition–control system can be achieved for an integrated process. However, to accomplish the biopharmaceutical DT, the development of real-time data acquisition, a dedicated data transferring system, an effective control and execution technique, robust simulation methods, anomaly detection, prediction tools, and easy access to secure the cloud server platform are still needed.

5. Conclusions

DTs are a crucial development of the close integration of manufacturing information and physical resources that raise much attention across industries. The critical parts of a fully developed DT include the physical and virtual components, and the interlinked data communication channels. Following the development of IoT technologies, there are many applications of DT in various industries, but the progress is lagging for pharmaceutical and biopharmaceutical manufacturing. This review paper summarizes the current state of DT in the two application scenarios, providing insights to stakeholders and highlighting possible challenges and solutions of implementing a fully integrated DT.

In pharmaceutical manufacturing, building blocks of a DT, including PAT methods, data management systems, unit operations, and flowsheet models, system analyses methods, and integration approaches have all been developed in the last few years, but gaps in PAT accuracy, real-time model computation, model maintenance capabilities, real-time data communication, as well as concerns in data security and confidentiality, are preventing the full integration of all the components. To solve these challenges, several insights are provided. The development of new tools such as NIRS and in-line UV spectroscopy, iterative optimization technologies, and different offline adaptive methodologies can help to resolve the existing issues in PAT methods. In order to reduce simulation time to achieve real-time computation, efficient algorithms, and reduced order modeling approaches should be further studied for process models. In terms of model maintenance, adaptive modeling methods with online streaming data are to be investigated further. To have a fully integrated and automated DT, the information flow from the virtual component to the physical plant also needs to be established. The virtual plant should be able to change system settings and control the physical plant to help to achieve an optimized process within the design space. Ideally, all these components should be placed under appropriate physical and virtual security protocols.

In biopharmaceutical manufacturing, similar constituting components of DTs have been discussed, as well as the implementation challenges in each block. In terms of process monitoring, the development of NIR or Raman spectroscopy, material calibration, and chemometric methods can help to obtain an accurate predicting/measurement result. Advanced data integration and synchronization technology should be in place. For process simulation, there is no robust model that captures CPPs and CQAs for all the unit operations in the integrated process due to the computational complexity. Pre-analysis to screen the CPPs and CQAs is a promising approach to reduce the computational burden. Process models to capture the auxiliary equipment and process contamination need to be further investigated. To achieve a fully integrated DT, real-time data acquisition methods, data transferring systems, effective control and execution techniques, robust simulation methods, and anomaly detection are still in need, with other supporting functions.

It is noted that given the rapid development and publication rate in this area, and that this paper is merely a narrative literature review, the authors are not able to list and review all studies in these areas in detail. The papers selected and problems described in the manuscript are only a nonholistic subset used to represent the capabilities and drawbacks of a method or technology. Since the manuscript is organized using a conceptual and topical frame, the authors recommend interested readers to go through cited references to explore additional details. In addition to the summarized research opportunities, further research directions can include the development of a demonstrative case study of DT in pharmaceutical and biopharmaceutical manufacturing and a systematic review of the field.

Author Contributions: Conceptualization, Y.C. and M.I.; writing—original draft preparation, Y.C., O.Y., C.S., P.B., R.R., and M.I.; writing—review and editing, Y.C., O.Y., C.S., P.B., R.R., and M.I.; supervision, R.R. and M.I.; project administration, M.I.; funding acquisition, M.I. and R.R. All authors have read and agreed to the published version of the manuscript.

Funding: This research was funded by US Food and Drug Administration, grant number DHHS-FDA-U01FD006487 and DHHS-FDA-R01FD006588.

Conflicts of Interest: The authors declare no conflict of interest.

References

1. Legner, C.; Eymann, T.; Hess, T.; Matt, C.; Böhmann, T.; Drews, P.; Mädche, A.; Urbach, N.; Ahlemann, F. Digitalization: Opportunity and Challenge for the Business and Information Systems Engineering Community. *Bus. Inf. Syst. Eng.* **2017**, *59*, 301–308. [CrossRef]
2. Kritzinger, W.; Karner, M.; Traar, G.; Henjes, J.; Sihn, W. Digital Twin in manufacturing: A categorical literature review and classification. *IFAC-PapersOnLine* **2018**, *51*, 1016–1022. [CrossRef]
3. Oztemel, E.; Gursev, S. Literature review of Industry 4.0 and related technologies. *J. Intell. Manuf.* **2018**, *31*, 127–182. [CrossRef]
4. Tao, F.; Cheng, J.; Qi, Q.; Zhang, M.; Zhang, H.; Sui, F. Digital twin-driven product design, manufacturing and service with big data. *Int. J. Adv. Manuf. Technol.* **2018**, *94*, 3563–3576. [CrossRef]
5. Venkatasubramanian, V. The promise of artificial intelligence in chemical engineering: Is it here, finally? *AIChE J.* **2019**, *65*, 466–478. [CrossRef]
6. Bao, J.; Guo, D.; Li, J.; Zhang, J. The modelling and operations for the digital twin in the context of manufacturing. *Enterp. Inf. Syst.* **2018**, *13*, 534–556. [CrossRef]
7. Tao, F.; Qi, Q.; Wang, L.; Nee, A.Y.C. Digital Twins and Cyber–Physical Systems toward Smart Manufacturing and Industry 4.0: Correlation and Comparison. *Engineering* **2019**, *5*, 653–661. [CrossRef]
8. Haag, S.; Anderl, R. Digital twin—Proof of concept. *Manuf. Lett.* **2018**, *15*, 64–66. [CrossRef]
9. Litster, J.; Bogle, I.D.L. Smart Process. Manufacturing for Formulated Products. *Engineering* **2019**, *5*, 1003–1009. [CrossRef]
10. Tourlomousis, F.; Chang, R.C. Dimensional Metrology of Cell-matrix Interactions in 3D Microscale Fibrous Substrates. *Procedia CIRP* **2017**, *65*, 32–37. [CrossRef]
11. Khan, M.; Wu, X.; Xu, X.; Dou, W. Big data challenges and opportunities in the hype of Industry 4.0. In Proceedings of the 2017 IEEE International Conference on Communications (ICC), Paris, France, 21–25 May 2017.
12. Li, X.; Li, D.; Wan, J.; Vasilakos, A.V.; Lai, C.-F.; Wang, S. A review of industrial wireless networks in the context of Industry 4.0. *Wirel. Netw.* **2015**, *23*, 23–41. [CrossRef]
13. Uhlemann, T.H.J.; Schock, C.; Lehmann, C.; Freiberger, S.; Steinhilper, R. The Digital Twin: Demonstrating the Potential of Real Time Data Acquisition in Production Systems. *Procedia Manuf.* **2017**, *9*, 113–120. [CrossRef]
14. Belanger, J.; Venne, P.; Paquin, J.-N. The What, Where and Why of Real-Time Simulation. *Planet RT.* **2010**, *1*, 37–49.
15. Roman-Ospino, A.D.; Singh, R.; Ierapetritou, M.; Ramachandran, R.; Mendez, R.; Ortega-Zuniga, C.; Muzzio, F.J.; Romanach, R.J. Near infrared spectroscopic calibration models for real time monitoring of powder density. *Int. J. Pharm.* **2016**, *512*, 61–74. [CrossRef] [PubMed]
16. Damiani, L.; Demartini, M.; Guizzi, G.; Revetria, R.; Tonelli, F. Augmented and virtual reality applications in industrial systems: A qualitative review towards the industry 4.0 era. *IFAC-PapersOnLine* **2018**, *51*, 624–630. [CrossRef]
17. Zühlke, D.; Gorecky, D.; Schmitt, M.; Loskyll, M. Human-machine-interaction in the industry 4.0 era. In Proceedings of the 2014 12th IEEE International Conference on Industrial Informatics (INDIN), Porto Alegre, Brazil, 27–30 July 2014.
18. Zhuang, C.; Liu, J.; Xiong, H. Digital twin-based smart production management and control framework for the complex product assembly shop-floor. *Int. J. Adv. Manuf. Technol.* **2018**, *96*, 1149–1163. [CrossRef]
19. Leng, J.; Zhang, H.; Yan, D.; Liu, Q.; Chen, X.; Zhang, D. Digital twin-driven manufacturing cyber-physical system for parallel controlling of smart workshop. *J. Ambient Intell. Humaniz. Comput.* **2018**, *10*, 1155–1166. [CrossRef]
20. Rosen, R.; von Wichert, G.; Lo, G.; Bettenhausen, K.D. About The Importance of Autonomy and Digital Twins for the Future of Manufacturing. *IFAC-PapersOnLine* **2015**, *48*, 567–572. [CrossRef]
21. Mayani, M.G.; Svendsen, M.; Oedegaard, S.I. Drilling Digital Twin Success Stories the Last 10 Years. In Proceedings of the SPE Norway One Day Seminar, Bergen, Norway, 18 April 2018.
22. Schleich, B.; Anwer, N.; Mathieu, L.; Wartzack, S. Shaping the digital twin for design and production engineering. *CIRP Ann.* **2017**, *66*, 141–144. [CrossRef]

23. Grieves, M. Digital Twin: Manufacturing Excellence through Virtual Factory Replication. *White Paper* **2014**, *1*, 1–7.
24. Grieves, M.; Vickers, J. Digital Twin: Mitigating Unpredictable, Undesirable Emergent Behavior in Complex Systems. In *Transdisciplinary Perspectives on Complex Systems*; Springer: Cham, Switzerland, 2017; pp. 85–113.
25. Stark, R.; Fresemann, C.; Lindow, K. Development and operation of Digital Twins for technical systems and services. *CIRP Ann.* **2019**, *68*, 129–132. [CrossRef]
26. Glaessgen, E.H.; Stargel, D.S. The Digital Twin Paradigm for Future NASA and U.S. Air Force Vehicles. In Proceedings of the 53rd AIAA/ASME/ASCE/AHS/ASC Structures, Structural Dynamics and Materials Conference—Special Session on the Digital Twin, Honolulu, HI, USA, 23–26 April 2012.
27. O'Connor, T.F.; Yu, L.X.; Lee, S.L. Emerging technology: A key enabler for modernizing pharmaceutical manufacturing and advancing product quality. *Int. J. Pharm.* **2016**, *509*, 492–498. [CrossRef] [PubMed]
28. Ding, B. Pharma Industry 4.0: Literature review and research opportunities in sustainable pharmaceutical supply chains. *Process Saf. Environ. Prot.* **2018**, *119*, 115–130. [CrossRef]
29. Barenji, R.V.; Akdag, Y.; Yet, B.; Oner, L. Cyber-physical-based PAT (CPbPAT) framework for Pharma 4.0. *Int. J. Pharm.* **2019**, *567*, 118445. [CrossRef] [PubMed]
30. Steinwandter, V.; Borchert, D.; Herwig, C. Data science tools and applications on the way to Pharma 4.0. *Drug Discov. Today* **2019**, *24*, 1795–1805. [CrossRef]
31. Lopes, M.R.; Costigliola, A.; Pinto, R.; Vieira, S.; Sousa, J.M.C. Pharmaceutical quality control laboratory digital twin—A novel governance model for resource planning and scheduling. *Int. J. Prod. Res.* **2019**, 1–15. [CrossRef]
32. Kumar, S.; Talasila, D.; Gowrav, M.; Gangadharappa, H. Adaptations of Pharma 4.0 from Industry 4.0. *Drug Invent. Today* **2020**, *14*, 405–415.
33. Reinhardt, I.C.; Oliveira, D.J.C.; Ring, D.D.T. Current Perspectives on the Development of Industry 4.0 in the Pharmaceutical Sector. *J. Ind. Inf. Integr.* **2020**, *18*, 100131. [CrossRef]
34. Zhang, C.; Xu, W.; Liu, J.; Liu, Z.; Zhou, Z.; Pham, D.T. A Reconfigurable Modeling Approach for Digital Twin-based Manufacturing System. *Procedia CIRP* **2019**, *83*, 118–125. [CrossRef]
35. Kabugo, J.C.; Jämsä-Jounela, S.-L.; Schiemann, R.; Binder, C. Industry 4.0 based process data analytics platform: A waste-to-energy plant case study. *Int. J. Electr. Power Energy Syst.* **2020**, *115*, 105508. [CrossRef]
36. González, I.; Calderón, A.J.; Figueiredo, J.; Sousa, J.M.C. A Literature Survey on Open Platform Communications (OPC) Applied to Advanced Industrial Environments. *Electronics* **2019**, *8*, 510. [CrossRef]
37. O'Donovan, P.; Leahy, K.; Bruton, K.; O'Sullivan, D.T.J. An industrial big data pipeline for data-driven analytics maintenance applications in large-scale smart manufacturing facilities. *J. Big Data* **2015**, *2*, 1–26.
38. Mandenius, C.-F.; Gustavsson, R. Mini-review: Soft sensors as means for PAT in the manufacture of bio-therapeutics. *J. Chem. Technol. Biotechnol.* **2015**, *90*, 215–227. [CrossRef]
39. Bosca, S.; Barresi, A.; Fissore, D. Use of a soft sensor for the fast estimation of dried cake resistance during a freeze-drying cycle. *Int. J. Pharm.* **2013**, *451*, 23–33. [CrossRef] [PubMed]
40. Ding, J.-G.; Qu, L.-L.; Hu, X.-L.; Liu, X.-H. Application of Temperature Inference Method Based on Soft Sensor Technique to Plate Production Process. *J. Iron Steel Res. Int.* **2011**, *18*, 24–27. [CrossRef]
41. Rogina, A.; Šiško, I.; Mohler, I.; Ujević, Ž.; Bolf, N. Soft sensor for continuous product quality estimation (in crude distillation unit). *Chem. Eng. Res. Des.* **2011**, *89*, 2070–2077. [CrossRef]
42. Kadlec, P.; Gabrys, B.; Strandt, S. Data-driven Soft Sensors in the process industry. *Comput. Chem. Eng.* **2009**, *33*, 795–814. [CrossRef]
43. Khalfe, N.; Lahiri, S.; Sawke, S. Soft sensor for better control of carbon dioxide removal process in ethylene glycol plant. *Chem. Ind. Chem. Eng. Q.* **2011**, *17*, 17–24. [CrossRef]
44. Teixeira, B.O.; Castro, W.S.; Teixeira, A.F.; Aguirre, L.A. Data-driven soft sensor of downhole pressure for a gas-lift oil well. *Control Eng. Pract.* **2014**, *22*, 34–43. [CrossRef]
45. Qin, S.J.; Yue, H.; Dunia, R. Self-validating inferential sensors with application to air emission monitoring. *Ind. Eng. Chem. Res.* **1997**, *36*, 1675–1685. [CrossRef]
46. Cao, H.; Mushnoori, S.; Higgins, B.; Kollipara, C.; Fermier, A.; Hausner, D.; Jha, S.; Singh, R.; Ierapetritou, M.; Ramachandran, R. A Systematic Framework for Data Management and Integration in a Continuous Pharmaceutical Manufacturing Processing Line. *Processes* **2018**, *6*, 53. [CrossRef]

47. Muñoz, E.; Capón-García, E.; Espuña, A.; Puigjaner, L. Ontological framework for enterprise-wide integrated decision-making at operational level. *Comput. Chem. Eng.* **2012**, *42*, 217–234. [CrossRef]
48. Židek, K.; Piteľ, J.; Adámek, M.; Lazorík, P.; Hošovský, A. Digital Twin of Experimental Smart Manufacturing Assembly System for Industry 4.0 Concept. *Sustainability* **2020**, *12*, 3658. [CrossRef]
49. Roblek, V.; Meško, M.; Krapež, A. A complex view of industry 4.0. *Sage Open* **2016**, *6*, 2158244016653987. [CrossRef]
50. Sanders, A.; Elangeswaran, C.; Wulfsberg, J. Industry 4.0 implies lean manufacturing: Research activities in industry 4.0 function as enablers for lean manufacturing. *J. Ind. Eng. Manag. (JIEM)* **2016**, *9*, 811–833. [CrossRef]
51. Wang, S.; Wan, J.; Zhang, D.; Li, D.; Zhang, C. Towards smart factory for industry 4.0: A self-organized multi-agent system with big data based feedback and coordination. *Comput. Netw.* **2016**, *101*, 158–168. [CrossRef]
52. Wang, Z.; Escotet-Espinoza, M.S.; Ierapetritou, M. Process analysis and optimization of continuous pharmaceutical manufacturing using flowsheet models. *Comput. Chem. Eng.* **2017**, *107*, 77–91. [CrossRef]
53. Grossmann, I.E.; Calfa, B.A.; Garcia-Herreros, P. Evolution of concepts and models for quantifying resiliency and flexibility of chemical processes. *Comput. Chem. Eng.* **2014**, *70*, 22–34. [CrossRef]
54. Eugene, E.A.; Gao, X.; Dowling, A.W. Learning and Optimization with Bayesian Hybrid Models. *arXiv* **2019**, arXiv:1912.06269. Available online: https://arxiv.org/abs/1912.06269 (accessed on 22 June 2020).
55. Von Stosch, M.; Oliveira, R.; Peres, J.; Feyo de Azevedo, S. Hybrid semi-parametric modeling in process systems engineering: Past, present and future. *Comput. Chem. Eng.* **2014**, *60*, 86–101. [CrossRef]
56. Bhosekar, A.; Ierapetritou, M. Advances in surrogate based modeling, feasibility analysis, and optimization: A review. *Comput. Chem. Eng.* **2018**, *108*, 250–267. [CrossRef]
57. Zendehboudi, S.; Rezaei, N.; Lohi, A. Applications of hybrid models in chemical, petroleum, and energy systems: A systematic review. *Appl. Energy* **2018**, *228*, 2539–2566. [CrossRef]
58. Bohlin, T. *Practical Grey-box Process Identification: Theory and Applications*, 1st ed.; Springer: London, UK, 2006.
59. Azarpour, A.; Borhani, T.N.; Alwi, S.R.W.; Manan, Z.A.; Mutalib, M.I.A. A generic hybrid model development for process analysis of industrial fixed-bed catalytic reactors. *Chem. Eng. Res. Des.* **2017**, *117*, 149–167. [CrossRef]
60. Laursen, S.Ö.; Webb, D.; Ramirez, W.F. Dynamic hybrid neural network model of an industrial fed-batch fermentation process to produce foreign protein. *Comput. Chem. Eng.* **2007**, *31*, 163–170. [CrossRef]
61. Schäfer, P.; Caspari, A.; Mhamdi, A.; Mitsos, A. Economic nonlinear model predictive control using hybrid mechanistic data-driven models for optimal operation in real-time electricity markets: In-silico application to air separation processes. *J. Process Control* **2019**, *84*, 171–181. [CrossRef]
62. Bikmukhametov, T.; Jäschke, J. Combining machine learning and process engineering physics towards enhanced accuracy and explainability of data-driven models. *Comput. Chem. Eng.* **2020**, *138*, 106834. [CrossRef]
63. Liu, H.; Wang, K.; Chen, Z.; Jordan, K.E.; Luo, J.; Deng, H. A Parallel Framework for Reservoir Simulators on Distributed-memory Supercomputers. In Proceedings of the SPE/IATMI Asia Pacific Oil & Gas. Conference and Exhibition, Society of Petroleum Engineers, Bali, Indonesia, 20–22 October 2015.
64. Prakash, A.V.; Chaudhury, A.; Barrasso, D.; Ramachandran, R. Simulation of population balance model-based particulate processes via parallel and distributed computing. *Chem. Eng. Res. Des.* **2013**, *91*, 1259–1271. [CrossRef]
65. Sampat, C.; Bettencourt, F.; Baranwal, Y.; Paraskevakos, I.; Chaturbedi, A.; Karkala, S.; Jha, S.; Ramachandran, R.; Ierapetritou, M. A parallel unidirectional coupled DEM-PBM model for the efficient simulation of computationally intensive particulate process systems. *Comput. Chem. Eng.* **2018**, *119*, 128–142. [CrossRef]
66. MathWorks. Simulation and Model-Based Design. 2020. Available online: https://www.mathworks.com/products/simulink.html (accessed on 19 June 2020).
67. COMSOL. Understand, Predict, and Optimize Physics-Based Designs and Processes with COMSOL Multiphysics. 2020. Available online: https://www.comsol.com/comsol-multiphysics (accessed on 19 June 2020).

68. PSE. gPROMS FormulatedProducts. 2020. Available online: https://www.psenterprise.com/products/gproms/formulatedproducts (accessed on 19 June 2020).
69. Aspentech. aspenONE Product Portfolio. 2020. Available online: https://www.aspentech.com/en/products/full-product-listing (accessed on 19 June 2020).
70. Siemens. Engineer Innovation with CFD- Focused Multiphysics Simulation. Available online: https://www.plm.automation.siemens.com/global/en/products/simcenter/STAR-CCM.html (accessed on 19 June 2020).
71. Pantelides, C. Digital Design, Digital Operations—The central role of modeling in digital world. In Proceedings of the PSE Advanced Process Modeling Forum, Tarrytown, NY, USA, 10–12 September 2019.
72. Siemens. Siemens PLM Software. MindSphere: The Cloud-Based, Open IoT Operating System for Digital Transformation. 2017. Available online: https://www.plm.automation.siemens.com/media/global/en/Siemens_MindSphere_Whitepaper (accessed on 27 May 2020).
73. GE Digital. Industrial Cloud Based Platform (PaaS). Available online: https://www.ge.com/digital/iiot-platform (accessed on 27 May 2020).
74. SEEQ. SEEQ Product Overview. Available online: https://www.seeq.com/product/overview (accessed on 19 June 2020).
75. TrendMiner. TrendMiner Self-Service Industrial Analytics. 2020. Available online: https://www.trendminer.com/software/ (accessed on 19 June 2020).
76. TIBCO. TIBCO Cloud: Connected Intelligence, Delivered. 2020. Available online: https://cloud.tibco.com/ (accessed on 19 June 2020).
77. Amazon. Start Building on AWS Today. 2020. Available online: https://aws.amazon.com/ (accessed on 19 June 2020).
78. Microsoft. Create Solutions Today that Adapt for Tomorrow. Invent with Purpose. 2020. Available online: https://azure.microsoft.com/en-us/ (accessed on 19 June 2020).
79. Google. Solve more with Google Cloud. 2020. Available online: https://cloud.google.com/ (accessed on 19 June 2020).
80. IBM. IBM Watson Products and Solutions. 2020. Available online: https://www.ibm.com/watson/products-services (accessed on 19 June 2020).
81. Subramanian, B. The disruptive influence of cloud computing and its implications for adoption in the pharmaceutical and life sciences industry. *J. Med. Mark. Device Diagn. Pharm. Mark.* **2012**, *12*, 192–203. [CrossRef]
82. Leukert, B.; Kubach, T.; Eckert, C.; Tsutsumi, K.; Crawford, M.; Vayssiere, N. IoT 2020: Smart and secure IoT platform. *IEC White Pap.* 2016, pp. 1–181. Available online: https://www.iec.ch/whitepaper/iotplatform/ (accessed on 25 July 2020).
83. Botta, A.; de Donato, W.; Persico, V.; Pescapé, A. Integration of Cloud computing and Internet of Things: A survey. *Future Gener. Comput. Syst.* **2016**, *56*, 684–700. [CrossRef]
84. Venkatasubramanian, V.; Zhao, C.; Joglekar, G.; Jain, A.; Hailemariam, L.; Suresh, P.; Akkisetty, P.; Morris, K.; Reklaitis, G.V. Ontological informatics infrastructure for pharmaceutical product development and manufacturing. *Comput. Chem. Eng.* **2006**, *30*, 1482–1496. [CrossRef]
85. Bray, T.; Paoli, J.; Sperberg-McQueen, C.M.; Eve Maler, F.Y. Extensible Markup Language (XML) 1.0 (Fifth Edition). 2008. Available online: https://www.w3.org/TR/2008/REC-xml-20081126/ (accessed on 28 May 2020).
86. Barbosa, D.; Bohannon, P.; Freire, J.; Kanne, C.-C.; Manolescu, I.; Vassalos, V.; Yoshikawa, M. XML Storage. In *Encyclopedia of Database Systems*; Liu, L., Özsu, M.T., Eds.; Springer: Boston, MA, USA, 2009; pp. 3627–3634.
87. Michels, J.; Hare, K.; Kulkarni, K.; Zuzarte, C.; Liu, Z.H.; Hammerschmidt, B.; Zemke, F. The New and Improved SQL: 2016 Standard. *ACM SIGMOD Rec.* **2018**, *47*, 51–60. [CrossRef]
88. Agrawal, R.; Ailamaki, A.; Bernstein, P.A.; Brewer, E.A.; Carey, M.J.; Chaudhuri, S.; Doan, A.; Florescu, D.; Franklin, M.J.; Garcia-Molina, H.; et al. The Claremont report on database research. *ACM SIGMOD Rec.* **2008**, *37*, 9–19. [CrossRef]
89. Jones, D.; Snider, C.; Nassehi, A.; Yon, J.; Ben, H. Characterising the Digital Twin: A systematic literature review. *CIRP J. Manuf. Sci. Technol.* **2020**, *29*, 36–52. [CrossRef]

90. Lund, A.M.; Mochel, K.; Lin, J.-W.; Onetto, R.; Srinivasan, J.; Gregg, P.; Bergman, J.E.; Hartling, K.D.; Ahmed, A.; Chotai, S. *Digital Twin Interface for Operating Wind Farms*; General Electric Co.: Boston, MA, USA, 2015.
91. Madni, A.; Madni, C.; Lucero, S. Leveraging Digital Twin Technology in Model—Based Systems Engineering. *Systems* **2019**, *7*, 7. [CrossRef]
92. GE Power Digital Solutions, GE Digital Twin—Analytic Engine for the Digital Power Plant. *White Pap.* 2016. Available online: https://www.ge.com/digital/sites/default/files/download_assets/Digital-Twin-for-the-digital-power-plant-.pdf (accessed on 25 July 2020).
93. Tao, F.; Zhang, M. Digital Twin Shop-Floor: A New Shop-Floor Paradigm towards Smart Manufacturing. *IEEE Access* **2017**, *5*, 20418–20427. [CrossRef]
94. Wagner, C.; Grothoff, J.; Epple, U.; Drath, R.; Malakuti, S.; Gruner, S.; Hoffmeister, M.; Zimermann, P. The role of the Industry 4.0 Asset Administration Shell and the Digital Twin during the life cycle of a plant. In Proceedings of the 2017 22nd IEEE International Conference on Emerging Technologies and Factory Automation (ETFA), Limassol, Cyprus, 12–15 September 2017.
95. Cheng, Y.; Zhang, Y.P.; Ji, P.; Xu, W.J.; Zhou, Z.D.; Tao, F. Cyber-physical integration for moving digital factories forward towards smart manufacturing: A survey. *Int. J. Adv. Manuf. Technol.* **2018**, *97*, 1209–1221. [CrossRef]
96. Siemens Switzerland Ltd. *The Digital Twin—Driving Business Value throughout the Building Life Cycle*; Siemens Switzerland Ltd.: Zug, Switzerland, September 2018.
97. Braun, K.; Laupp, G.; Leich, R.; Saur, W.; Scheifele, H.; Schick, J. Method for filling packaging containers by weight. U.S. Patent 4385670A, 31 May 1983.
98. Guo, F.Y.; Zou, F.; Liu, J.H.; Wang, Z.Q. Working mode in aircraft manufacturing based on digital coordination model. *Int. J. Adv. Manuf. Technol.* **2018**, *98*, 1547–1571. [CrossRef]
99. Tuegel, E.J.; Ingraffea, A.R.; Eason, T.G.; Spottswood, S.M. Reengineering Aircraft Structural Life Prediction Using a Digital Twin. *Int. J. Aerosp. Eng.* **2011**, *2011*, 154798. [CrossRef]
100. Gockel, B.; Tudor, A.; Brandyberry, M.; Penmetsa, R.; Tuegel, E. Challenges with Structural Life Forecasting Using Realistic Mission Profiles. In Proceedings of the 53rd AIAA/ASME/ASCE/AHS/ASC Structures, Structural Dynamics and Materials Conference, Honolulu, HI, USA, 23–26 April 2012.
101. Glaessgen, E.; Biegel, B.; Chandler, F.; Crichton, D.; LeMoigne, J.; Little, M.; Null, C.; Peters, W.; Ransom, J.; Wang, L. *NASA Technology Roadmaps TA11: Modeling, Simulation, Information Technology, and Processing*; NASA Office of the Chief Technologist: Washington, DC, USA, 2015.
102. Qi, Q.; Tao, F.; Hu, T.; Answer, N.; Liu, A.L.; Wei, A.; Wang, L.; Nee, A.Y.C. Enabling technologies and tools for digital twin. *J. Manuf. Syst.* **2019**. [CrossRef]
103. Toru Ishida, K.I.E. *Digital Cities: Technologies, Experiences, and Future Perspectives*, 1st ed.; Springer: Berlin/Heidelberg, Germany, 2000.
104. Parris, C. Meet a Digital Twin. In *Minds + Machines*; GE Digital: San Francisco, CA, USA, 2017.
105. Predictive Insights Aid Aircraft Landing Gear Performance | GE Digital. ge.com. 2020. Available online: https://www.ge.com/digital/customers/predictive-insights-aid-aircraft-landing-gear-performance (accessed on 25 July 2020).
106. Seshadri, B.R.; Krishnamurthy, T. Structural Health Management of Damaged Aircraft Structures Using the Digital Twin Concept. In Proceedings of the 25th AIAA/AHS Adaptive Structures Conference, Grapevine, TX, USA, 9–13 January 2017.
107. Damjanovic-Behrendt, V. A digital twin based privacy enhancement mechanism for the automative industry. In Proceedings of the 2018 International Conference on Intelligent Systems (IS), Funchal—Madeira, Portugal, 25–27 September 2018.
108. Yoo, Y.; Boland, R.J., Jr.; Lyytinen, K.; Majchrzak, A. Organizing for Innovation in the Digitized World. *Organ. Sci.* **2012**, *23*, 1398–1408. [CrossRef]
109. Wannenburg, J.; Malekian, R. Body Sensor Network for Mobile Health Monitoring, a Diagnosis and Anticipating System. *IEEE Sens. J.* **2015**, *15*, 6839–6852. [CrossRef]
110. Bruynseels, K.; de Sio, F.S.; van den Hoven, J. Digital Twins in Health Care: Ethical Implications of an Emerging Engineering Paradigm. *Front. Genet.* **2018**, *9*, 31. [CrossRef] [PubMed]

111. Baillargeon, B.; Rebelo, N.; Fox, D.D.; Taylor, R.L.; Kuhl, E. The Living Heart Project: A robust and integrative simulator for human heart function. *Eur. J. Mech. A-Solids* **2014**, *48*, 38–47. [CrossRef] [PubMed]
112. Francisco, A.; Mohammadi, N.; Taylor, J.E. Smart City Digital Twin-Enabled Energy Management: Toward Real-Time Urban. Building Energy Benchmarking. *J. Manag. Eng.* **2020**, *36*, 04019045. [CrossRef]
113. Silva, B.N.; Khan, M.; Han, K. Towards sustainable smart cities: A review of trends, architectures, components, and open challenges in smart cities. *Sustain. Cities Soc.* **2018**, *38*, 697–713. [CrossRef]
114. Komninos, N. The Architecture of Intelligent Cities. In Proceedings of the 2nd International Conference on Intelligent Environments 2006. Institute of Engineering and Technology, Athens, Greece, 5–6 July 2006.
115. Datta, S.P.A. Emergence of Digital Twins—Is this the march of reason? *J. Innov. Manag.* **2017**, *5*, 14–33. [CrossRef]
116. Gunes, V.; Peter, S.; Givargis, T.; Vahid, F. A Survey on Concepts, Applications, and Challenges in Cyber-Physical Systems. *KSIIS Trans. Internet Inf. Syst.* **2014**, *8*. [CrossRef]
117. Eckhart, M.; Ekelhart, A. Towards Security-Aware Virtual Environments for Digital Twins. In Proceedings of the 4th ACM Workshop on Cyber-Physical System Security—CPSS '18, Incheon, Korea, 4–8 June 2018; pp. 61–72.
118. Rasheed, A.; San, O.; Kvamsdal, T. Digital Twin: Values, Challenges and Enablers from a Modeling Perspective. *IEEE Access* **2020**, *8*, 21980–22012. [CrossRef]
119. Alam, K.M.; El Saddik, A. C2ps: A Digital Twin Architecture Reference Model for the Cloud-Based Cyber-Physical Systems. *IEEE Access* **2017**, *5*, 2050–2062. [CrossRef]
120. Park, H.; Easwaran, A.; Andalam, S. *Challenges in Digital Twin Development for Cyber-Physical Production Systems*; Springer International Publishing: Cham, Switzerland, 2019; pp. 28–48.
121. Knapp, E.; Langill, J. *Industrial Network Security Securing Critical Infrastructure Networks for Smart Grid, SCADA, and Other Industrial Control. Systems*, 2nd ed.; Elsevier: Waltham, MA, USA, 2015.
122. Elkaseer, A.; Salem, M.; Ali, H.; Scholz, S. Approaches to a Practical Implementation of Industry 4.0. In Proceedings of the 11th International Conference on Advances in Computer-Human Interactions, Rome, Italy, 25–29 March 2018.
123. Barrasso, D. Developing and applying digital twins for Continuous Drug Product Manufacturing. In Proceedings of the PSE Advanced Process Modeling Forum, Tarrytown, NY, USA, 10–12 September 2019.
124. Ierapetritou, M.; Muzzio, F.; Reklaitis, G. Perspectives on the continuous manufacturing of powder-based pharmaceutical processes. *AIChE J.* **2016**, *62*, 1846–1862. [CrossRef]
125. Boukouvala, F.; Niotis, V.; Ramachandran, R.; Muzzio, F.J.; Ierapetritou, M.G. An integrated approach for dynamic flowsheet modeling and sensitivity analysis of a continuous tablet manufacturing process. *Comput. Chem. Eng.* **2012**, *42*, 30–47. [CrossRef]
126. Kamble, R.; Sharma, S.; Varghese, V.; Mahadik, K. Process analytical technology (PAT) in pharmaceutical development and its application. *Int. J. Pharm. Sci. Rev. Res.* **2013**, *23*, 212–223.
127. Simon, L.L.; Pataki, H.; Marosi, G.; Meemken, F.; Hungerbühler, K.; Baiker, A.; Tummala, S.; Glennon, B.; Kuentz, M.; Steele, G.; et al. Assessment of Recent Process. Analytical Technology (PAT) Trends: A Multiauthor Review. *Org. Process Res. Dev.* **2015**, *19*, 3–62. [CrossRef]
128. Yu, Z.; Chew, J.; Chow, P.; Tan, R. Recent advances in crystallization control: An industrial perspective. *Chem. Eng. Res. Des.* **2007**, *85*, 893–905. [CrossRef]
129. Sierra-Vega, N.O.; Román-Ospino, A.; Scicolone, J.; Muzzio, F.J.; Romañach, R.J.; Méndez, R. Assessment of blend uniformity in a continuous tablet manufacturing process. *Int. J. Pharm.* **2019**, *560*, 322–333. [CrossRef]
130. Goodwin, D.J.; van den Ban, S.; Denham, M.; Barylski, I. Real time release testing of tablet content and content uniformity. *Int. J. Pharm.* **2018**, *537*, 183–192. [CrossRef]
131. De Beer, T.R.M.; Bodson, C.; Dejaegher, B.; Walczak, B.; Vercruysse, P.; Burggraeve, A.; Lemos, A.; Delattre, L.; Heyden, Y.V.; Remon, J.P.; et al. Raman spectroscopy as a process analytical technology (PAT) tool for the in-line monitoring and understanding of a powder blending process. *J. Pharm. Biomed. Anal.* **2008**, *48*, 772–779. [CrossRef]

132. Singh, R.; Sahay, A.; Muzzio, F.; Ierapetritou, M.; Ramachandran, R. A systematic framework for onsite design and implementation of a control system in a continuous tablet manufacturing process. *Comput. Chem. Eng.* **2014**, *66*, 186–200. [CrossRef]
133. Baranwal, Y.; Román-Ospino, A.D.; Keyvan, G.; Ha, J.M.; Hong, E.P.; Muzzio, F.J.; Ramachandran, R. Prediction of dissolution profiles by non-destructive NIR spectroscopy in bilayer tablets. *Int. J. Pharm.* **2019**, *565*, 419–436. [CrossRef] [PubMed]
134. Shekunov, B.Y.; Chattopadhyay, P.; Tong, H.H.; Chow, A.H. Particle size analysis in pharmaceutics: Principles, methods and applications. *Pharm. Res.* **2007**, *24*, 203–227. [CrossRef]
135. Wu, H.; White, M.; Khan, M. Quality-by-Design (QbD): An integrated process analytical technology (PAT) approach for a dynamic pharmaceutical co-precipitation process characterization and process design space development. *Int. J. Pharm.* **2011**, *405*, 63–78. [CrossRef]
136. Meng, W.; Román-Ospino, A.D.; Panikar, S.S.; O'Callaghan, C.; Gilliam, S.J.; Ramachandran, R.; Muzzio, F.J. Advanced process design and understanding of continuous twin-screw granulation via implementation of in-line process analytical technologies. *Adv. Powder Technol.* **2019**, *30*, 879–894. [CrossRef]
137. Ostergaard, I.; Szilagyi, B.; de Diego, H.L.; Qu, H.; Nagy, Z.K. Polymorphic Control and Scale-up Strategy for Antisolvent Crystallization Using Direct Nucleation Control. *Cryst. Growth Des.* **2020**, *20*, 2683–2697. [CrossRef]
138. U.S. Department of Health and Human Services, F.D.A. *PAT-A Framework for Innovative Pharmaceutical Development, Manufacturing and Quality Assurance*; U.S. Department of Health and Human Services, F.D.A.: Rockville, MD, USA, 2004.
139. Bakeev, K.A. *Process Analytical Technology: Spectroscopic Tools and Implementation Strategies for the Chemical and Pharmaceutical Industries*, 2nd ed.; John Wiley & Sons: Hoboken, NJ, USA, 2010.
140. James, M.; Stanfield, C.F.; Bir, G. A Review of Process. Analytical Technology (PAT) in the U.S. Pharmaceutical Industry. *Curr. Pharm. Anal.* **2006**, *2*, 405–414.
141. Nagy, Z.K.; Fevotte, G.; Kramer, H.; Simon, L.L. Recent advances in the monitoring, modelling and control of crystallization systems. *Chem. Eng. Res. Des.* **2013**, *91*, 1903–1922. [CrossRef]
142. Simon, L.L.; Kiss, A.A.; Cornevin, J.; Gani, R. Process engineering advances in pharmaceutical and chemical industries: Digital process design, advanced rectification, and continuous filtration. *Curr. Opin. Chem. Eng.* **2019**, *25*, 114–121. [CrossRef]
143. Papadakis, E.; Woodley, J.M.; Gani, R. Perspective on PSE in pharmaceutical process development and innovation. In *Process. Systems Engineering for Pharmaceutical Manufacturing*; Elsevier: Amsterdam, The Netherlands, 2018; pp. 597–656.
144. Pandey, P.; Bharadwaj, R.; Chen, X. Modeling of drug product manufacturing processes in the pharmaceutical industry. In *Predictive Modeling of Pharmaceutical Unit Operations*; Woodhead Publishing: Sawston, Cambridge, UK, 2017; pp. 1–13.
145. Escotet-Espinoza, M.S.; Foster, C.J.; Ierapetritou, M. Discrete Element Modeling (DEM) for mixing of cohesive solids in rotating cylinders. *Powder Technol.* **2018**, *335*, 124–136. [CrossRef]
146. Toson, P.; Siegmann, E.; Trogrlic, M.; Kureck, H.; Khinast, J.; Jajcevic, D.; Doshi, P.; Blackwood, D.; Bonnassieux, A.; Daugherity, P.D.; et al. Detailed modeling and process design of an advanced continuous powder mixer. *Int. J. Pharm.* **2018**, *552*, 288–300. [CrossRef]
147. Bhalode, P.; Ierapetritou, M. Discrete element modeling for continuous powder feeding operation: Calibration and system analysis. *Int. J. Pharm.* **2020**, *585*, 119427. [CrossRef] [PubMed]
148. Rantanen, J.; Khinast, J. The Future of Pharmaceutical Manufacturing Sciences. *J. Pharm. Sci.* **2015**, *104*, 3612–3638. [CrossRef] [PubMed]
149. Sajjia, M.; Shirazian, S.; Kelly, C.B.; Albadarin, A.B.; Walker, G. ANN Analysis of a Roller Compaction Process. in the Pharmaceutical Industry. *Chem. Eng. Technol.* **2017**, *40*, 487–492. [CrossRef]
150. Pandey, P.; Katakdaunde, M.; Turton, R. Modeling weight variability in a pan coating process using Monte Carlo simulations. *AAPS PharmSciTech* **2006**, *7*, E2–E11. [CrossRef] [PubMed]
151. Metta, N.; Verstraeten, M.; Ghijs, M.; Kumar, A.; Schafer, E.; Singh, R.; De Beer, T.; Nopens, I.; Cappuyns, P.; Van Assche, I.; et al. Model development and prediction of particle size distribution, density and friability of a comilling operation in a continuous pharmaceutical manufacturing process. *Int. J. Pharm.* **2018**, *549*, 271–282. [CrossRef]

152. Barrasso, D.; Tamrakar, A.; Ramachandran, R. Model Order Reduction of a Multi-scale PBM-DEM Description of a Wet Granulation Process via ANN. *Procedia Eng.* **2015**, *102*, 1295–1304. [CrossRef]
153. Bostijn, N.; Dhondt, J.; Ryckaert, A.; Szabó, E.; Dhondt, W.; Van Snick, B.; Vanhoorne, V.; Vervaet, C.; De Beer, T. A multivariate approach to predict the volumetric and gravimetric feeding behavior of a low feed rate feeder based on raw material properties. *Int. J. Pharm.* **2018**, *557*, 342–353. [CrossRef]
154. Van Snick, B.; Grymonpre, W.; Dhondt, J.; Pandelaere, K.; Di Pretoro, G.; Remon, J.P.; De Beer, T.; Vervaet, C.; Vanhoorne, V. Impact of blend properties on die filling during tableting. *Int. J. Pharm.* **2018**, *549*, 476–488. [CrossRef]
155. Escotet-Espinoza, M.S.; Moghtadernejad, S.; Oka, S.; Wang, Y.; Roman-Ospino, A.; Schäfer, E.; Cappuyns, P.; Van Assche, I.; Futran, M.; Ierapetritou, M.; et al. Effect of tracer material properties on the residence time distribution (RTD) of continuous powder blending operations. Part. I of II: Experimental evaluation. *Powder Technol.* **2019**, *342*, 744–763. [CrossRef]
156. Escotet-Espinoza, M.S.; Moghtadernejad, S.; Oka, S.; Wang, Z.; Wang, Y.; Roman-Ospino, A.; Schäfer, E.; Cappuyns, P.; Van Assche, I.; Futran, M.; et al. Effect of material properties on the residence time distribution (RTD) characterization of powder blending unit operations. Part. II of II: Application of models. *Powder Technol.* **2019**, *344*, 525–544. [CrossRef]
157. Escotet-Espinoza, M.S.; Vadodaria, S.; Singh, R.; Muzzio, F.J.; Ierapetritou, M.G. Modeling the effects of material properties on tablet compaction: A building block for controlling both batch and continuous pharmaceutical manufacturing processes. *Int. J. Pharm.* **2018**, *543*, 274–287. [CrossRef]
158. Rogers, A.; Hashemi, A.; Ierapetritou, M. Modeling of Particulate Processes for the Continuous Manufacture of Solid-Based Pharmaceutical Dosage Forms. *Processes* **2013**, *1*, 67–127. [CrossRef]
159. Metta, N.; Ghijs, M.; Schäfer, E.; Kumar, A.; Cappuyns, P.; Assche, I.V.; Singh, R.; Ramachandran, R.; Beer, T.D.; Ierapetritou, M.; et al. Dynamic Flowsheet Model Development and Sensitivity Analysis of a Continuous Pharmaceutical Tablet Manufacturing Process Using the Wet Granulation Route. *Processes* **2019**, *7*, 234. [CrossRef]
160. Wang, Z.; Escotet-Espinoza, M.S.; Singh, R.; Ierapetritou, M. Surrogate-based Optimization for Pharmaceutical Manufacturing Processes. In *Computer Aided Chemical Engineering*; Espuña, A., Graells, M., Puigjaner, L., Eds.; Elsevier: Amsterdam, The Netherlands, 2017; pp. 2797–2802.
161. U.S. Department of Health and Human Services, F.D.A. *Data Integrity and Compliance with Drug CGMP*; U.S. Department of Health and Human Services, F.D.A.: Silver Spring, MD, USA, 2018.
162. Su, Q.; Bommireddy, Y.; Shah, Y.; Ganesh, S.; Moreno, M.; Liu, J.; Gonzalez, M.; Yazdanpanah, N.; O'Connor, T.; Reklaitis, G.V.; et al. Data reconciliation in the Quality-by-Design (QbD) implementation of pharmaceutical continuous tablet manufacturing. *Int. J. Pharm.* **2019**, *563*, 259–272. [CrossRef]
163. Ganesh, S.; Moreno, M.; Liu, J.; Gonzalez, M.; Nagy, Z.; Reklaitis, G. Sensor Network for Continuous Tablet Manufacturing. In *13th International Symposium on Process. Systems Engineering (PSE 2018)*; Elsevier: Amsterdam, The Netherlands, 2018; pp. 2149–2154.
164. Ganesh, S. Continuous Pharmaceutical Manufacturing: Systems Integration for Process Operations Management. Ph.D. Thesis, Purdue University Graduate School, West Lafayette, IN, USA, 2020.
165. Singh, R.; Sahay, A.; Karry, K.M.; Muzzio, F.; Ierapetritou, M.; Ramachandran, R. Implementation of an advanced hybrid MPC–PID control system using PAT tools into a direct compaction continuous pharmaceutical tablet manufacturing pilot plant. *Int. J. Pharm.* **2014**, *473*, 38–54. [CrossRef]
166. Hailemariam, L.; Venkatasubramanian, V. Purdue Ontology for Pharmaceutical Engineering: Part I. Conceptual Framework. *J. Pharm. Innov.* **2010**, *5*, 88–99. [CrossRef]
167. Hailemariam, L.; Venkatasubramanian, V. Purdue Ontology for Pharmaceutical Engineering: Part II. Applications. *J. Pharm. Innov.* **2010**, *5*, 139–146. [CrossRef]
168. Zhao, C.; Jain, A.; Hailemariam, L.; Suresh, P.; Akkisetty, P.; Joglekar, G.; Venkatasubramanian, V.; Reklaitis, G.V.; Morris, K.; Basu, P. Toward intelligent decision support for pharmaceutical product development. *J. Pharm. Innov.* **2006**, *1*, 23–35. [CrossRef]
169. Muñoz, S.G.; Torres, E.H. Supervised Extended Iterative Optimization Technology for Estimation of Powder Compositions in Pharmaceutical Applications: Method and Lifecycle Management. *Ind. Eng. Chem. Res.* **2020**, *59*, 10072–10081. [CrossRef]

170. Shi, Z.; Hermiller, J.; Muñoz, S.G. Estimation of mass-based composition in powder mixtures using Extended Iterative Optimization Technology (EIOT). *AIChE J.* **2019**, *65*, 87–98. [CrossRef]
171. Kadlec, P.; Grbić, R.; Gabrys, B. Review of adaptation mechanisms for data-driven soft sensors. *Comput. Chem. Eng.* **2011**, *35*, 1–24. [CrossRef]
172. Flåten, G.R. Model Maintenance. In *Multivariate Analysis in the Pharmaceutical Industry*; Academic Press: Cambridge, MA, USA, 2018; pp. 313–321.
173. Chan, K.H.; Dozal-Mejorada, E.J.; Cheng, X.; Kephart, R.; Ydstie, B.E. Predictive control with adaptive model maintenance: Application to power plants. *Comput. Chem. Eng.* **2014**, *70*, 91–103. [CrossRef]
174. Chen, K.; Castillo, I.; Chiang, L.H.; Yu, J. Soft Sensor Model Maintenance: A Case Study in Industrial Processes. *IFAC-PapersOnLine* **2015**, *48*, 427–432. [CrossRef]
175. Gama, J.; Žliobaitė, I.; Bifet, A.; Pechenizkiy, M.; Bouchachia, A. A survey on concept drift adaptation. *ACM Comput. Surv.* **2014**, *46*, 1–37. [CrossRef]
176. Janardan, S.M. Concept Drift Streaming Data Classification Algorithms Platforms and Issues. *Procedia Comput. Sci.* **2017**, *122*, 804–811. [CrossRef]
177. Webb, G.I.; Hyde, R.; Cao, H.; Nguyen, H.L.; Petitjean, F. Characterizing concept drift. *Data Min. Knowl. Discov.* **2016**, *30*, 964–994. [CrossRef]
178. Kadwe, Y.; Suryawanshi, V. A Review on Concept Drift. *Iosr J. Comput. Eng.* **2015**, *17*, 20–26.
179. Sun, Y.; Zhang, J.; Xiong, Y.; Zhu, G. Data Security and Privacy in Cloud Computing. *Int. J. Distrib. Sens. Netw.* **2014**, *10*, 190903. [CrossRef]
180. O'Connor, T. Opportunities and Challenges for the Application of Process. Modeling and Simulation for Product Quality Risk Management. In Proceedings of the Advanced Process Modeling Forum, Tarrytown, NY, USA, 10–12 September 2019.
181. Badr, S.; Sugiyama, H. A PSE perspective for the efficient production of monoclonal antibodies: Integration of process, cell, and product design aspects. *Curr. Opin. Chem. Eng.* **2020**, *27*, 121–128. [CrossRef]
182. Lin-Gibson, S.; Srinivasan, V. Recent Industrial Roadmaps to Enable Smart Manufacturing of Biopharmaceuticals. *IEEE Trans. Autom. Sci. Eng.* **2019**, *2019*, 1–8. [CrossRef]
183. Narayanan, H.; Luna, M.F.; von Stosch, M.; Cruz Bournazou, M.N.; Polotti, G.; Morbidelli, M.; Butte, A.; Sokolov, M. Bioprocessing in the Digital Age: The Role of Process Models. *Biotechnol. J.* **2020**, *15*, e1900172. [CrossRef] [PubMed]
184. Read, E.K.; Park, J.T.; Shah, R.B.; Riley, B.S.; Brorson, K.A.; Rathore, A.S. Process analytical technology (PAT) for biopharmaceutical products: Part I. concepts and applications. *Biotechnol. Bioeng.* **2010**, *105*, 276–284. [CrossRef] [PubMed]
185. Biechele, P.; Busse, C.; Solle, D.; Scheper, T.; Reardon, K. Sensor systems for bioprocess monitoring. *Eng. Life Sci.* **2015**, *15*, 469–488. [CrossRef]
186. Zhao, L.; Fu, H.-Y.; Zhou, W.; Hu, W.-S. Advances in process monitoring tools for cell culture bioprocesses. *Eng. Life Sci.* **2015**, *15*, 459–468. [CrossRef]
187. Roch, P.; Mandenius, C.-F. On-line monitoring of downstream bioprocesses. *Curr. Opin. Chem. Eng.* **2016**, *14*, 112–120. [CrossRef]
188. Guerra, A.; von Stosch, M.; Glassey, J. Toward biotherapeutic product real-time quality monitoring. *Crit. Rev. Biotechnol.* **2019**, *39*, 289–305. [CrossRef]
189. Pais, D.A.M.; Carrondo, M.J.T.; Alves, P.M.; Teixeira, A.P. Towards real-time monitoring of therapeutic protein quality in mammalian cell processes. *Curr. Opin. Biotechnol.* **2014**, *30*, 161–167. [CrossRef]
190. Classen, J.; Aupert, F.; Reardon, K.F.; Solle, D.; Scheper, T. Spectroscopic sensors for in-line bioprocess monitoring in research and pharmaceutical industrial application. *Anal. Bioanal. Chem.* **2017**, *409*, 651–666. [CrossRef]
191. Berry, B.N.; Dobrowsky, T.M.; Timson, R.C.; Kshirsagar, R.; Ryll, T.; Wiltberger, K. Quick generation of Raman spectroscopy based in-process glucose control to influence biopharmaceutical protein product quality during mammalian cell culture. *Biotechnol. Prog.* **2016**, *32*, 224–234. [CrossRef]

192. Mehdizadeh, H.; Lauri, D.; Karry, K.M.; Moshgbar, M.; Procopio-Melino, R.; Drapeau, D. Generic Raman-based calibration models enabling real-time monitoring of cell culture bioreactors. *Biotechnol. Prog.* **2015**, *31*, 1004–1013. [CrossRef] [PubMed]
193. Abu-Absi, N.R.; Martel, R.P.; Lanza, A.M.; Clements, S.J.; Borys, M.C.; Li, Z.J. Application of spectroscopic methods for monitoring of bioprocesses and the implications for the manufacture of biologics. *Pharm. Bioprocess.* **2014**, *2*, 267–284. [CrossRef]
194. Rathore, A.S.; Kateja, N.; Kumar, D. Process integration and control in continuous bioprocessing. *Curr. Opin. Chem. Eng.* **2018**, *22*, 18–25. [CrossRef]
195. Wasalathanthri, D.P.; Rehmann, M.S.; Song, Y.; Gu, Y.; Mi, L.; Shao, C.; Chemmalil, L.; Lee, J.; Ghose, S.; Borys, M.C.; et al. Technology outlook for real-time quality attribute and process parameter monitoring in biopharmaceutical development—A review. *Biotechnol. Bioeng.* **2020**, 117. [CrossRef]
196. Tang, P.; Xu, J.; Louey, A.; Tan, Z.; Yongky, A.; Liang, S.; Li, Z.J.; Weng, Y.; Liu, S. Kinetic modeling of Chinese hamster ovary cell culture: Factors and principles. *Crit. Rev. Biotechnol.* **2020**, *40*, 265–281. [CrossRef]
197. Farzan, P.; Mistry, B.; Ierapetritou, M.G. Review of the important challenges and opportunities related to modeling of mammalian cell bioreactors. *AIChE J.* **2017**, *63*, 398–408. [CrossRef]
198. Baumann, P.; Hubbuch, J. Downstream process development strategies for effective bioprocesses: Trends, progress, and combinatorial approaches. *Eng. Life Sci.* **2017**, *17*, 1142–1158. [CrossRef]
199. Smiatek, J.; Jung, A.; Bluhmki, E. Towards a Digital Bioprocess. Replica: Computational Approaches in Biopharmaceutical Development and Manufacturing. *Trends Biotechnol.* **2020**. [CrossRef]
200. Olughu, W.; Deepika, G.; Hewitt, C.; Rielly, C. Insight into the large-scale upstream fermentation environment using scaled-down models. *J. Chem. Technol. Biotechnol.* **2019**, *94*, 647–657. [CrossRef]
201. Li, X.; Scott, K.; Kelly, W.J.; Huang, Z. Development of a Computational Fluid Dynamics Model for Scaling-up Ambr Bioreactors. *Biotechnol. Bioprocess Eng.* **2018**, *23*, 710–725. [CrossRef]
202. Farzan, P.; Ierapetritou, M.G. A Framework for the Development of Integrated and Computationally Feasible Models of Large-Scale Mammalian Cell Bioreactors. *Processes* **2018**, *6*, 82. [CrossRef]
203. Menshutina, N.V.; Guseva, E.V.; Safarov, R.R.; Boudrant, J. Modelling of hollow fiber membrane bioreactor for mammalian cell cultivation using computational hydrodynamics. *Bioprocess Biosyst. Eng.* **2020**, *43*, 549–567. [CrossRef] [PubMed]
204. Xu, J.; Tang, P.; Yongky, A.; Drew, B.; Borys, M.C.; Liu, S.; Li, Z.J. Systematic development of temperature shift strategies for Chinese hamster ovary cells based on short duration cultures and kinetic modeling. *MAbs* **2019**, *11*, 191–204. [CrossRef] [PubMed]
205. Sokolov, M.; Ritscher, J.; MacKinnon, N.; Souquet, J.; Broly, H.; Morbidelli, M.; Butte, A. Enhanced process understanding and multivariate prediction of the relationship between cell culture process and monoclonal antibody quality. *Biotechnol. Prog.* **2017**, *33*, 1368–1380. [CrossRef] [PubMed]
206. Villiger, T.K.; Scibona, E.; Stettler, M.; Broly, H.; Morbidelli, M.; Soos, M. Controlling the time evolution of mAb N-linked glycosylation—Part II: Model-based predictions. *Biotechnol. Prog.* **2016**, *32*, 1135–1148. [CrossRef] [PubMed]
207. Kotidis, P.; Jedrzejewski, P.; Sou, S.N.; Sellick, C.; Polizzi, K.; del Val, I.J.; Kontoravdi, C. Model-based optimization of antibody galactosylation in CHO cell culture. *Biotechnol. Bioeng.* **2019**, *116*, 1612–1626. [CrossRef]
208. Radhakrishnan, D.; Robinson, A.S.; Ogunnaike, B. Controlling the Glycosylation Profile in mAbs Using Time-Dependent Media Supplementation. *Antibodies* **2017**, *7*, 1. [CrossRef]
209. Karst, D.J.; Steinebach, F.; Soos, M.; Morbidelli, M. Process performance and product quality in an integrated continuous antibody production process. *Biotechnol. Bioeng.* **2017**, *114*, 298–307. [CrossRef]
210. Shirahata, H.; Diab, S.; Sugiyama, H.; Gerogiorgis, D.I. Dynamic modelling, simulation and economic evaluation of two CHO cell-based production modes towards developing biopharmaceutical manufacturing processes. *Chem. Eng. Res. Des.* **2019**, *150*, 218–233. [CrossRef]
211. Xing, Z.; Kenty, B.; Koyrakh, I.; Borys, M.; Pan, S.-H.; Li, Z.J. Optimizing amino acid composition of CHO cell culture media for a fusion protein production. *Process Biochem.* **2011**, *46*, 1423–1429. [CrossRef]

212. Spahn, P.N.; Hansen, A.H.; Hansen, H.G.; Arnsdorf, J.; Kildegaard, H.F.; Lewis, N.E. A Markov chain model for N-linked protein glycosylation–towards a low-parameter tool for model-driven glycoengineering. *Metab. Eng.* **2016**, *33*, 52–66. [CrossRef] [PubMed]
213. Hutter, S.; Villiger, T.K.; Brühlmann, D.; Stettler, M.; Broly, H.; Soos, M.; Gunawan, R. Glycosylation flux analysis reveals dynamic changes of intracellular glycosylation flux distribution in Chinese hamster ovary fed-batch cultures. *Metab. Eng.* **2017**, *43*, 9–20. [CrossRef] [PubMed]
214. Nolan, R.P.; Lee, K. Dynamic model of CHO cell metabolism. *Metab. Eng.* **2011**, *13*, 108–124. [CrossRef] [PubMed]
215. Bayrak, E.S.; Wang, T.; Cinar, A.; Undey, C. Computational Modeling of Fed-Batch Cell Culture Bioreactor: Hybrid Agent-Based Approach. *IFAC-PapersOnLine* **2015**, *48*, 1252–1257. [CrossRef]
216. Kiparissides, A.; Pistikopoulos, E.N.; Mantalaris, A. On the model-based optimization of secreting mammalian cell (GS-NS0) cultures. *Biotechnol. Bioeng.* **2015**, *112*, 536–548. [CrossRef]
217. Kotidis, P.; Demis, P.; Goey, C.H.; Correa, E.; McIntosh, C.; Trepekli, S.; Shah, N.; Klymenko, O.V.; Kontoravdi, C. Constrained global sensitivity analysis for bioprocess design space identification. *Comput. Chem. Eng.* **2019**, *125*, 558–568. [CrossRef]
218. Narayanan, H.; Sokolov, M.; Morbidelli, M.; Butte, A. A new generation of predictive models: The added value of hybrid models for manufacturing processes of therapeutic proteins. *Biotechnol. Bioeng.* **2019**, *116*, 2540–2549. [CrossRef]
219. Psichogios, D.C.; Ungar, L. A hybrid neural network-first principles approach to process modeling. *AIChE J.* **1992**, *38*, 1499–1511. [CrossRef]
220. Von Stosch, M.; Hamelink, J.-M.; Oliveira, R. Hybrid modeling as a QbD/PAT tool in process development: An industrial E. coli case study. *Bioprocess Biosyst. Eng.* **2016**, *39*, 773–784. [CrossRef]
221. Zalai, D.; Koczka, K.; Párta, L.; Wechselberger, P.; Klein, T.; Herwig, C. Combining mechanistic and data-driven approaches to gain process knowledge on the control of the metabolic shift to lactate uptake in a fed-batch CHO process. *Biotechnol. Prog.* **2015**, *31*, 1657–1668. [CrossRef]
222. Selvarasu, S.; Kim, D.Y.; Karimi, I.A.; Lee, D.-Y. Combined data preprocessing and multivariate statistical analysis characterizes fed-batch culture of mouse hybridoma cells for rational medium design. *J. Biotechnol.* **2010**, *150*, 94–100. [CrossRef] [PubMed]
223. Lienqueo, M.E.; Mahn, A.; Salgado, J.C.; Shene, C. Mathematical Modeling of Protein Chromatograms. *Chem. Eng. Technol.* **2012**, *35*, 46–57. [CrossRef]
224. Shi, C.; Gao, Z.-Y.; Zhang, Q.-L.; Yao, S.-J.; Slater, N.K.H.; Lin, D.-Q. Model-based process development of continuous chromatography for antibody capture: A case study with twin-column system. *J. Chromatogr. A* **2020**, *1619*, 460936. [CrossRef] [PubMed]
225. Wang, G.; Briskot, T.; Hahn, T.; Baumann, P.; Hubbuch, J. Estimation of adsorption isotherm and mass transfer parameters in protein chromatography using artificial neural networks. *J. Chromatogr. A* **2017**, *1487*, 211–217. [CrossRef]
226. Baur, D.; Angarita, M.; Müller-Späth, T.; Morbidelli, M. Optimal model-based design of the twin-column CaptureSMB process improves capacity utilization and productivity in protein A affinity capture. *Biotechnol. J.* **2016**, *11*, 135–145. [CrossRef]
227. Huter, J.M.; Strube, J. Model-Based Design and Process. Optimization of Continuous Single Pass Tangential Flow Filtration Focusing on Continuous Bioprocessing. *Processes* **2019**, *7*, 317. [CrossRef]
228. Mandenius, C.-F.; Brundin, A. Bioprocess optimization using design-of-experiments methodology. *Biotechnol. Prog.* **2008**, *24*, 1191–1203. [CrossRef]
229. Pleitt, K.; Somasundaram, B.; Johnson, B.; Shave, E.; Lua, L.H.L. Evaluation of process simulation as a decisional tool for biopharmaceutical contract development and manufacturing organizations. *Biochem. Eng. J.* **2019**, *150*, 107252. [CrossRef]
230. Arnold, L.; Lee, K.; Rucker-Pezzini, J.; Lee, J.H. Implementation of Fully Integrated Continuous Antibody Processing: Effects on Productivity and COGm. *Biotechnol. J.* **2019**, *14*, 1800061. [CrossRef]
231. Pollock, J.; Coffman, J.; Ho, S.V.; Farid, S.S. Integrated continuous bioprocessing: Economic, operational, and environmental feasibility for clinical and commercial antibody manufacture. *Biotechnol. Prog.* **2017**, *33*, 854–866. [CrossRef]

232. Yang, O.; Prabhu, S.; Ierapetritou, M. Comparison between Batch and Continuous Monoclonal Antibody Production and Economic Analysis. *Ind. Eng. Chem. Res.* **2019**, *58*, 5851–5863. [CrossRef]
233. Walther, J.; Godawat, R.; Hwang, C.; Abe, Y.; Sinclair, A.; Konstantinov, K. The business impact of an integrated continuous biomanufacturing platform for recombinant protein production. *J. Biotechnol.* **2015**, *213*, 3–12. [CrossRef] [PubMed]
234. Pirrung, S.M.; van der Wielen, L.A.M.; van Beckhoven, R.F.W.C.; van de Sandt, E.J.A.X.; Eppink, M.H.M.; Ottens, M. Optimization of biopharmaceutical downstream processes supported by mechanistic models and artificial neural networks. *Biotechnol. Prog.* **2017**, *33*, 696–707. [CrossRef] [PubMed]
235. Zobel-Roos, S.; Schmidt, A.; Mestmäcker, F.; Mouellef, M.; Huter, M.; Uhlenbrock, L.; Kornecki, M.; Lohmann, L.; Ditz, R.; Strube, J. Accelerating Biologics Manufacturing by Modeling or: Is Approval under the QbD and PAT Approaches Demanded by Authorities Acceptable Without a Digital-Twin? *Processes* **2019**, *7*, 94. [CrossRef]
236. Sencar, J.; Hammerschmidt, N.; Jungbauer, A. Modeling the Residence Time Distribution of Integrated Continuous Bioprocesses. *Biotechnol. J.* **2020**, e2000008. [CrossRef]
237. Gomis-Fons, J.; Schwarz, H.; Zhang, L.; Andersson, N.; Nilsson, B.; Castan, A.; Solbrand, A.; Stevenson, J.; Chotteau, V. Model–based design and control of a small-scale integrated continuous end-to-end mAb platform. *Biotechnol. Prog.* **2020**, e2995. [CrossRef]
238. Zahel, T.; Hauer, S.; Mueller, E.M.; Murphy, P.; Abad, S.; Vasilieva, E.; Maurer, D.; Brocard, C.; Reinisch, D.; Sagmeister, P.; et al. Integrated Process Modeling-A Process Validation Life Cycle Companion. *Bioengineering* **2017**, *4*, 86. [CrossRef]
239. Borys, B.S.; Roberts, E.L.; Le, A.; Kallos, M.S. Scale-up of embryonic stem cell aggregate stirred suspension bioreactor culture enabled by computational fluid dynamics modeling. *Biochem. Eng. J.* **2018**, *133*, 157–167. [CrossRef]
240. Sou, S.N.; Jedrzejewski, P.M.; Lee, K.; Sellick, C.; Polizzi, K.M.; Kontoravdi, C. Model–based investigation of intracellular processes determining antibody Fc-glycosylation under mild hypothermia. *Biotechnol Bioeng* **2017**, *114*, 1570–1582. [CrossRef]
241. Agarwal, N.; Mason, A.; Pradhan, R.; Kemper, J.; Bosley, A.; Serfiotis-Mitsa, D.; Wang, J.; Lindo, V.; Ahuja, S.; Hatton, D.; et al. Kinetic modeling as a tool to understand the influence of cell culture process parameters on the glycation of monoclonal antibody biotherapeutics. *Biotechnol. Prog.* **2019**, *35*, e2865. [CrossRef]
242. Sha, S.; Huang, Z.; Wang, Z.; Yoon, S. Mechanistic modeling and applications for CHO cell culture development and production. *Curr. Opin. Chem. Eng.* **2018**, *22*, 54–61. [CrossRef]
243. Zurcher, P.; Sokolov, M.; Bruhlmann, D.; Ducommun, R.; Stettler, M.; Souquet, J.; Jordan, M.; Broly, H.; Morbidelli, M.; Butte, A. Cell culture process metabolomics together with multivariate data analysis tools opens new routes for bioprocess development and glycosylation prediction. *Biotechnol. Prog.* **2020**, e3012. [CrossRef] [PubMed]
244. Sokolov, M.; Morbidelli, M.; Butte, A.; Souquet, J.; Broly, H. Sequential Multivariate Cell Culture Modeling at Multiple Scales Supports Systematic Shaping of a Monoclonal Antibody toward a Quality Target. *Biotechnol. J.* **2018**, *13*, e1700461. [CrossRef] [PubMed]
245. Behere, K.; Yoon, S. Chromatography bioseparation technologies and in-silico modelings for continuous production of biotherapeutics. *J Chromatogr A* **2020**, *1627*, 461376. [CrossRef]
246. Kumar, V.; Lenhoff, A.M. Mechanistic Modeling of Preparative Column Chromatography for Biotherapeutics. *Annu. Rev. Chem. Biomol. Eng.* **2020**, *11*, 235–255. [CrossRef]
247. Krippl, M.; Dürauer, A.; Duerkop, M. Hybrid modeling of cross-flow filtration: Predicting the flux evolution and duration of ultrafiltration processes. *Sep. Purif. Technol.* **2020**, *248*, 117064. [CrossRef]
248. Lohmann, L.J.; Strube, J. Accelerating Biologics Manufacturing by Modeling: Process Integration of Precipitation in mAb Downstream Processing. *Processes* **2020**, *8*, 58. [CrossRef]
249. Pirrung, S.M.; Berends, C.; Backx, A.H.; van Beckhoven, R.F.W.C.; Eppink, M.H.M.; Ottens, M. Model–based optimization of integrated purification sequences for biopharmaceuticals. *Chem. Eng. Sci. X* **2019**, *3*, 100025. [CrossRef]
250. Sachidananda, M.; Erkoyuncu, J.; Steenstra, D.; Michalska, S. Discrete Event Simulation Modelling for Dynamic Decision Making in Biopharmaceutical Manufacturing. *Procedia CIRP* **2016**, *49*, 39–44. [CrossRef]

251. Petrides, D.; Carmichael, D.; Siletti, C.; Koulouris, A. Biopharmaceutical Process Optimization with Simulation and Scheduling Tools. *Bioengineering* **2014**, *1*, 154–187. [CrossRef]
252. Taras, S.; Woinaroschy, A. Simulation and Multi-objective Optimization of Bioprocesses with Matlab and Superpro Designer Using a Client-server Interface. *Chem. Eng. Trans.* **2011**, *25*, 207–212.
253. Gangadharan, N.; Turner, R.; Field, R.; Oliver, S.G.; Slater, N.; Dikicioglu, D. Metaheuristic approaches in biopharmaceutical process development data analysis. *Bioprocess Biosyst. Eng.* **2019**, *42*, 1399–1408. [CrossRef] [PubMed]
254. Casola, G.; Siegmund, C.; Mattern, M.; Sugiyama, H. Data mining algorithm for pre-processing biopharmaceutical drug product manufacturing records. *Comput. Chem. Eng.* **2019**, *124*, 253–269. [CrossRef]
255. Lee, H.W.; Christie, A.; Xu, J.; Yoon, S. Data fusion-based assessment of raw materials in mammalian cell culture. *Biotechnol. Bioeng.* **2012**, *109*, 2819–2828. [CrossRef] [PubMed]
256. Herold, S.; King, R. Automatic identification of structured process models based on biological phenomena detected in (fed-)batch experiments. *Bioprocess Biosyst. Eng.* **2014**, *37*, 1289–1304. [CrossRef]
257. Luna, M.F.; Martínez, E.C. Iterative modeling and optimization of biomass production using experimental feedback. *Comput. Chem. Eng.* **2017**, *104*, 151–163. [CrossRef]
258. Feidl, F.; Vogg, S.; Wolf, M.; Podobnik, M.; Ruggeri, C.; Ulmer, N.; Wälchli, R.; Souquet, J.; Broly, H.; Butté, A.; et al. Process–wide control and automation of an integrated continuous manufacturing platform for antibodies. *Biotechnol. Bioeng.* **2020**, *117*, 1367–1380. [CrossRef]
259. Fahey, W.; Jeffers, P.; Carroll, P. A business analytics approach to augment six sigma problem solving: A biopharmaceutical manufacturing case study. *Comput. Ind.* **2020**, *116*, 103153. [CrossRef]
260. Portela, R.M.C.; Varsakelis, C.; Richelle, A.; Giannelos, N.; Pence, J.; Dessoy, S.; von Stosch, M. *When Is an In Silico Representation a Digital Twin? A Biopharmaceutical Industry Approach to the Digital Twin Concept*; Springer: Berlin/Heidelberg, Germany, 2020; pp. 1–21.

processes

Opinion

Emerging Challenges and Opportunities in Pharmaceutical Manufacturing and Distribution

Miriam Sarkis, Andrea Bernardi, Nilay Shah and Maria M. Papathanasiou *

Centre for Process Systems Engineering, Department of Chemical Engineering, Imperial College London, South Kensington Campus, London SW7 2AZ, UK; miriam.sarkis16@imperial.ac.uk (M.S.); a.bernardi13@imperial.ac.uk (A.B.); n.shah@imperial.ac.uk (N.S.)
* Correspondence: maria.papathanasiou11@imperial.ac.uk

Abstract: The rise of personalised and highly complex drug product profiles necessitates significant advancements in pharmaceutical manufacturing and distribution. Efforts to develop more agile, responsive, and reproducible manufacturing processes are being combined with the application of digital tools for seamless communication between process units, plants, and distribution nodes. In this paper, we discuss how novel therapeutics of high-specificity and sensitive nature are reshaping well-established paradigms in the pharmaceutical industry. We present an overview of recent research directions in pharmaceutical manufacturing and supply chain design and operations. We discuss topical challenges and opportunities related to small molecules and biologics, dividing the latter into patient- and non-specific. Lastly, we present the role of process systems engineering in generating decision-making tools to assist manufacturing and distribution strategies in the pharmaceutical sector and ultimately embrace the benefits of digitalised operations.

Keywords: pharmaceutical manufacturing; process systems engineering; Industry 4.0; digitalisation

Citation: Sarkis, M.; Bernardi, A.; Shah, N.; Papathanasiou, M.M. Emerging Challenges and Opportunities in Pharmaceutical Manufacturing and Distribution. *Processes* **2021**, *9*, 457. https://doi.org/10.3390/pr9030457

Academic Editors: Luis Puigjaner and Bhavik Bakshi

Received: 11 January 2021
Accepted: 25 February 2021
Published: 3 March 2021

Publisher's Note: MDPI stays neutral with regard to jurisdictional claims in published maps and institutional affiliations.

1. Introduction

Complexity in pharmaceutical manufacturing and distribution is highly dependent on the product nature. Therapeutic drugs can be classified into two broad categories: (a) small molecules, (b) biologics. The former refers to chemically synthesised drugs, while the latter refers to products that involve components extracted from or produced by a living organism [1]. Biologics include monoclonal antibodies (mAbs), vaccines, blood products, and advanced therapy medicinal products (ATMPs). Figure 1 illustrates the drug categories considered here. Each of these products is characterised by key specifications and/or formulation that dominate decisions related to its manufacturing and supply chain. Small molecules are pharmaceuticals based on chemical components and characterised by large scale manufacturing. On the other hand, manufacturing of biologics involves cell-based production systems and complex downstream separation trains, largely performed in batch/semi-batch mode [2,3]. This often presents challenges in the optimisation and scale up of unit operations.

Enhanced clinical disease understanding has led the pharmaceutical industry to move from one-size-fits-all approaches and develop targeted therapeutics such as ATMPs. Their production process differs significantly from small molecules or mAbs as it involves a series of product- and often patient-specific steps [4]. Their patient-specific nature may challenge scale up and distribution and has led to a shift in the manufacturing and supply chain status quo, highlighting the need for smaller, more agile, and often regional manufacturing units that translate into distributed networks closer to the patient. In addition, such products are coupled with stringent distribution timelines and tight storage constraints that need to be satisfied. As a result, questions related to optimal number and location of facilities arise, as well as how can one design a robust investment planning model. Furthermore, network and task coordination become of primary importance as the supply chain becomes more complex. Once the network has been designed, manufacturers need to ensure that distribution and storage conditions are met and maintained throughout the product journey, in order to reduce losses due to product degradation that can lead to drug shortages or reduced quality.

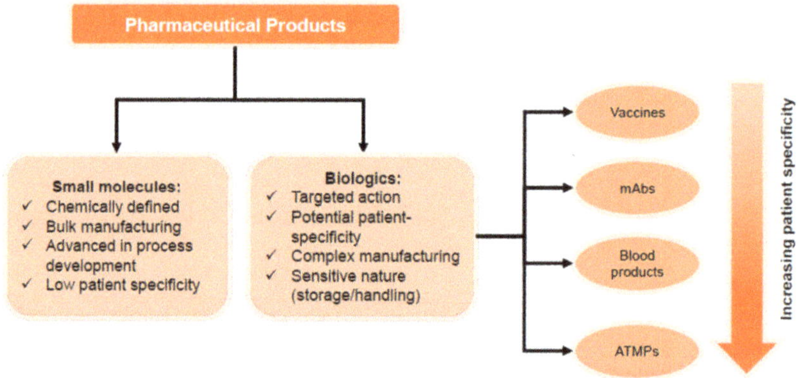

Figure 1. Schematic of simplified pharmaceutical product categories.

In this paper, we discuss how the nature of therapeutics may impact the design of suitable manufacturing processes and supply chain networks. We have performed a literature review and we summarise some of the latest initiatives taken to assist the decision-making process in the pharmaceutical industry. We also discuss how process systems engineering has been aiding innovation in this space. In the last part of this paper, we present a perspective on current and future developments in this space.

2. Engineering Challenges and Opportunities in Pharmaceutical Manufacturing and Supply Chain

Recently, the term Pharma 4.0 has been introduced, referring to the adaptation of digital strategies and tools of Industry 4.0 principles, and their application to pharmaceutical manufacturing and supply chain practices [5,6]. In this context, digital tools and orchestration platforms are being developed under Industry 4.0/5.0 principles [7]. The term refers to manufacturing digitalisation and automation of processes, introducing autonomous, computerised systems. It utilises different types of mathematical models (e.g., statistical, kinetic) and Internet of Things to facilitate and maintain internal communication within and across the factories. Application of Industry 4.0/5.0 principles aims to facilitate: (a) data collection, analysis, and interpretation, (b) man-machine co-operation, (c) online monitoring and control, and (d) intra- and inter-facility data sharing. In the last few years, we have seen the emergence of cloud-based applications coming to assist decision-making in the pharmaceutical industry. Several industrial players have embraced Pharma 4.0 either through the development of digital platforms to be used by manufacturers (e.g., Siemens) or by integrating digitalisation into their manufacturing processes (e.g., ChemeCon GmbH) [8].

2.1. Manufacturing

Pharmaceutical manufacturing is divided in two main parts: firstly, the pharmaceutical ingredient or drug (active pharmaceutical ingredient (API)/drug substance) is being produced, while the second step is focused on making this product suitable for administration to the patients (drug product). Common process steps usually involve drug formulation-specific and therefore differ across drug types. Often, small molecule primary manufacturing involves chemical synthesis and purification steps, while secondary manufacturing starts with the mixing of the API with excipients, followed by granulation, compression, coating, and packaging. On the other hand, biologics involve the production of either the API or parts of the drug product by a living organism. For example, mAbs are produced in mammalian cell culture systems using bioreactors, a process referred to also as upstream (USP) [2]. Following USP, the product undergoes a series of separation/purification steps, including filtration and chromatography to ensure that impurities are removed from the final formulation. Different to all other categories, ATMPs, such as chimeric antigen receptor T (CAR-T) cells, often involve one or more patient-specific steps [9]. Autologous CAR-T cells are a representative example as their manufacturing is based on T cells that have been extracted from the patient's blood stream [10,11].

Pharmaceutical manufacturers are focused on delivering efficacious and safe products at quantities that meet the global demand. In addition, process and product standardisation are primary goals to ensure batch-to-batch variability is minimised. In parallel, production processes need to be economically viable, adding to the complexity of identifying the best candidate design(s). These are often conflicting objectives (Table 1) that require systematic procedures for the identification of the most suitable operating units and modes that will meet product specifications, while yielding a profitable process. In an effort towards process improvement and modernisation, the pharmaceutical industry has pioneered by creating new and/or adapting existing innovations. Here, we present some of them and discuss the challenges that remain open in each space.

Table 1. Summary of the key challenges and opportunities in pharmaceutical manufacturing and distribution. The tick sign highlights the relevance of the identified issues and solutions to each drug product category.

Decisions	Challenges and Issues	Small Molecules	Conventional Biologics	Personalised Products	Solutions and Opportunities
			Manufacturing		
Product Portfolio	Identification of product profile		✔	✔	Understand QTTP; Outsource manufacturing and development to contractors
Process Design and Operations	Long approval times	✔	✔	✔	Standardisation; Single-use technology; Multiproduct facilities
	Batch solutions are well-established	✔	✔		Incentivise investment in continuous manufacturing
	Batch-to-batch variability and shortages		✔		Continuous manufacturing; QbD
	Identification of optimal operating units and modes	✔	✔	✔	Process optimisation tools; QbD
	Measurements availability and lack of process understanding		✔	✔	Develop online PAT tools; Digital twins of process; QbD
Capacity Planning	Long lead times for scale up	✔	✔		Scale up existing suites; Scale out through new suites instalment; Single use equipment
	Patient-specific products and process			✔	Scale out and set up of parallel suites
	Uncertainty of long-term demand	✔	✔	✔	Decision-making tools for long-term investment strategies; Multiproduct facilities; Outsource production to CMOs and CDMOs

Table 1. *Cont.*

Decisions	Challenges and Issues	Small Molecules	Conventional Biologics	Personalised Products	Solutions and Opportunities
Production Planning and Scheduling	Adaptability to short-term demand fluctuations	✔	✔	✔	Real time demand forecasts; Continuous manufacturing for flexible campaigns; Single use equipment to reduce changeover times Planning and scheduling decision-making tools
	Time constraint and patient specificity			✔	COI and COC; Monitor patient schedule; Planning and scheduling decision-making tools
Quality Control	Quality assurance tasks lead times	✔	✔	✔	Continuous manufacturing; QbD
	Unavailability of measurements	✔	✔	✔	Digital twins for real-time monitoring
Distribution					
Inventory Planning	Prevention of shortages	✔	✔		Real-time sharing of stock data, inventory levels and forecasted demand
	Monitor CQAs and CPPs		✔	✔	Track & Trace tools; Outsource distribution to contract logistics providers
	Time constraint and patient specificity			✔	COI and COC; Track & Trace tools
Network Structure	Compliance and coordination of stakeholders.	✔	✔	✔	Track & Trace tools; Data sharing
	Time constraint and patient specificity			✔	Decentralised supply chain closer to the patient
Transport Modes and Connections	Monitor CQAs and CPPs		✔	✔	Track & Trace tools; Outsource distribution to contract logistics providers
	Counterfeit drugs entering supply chain	✔	✔		Track & Trace tools
	Time constraint and patient specificity		✔	✔	Track & Trace tools; Outsource distribution to contract logistics providers

2.1.1. Quality by Design

The emergence of biologically derived drugs has underlined the necessity for thorough system understanding that includes detailed mapping of how process conditions may affect product quality. Quality by design (QbD) was firstly discussed by Juran [12] in 1992 and refers to the integration of quality into the process and product. In other words, all design and operation decisions are taken aiming to meet a predefined product quality. In the pharmaceutical industry, QbD has been increasingly endorsed by regulators and adapted by manufacturers [13–16], while in recent years it has become an integral part of approval submission dossiers. QbD suggests that firstly the quality target product profile (QTTP) needs to be decided, followed by the identification of the critical quality attributes (CQAs) [16,17]. CQAs are defined as product properties and/or characteristics that need to be within certain limits. The process is then designed, aiming to meet the pre-defined QTTP, while maintaining CQAs within the allowed threshold. This is achieved by manipulating those process parameters that directly affect CQA performance, known as critical process parameters (CPPs). QbD offers a systematic procedure for the development of processes based on thorough system understanding and prior knowledge integrated to the design and operation. Efforts have been made to integrate mathematical models with QbD principles to explore CPP-CQA interplay. Such understanding enables the determination of a set of *feasible* points within the space of the operating conditions that assure that the CQAs are within specifications, known also as "design space" (DS). This has allowed manufactures to move away from *uniquely optimal* operating profiles and adopt a more flexible strategy, whereby the manufacturing process is approved to operate within the DS and allowing greater flexibility for post-approval improvements within the DS.

Despite the wide application of QbD in mAbs and lately in vaccines, when it comes to ATMPs, QbD-driven processes remain an open challenge [18]. The often patient-/donor-specific nature of the starting material renders systematic CQA identification impossible to perform. In addition, the manufacturing performance of cell-based therapies is highly dependent on the quality of the extracted cells, leading therefore to a highly variable CPP-CQA interplay. As ATMP manufacturing matures and more understanding on the optimal portfolio of conditions is gained, QbD principles can be adapted to incorporate patient profile and incoming materials as key CPPs and map their impact on the process and product performance.

2.1.2. Continuous Manufacturing

Process performance has been the driver for many of the latest advances in the pharmaceutical industry, such as the one of continuous manufacturing (CM). CM offers the possibility for robust processes that involve smaller equipment size. In addition, by running longer and producing higher product yields, CM processes can lead to decreased batch-to-batch variability and therefore minimise the risk of drug shortages due to unmet quality specifications [19,20]. On the other hand, operating in continuous mode is translated into a must-have requirement of rapid, online measurements and a high level of process understanding to allow the operator to ensure that the product will meet specifications. This is of utmost importance in CM as its *plug-and-play* profile means that an intermediate intervention is not possible which translates into a significant financial and shortage risk if the process deviates from the optimal significantly. CM is one of the most discussed trends and innovations of the latest years in the pharmaceutical industry, endorsed by regulators [21]. Promising eco-efficient processes of higher productivity, CM has been successfully applied in many existing production processes leading to significant improvements [19]. Small molecules have seen applications of CM early on with the initiatives from Novartis-MIT on continuous crystallisation [22] and the GSK-Pfizer partnership for the development of continuous processing technology for oral solid dosage (OSD) drugs [23]. Innovation has been demonstrated in the space of biologics as well with Genzyme and Bayer as leading adapters of perfusion and other continuous manufacturing processes [24], while Novasep, GE Healthcare, Knauer, and ChromaCon are some of the equipment manufacturers offering small- and pilot-scale continuous chromatography systems. Warikoo et al. [25] demonstrated one of the first fully continuous pilot-scale bioprocesses for the production of a mAb and a recombinant human enzyme. They designed and used a system composed by a 12 L perfusion bioreactor connected to 4-column periodic counter-current chromatography and they successfully demonstrated the production and purification of the desired products. Godawat et al. [26] showcased an end-to-end continuous bioprocess using a perfusion bioreactor connected to an ATF cell retention device. The upstream mixture was then processed by two 4-column PCC systems. Additionally, Karst et al. [27] presented a lab-scale continuous mAb production process using a perfusion cell culture, a surge tank, and a continuous capture process.

Despite the success of CM in small molecules, challenges still exist that prevent biologics from reaching a fully continuous process at scale. A significant percentage of this slower adaptation can be attributed to system complexity. Relying on living organisms as production systems, biologics are coupled with complex process dynamics that challenge the identification and maintenance of the optimal operating profile. Although, CM promises more stable processes and decreased batch-to-batch variability, it requires increased certainty that the optimal operating conditions will be maintained throughout the process. This is to ensure that the desired product will meet specifications and reduce financial and supply risks associated to out-of-spec batches. To enable the design of robust processes that are continuously monitored requires suitable analytics to be in place. Despite advances in the field of continuous online measurements [28–32], process analytical technologies (PATs) are yet to be further developed in order for uninterrupted CM to be realised. Focusing on biologics and specifically mAbs, another limiting step that hinders end-to-end continuous processing is upstream/downstream (USP/DSP) integration. Process intensification via process integration in mAbs is a challenge, firstly as DSP units are not at the scale to handle the volumes produces by the USP counterpart. A way to mitigate this would be scaling up DSP equipment, risking increasing the already high DSP cost (80% of the end-to-end process).

Aiming to tackle this, initiatives have been made towards the development of smaller scale separation units, operating in continuous mode, increasing therefore their volume processing capabilities [26,33,34]. Another alternative could be to scale out the DSP step, offering also higher operating flexibility. Some of the remaining challenges are currently being tackled through the development of computer-modelling platforms as discussed later in the manuscript. Manufacturing challenges increase as products become more specialised. For example, CAR-T cells (ATMPs) are manufactured using *closed-box* production platforms that do not allow for task parallelisation or scale up [35,36]. This translates into integrated lines of unit operations being occupied for the entire manufacturing (>10 days) duration of a single therapy before they can become available to receive the next one. As ATMPs gain momentum, manufacturers will be required to increase their capacity. Given that volumetric scale up is not possible, other possibilities can be explored, such as scale-out, referring to multiple suites running in parallel or a completely granular manufacturing procedure where every step is performed in a separate unit, allowing therefore for sequential manufacturing with decreased waiting times. The latter model could greatly benefit from process intensification initiatives as it resembles the well-known model of biologics.

2.2. Supply Chains

Supply chain design decisions, strategies, and operations are highly dependent on the product that is delivered to the patient. With increasingly complex portfolios and stringent regulations to deliver an effective and safe therapy to end-users, pharmaceutical supply chains costs are on the rise. The nature of the product type, from the chemically-derived small molecules to highly targeted biologics, such as mAbs and ATMPs, entails different distribution and storage challenges [37]. Table 1 summarises the main challenges faced in pharmaceutical supply chains and related innovations.

2.2.1. Demand Scales

The pharmaceutical industry is inherently global, and its supply chains comprise a network of manufacturers (primary and secondary), which include in-house or external contractors, packaging facilities, regional distribution centres (wholesalers), and final healthcare providers, such as hospital and pharmacies. Off-the shelf products, prescription drugs and vaccines can be produced on a large scale, with single manufactured batches delivering numerous patient non-specific doses, following a one-size-fits-all distribution approach. This strategy is preserved in the case of emerging specialty drug products as well. Demands for these products, which are often biologics, including mAbs, can be predicted to be smaller in scale as they provide treatment of rare and complex chronic diseases, which only certain patient subgroups present. However, as the complexity of the treatment increases, it becomes increasingly difficult to synthesise a product that is compatible with the entire patient cohort. In the case of ATMPs, distribution has been envisioned through two channels: allogeneic and autologous. Allogeneic therapies are manufactured in larger batches from unrelated donor tissues [38]. Off-the-shelf production offered by the allogenic route is presenting several donor-patient matching challenges which have slowed down the success of these therapies in clinical trials. By contrast, autologous ATMPs have thus far been more successful clinically [9] and have the potential to reconfigure standard supply chain structures, as they represent a turning point in the feasibility of personalised medicines. Figure 2 illustrates the general supply chain structure for batch-produced drugs and patient-specific therapeutics. In the instance of CAR-T cell therapies, a sample of cells is extracted from the patient, shipped, modified, and administered to the patient, with a minimised cycle time (17–19 days return time for leading commercial products) [39–42]. The supply chain for these therapies is closer to the customer and the need for a *1:1 business model* emerges, where the product released by a single batch is patient-specific. Opportunities of scale up are limited and decentralisation of manufacturing is a promising approach [43]. Companies, expanding their primary and specialty drugs portfolios to personalised therapeutics, are expected to deal with a spectrum of decoupled demands simultaneously, which require extensive coordination of the stakeholders in the supply chain [44].

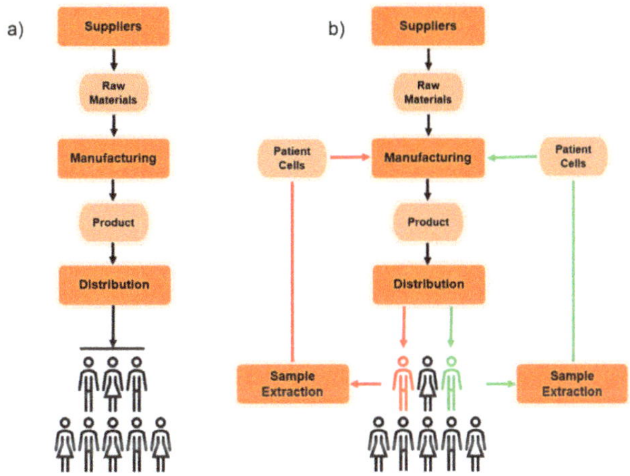

Figure 2. Pharmaceutical supply chain for (**a**) batch-produced drugs and (**b**) patient-specific therapeutics.

2.2.2. New Players

The pharmaceutical ecosystem comprises large R&D multinationals, local companies, generic manufactures, contract development, and manufacturing organizations (contract manufacturing organisations, CMOs, and contract and development manufacturing organisations, CDMOs) and biotechnology companies [45]. Large R&D multinationals are the key players in the marketplace, with presence in branded products and manufacturing sites across many locations. In recent years, their research focus has shifted to unmet needs of smaller patient populations, such as prevention and cure of rare diseases [46]. The increasing complexity of novel targeted therapeutics and lack of in-house manufacturing expertise of large multinationals in these contexts have determined an increased in mergers and acquisition (M&A) and outsourcing strategies through CMOs or CDMOs [46]. For instance, CellforCure was acquired by Novartis, expanding the company's manufacturing capabilities in CAR-T cell therapies, Hitachi acquired Aptech to increase manufacturing capabilities in Europe, Thermo Fisher acquire CMO Brammer Bio for $1.7 bn and GE Healthcare was acquired by Danaher ($21.4 bn) [43]. CMOs and CDMOs are equivalently attractive for biotechnology companies, which are often the main innovators in genetically engineered therapeutics, but lack of manufacturing resources and liquidity for in-house manufacturing. As the number of stakeholders involved in clinical and commercial supply chains increases, end-to-end monitoring of CQAs becomes increasingly difficult [47]. Outsourcing distribution and handling to specialised contract logistic providers is an appealing option to assure safe and secure delivery of complex biological drug products; however, management and coordination between the multiple agents becomes the key challenge.

2.2.3. Logistics Considerations

Manufacturers indicate on the product label the stability conditions for the product, which must be maintained throughout the whole supply chain. Small molecule drug products can typically be stored at 25 °C [43]. By contrast, the stability of bioproducts is highly compromised by temperature excursions and shocks. For instance, blood products, conventional vaccines (e.g., live-attenuated viruses) and monoclonal antibodies must be transported and stored under refrigeration conditions of 2–9 °C [48]. If cold chain logistics introduce additional costs in the supply chain, these are further exacerbated when handling ATMPs. CAR-Ts can be stored and transported either fresh (−80 °C) or cryopreserved (−180 °C), depending on the manufacturing practice, noting that they are also highly sensitive to shear stress and vibrations, because of their cell-based nature [9]. This ensures stability, maintains viability, and prevents genetic changes. Other genetically engineered products, such as mRNA vaccines, must be stored and handled under similar conditions (−70 °C) [49]. Monitoring the CQAs in relation to storage and transport environment conditions and ensuring timely delivery of therapies becomes increasingly crucial as the product structure and scope increases in complexity. Whether distribution is tackled in-house or outsourced, transparency of manufacturing and logistics operations facilitates quality assurance and effectiveness of the entire supply chain [47].

2.2.4. End-to-End Monitoring

CAR-Ts and personalised therapies offer a new perspective on the importance of track and trace capabilities for supply chain management and real-time monitoring. In these supply chains, chain of identity (COI) and tracking is crucial in order to ensure return of the therapy to the right patient by the end of the product cycle [9]. In addition, chain of custody (COC) principles must be applied with the aim of recording data related to handling, collection, and performed actions on the sample, thus monitoring the patient-specific product profile closely. It is worth noting that potential success of off-the-shelf ATMPs will equally require donor information to be tracked throughout the supply chain to ensure compatibility and aid effective donor-patient matching. Patients will also need to be monitored for several years after receipt of therapy; this information should be used to improve therapy design wherever possible. Initiatives to improve end-to-end visibility of supply chains are emerging in the fields of conventional non-specific products as well, such as Merck KGaA's commitment to utilise data analytics to predict and prevent drug shortages [50]. Companies are in fact becoming more aware of the improved supply-and-demand forecasting that traceability offers, including its potential to prevent API stock-outs and counterfeit drugs from entering the supply chain. As discussed above, drugs, including targeted biologics and small molecules, can also greatly benefit from real-time monitoring of CQAs, as the risk of failing to comply with labelled requirements can be reduced.

As highlighted by Papathanasiou [51], cloud-based platforms can facilitate communication and seamless connection between stakeholders. Maintaining and upgrading data security will though become a constant requirement for reducing vulnerabilities to the increasing sophistication of cyber-attacks. Particularly, secure safeguarded systems to protect data will become central to foster patient trust for data sharing and conduct the research needed to drive personalised medicine [52]. Alongside cloud-based solutions, blockchain-based alternatives are being developed in recent years. In a nutshell, blockchain is part of the broader category of distributed ledger technologies (DLTs) and it is based in the participation of a network of devices, called nodes, that keep a copy of the database [53]. A distinct advantage of the blockchain is that it does not require a central trusted party to verify the validity of the data but it relies on a consensus protocol which is publicly available and agreed upon by all the participants [54]. The information stored in the blockchain is public, immutable, and tamper-proof, while the security of the sensible data is assured by the utilization of strong state-of-the-art cryptographic algorithms. By adopting the blockchain a unified distributed health records database that can be accessed by every stakeholder along the supply chain, from the raw material providers to the final patient, with different levels of access to the information. For instance, information about the QC and storage conditions of a therapy along the supply chain could be accessed by everyone at any time by scanning a QR code attached to the therapy, while patient-specific data would be cryptographically sealed for most of the stakeholders except the patient himself and the hospital. An extensive review of blockchain solutions in the healthcare sector is out of the scope of this work and can be found elsewhere [55]. Despite the great potential, scalability of blockchain application remains an issue and is yet to be demonstrated. An interesting use-case is the recent partnership between NHS England and Hedera Hashgraph, a company providing blockchain-based solutions, in an attempt to use blockchain for enabling cold chain monitoring of COVID-19 vaccines for a selected group of facilities [56]. Other examples of blockchain-based tools for real-time monitoring of storage conditions of sensitive goods and traceability solutions are being developed by Modum.io [57] and is under investigation in a leading Italian company of the ophthalmic sector [53].

2.2.5. Production Planning and Scheduling

Despite the exciting opportunities brought by digitalisation and advanced monitoring, well-established technologies still present a large margin of improvement in terms of adaptability of production levels to demand. One of the main bottlenecks of current manufacturing and distribution networks, for both small molecules and conventional biologics is the planning and scheduling of production in response to short-term demand fluctuations [45]. Primary manufacturing sites usually comprise multipurpose batch equipment setups to distribute the capital cost over a spectrum of products. In the instance of biopharmaceutical manufacturing, such as manufacturing of mAbs, perfusion and fed-batch modes are preferred modes of operation due to improved fermentation titres [58]. Significant losses in revenues can result from downtime due to changeovers and required extensive cleaning tasks to prevent contamination. This pushes manufacturers to operate the site in long product campaigns, which ensure profitable utilisation of the plant throughout the time horizon [45,59]. Small molecule drug substances exiting the primary sites are stored up to 1 year and can be further processed in secondary manufacturing sites upon demand. Simpler tasks of fill and finish and packaging taking place in this secondary stage allow more flexible scheduling of operations and supply products to distribution centres.

The intermediate storage installations between drug substance and drug product manufacturing can act as a buffer to tackle variations in market dynamics: the customer-facing end (hospitals and pharmacies) place orders on wholesalers, carry out an assessment on inventory levels and if necessary, place orders upstream. In the event of an API shortage, the lack of responsiveness of primary manufacturing long campaigns emerges, which can then lead to drug product shortages and impact patients in need of the therapy. Stockpiling has been a profitable option for well-established chemically synthesised drug products; however, it is not always the best-suited solution for more complex and expensive biologics with short shelf lives. The high value of these products constrains the size of product inventory held as this might constitute tying up working capital [60]. Off-the-shelf production has followed the above planning paradigm for years, but patient-specific therapies come to reshape this approach. Scheduling production becomes *patient scheduling*, where each batch contains solely a dose of therapy that is specific to the patient [9]. The business model changes radically and adaptation to demand dynamics becomes increasingly important as operations are now constrained by return times between collection of the sample at the start of the supply chain, manufacturing, product release, and re-infusion.

2.2.6. Capacity and Investment Planning

Investment planning into expansions, establishment, and shutdown of facilities would have to be carried out under high uncertainty of demand of pipeline products and drugs under development. In order to avoid financial losses related to poor forecasting and suboptimal utilisation of facilities, R&D companies are externalising development and manufacturing of novel entries in their portfolio to contractors. The problem of capacity management is outsourced to the CMO, which is able to better balance utilisation by making products for multiple innovators [43]. Stainless-steel plants are well-established production facilities for conventional vaccines, mAbs, small molecule products and are suitable for large scale production. The capital investment for these facilities can extend from $500 M to $1 bn [61], highlighting the financial losses that can derive from underutilisation of the facility. The process of setting up entirely new facilities can extend up to 5–10 years, which once more hinders flexibility and responsiveness to varying therapeutic needs of the population. It is often the case that capacity within the facility is expanded by either setting up suites in parallel or scaling up existing ones with larger equipment [62].

Nevertheless, the operational burden of cleaning tasks, contamination concerns, and the ever-present need of more flexible production as biopharmaceutical products become more advanced and complex, is pushing many companies to utilise single-use production technologies. This trend in manufacturing offers multiple advantages in terms of savings in instalment, which fall in the range of $20–$100 M (2–20% of the capital investment), and operational costs. Set up times for new facilities are shorter (1.5 years) and the advantage of parallelising production with suites is preserved, in order to cope with short-term demand changes. Interestingly, COVID-19 vaccine producers choose to rely on flexible single-use systems as opposed to traditional commercial-large scale bioreactors and fermenters, valuing the potential to install manufacturing capacity at a higher speed, which is crucial during a global health crisis [63]. The advantage of single-use equipment is seen also in the space of personalised medicines, where cross-contamination risk between products can cause loss of patient specificity and have detrimental effects on the patient's health. Changeover time within each suite is decreased from 1 month to 0.5 days [61] as equipment components no longer require, cleaning, but are rather disposed, substituted, with an adaptable capacity to the incoming patient schedule. The environmental drawbacks of utilising high purity water and heat to clean and sterilise stainless steel equipment are removed. Disposal routes of single-use technologies is, however, still an issue to be considered. Used components are typically bio-hazardous, which entails that waste treatment tasks have to be carried out on-site prior to landfill disposal. Another option is to send used components to geographically separate waste-to-energy facilities for incineration and recovery of electricity. Latterly, initiatives, such as the Biopharma Recycling Program, are investigating recycling strategies to further reduce the environmental footprint of plastic single-use equipment and exploit the full benefits of flexible manufacturing [64].

3. Assisting Digitalisation in Pharmaceutical Industry via Process Systems Engineering (PSE)

Process systems engineering (PSE) has been traditionally assisting decision making in the pharmaceutical industry [46,65–68]. The adoption of digitalisation in pharmaceutical manufacturing and the supply chain will be key for seamless data exchange across manufacturing facilities and supply chain networks, as it will allow connectivity of processes, products, and people. As highlighted previously in this manuscript, there is a wide range of opportunities to improve the strategies and operations of the entire pharmaceutical supply chain in order to meet the evolving therapeutic needs of the population. In this space, the concept of enterprise-wide optimisation [69,70] is at the core of a coordination between R&D, supply of materials, manufacturing, and distribution of pharmaceutical products. The main objectives of an enterprise-wide approach are to maximise profits, responsiveness to customer needs, resource utilisation and minimisation of costs, stock levels, and environmental footprints. This is achieved while accounting for the complex interactions between the many stakeholders of the supply chain. Instant flow of information and data sharing can also be achieved via the adoption of *transactional* IT tools, such as track and trace tools, cloud-based and blockchain platforms. Such tools can improve communication and operations across the supply chain, specifically in the evolving pharmaceutical ecosystem that comprises multiple players. Nevertheless, these digital tools do not provide comprehensive frameworks to support the decision-making process. Therefore, it is of paramount importance to develop *analytical* IT tools to explore, analyse alternatives, and predict actions for the design, planning, and operation of each of the components of the supply chain. The PSE community has largely focused on the development of sophisticated optimisation and decision-support tools so as to yield optimum performance and ensure customer satisfaction. Figure 3 presents a summary of key considerations and challenges currently faced by the pharmaceutical industry, as well as some of the most remarkable computational and other innovations that can assist the decision making.

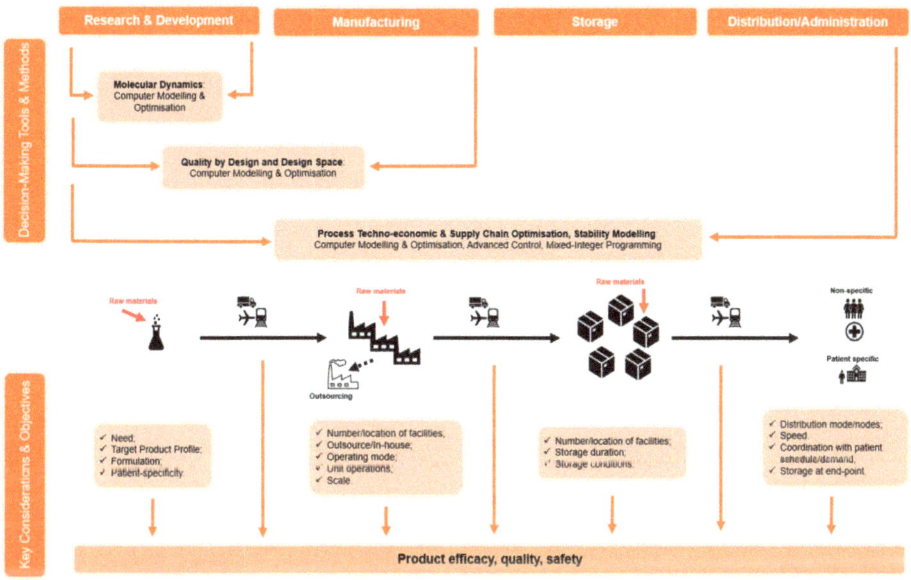

Figure 3. Pharmaceutical manufacturing and supply chain ecosystem.

In the pharmaceutical sector, changes introduced to approved processes and/or products need to be registered re-approved by regulators. This poses an additional challenge to the adaptation of new methods and technologies. Initiatives such as quality by design and design space identification that allow thorough process and product understanding can lead to more flexible processes and therefore faster approval procedures. In that respect, computer-based modelling initiatives and tools demonstrate significant potential as they offer lower-cost experimentation platforms and the ability to identify the optimal process operating profiles offline. Such models and platforms, also known as *digital twins*, find applications across a variety of activities in pharmaceutical manufacturing. Computer-based modelling in product development and manufacturing has demonstrated significant potential, offering low-cost experimentation platforms to assess CQA-CPP interplay under a vast spectrum of conditions [71–73]. In a similar fashion, PSE researchers are also looking to quantify parameter uncertainty and its impact on product and process performance [74,75]. From an operational standpoint, many groups are developing using digital twins for the design of optimal operating setups, optimisation profiles [76–79], and smart controllers [80–82] that can operate bypassing measurement unavailability.

In the field of supply chain management, optimisation-based approaches have improved strategies and operations of pharmaceutical and bio-pharmaceutical processes and distribution in multiple ways. Computational tools have been developed to seek optimal long-term strategic plans, considering the problem supply chain design and capacity planning, as well mid- and short-term decisions, addressing the problem of production planning and scheduling [69]. Supply chain design, capacity, and investment optimisation models focus on the long-term decisions regarding the strategic locations of the plants, storage, and sourcing of raw materials, contracts with CMOs and CDMOs and logistics providers as well as future investments in new capacity over the years [58,62,83–87]. Tools to optimise production planning and scheduling offer great potential in assisting operational, day-to-day decision-making. Systematic approaches in production planning yield estimates of production targets, inventory levels, and material flows across the supply chain over a horizon of several months [59,60,88]. Scheduling tools, instead, provide detailed sequencing of tasks and operations, fulfilling orders and meeting the production targets, relying on a more granular description of the manufacturing and distribution processes and accounting for resource constraints [89,90].

Integrating the different levels of decision-making across many time scales is an issue of great interest in research [91,92], alongside coordination between multiple geographically distributed manufacturing and storage facilities comprising the supply chain. The size of the optimisation problem becomes challenging to solve by commercially available solvers and numerous approaches, including rolling horizon, spatial [88], and temporal decomposition [88,89] schemes have been proposed in literature. Solving the above problems under uncertainty remains an open challenge [93]. Uncertainty introduced by demand fluctuations, on-going global competition and pending clinical trial results, challenges the long-term strategic decision-making. Most of the frameworks proposed in literature use stochastic programming and scenario-based approaches [85,86,94], often coupled with decomposition strategies [95] to tackle computationally intractable formulations. Case studies have mainly focused on manufacturing and distribution of chemically derived drugs and conventional biologics. Therefore, the novelty of patient-specific products and ATMPs is a fertile ground for PSE tools to support investment planning exercises and establish supply chains that can cope with the predicted demand and success of these products. Similarly, planning and scheduling tools can aid the decision-making and tackle the operational challenges brought by patient specificity and time constraints of the therapy cycle [96,97].

4. Conclusions and Outlook

In the latest years, pharmaceutical products have evolved towards disease- and patient-specific therapeutics, involving meticulous manufacturing steps. In addition, cell-based therapeutics and vaccines present high sensitivity to environmental and transport conditions, complicating supply chain logistics. Increased drug specificity and demand uncertainty are adding a further level of complexity when it comes to the design and operation of robust manufacturing processes and distribution networks. As discussed in this paper, the pharmaceutical industry has taken significant steps towards the improvement of existing and/or the development of novel processes that promise agile, responsive, and reproducible manufacturing. Similarly, distribution networks in the pharmaceutical sector are undergoing a paradigm shift, exploring the capabilities of decentralised models.

Such developments are accompanied by digital innovation in the pharmaceutical industry that comes to enable seamless communication between process units, production plants, and distribution nodes. As discussed earlier, process systems engineering has been at the forefront of enabling digitalisation through the development of computer modelling tools. The latter can assist with real-time monitoring of storage conditions that are critical for sensitive pharmaceutical products with short shelf-life, thus increasing drug safety. One of the main challenges hindering fast exploitation of Industry 4.0 principles in pharmaceutical manufacturing is the change of mindset. Practitioners should embrace the benefits arising from the realisation of Pharma 4.0 towards replacing paper-based systems with cloud-based servers. This will allow significantly improved agility and productivity in the operations of the pharmaceutical sector.

Author Contributions: Conceptualization, M.S., A.B., N.S. and M.M.P.; methodology, M.S., A.B. and M.M.P.; investigation, M.S., A.B. and M.M.P.; images/tables, M.S.; writing—original draft preparation, M.S., A.B. and M.M.P.; writing—review and editing, M.S., A.B., N.S. and M.M.P.; All authors have read and agreed to the published version of the manuscript.

Funding: Funding from the UK Engineering & Physical Sciences Research Council (EPSRC) for the Future Targeted Healthcare Manufacturing Hub hosted at University College London with UK university partners is gratefully acknowledged (Grant Reference: EP/P006485/1).

Institutional Review Board Statement: Not applicable.

Informed Consent Statement: Not applicable.

Acknowledgments: The authors would like to acknowledge expert opinion received through multiple conversations with the User Steering Committee of the Future Targeted Healthcare Manufacturing Hub. Financial and in-kind support from the consortium of industrial users and sector organisations is also acknowledged.

Conflicts of Interest: The authors declare no conflict of interest.

References

1. European Medicines Agency (EMA). Biological Medicine European Medicines Agency. Available online: https://www.ema.europa.eu/en/glossary/biological-medicine (accessed on 28 December 2020).
2. Papathanasiou, M.M.; Kontoravdi, C. Engineering challenges in therapeutic protein product and process design. *Curr. Opin. Chem. Eng.* **2020**, *27*, 81–88. [CrossRef]
3. Shukla, A.A.; Wolfe, L.S.; Mostafa, S.S.; Norman, C. Evolving trends in mAb production processes. *Bioeng. Transl. Med.* **2017**, *2*, 58–69. [CrossRef]
4. Iancu, E.M.; Kandalaft, L.E. Challenges and advantages of cell therapy manufacturing under Good Manufacturing Practices within the hospital setting. *Curr. Opin. Biotechnol.* **2020**, *65*, 233–241. [CrossRef]
5. ISPE. Pharma 4.0. Available online: https://ispe.org/initiatives/pharma-4.0 (accessed on 15 February 2021).
6. ISPE. Pharma 4.0: Hype or Reality. Available online: https://ispe.org/pharmaceutical-engineering/july-august-2018/pharma-40tm-hype-or-reality (accessed on 15 February 2021).
7. Nahavandi, S. Industry 5.0—A Human-Centric Solution. *Sustainability* **2019**, *11*, 4371. [CrossRef]
8. SIEMENS. Embracing the Digital Transformation. Available online: https://assets.new.siemens.com/siemens/assets/api/uuid:f06590be-f311-4966-94df-f490abcc7d40/siemens-pharma-digi-consulting-en.pdf (accessed on 15 February 2021).
9. Papathanasiou, M.M.; Stamatis, C.; Lakelin, M.; Farid, S.; Titchener-Hooker, N.; Shah, N. Autologous CAR T-cell therapies supply chain: Challenges and opportunities? *Cancer Gene.* **2020**, *27*, 1–11. [CrossRef]
10. Levine, B.L.; Miskin, J.; Wonnacott, K.; Keir, C. Global Manufacturing of CAR T Cell Therapy. *Mol. Methods Clin. Dev.* **2017**, *4*, 92–101. [CrossRef] [PubMed]
11. Levine, B.L. Performance-enhancing drugs: Design and production of redirected chimeric antigen receptor (CAR) T cells. *Cancer Gene.* **2015**, *22*, 79–84. [CrossRef] [PubMed]
12. Juran, J.M. *Juran on Quality by Design: The New Steps for Planning Quality into Goods and Services*; The Free Press: New York, NY, USA, 1992; ISBN 9780029166833.
13. Warren, G. Quality by Design (QbD) Overview. 2015. Available online: https://www.pda.org/docs/default-source/website-document-library/chapters/presentations/australia/quality-by-design-(qbd)-overview.pdf?sfvrsn=f022b28e_6 (accessed on 1 February 2021).
14. Yu, L.X.; Amidon, G.; Khan, M.A.; Hoag, S.W.; Polli, J.; Raju, G.K.; Woodcock, J. Understanding pharmaceutical quality by design. *AAPS J.* **2014**, *16*, 771–783. [CrossRef] [PubMed]
15. Moore, C.M.V. Quality by Design—FDA Lessons Learned and Challenges for International Harmonization. 2012. Available online: https://www.fda.gov/media/85369/download (accessed on 1 March 2021).
16. PharmTech. EMA and FDA Release QbD Guidance. Available online: https://www.pharmtech.com/view/ema-and-fda-release-qbd-guidance (accessed on 28 December 2020).
17. Eon-Duval, A.; Broly, H.; Gleixner, R. Quality attributes of recombinant therapeutic proteins: An assessment of impact on safety and efficacy as part of a quality by design development approach. *Biotechnol. Prog.* **2012**, *28*, 608–622. [CrossRef] [PubMed]
18. Detela, G.; Lodge, A. Manufacturing process development of ATMPs within a regulatory framework for EU clinical trial & marketing authorisation applications. *Cell Gene Insights* **2016**, *2*, 425–452. [CrossRef]
19. Konstantinov, K.B.; Cooney, C.L. White paper on continuous bioprocessing 20–21 May 2014: Continuous Manufacturing Symposium. *J. Pharm Sci.* **2015**, *104*, 813–820. [CrossRef]
20. Hummel, J.; Pagkaliwangan, M.; Gjoka, X.; Davidovits, T.; Stock, R.; Ransohoff, T.; Gantier, R.; Schofield, M. Modeling the Downstream Processing of Monoclonal Antibodies Reveals Cost Advantages for Continuous Methods for a Broad Range of Manufacturing Scales. *Biotechnol. J.* **2019**, 14. [CrossRef] [PubMed]
21. Chatterjee, S. FDA Perspective on Continuous Manufacturing. *IFPAC Annu. Meet.* **2012**, *26*, 34–42.
22. Quon, J.L.; Zhang, H.; Alvarez, A.; Evans, J.; Myerson, A.S.; Trout, B.L. Continuous crystallization of Aliskiren hemifumarate. *Cryst. Growth Des.* **2012**, *12*, 3036–3044. [CrossRef]

23. Taylor, P. Outsourcing PharmA GSK and Pfizer Team up on Continuous Manufacturing Project. Available online: https://www.outsourcing-pharma.com/Article/2015/11/03/GSK-and-Pfizer-team-up-on-continuous-manufacturing-project (accessed on 25 December 2020).
24. Langer, E.S.; BioProcess International. Trends Perfusion Bioreactors the Next Revolution Bioprocessing? Available online: https://bioprocessintl.com/upstream-processing/bioreactors/trends-in-perfusion-bioreactors-323459/ (accessed on 23 December 2020).
25. Warikoo, V.; Godawat, R.; Brower, K.; Jain, S.; Cummings, D.; Simons, E.; Johnson, T.; Walther, J.; Yu, M.; Wrigth, B.; et al. Integrated continuous production of recombinant therapeutic proteins. *Biotechnol. Bioeng.* **2012**, *109*, 3018–3029. [CrossRef] [PubMed]
26. Godawat, R.; Konstantinov, K.; Rohani, M.; Warikoo, V. End-to-end integrated fully continuous production of recobinant monoclonal antibodies. *J. Biotechnol.* **2015**, *213*, 13–19. [CrossRef] [PubMed]
27. Karst, D.J.; Steinebach, F.; Soos, M.; Morbidelli, M. Process performance and product quality in an integrated continuous antibody production process. *Biotechnol. Bioeng.* **2017**, *114*, 298–307. [CrossRef]
28. André, S.; Cristau, L.S.; Gaillard, S.; Devos, O.; Calvosa, É.; Duponchel, L. In-line and real-time prediction of recombinant antibody titer by in situ Raman spectroscopy. *Anal. Chim. Acta* **2015**, *892*, 148–152. [CrossRef]
29. Li, M.Y.; Ebel, B.; Paris, C.; Chauchard, F.; Guedon, E.; Marc, A. Real-time monitoring of antibody glycosylation site occupancy by in situ Raman spectroscopy during bioreactor CHO cell cultures. *Biotechnol. Prog.* **2018**, *34*, 486–493. [CrossRef]
30. Shang, Y.; Zeng, Y.; Zeng, Y. Integrated Microfluidic Lectin Barcode Platform for High-Performance Focused Glycomic Profiling. *Sci. Rep.* **2016**, *6*, 20297. [CrossRef]
31. Rüdt, M.; Brestrich, N.; Rolinger, L.; Hubbuch, J. Real-time monitoring and control of the load phase of a protein A capture step. *Biotechnol. Bioeng.* **2017**, *114*, 368–373. [CrossRef]
32. Großhans, S.; Rüdt, M.; Sanden, A.; Brestrich, N.; Morgenstern, J.; Heissler, S.; Hubbuch, J. In-line Fourier-transform infrared spectroscopy as a versatile process analytical technology for preparative protein chromatography. *J. Chromatogr. A* **2018**, *1547*, 37–44. [CrossRef]
33. Krättli, M.; Steinebach, F.; Morbidelli, M. Online control of the twin-column countercurrent solvent gradient process for biochromatography. *J. Chromatogr A* **2013**, *1293*, 51–59. [CrossRef] [PubMed]
34. Godawat, R.; Brower, K.; Jain, S.; Konstantinov, K.; Riske, F.; Warikoo, V. Periodic counter-current chromatography—design and operational considerations for integrated and continuous purification of proteins. *Biotechnol. J.* **2012**, *7*, 1496–1508. [CrossRef] [PubMed]
35. Vormittag, P.; Gunn, R.; Ghorashian, S.; Veraitch, F.S. A guide to manufacturing CAR T cell therapies. *Curr. Opin. Biotechnol.* **2018**, *53*, 164–181. [CrossRef] [PubMed]
36. Mock, U.; Nickolay, L.; Philip, B.; Cheung, G.W.K.; Zhan, H.; Johnston, I.C.D.; Kaiser, A.D.; Peggs, K.; Pule, M.; Thrasher, A.J.; et al. Automated manufacturing of chimeric antigen receptor T cells for adoptive immunotherapy using CliniMACS prodigy. *Cytotherapy* **2016**, *18*, 1002–1011. [CrossRef] [PubMed]
37. Jaffer, G. Pharma's Almanac. Keys to Successful Storage, Management and Transport of Biological Materials. Available online: https://www.pharmasalmanac.com/articles/keys-to-successful-storage-management-and-transport-of-biological-materials (accessed on 28 December 2020).
38. Ruella, M.; Kenderian, S.S. Next-Generation Chimeric Antigen Receptor T-Cell Therapy: Going off the Shelf. *BioDrugs* **2017**, *31*, 473–481. [CrossRef] [PubMed]
39. U.S. Food and Drug Administration. *Summary Basis for Regulatory Action*; U.S. Food and Drug Administration: Silver Spring, MD, USA, 2017; pp. 1–16.
40. Novartis. Kymriah®(Tisagenlecleucel), First-in-Class CAR-T Therapy from Novartis, Receives Second FDA Approval to treat Appropriate r/r Patients with Large B-Cell Lymphoma. Available online: https://www.novartis.com/news/media-releases/kymriah-tisagenlecleucel-first-class-car-t-therapy-from-novartis-receives-second-fda-approval-treat-appropriate-rr-patients-large-b-cell-lymphoma#:~{):text=Quantitative%20Sciences%20Hackathon-,Kymriah%C2%AE%20(tisagenlecleucel)%2C%20first%2Din%2Dclass%20CAR,with%20large%20B%2Dcell%20lymphoma (accessed on 5 January 2021).
41. Novartis. Kymriah®(Tisagenlecleucel): Treatment Process, Dosing, Administration HCP. Available online: https://www.hcp.novartis.com/products/kymriah/acute-lymphoblastic-leukemia-children/treatment-process/ (accessed on 5 January 2021).
42. Kite PharmA First CAR T Therapy for Certain Types of Relapsed or Refractory B-Cell Lymphoma. Available online: https://www.kitepharma.com/news/press-releases/2020/7/us-fda-approves-kites-tecartus-the-first-and-only-car-t-treatment-for-relapsed-or-refractory-mantle-cell-lymphoma (accessed on 20 November 2020).
43. Lehmicke, M.; In Vivo. Manufacturing Cures: Infrastructure Challenges Facing Cell and Gene. *Therapy Developers*. Available online: https://invivo.pharmaintelligence.informa.com/IV124277/Manufacturing-Cures-Infrastructure-Challenges-Facing-Cell-And-Gene-Therapy-Developers (accessed on 15 December 2020).
44. Ebel, T.; George, K.; Larsen, E.; Shah, K.; Ungerman, D. *Pharmaceuticals and Medical Products Operations*; Mckinsey Co.: New York, NY, USA, 2013.
45. Shah, N. Pharmaceutical supply chains: Key issues and strategies for optimisation. *Comput. Chem. Eng.* **2004**, *29*, 929–941. [CrossRef]

46. Marques, C.M.; Moniz, S.; de Sousa, J.P.; Barbosa-Póvoa, A.P.; Reklaitis, G. Decision-support challenges in the chemical-pharmaceutical industry: Findings and future research directions. *Comput. Chem. Eng.* **2020**, *134*, 106672. [CrossRef]
47. Srai, J.S.; Badman, C.; Krumme, M.; Futran, M.; Johnston, C. Future supply chains enabled by continuous processing-opportunities and challenges May 20-21, 2014 Continuous Manufacturing Symposium. *J. Pharm. Sci.* **2015**, *104*, 840–849. [CrossRef]
48. Sykes, C. Time- and temperature-controlled transport: Supply chain challenges and solutions. *Pharm. Ther.* **2018**, *43*, 154–170.
49. U.S. Food and Drug Administration. Vaccines and Related Biological Products Advisory Committee. December 10. 2020 Meeting Briefing Document. Available online: https://www.fda.gov/advisory-committees/vaccines-and-related-biological-products-advisory-committee/2020-meeting-materials-vaccines-and-related-biological-products-advisory-committee (accessed on 28 December 2020).
50. Castellanos, S.; The Wall Street Journal. Drugmaker to Test Machine Learning to Prevent Drug Shortages. Available online: https://www.wsj.com/articles/drugmaker-to-test-machine-learning-to-prevent-drug-shortages-11571079794#:~:text=Merck%20KGaA%20plans%20to%20use,before%20they%20can%20be%20used (accessed on 28 December 2020).
51. Papathanasiou, M. Advances in Enabling Smart Technologies across the Cell Therapy Supply Chain. *Cell Gene. Insights* **2018**, *4*, 495–500. [CrossRef]
52. Raza, S.; Blackburn, L.; Moortie, S.; Cook, S.; Johnson, E.; Gaynor, L.; Kroese, M. *The Personalised Medicine Technology Landscape*; PHG Found: Cambridge, UK, 2018; ISBN 978-1-907198-31-1.
53. Chiaccio, F.; D'Urso, D.; Compagno, L.; Chiarenza, M.; Velardita, L. Towards a Blockchain Based Traceability Process: A Case Study from Pharma Industry. *IFIP Adv. Inf. Commun. Technol.* **2019**, *566*, 451–457. [CrossRef]
54. Nakamoto, S. Bitcoin: A Peer-to-Peer Electronic Cash System. Available online: https://bitcoin.org/bitcoin.pdf (accessed on 3 February 2021).
55. Agbo, C.C.; Mahmoud, Q.H.; Eklund, J.M. Blockchain Technology in Healthcare: A Systematic Review. *Healtcare* **2019**, *7*, 56. [CrossRef] [PubMed]
56. CISION, PR Newswire. Everyware and Hedera Hashgraph Enabling Cold ChaMonitoring of COVID-19 Vaccine for NHS Fcilities. Available online: https://www.prnewswire.com/news-releases/everyware-and-hedera-hashgraph-enabling-cold-chain-monitoring-of-covid-19-vaccine-for-nhs-facilities-301209642.html (accessed on 3 February 2021).
57. Modum. Success Stories. Available online: https://www.modum.io/industries/success-stories (accessed on 3 February 2021).
58. Siganporia, C.C.; Ghosh, S.; Daszkowski, T.; Papageorgiou, L.G.; Farid, S.S. Capacity planning for batch and perfusion bioprocesses across multiple biopharmaceutical facilities. *Biotechnol. Prog.* **2014**, *30*, 594–606. [CrossRef] [PubMed]
59. Vieira, M.; Pinto-Varela, T.; Moniz, S.; Barbosa-Póvoa, A.P.; Papageorgiou, L.G. Optimal planning and campaign scheduling of biopharmaceutical processes using a continuous-time formulation. *Comput. Chem. Eng.* **2016**, *91*, 422–444. [CrossRef]
60. Lakhdar, K.; Zhou, Y.; Savery, J.; Titchener-Hooker, N.J.; Papageorgiou, L.G. Medium term planning of biopharmaceutical manufacture using mathematical programming. *Biotechnol. Prog.* **2005**, *21*, 1478–1489. [CrossRef]
61. Millipore SigmA Bioprocess Online. Flexible Manufacturing of Vaccines. Available online: https://www.bioprocessonline.com/doc/flexible-manufacturing-of-vaccines-0001 (accessed on 20 December 2020).
62. Papageorgiou, L.G.; Rotstein, G.E.; Shah, N. Strategic supply chain optimization for the pharmaceutical industries. *Ind. Eng. Chem. Res.* **2001**, *40*, 275–286. [CrossRef]
63. MacDonald, G.J.; Genetic Engineering & Biotechnology News. Single-Use Tech Key for Capacity Hike Required for COVID-19. Available online: https://www.genengnews.com/topics/bioprocessing/single-use-tech-key-for-capacity-hike-required-for-covid-19/ (accessed on 29 December 2020).
64. Markarian, J. Supply Chain Challenges for Single-Use Systems. *Biopharm. Int.* **2019**, *32*, 16–19.
65. Gernaey, K.V.; Cervera-Padrell, A.E.; Woodley, J.M. A perspective on process systems engineering in pharmaceutical process development and innovation. *Comput. Chem. Eng.* **2012**, *42*, 15–29. [CrossRef]
66. Papadakis, E.; Woodley, J.M.; Gani, R. Perspective on PSE in pharmaceutical process development and innovation. *Comput. Aided Chem. Eng.* **2018**, *41*, 597–656. [CrossRef]
67. Barbosa-Póvoa, A.P.; Pinto, J.M. Process supply chains: Perspectives from academia and industry. *Comput. Chem. Eng.* **2020**, *132*, 106606. [CrossRef]
68. Chen, Y.; Yang, O.; Sampat, C.; Bhalode, P.; Ramachandran, R.; Ierapetritou, M. Digital Twins in Pharmaceutical and Biopharmaceutical Manufacturing: A Literature Review. *Processes* **2020**, *8*, 1088. [CrossRef]
69. Grossmann, I.E. Advances in mathematical programming models for enterprise-wide optimization. *Comput. Chem. Eng.* **2012**, *47*, 2–18. [CrossRef]
70. Lainez, J.M.; Schaefer, E.; Reklaitis, G.V. Challenges and opportunities in enterprise-wide optimisation in the pharmaceutical industry. *Comput. Chem. Eng.* **2012**, *47*, 19–28. [CrossRef]
71. Kotidis, P.; Demis, P.; Goey, C.H.; Correa, E.; McIntosh, C.; Trepekli, S.; Shah, N.; Klymenko, O.V.; Kontoravdi, C. Constrained global sensitivity analysis for bioprocess design space identification. *Comput. Chem. Eng.* **2019**, *125*, 558–568. [CrossRef]
72. Böhl, O.J.; Schellenberg, J.; Bahnemann, J.; Hitzmann, B.; Scheper, T.; Solle, D. Implementation of QbD strategies in the inoculum expansion of a mAb production process. *Eng. Life Sci.* **2020**. [CrossRef]
73. Bano, G.; Wang, Z.; Facco, P.; Bezzo, F.; Barolo, M.; Ierapetritou, M. A novel and systematic approach to identify the design space of pharmaceutical processes. *Comput. Chem. Eng.* **2018**, *115*, 309–322. [CrossRef]

74. Xie, X.; Schenkendorf, R. Robust Process Design in Pharmaceutical Manufacturing under Batch-to-Batch Variation. *Processes* **2019**, *7*, 509. [CrossRef]
75. Mortier, S.T.F.C.; Van Bockstal, P.J.; Corver, J.; Nopens, I.; Gernaey, K.V.; De Beer, T. Uncertainty analysis as essential step in the establishment of the dynamic Design Space of primary drying during freeze-drying. *Eur. J. Pharm. Biopharm.* **2016**, *103*, 71–83. [CrossRef]
76. Ganesh, S.; Su, Q.; Vo, L.B.D.; Pepka, N.; Rentz, B.; Vann, L.; Yazdanpanah, N.; O'Connor, T.; Nagy, Z.K.; Reklaitis, G.V. Design of condition-based maintenance framework for process operations management in pharmaceutical continuous manufacturing. *Int. J. Pharm.* **2020**, *587*, 119621. [CrossRef] [PubMed]
77. Diab, S.; McQuade, D.T.; Gupton, B.F.; Gerogiorgis, D.I. Process Design and Optimization for the Continuous Manufacturing of Nevirapine, an Active Pharmaceutical Ingredient for HIV Treatment. *Org. Process Res. Dev.* **2019**, *23*, 320–333. [CrossRef]
78. Giridhar, A.; Reklaitis, G.V. Real-Time Optimization: How to Change Setpoints in Pharmaceutical Manufacturing. In *AAPS Advances the Pharmaceutical Sciences Series*; Nagy, Z.K., El Hagrasy, A., Litster, J., Eds.; Springer: Cham, Switzerland, 2020; Volume 42, pp. 429–440. ISBN 978-3-030-41524-2_12.
79. Kotidis, P.; Jedrzejewski, P.; Sou, S.N.; Sellick, C.; Polizzi, K.; del Val, I.J.; Kontoravdi, C. Model-based optimization of antibody galactosylation in CHO cell culture. *Biotechnol. Bioeng.* **2019**, *116*, 1612–1626. [CrossRef] [PubMed]
80. Papathanasiou, M.M.; Burnak, B.; Katz, J.; Shah, N.; Pistikopoulos, E.N. Assisting continuous biomanufacturing through advanced control in downstream purification. *Comput. Chem. Eng.* **2019**, *125*, 232–248. [CrossRef]
81. Feidl, F.; Vogg, S.; Wolf, M.; Podobnik, M.; Ruggeri, C.; Ulmer, N.; Wachli, R.; Souquet, J.; Broly, H.; Butté, A.; et al. Process-wide control and automation of an integrated continuous manufacturing platform for antibodies. *Biotechnol. Bioeng.* **2020**, *117*, 1367–1380. [CrossRef]
82. Su, Q.; Moreno, M.; Ganesh, S.; Reklaitis, G.V.; Nagy, Z.K. Resilience and risk analysis of fault-tolerant process control design in continuous pharmaceutical manufacturing. *J. Loss. Prev. Process. Ind.* **2018**, *55*, 411–422. [CrossRef]
83. De Carvalho, M.I.; Ribeiro, D.; Barbosa-Póvoa, A.P. Design and Planning of Sustainable Vaccine Supply Chain. In *Pharmaceutical Supply Chains Medicine Shortages*; Barbosa-Póvoa, A.P., Jenzer, H., De Miranda, J.L., Eds.; Springer: Cham, Switzerland, 2019; pp. 23–55. ISBN 978-3-030-15398-4.
84. Liu, S.; Papageorgiou, L.G. Multi-objective optimisation for production, distribution and capacity planning of global supply chains in the process industry. *Omega* **2014**, *41*, 369–382. [CrossRef]
85. Gatica, G.; Papageorgiou, L.G.; Shah, N. Capacity Planning Under Uncertainty for the Pharmaceutical Industry. *Trans. Icheme* **2003**, *81*, 665–678. [CrossRef]
86. Tsang, K.H.; Samsatli, N.J.; Shah, N. Capacity Investment Planning for Multiple Vaccines Under Uncertainty 1: Capacity Planning. *Inst. Chem. Eng.* **2007**, *85*, 120–128. [CrossRef]
87. Liu, S.; Papageorgiou, L.G.; Shah, N. Optimal design of low-cost supply chain networks on the benefits of new product formulations. *Comput. Ind. Eng.* **2020**, *139*, 1061893. [CrossRef]
88. Sousa, R.T.; Liu, S.; Papageorgiou, L.G.; Shah, N. Global supply chain planning for pharmaceuticals. *Chem. Eng. Res. Des.* **2011**, *89*, 2396–2409. [CrossRef]
89. Stefansson, H.; Jensson, P.; Shah, N. Multiscale Planning and Scheduling in the Secondary Pharmaceutical Industry. *Am. Inst. Chem. Eng.* **2006**, *52*, 4133–4149. [CrossRef]
90. Amaro, A.C.S.; Barbosa-Póvoa, A.P. Planning and scheduling of industrial supply chains with reverse flows: A real pharmaceutical case study. *Comput. Chem. Eng.* **2008**, *32*, 2606–2625. [CrossRef]
91. Rossi, F.; Casas-Orozco, D.; Reklaitis, G.; Manenti, F.; Buzzi-Ferraris, G. A computational framework for integrating campaign scheduling, dynamic optimization and optimal control in multi-unit batch processes. *Comput. Chem. Eng.* **2017**, *107*, 184–220. [CrossRef]
92. Dias, L.S.; Ierapetritou, M.G. From process control to supply chain management: An overview of integrated decision making strategies. *Comput. Chem. Eng.* **2017**, *106*, 826–835. [CrossRef]
93. Jankauskas, K.; Papageorgiou, L.G.; Farid, S.S. Fast genetic algorithm approaches to solving discrete-time mixed integer linear programming problems of capacity planning and scheduling of biopharmaceutical manufacture. *Comput. Chem. Eng.* **2019**, *121*, 212–223. [CrossRef]
94. Franco, C.; Alfonso-Lizarazo, E. Optimization under uncertainty of the pharmaceutical supply chain in hospitals. *Comput. Chem. Eng.* **2020**, *135*, 106689. [CrossRef]
95. Levis, A.A.; Papageorgiou, L.G. A Hierarchial Solution Approach for Multi-Site Capacity Planning Under Uncertainty in the Pharmaceutical Industry. *Comput. Chem. Eng.* **2004**, *28*, 707–725. [CrossRef]
96. Wang, X.; Kong, Q.; Papathanasiou, M.M.; Shah, N. Precision healthcare supply chain design through multi-objective stochastic programming. *Comput. Aided Chem. Eng.* **2018**, *44*, 2137–2142. [CrossRef]
97. Moschou, D.; Papathanasiou, M.M.; Lakelin, M.; Shah, N. Investment Planning in Personalised Medicine. *Comput. Aided Chem. Eng.* **2020**, *48*, 49–54. [CrossRef]

processes

MDPI

Perspective

Why Is Batch Processing Still Dominating the Biologics Landscape? Towards an Integrated Continuous Bioprocessing Alternative

Ashish Kumar [1], Isuru A. Udugama [2], Carina L. Gargalo [2] and Krist V. Gernaey [2,*]

[1] Pharmaceutical Engineering Research Group (PharmaEng), Department of Pharmaceutical Analysis, Ghent University, 9000 Gent, Belgium; Ashish.Kumar@UGent.be

[2] Process and Systems Engineering Center (PROSYS), Department of Chemical and Biochemical Engineering, Technical University of Denmark, 2800 Kongens Lyngby, Denmark; isud@kt.dtu.dk (I.A.U.); carlour@kt.dtu.dk (C.L.G.)

* Correspondence: kvg@kt.dtu.dk

Received: 9 November 2020; Accepted: 10 December 2020; Published: 12 December 2020

Abstract: Continuous manufacturing of biologics (biopharmaceuticals) has been an area of active research and development for many reasons, ranging from the demand for operational streamlining to the requirement of achieving obvious economic benefits. At the same time, biopharma strives to develop systems and concepts that can operate at similar scales for clinical and commercial production—using flexible infrastructures, such as single-use flow paths and small surge vessels. These developments should simplify technology transfer, reduce footprint and capital investment, and will allow to react readily to changing market pressures while maintaining quality attributes. Despite a number of clearly identified benefits compared to traditional batch processes, continuous bioprocessing is still not widely adopted for commercial manufacturing. This paper details how industry-specific technological, organizational, economic, and regulatory barriers that exist in biopharmaceutical manufacturing are hindering the adoption of continuous production processes. Based on this understanding, the roles of process systems engineering (PSE), process analytical technologies, and process modeling and simulation are highlighted as key enabling tools in overcoming these multi-faceted barriers in today's manufacturing environment. Of course, we do recognize that there is also a need for a clear set of regulations to guide a transition of biologics manufacturing towards continuous processing. Furthermore, the role played by the emerging fields of process integration and automation as well as digitalization is explored, as these are the tools of the future to facilitate this transition from batch to continuous production. Finally, an outlook focusing on technology, management, and regulatory aspects is presented to identify key concerted efforts required to drive the broad adaptation of continuous manufacturing in biopharmaceutical processes.

Keywords: continuous manufacturing; bioprocessing; process systems engineering; single-use technology

1. Introduction: Where Are We Now

There is a growing interest in continuous biologics manufacturing in the pharmaceutical sector, encouraged by regulatory entities such as the Food and Drug Administration (FDA) and the International Council for Harmonization (ICH) [1–4]. Continuous manufacturing (CM) constitutes the critical aspect of process intensification which endeavors to reduce the time, cost, and complexities of bioprocess development and manufacturing. Several potential benefits are to be expected from its implementation, such as sustained and steady-state operation, the promise of more consistent product quality and high productivity, reduced equipment size, and streamlined process flow, thus lowering operating and capital costs [5,6]. An overall increase in sustainability can, in principle, be achieved by switching to continuous manufacturing, in line with the UN's Sustainable Development Goals to adopt better and more sustainable processes. However, despite the immense opportunities of continuous over batch operation, technical and operational challenges need to be addressed and overcome for integrated continuous bioprocessing to become a reality. Like any other disruptive "new" technology, besides the scientific element, there are also business and human aspects involved in decision-making that need attention. Of course, some changes are already happening, and people are currently working to develop more of the continuous biomanufacturing technology. However, such development is generally occurring at a much slower pace than expected. The main question to be answered is then why adopting continuous biomanufacturing technology is, in general, extremely slow, despite the obvious potential benefits that can be achieved by its implementation.

This manuscript aims to explore this unknown conundrum behind the slow adoption of continuous processing in commercial biologics manufacturing despite known business benefits, innovations, technical support, and regulatory push (Figure 1).

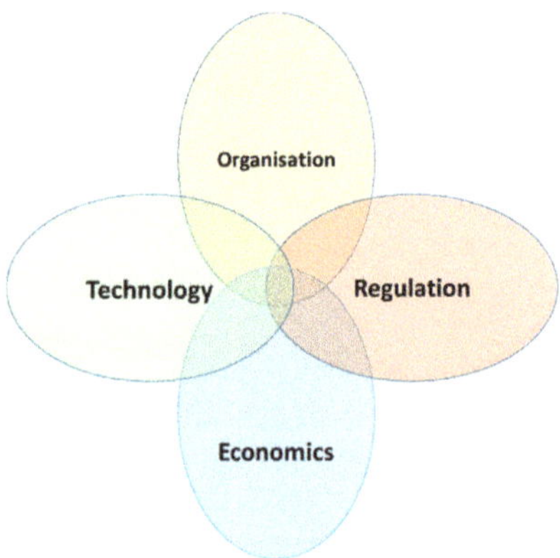

Figure 1. Key forces that must be taken into consideration when introducing process technologies to biologics manufacturing.

1.1. Critical Views on Current Practices

On the one hand, the manufacturing of biologics is the epitome of the contemporary industry with cutting-edge research and development, resulting in the steady discovery of novel molecules. On the other hand, there are the manufacturing processes of these biologics, which are, even today, to a large extent, based on inefficient batch production platforms. Since biotech production started developing gradually, the first reactors were typically relatively small, inspired, e.g., by the dairy industry. However, throughout the years, the need for reactor capacity and control has been growing steadily, which has resulted in bioreactor vessels that are often up to several hundred cubic meters in capacity. Operating these processes in a batch mode has its advantages. Namely, batch processes require less precise and robust controls to make the desired product than continuous operation. At the same time, the key disadvantages are the inherent inefficiencies of batch fermentation processes. For example, several studies [7,8] have shown that continuous, or for that matter, even semi-continuous, fermentation processes are more efficient than batch processes due to promoting gains both in space-time yield and raw material usage.

To this end, both academia and industry have been deliberating the adoption of continuous biologics manufacturing for over half a century. Challenges with regards to various aspects of operation and control of the biologics process in a continuous mode were identified in this period. A significant number of mechanistic insights translated into several technological advancements. On the other hand, regulatory agencies have also been evolving in support of continuous manufacturing from the FDA guidance on process analytical technologies (PATs) in 2004 for the adoption of a science- and risk-based approach allowing quality by design (QbD) [9] to the recent draft guidance dedicated to continuous manufacturing [2,10]. Despite explicit knowledge, technological advancement, and regulatory support, the implementation of continuous bioprocessing is and has been slow [11]. At best, few unit operations/steps in upstream and downstream have been implemented as continuous processes by a minority of facilities. This, too, is limited to some large-scale biopharmaceutical products, such as antibodies [12], insulin [13], Factor VIII, and coagulation factors [14], which essentially require moderate processing conditions in perfusion systems and external alternating tangential flow (ATF) filtration systems for cell retention [5,15]. Some bio-manufacturers, such as Genzyme, Bayer, and Amgen, with considerable experience using perfusion and single-use systems for commercial products manufacture are leaders in continuous bioprocessing implementation. Several others are still evaluating the commercial feasibility of the technology. Commercial continuous downstream processing, particularly chromatography operations, still remains rare. Optimistically, only one or a few out of the multiple chromatographic steps in multi-step downstream processing having been implemented as continuous operations. It is noteworthy that some processes claim to be "continuous" manufacturing operations. Upon closer inspection, they are found to be parallelized upstream and downstream unit operations that give only the "illusion" of being continuous operations. These processes are cyclical operations with multiple batch units being sequentially scheduled to provide a consistent process flow. In contrast, continuous bioprocessing is expected to contribute to process intensification by offering an opportunity to simplify the processes and, thus, replace complex cyclical batch operations.

1.2. Academic Engagements Supporting the Advancement

The first challenges towards adopting continuous manufacturing in biologics were posed in terms of knowledge gaps—pointing towards long-term stable and sterile cultivation of cells. While continuous processing is beneficial for unstable products that cannot be left in the culture medium over the entire cultivation period, it is not suitable for the vectors' genetic stability and the producer cell lines. Studies to understand cells' microenvironment, functioning, and long-term genetic stability within the cells were performed in great detail [16,17]. Similarly, several technological developments lowered the risk of contamination during in situ product removal as in continuous bioprocessing [18,19]. Besides resolving the risks during continuous processing, increasing the historically low upstream drug substance titer in batch biologics manufacturing (<1 g/L) has been an important underlying driver for the exploration of continuous processing in biologics [20]. A solution came with perfusion processes that kept cell counts constant while exchanging culture supernatants with fresh media at regular intervals. However, this solution came with several process systems engineering challenges such as (i) inhomogeneity in the continuous bioreactor vessel, including nutrient shortages or regions of cell accumulation; (ii) maintenance of sterility in the long run; (iii) increase in process development cost and time due to complex continuous culture systems in labs; and (iv) regulatory expectation towards developing a complex control strategy addressing time invariance in a continuous run.

As mentioned, the presence of heterogeneities in the continuous bioreactor vessel continues to be one of the main challenges in the manufacturing of biologics. However, not easy to tackle, new technologies are being developed to investigate and improve this. For example, recently, computational fluid dynamics (CFD) has been the focus of growing interest in the study of mixing, mass transfer, and substrate gradients in (bio)reactors [21]. Furthermore, CFD, when coupled with compartmental modeling, has been shown to be a powerful tool for process design and optimization [22,23]. Another promising approach for the monitoring of real gradients inside the bioreactor is, for example, the use of free-floating sensors [24]. Since in long perfusion processes, membrane fouling and blocking were observed, newer technologies for external cell retention such as ATF filtration were developed and widely adopted [15]. In an ATF system, a feed stream through hollow fibers is regularly reversed to wash off material that can clog the pores. Similarly, with respect to agility in process development and handling the complexity of continuous equipment, innovative PAT techniques (Section 3.2) and process control strategies (including process modeling and advanced process control approaches) (Section 3.3) and robotics and automation (Section 3.2) have been the key focus in academic research bringing new solutions [25,26]. These innovations are expected to simplify operation, improve process robustness, and reduce handling efforts significantly.

Furthermore, due to the inherent multi-step nature of the process, the implementation of continuous operations is complicated in downstream processing. However, downstream processing of biologics usually represents 50–80% of the manufacturing cost, making commercial batch manufacturing very inefficient and expensive [26]. Thus, significant research efforts on the development of continuous downstream processes were made in recent decades to offer improved use of the column capacity and increased resolution efficiency. A better resin capacity use and column lifetime exploitation are particularly essential when expensive protein and resins are involved in the protein capture step of downstream processing. Most of the current studies on continuous downstream processing have been carried out on monoclonal antibodies (mAbs) for two key reasons: first, mAbs comprise the most significant segment of biopharmaceutical products under development and getting ready for commercial manufacturing; second, the use of three chromatography steps—Protein A affinity, cation-exchange (CEX), and anion-exchange (AEX)—already provides a standardized route for achieving high purities and high recoveries, but it is in need of a fresh intensification focus. Therefore, several methods and multi-column systems for continuous liquid chromatography have been developed in recent years, several of which translated into commercially available solutions, such as ÄKTA PCC (GE Healthcare), BioSC® (Novasep), BioSMB® (Pall Life Sciences), and Contichrom® (ChromaCon® AG) [27]. Next to innovation in the established production route, innovative separation technologies have been proposed using computational modeling and simulation. For example, Castilho et al. [28] developed Protein A membrane adsorbers and utilized computational fluid dynamics to design a rotating disk filter.

Besides individual research in academia and industry, several collaborative platforms also aim to address these challenges. A non-exhaustive list of such collaborative platforms enabling development and adoption of continuous manufacturing technologies in bioprocessing is given in Table 1. Some of the key focus points for these groups include accelerating the biopharmaceutical innovation, developing adaptive process control and advanced sensing for robust quality in biomanufacturing, demonstrating the commercial feasibility of new technologies, and enabling the biopharmaceutical manufacturing workforce for new and future manufacturing technologies. Such collaborations are playing an increasingly important role in pushing the accepted state of the art on technology and its knowledge, as well as helping in developing a well-educated workforce that can address the challenges related to the introduction of novel production technology.

Table 1. Some of the collaborative platforms that enable the development and adoption of new manufacturing technologies such as continuous bioprocessing.

Collaborative Platforms	Kind/Lead	Purpose
Biomanufacturing consortium (BioMAN)	Industry–academia collaboration lead by Massachusetts Institute of Technology (MIT)	Several stakeholders work together to develop new knowledge, science, technologies, and strategies to improve biomanufacturing [29].
BioPhorum consortium	Cross-industry collaboration	Connecting most big biopharmaceutical manufacturers and suppliers collaboratively to produce technical documents to explore, propose, and define industry best practices on the topics mentioned earlier [30].

Table 1. *Cont.*

Collaborative Platforms	Kind/Lead	Purpose
National Institute for Innovation in Manufacturing Biopharmaceuticals (NIIMBL)	Public–private partnership, manufacturing innovation institutes funded through a cooperative agreement with the National Institute of Standards and Technology (NIST)	To achieve a public–private partnership to enable more efficient and rapid manufacturing capabilities and biopharmaceutical manufacturing workforce to accelerate biopharmaceutical innovation.
BIOPRO cluster	Industry–academia collaboration lead by Technical University of Denmark (DTU)	Developing new ways of making bio-based production more efficient and sustainable by reducing the consumption of energy and raw materials while improving yields [31].

2. What Is Creating This Disjoint

The initial implementation of continuous bioprocessing was slow due to technical challenges. Similar to other technological advancements and new solutions development, it could be argued that the benefits of continuous processing outweigh its challenges and constraints. Hence, even though many biologics firms have been successfully operating continuous manufacturing plants for years, it is esoteric that continuous bioprocessing has not implemented a typical technology lifecycle paradigm. The negative assessment from within the industry, despite known business and quality benefits, lies in many different dimensions of decision making for technology adoption (Figure 1). From a practical point of view, a migration to continuous manufacturing, both during clinical trials and mass manufacturing, requires approval at a senior executive level. However, with the need to maintain quarterly and yearly performance figures, it means that a drive towards continuous manufacturing, which requires a more long-term time horizon to be truly beneficial, becomes a tough sell. As expected, business priorities and decision making are often driven by short-term goals, visible challenges, and deeply entrenched mental models rooted in batch manufacturing technologies (Figure 2).

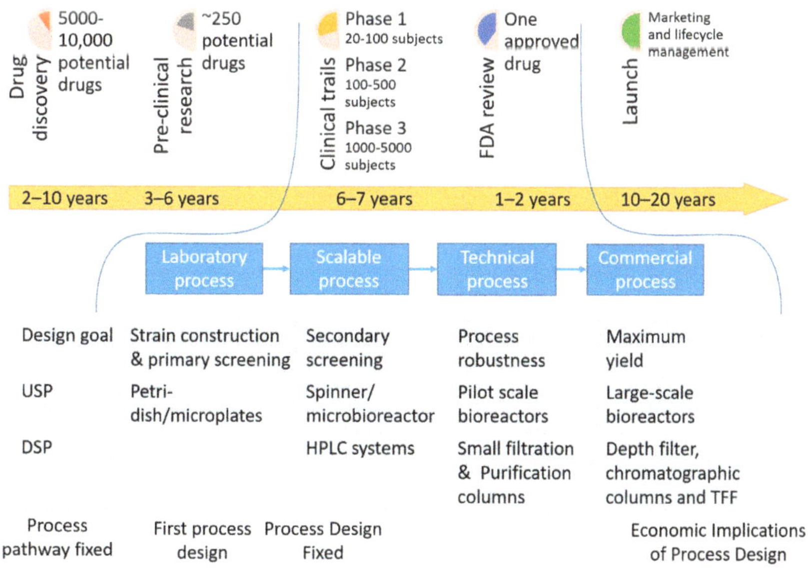

Figure 2. Lifecycle of a drug from discovery to mass market availability. Abbreviations: DSP (Downstream processing), HPLC (High performance liquid chromatography), TFF (Tangential-flow filtration system), USP (Upstream processing).

In the following section, we look into various dimensions as the cause for the slow adoption of industry's continuous biologics manufacturing.

2.1. Are Costs or Regulators to Blame

It is often argued that the biopharmaceutical industry will eventually evolve towards continuous processing, similar to the trends observed in other industrial sectors such as oil, food, and paper manufacturing. That is, however, not really happening. Such arguments ignore the considerable differences in the required processing capacity and product value in the manufacturing of biologics. It is safe to say that a biologics plant that processes anywhere near the per-day processing capacity of plants in the aforementioned industrial sectors will never be needed. Thus, a business case never truly stands if driven by the need to be continuous on a capacity basis [5]. In general, much of the cost advantage from continuous processing comes from intensifying the process, resulting in smaller equipment footprints (compared with batch and fed-batch systems) and, consequently, lower capital investments. Simultaneously, smaller facilities are more comfortable to expand or replicate as back-up systems at multiple sites. Moreover, it is essential to realize that biopharmaceutical manufacturing is currently challenged by questions that are very different from the questions faced by the biopharma industry of the past, built on the heritage of blockbuster drugs, and faced during its development and manufacturing. Under development, the current biopharma portfolio is a collection of smaller targeted treatments, requiring a flexible and economical production option for these products instead of hitting a capacity dead-end. Thus, it is crucial to make a correct comparison while building the business case for the continuous manufacturing of biologics.

The early discourse presented regulatory aspects as the biggest hurdle on the road toward the actual implementation of continuous bioprocessing. However, over the past few years, regulatory agencies such as the FDA appeared to be a strong supporter of continuous bioprocessing as it reduces manual interference with product streams and improves process control and quality. In this context, the agency also recommends using QbD principles to establish sufficient process robustness [9]. However, it is still challenging to develop proper scale-down models, which are prerequisites for this approach. While running experiments for this purpose, the prohibitively expensive experimental campaigns associated with media costs alone restrict this [32]. Different strategies are being investigated to avert these costs potentially (Figure 3). For example, Tajsoleiman et al. (2019) [22] explored the use of compartmental modeling by the automatic conversion of fully developed CFD (bio)reactor models into compartment models. This approach is highly efficient and can be used as a strategy to develop a scale-down model of the process/equipment [33].

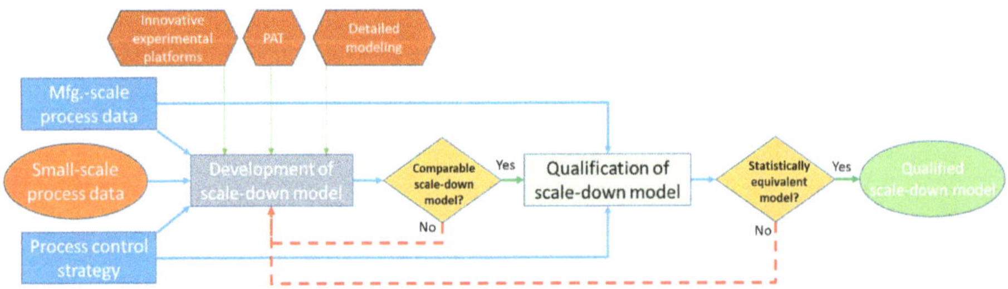

Figure 3. Scale-down modeling workflow supported by innovative solutions.

Furthermore, for the same purpose, the application of continuous hollow-fiber bioreactors and miniaturized stirred bioreactors (MSBRs) (or small scale reactors in general) as well as microfluidics is currently being explored [34]. While a hollow-fiber system can be as small as 2.5 mL, it is not easy to measure process parameters. On the other side, although easier to measure parameters, the minimum volume for commercially available MSBRs is finally scaling-down to actual low volumes, such as 15 mL and 250 mL volume Ambr® systems [35]. Similarly, a few commercial microfluidic perfusion systems without active mixing are also currently available. Most of these systems struggle with challenges such as the inability to incorporate control of pH and dissolved gases. Tajsoleiman et al. [33] reviewed opportunities, critical parameters, and challenges of implementing MSBRs for scale-down purposes.

2.2. Realization of Flexible and Intensified Manufacturing

Success for a modern biopharma company strongly depends on operational flexibility to adjust production capacity in response to rapidly changing demand forecast, unexpected failures in late-stage clinical trials, or faster than expected clinical adoption. With smart batch definitions, scale-up could be enabled simply by extending a perfusion process' duration in continuous processing. Capacity use is improved when only a small number of long-duration campaigns are performed, reducing the number of required product change-overs [36]. Moreover, continuous bioprocessing plants are often made specifically for a single process/product, with little or no intention of adapting the same equipment for a different process. However, this translates into an inherently low facility flexibility level with long continuous operation run times. It also limits the degree of flexibility a company has to switch quickly between different products. Process intensification to achieve final titers similar to conventional upstream processes in a shorter timeframe and a modular biomanufacturing platform have been proposed as strategies to regain the degree of flexibility.

Early-stage process development and clinical manufacturing are often critical to getting a new product into clinical trials. Clinical delays due to process development or manufacturing issues can be very costly. Although many firms have successfully operated continuous perfusion for years, it is generally not seen as a more reliable production method than (fed-)batch, especially at the early stage. On the other hand, standard fed-batch culture services are offered by many biologics contract manufacturers. Perfusion culture services are far less common and, even when found, are readily available only with the particular cell retention method already in use by that vendor. The compatibility of that method for a specific process may take time to resolve. Such early-stage challenges often impact early decisions about the most appropriate way to use the available manufacturing technology to meet the demand and timeline. Even at the commercial scale, two significant biopharmaceutical shortages occurred due to commercial production problems using continuous perfusion systems. Therefore, it is crucial to improve design, characterization, and protocol standardization methods to improve continuous bioprocessing predictability and confidence.

2.3. The Dilemma of Technology Evolution—Continuous Stainless Steel vs. Single-Use

Continuous processing equipment manufacturers and users report that many of the problems long associated with perfusion and continuous bioprocessing have been resolved in recent years by applying innovative technologies, including new developments in single-use equipment. All the necessary building blocks for a fully disposable continuous value stream exist—bioreactors, cell-retention devices, prepacked columns, and filters. However, for very high cell densities aimed for titer increase, disposable/single-use bioreactors are not recommended. At some point, in single-use bioreactors, the dissolved oxygen (DO) level might become a limiting factor for high cell densities. In such cases, continuous processing at a sufficient scale is not possible due to limits of oxygen transfer rates and pressure limits of welded seams in the plastic bags. A high cell concentration also generates heat that is difficult to transfer across insulating plastic layers.

While single-use or disposable systems provide much more flexibility, an inevitable concern appears regarding these systems' environmental sustainability. In a world where people are worried about grocery bags and disposable cups, justifying an industry's transition from multi-use stainless steel to disposable plastic adds another dimension of concern. Although the waste output from biomanufacturing is only a fraction, appropriate waste disposal procedures are crucial to adopting these systems, especially towards scale-up uses. The process intensification supported by continuous manufacturing can, in fact, help to reduce the footprint and improve sustainability. Thus, although the two technology trends look to be competing, they are mutually enabling the future of bioprocessing. Table 2 compares the benefits and challenges of the two manufacturing platforms—continuous stainless steel vs. single-use systems.

Table 2. Benefits and challenges that single-use systems can offer compared to stainless steel-based manufacturing platforms.

Multi-Use Stainless Steel Systems	Single-Use Systems
Benefits	
• Well-known systems with standardization • Less disposable waste • Available in large capacities • More advanced sampling and process control	• Capacity flexibility • Can be easily scaled out • Easy product/facility changeover • Reduced cleaning validation • Lower energy demand and water use • Lower risk of contamination
Challenges	
• Inflexible infrastructure • Challenging in scale-up • Higher maintenance requirements, utilities • Cleaning validation required • Higher risk of contamination • Higher capital expenditure (CAPEX) required to start	• Lack of mechanical strength, i.e., difficult scale-up • Waste generation • Risk of extractables and leachables • Lack of standardization • Requirement of sustainable supply of consumables • More dependence on automation and a very skilled operator needed

2.4. Organisational Readiness

From a management perspective, it should be noted that multiple stakeholders are involved in successfully implementing continuous biologics manufacturing—for both clinical trials and, more importantly, consumers. One key stakeholder that is often overlooked is the plant operations department. With the inherent variations present in raw materials and processes, an argument can be made that plant operations would instead prefer to operate batch units (where these variations can be more easily dealt with) instead of dealing with the precise operations required for continuous manufacturing. Besides, operating a new type of process requires plant operations to develop appropriate strategies and train operators to understand and respond to a new set of operational challenges. Thus, onboarding plant operations at an early stage of process design and reducing their burden of a late-stage switch from batch to continuous operations play a major role in gathering a critical mass of support to implement continuous production technologies.

Considering an alternative perspective, it can be argued that for biologic firms focused on drug discovery as the main form of innovation, the introduction of novel production methodologies can be seen as a distraction. This is because the primary "value creation", i.e., the economic benefit, happens through drug discovery. From an organization's point of view, the production process is just the "vehicle" used to mass-produce these drugs. It may be argued in such a case that any "economic benefit" that can be achieved by improved production is minor in comparison to drug discovery, and the pursuit of such improvements can "jeopardize" the focus on more significant sources of economic benefit. To this end, generic and control biologic manufacturers might have a greater willingness to consider novel technologies such as continuous manufacturing processes, as their core business model is founded on "cost-efficient production".

3. What Can Be Done to Improve the Situation

3.1. Need for More PSE Case Studies to Build Confidence

There have been significant knowledge-driven advancements in continuous bioprocessing technologies. Perfusion processing is now significantly less complex, less prone to contamination, and more readily scalable than previously. Similarly, multiple solutions for continuous downstream processing have intensified the process to accommodate the increased titers and processing rate. A cost-of-goods analysis on consumables has shown a 6–10-fold cost reduction from the conventional batch to the intensified continuous process [8]. Thus, the negative assessments from within the industry of continuous perfusion and downstream processing may, overall, reflect a lack of direct exposure or experience with continuous technology combined with a relative lack of case studies documenting success [11]. Case studies and reports of superior performance, when compared to existing batch platforms, will further help the rapid adoption of continuous bioprocessing. Moreover, such case studies can reflect business drivers' alignment during research and development of a biologics drug and its commercial manufacturing to create a supportive environment (Figure 4). The business drivers during research and development are focused on maximizing knowledge acquisition with minimal expenditure. As "kiss" or "kill" is an approach regarding either continuation or termination heavily used in early stage of drug development, a knowledge spill-over across projects is a supporting driver. In fact, nine times out of 10, candidate molecules that show promise at the early stage of development fail at later trials. Therefore, while accuracy in such a decision is critical for ultimate success, a delay in search of accuracy comes with a heavy cost due to the fact that there are many candidate molecules at the early stage. At the later stages, the speed and flexibility in process technology for achieving capacity for clinical supplies and later-stage market demand drive the development strategy. On the commercial side, business drivers focus on maximizing supplies and savings while maintaining the promised quality. This involves improving the process' robustness by continued process validation programs, continuous improvement projects for lean manufacturing, and efficient product facility change-overs.

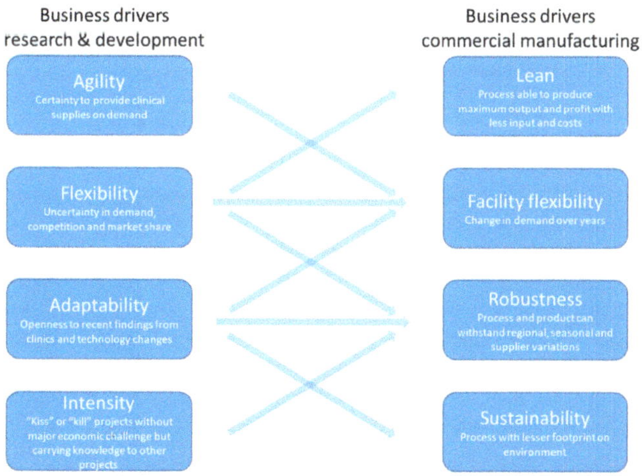

Figure 4. Alignment between interlinked business drivers in industrial research and development (R&D) unit and commercial manufacturing is key to success.

Furthermore, analyzing the inherently negative perception of new process technologies by plant operations and management, in general, can be traced to the nature in which the performance of these individuals is assessed. In particular, the primary objective of management and plant operations is to achieve a high level of plant "uptime" while guaranteeing "on-specification" production, with minimal operator intervention. In other words, the obvious choice is, then, to operate technologies that are robust and well understood by the operators. From a systems thinking standpoint, these two risk factors can be categorized into (i) technology risk (the likelihood that a new technology fails); and (ii) plant-wide operational risk (the reduction in operational flexibility due to the new technology).

Process systems engineering methods can be employed to systematically assess (and thus manage) these two risks during technology development and implementation. One such method is the concept of technology readiness level (TRL). The TRL was originally developed by National Aeronautics and Space Administration (NASA) to assess space technology development but has now been adopted in other areas, including bio-based production [37]. The TRL provides a systematic framework in which technology readiness can be easily estimated based on literature and publicly available information. This framework also provides a guide on how to factor in the TRL's economic consequences of new technology. While these frameworks are not explicitly developed to rate the technology development in biologics production, they can be adopted to make informed decisions based on "facts and figures" rather than on perception. Thus, the TRL allows technology developers to identify key areas of improvements throughout the design process.

Similarly, the plant-wide operational risk introduced by new technology can be systematically evaluated (and thus managed) based on a systematic assessment. For example, the framework presented in [38], which was initially intended to rate process control configurations, can be adopted to systematically assess the economic consequences of reducing operational flexibility due to the implementation of new technologies. In this particular framework, the concepts of layer of protection analysis (LOPA) and net present value (NPV) analysis are employed to quantify the consequences. This type of analysis allows both the technology developer and plant operations to chart a minimal risk path to implementation. This can include steps such as the development of validated plant-wide process models or the use of modified benchmark simulations to demonstrate the plant-wide ramification that a new process technology will bring. Moreover, the development of built-in redundancies, as well as implementation strategies, minimize plant-wide ramifications.

3.2. PAT Solutions and Robotics for Better Control

From a pure process control point of view, both continuous and batch production processes have unique characteristics that make their control difficult. One of the most important differences between continuous bioprocessing and batch or even fed-batch operations is in their required level of control sophistication [39]. For a fed-batch process, the objective of process control is mainly to achieve a given feeding profile while ensuring that variables such as pH and temperature are kept in check. The fact that the content within a vessel (e.g., fermentation) changes over time means that there is a need to adjust the controls' behavior to account for these variations. In academia, an extensive body of work can be found on batch process optimization and control, including the use of machine learning techniques, such as Kalman filters, neural networks [40], adaptive neuro-fuzzy inference systems (ANFISs) [41], or evolutionary algorithms [42]. However, in practice, industrial batch and fed-batch operations imply a minimum degree of automation and, in many instances, require operators to carry out manual tasks (e.g., nutrient addition). For example, in fermentation operations, pH and temperature controllers are present, while fed-batch operations may follow a pre-determined feeding regime. In batch bioprocessing, this is an acceptable practice as operators can prolong or shorten the fermentation processes to counter all other variations.

On the other hand, continuous bioprocesses have a simpler technical requirement as the objective of the control structure is to maintain a set point where, in principle, the content within the vessel/unit operation stays the same as raw materials are added continuously and products are removed continuously. Nonetheless, in practice, the control of continuous processes is far more complex. First of all, available measurement devices are unable to measure state variables of interest directly. Thus, it requires inferences to be made to determine key variables (e.g., the concentration of product in a continuous fermentation). In situations where key state variables can be measured, there is a limitation in the number of measurements that can be performed in a given time frame and in the fact that significant time is needed to carry out the analysis. Furthermore, unlike batch processes, where variables (e.g., pH and temperature) must be tightly controlled, in continuous manufacturing, process variations must be detected by monitoring key state variables, and appropriate real-time control action is to be taken to counteract these variations. While these objectives can be accomplished from a base layer control structure, a hierarchical advanced regulatory control structure must be ensured for the purpose of within-specification operations. To achieve this and, thus, introduce an integrated process control approach, one of the strategies is to embed novel sensors to monitor the critical control parameters of key processes in real-time.

As emphasized by the FDA [43], novel control and monitoring approaches are necessary to improve and ensure product quality in continuous biologics manufacturing. These strategies, such as PAT solutions [44] and the introduction of an industrial automation hierarchy, play a detrimental role in adapting to future demands and are, thus, essential for the robust and dependable monitoring and control of bioprocesses. However, depending on the TRL of advanced sampling methods, the time needed for process monitoring and control might be reduced, which is highly desired.

Among other developments in terms of sensing tools, the development and application of smart sensors seem to pave the way for the future, facilitating the adoption of continuous manufacturing and promoting development towards the factory of the future (smart manufacturing, also known as Industry 4.0). Free-floating wireless sensors, advanced image analysis, spectroscopic and soft sensors, and the use of chemometrics are some examples of smart sensing technology. They facilitate collecting more and improved data (information-rich), thus enhancing monitoring and control tasks. In particular, these developments break away from the traditional pressure, flow, temperature, and pH sensors that are usually available in a manufacturing process and rather focus on generating data related to the state of development of a process. Free-floating wireless sensors are a quite novel and bold sensing technology, based upon non-invasive instrumented particles [45], which can provide access to process data in a harsh environment inside an agitated bioreactor [46]

Another recent development is the use of advanced image analysis for process monitoring. For example, current studies have demonstrated that imaging and advanced image analysis, coupled with chemometrics and state-of-the-art machine learning algorithms, are promising to monitor fermentation [47].

On the other hand, spectroscopic sensors have been used and tested for a longer period. In theory, they can detect several compounds simultaneously without time delay by processing the spectra with chemometrics [48]. For example, new studies include developing infrared spectrometers [49] and Raman spectroscopy [50], which are being modified to allow for application in the demanding manufacturing environment.

Soft sensors are another sensing technology that has shown great promise for on-line monitoring. They are an alternative to traditional approaches which allow for the monitoring of state variables that affect the bioprocesses but cannot be sensed in real time otherwise [51–54]. As an advanced process monitoring technology, they use algorithms to perform an on-line analysis to collect information on unmeasured process state variables [44,55]. The interest in using these sensors has spiked in recent years due to the growing computer capacity and the numerous advances in signal processing (artificial intelligence and machine learning algorithms), which are, without a doubt, the great enablers of their successful implementation. The opportunities they bring align with the PAT initiatives and, thus, agree with the intended shift towards smart manufacturing and, consequently, the implementation of continuous manufacturing [49].

The use of robotics is also regarded as a promising strategy to facilitate the transition to continuous production lines. Great examples are the use of mechanical arms for the final production steps, such as filling millions of vials per year in big pharmaceutical companies' filling operations. Furthermore, robots are also being used to transport samples across factories for analysis and gradually replace repetitive manual tasks in the production line, tasks that were, until recently, mainly carried out by process operators. However, despite the enticing and demonstrated usefulness, we are still in the first stages of adopting this technology.

3.3. Modeling and Simulation

The successful shift from batch to continuous operation calls for a systematic modeling and simulation framework to explore, test, and evaluate a set of scenarios to narrow down the list of process candidates. Testing different scenarios on a full scale, pilot, or even lab bench is extremely resource-demanding, especially in the biopharma industry. For example, integrated continuous API production is barely investigated due to challenges such as the lack of process understanding and technology readiness [56]. Thus, it seems necessary and highly beneficial to apply modeling and simulations tools to design, control, and optimize processes when testing continuous process alternatives. These tools have been evaluated in the chemical and fine chemical industry and academia for many years.

Furthermore, looking at the many inherent challenges, the development of reliable in silico plant-wide simulation models that represent, as well as make possible, the potential continuous production of biopharmaceuticals would tremendously empower and encourage this industry to take the next steps. Plant-wide simulation models are a comprehensive modeling approach that requires a plug-and-play type of model development [56]. There are different types of modeling strategies currently available [23] which vary in their degree of accuracy, from first principles models to data-driven and even hybrid modeling. To the interested readers, Gargalo et al. (2020) [23] presented a detailed review of the different types of modeling approaches that are available, especially for bioprocess modeling.

Besides process understanding, optimization, and control, modeling has a significant role in the techno-economic and sustainability evaluation of emerging platforms. For example, in this regard, several studies have compared the impact on the overall sustainability, economics, and environmental aspects of switching from classic stainless steel batch options to single-use equipment in continuous biomanufacturing [57–59].

3.4. The Clarity in Regulation for Continuous Biologics Manufacturing

Beside scientific, technical, and business drivers, the adoption of industrial-scale continuous biologics manufacturing strongly demands clear scientific and regulatory considerations for their development, implementation, operation, and lifecycle management. The scientific guidelines, both by the FDA and ICH, ever since the publication of FDA's PAT guidance in 2004, have been providing principles and concepts which apply to continuous manufacturing. Specifically, building upon the principles and concepts described in the already published ICH guidelines, this guideline offers clarification on CM concepts, describes scientific approaches, and presents regulatory considerations unique to CM of drug substances and drug products. However, the need for a more clearly identified path was more recently identified by regulators. The FDA established an Emerging Technology Program in 2014 that works collaboratively with companies to support the adoption of a continuous manufacturing platform. In 2019, a draft guidance on Quality Considerations for Continuous Manufacturing was presented to clarify the FDA's current thinking regarding innovative CM approaches and to resolve potential issues some companies have as they consider implementation. However, the guidance only provides recommendations for small-molecule, oral, solid drug products linked with (abbreviated-)new drug application (ANDA/NDA) and explicitly notes that the recommendations do not apply to biological products submitted under a biologics license application (BLA). The ICH is, at the same time, working on the Q13 guideline on continuous manufacturing in a broader sense. This guideline will apply to CM of drug substances and drug products for both chemical entities and therapeutic proteins. Moreover, this guideline will apply to CM processes for new products (e.g., new drugs, generic drugs, and biosimilars) and the conversion of batch manufacturing to CM processes for existing products. While such regulatory efforts are very encouraging, it is also essential to ensure that continuous manufacturing standards are harmonized.

4. Sustainability Considerations of Continuous Manufacturing of Biopharmaceuticals: Process Integration and Automation

Currently, all companies are pursuing viable strategies to stand by green chemistry principles and achieve overall sustainability. As early as 2007, the pharmaceutical roundtable founded by the American Chemical Society Green Chemistry Institute (GCI) and global pharmaceutical companies recommended continuous manufacturing as the number one key green research field to boost green and sustainable production [60,61]. Continuous bioprocessing requires automation, instrumentation, and process integration, for example, by using process intensification. Integrated analytics and automation systems open the door for on-line detection and deviation handling to make corrections without human intervention. However, a fully integrated system would require high analytical precision, robust control, execution, and well-defined protocols to handle the potential deviations. The next generation of automation for bioprocessing platforms should be capable of connecting upstream and downstream, creating an integrated continuous biomanufacturing platform. Nonetheless, a clear disadvantage is that the more we connect and automate, the lower the systems' flexibility is. Thus, supervisory automation with the flexibility to switch or bypass unit operations in a lab setting to evaluate different and experimental bioprocessing approaches is desired [62]. In a Good Manufacturing Practices (GMP) environment where such flexibility within an approved workflow is not needed, such features can be controlled through an access control system, allowing operational flexibility in new manufacturing facilities. Thus, automation, instrumentation, and process integration enable more effective scale-up, better product quality, safety, and greater efficiency in terms of economy and productivity. This leads to higher raw material utilization levels, waste minimization, energy efficiency, and better facility utilization compared to a similar batch process.

Furthermore, it is important to note that continuous manufacturing processes are more suited for implementing smart manufacturing concepts (Industry 4.0). These concepts also promise further gains in process efficiency and sustainability. Batch processes, due to the dynamic nature, are more dependent on human operators to carry out complex tasks and are limited in their capabilities to carry out timely actions. In contrast, continuous bioprocessing, due to the steady-state nature, allows employing a higher degree of process automation for streamlining operations by continuous monitoring of the state of control and, thus, overall sustainability.

However, one should also acknowledge that continuous production in the biomanufacturing industry usually comes with introducing single-use and/or disposable systems. Even though they provide flexibility, an inevitable consequence is compromising the environmental sustainability of such processes. In a world where we try to avoid single-use plastics, such as plastic bags and bottles, justifying the transition from multi-use stainless steel to disposable systems is a real concern. Hence, appropriate waste disposal procedures are vital in adopting these systems, especially towards scale-up uses [63,64].

5. Digitalization

Biologics manufacturing, like many other manufacturing industries, is being influenced by the Industry 4.0 push. However, from the numerous concepts currently explored under the digitalization umbrella, which spans across multiple sectors, three key advances are likely to impact the biologics industry and create further incentives to adopt continuous manufacturing. These are smart sensors, Big Data, and Digital Twins.

5.1. From Smart Sensors to Big Data

Smart sensors, as discussed in Section 3.2, as part of the general digitalization drive, are indeed a promising tool to obtain large and improved datasets that better reflect the state of development of a process. Leveraging these data is the other half of the digitalization movement. In this case, the objective is to take the large amount of data collected and convert it into actionable information. Initially, the aim is to monitor a process with the intention of later using this information to change/control the process [65]. In batch production, these concepts have already been employed to predict batch end-time from seemingly information-poor variables, such as temperature and pH. Concepts such as predictive maintenance also leverage data to predict and schedule maintenance operations [65].

5.2. Digital Twins

Like Big Data, Digital Twins form one of the key pillars in Industry 4.0 and the digitalization drive [23,49]. While the Digital Twins concept is somewhat vaguely defined for bio-based manufacturing, there are two types of digital twins that can aid future development. This first type of digital twins deals with operational support and control. For example, the use of digital twins to forecast the evolution of a fermentation process, such as in [66]. On the other hand, a digital twin can be a digital representation of a future production process, where it acts as a validated test-bed which can be used to refine and build confidence in a process design prior to construction.

All these developments directly impact further strengthening of the case of continuous production and its business case. The development of novel sensors enables collecting datasets that are needed for "real-time" tracking of key state variables in continuous production. Simultaneously, the big data-based process monitoring and control methods are required to ensure that the data gathered can be turned into actionable information for the process operators and/or perform closed-loop control action. Digital twins, on the one hand, play a similar role in producing actionable information and providing the ability to implement closed-loop controls.

A knock-on effect of the operation supports that these elements provide, together with the high degree of process automation, is needed for continuous production processes. This brings the need for reduced staff for given production output. In addition, the operators who are working on continuous production processes would operate the process through more automated operations, avoiding manual tasks as much as possible, which are the norm in batch production.

5.3. Application of Modeling in Regulatory Decision

A unique and challenging aspect of introducing changes to pharmaceutical processes' design and operation is the need for regulatory bodies to approve any specific changes. To this end, a concept such as "Digital Twins", which, in essence, opens up towards shifting validations and testing on a process into an in-silico environment, needs to be accepted by the regulatory bodies. Quality by design is one such framework. It has been adopted and endorsed by the FDA, which indicates their willingness to acknowledge the need for more in-silico-based studies to improve the precision and speed at which process designs can be created and tested. However, the question remains whether multiple designs (process paths) can be validated by employing the digital twin concept. At the very least, these digital twins can be applied for building a multivariate design space and scale-down models for commercial-scale systems, reducing the economic burden on the R&D department without compromising the quality.

5.4. Leveraging Process Data

With the development and pilot-scale operation of hundreds of process designs in a year and dozens of commercial production processes, pharmaceutical companies can collect a large amount of diverse process data from operations. With the recent advances made in data-driven analytics and the availability of "big data", pharmaceutical companies can leverage these data to identify common failures and successes rapidly. In turn, these situations can be further analyzed by subject matter experts to develop process insights for an improved design and operation of future production processes. As such, data analytics will allow pharmaceutical companies to rapidly improve process designs as opposed to the more subdued pace at which changes usually occur. It is noteworthy that data analytics acts as an enabler for experts to analyze hundreds, if not thousands, of datasets effectively, thus facilitating the gathering of insights and pro-active planning of operational changes.

6. Hybrid Facilities: Acknowledging the Best of Both Worlds

Depending on the demand and manufacturing stage, it becomes easier or more challenging to work with single-use/disposable systems or multi-use stainless steel systems. A hybrid facility that uses disposables and reusable systems potentially combines both systems' benefits and could be the path forward depending on the required capacity. This has often come to mean single-use technologies in seed-train development. However, for upstream production, the use of stainless steel-based perfusion bioreactors is more common, and then again using single-use systems in downstream processing for in-process holding and filtration units. Such flexibility will automatically enable the flexibility of connected unit operations and, thus, continuous operations while keeping the risk of contamination to a minimal level [36].

7. Summary and Outlook

Despite the benefits of continuous over batch bioprocessing, its adoption has lagged, with few exceptions. However, the batch manufacturing paradigm's dominance in the industry for reasons such as "batch processing is familiar and works very well" cannot be sustained in the long term given the new biomanufacturing challenges. The industry-held perception of complexity in continuous bioprocessing is becoming obsolete, as more and more new technologies and solutions are continually improving the situation. Several academic- and industry-led consortia are working to improve the perception regarding continuous bioprocessing by bringing the questions to the correct stakeholders who can address them. The training provided by these initiatives to the top management of the companies is playing an essential role in changing the perception and, at the same time, also creating new scientists and operators that can understand and respond to a new set of operational challenges. However, wider adoption of continuous bioprocessing will only be possible if the gaps at the technical, management, and regulatory levels are acknowledged. As discussed in this paper, concerted efforts are being made to abridge them. These include:

Technical:

i. Improvement in automation to allow flexibility in the design and control of continuous processes;
ii. Adoption of a "digital twin" of processes to reduce the costs and risks linked with a decision made based on a limited set of experimental results;
iii. Application of detailed modeling and expert systems to support the development and regulatory requirements, such as scale-down modeling;
iv. Working on hybrid approaches such as single-use and multi-use to obtain best-of-all outcomes, thus enabling continuous manufacturing;
v. Application of "big data" to support process development, control, and regulatory filing of a project.

Management:

i. Training on realistic situations highlighting risks and benefits of continuous manufacturing will allow removing the barrier caused by preexisting perceptions;
ii. Acquisition of trained staff who can support the adoption of new technology;
iii. Identifying key stakeholders from across the organization and getting them involved in the migration processes;
iv. Early alignment of R&D and commercial manufacturing business drivers to realize the extensive benefits of continuous biologics manufacturing.

Regulatory:

i. Clarity on the regulation of continuous vs. batch definitions under newer integrated and hybrid biomanufacturing process designs will be very useful;
ii. Increase in acceptability of digital twins as evidence for regulatory clearance;
iii. A joint effort by regulatory agencies and industries to develop a possible roadmap for the integrated continuous manufacturing will be highly beneficial for the biologics sector;
iv. Harmonization of continuous manufacturing standards and regulations.

Author Contributions: Conceptualization, A.K., C.L.G., I.A.U. and K.V.G.; writing—original draft preparation, A.K., C.L.G., I.A.U.; writing—review and editing, A.K., C.L.G., I.A.U. and K.V.G.; visualization, A.K., C.L.G., I.A.U.; supervision, K.V.G.; project administration, K.V.G.; funding acquisition, K.V.G.; All authors have read and agreed to the published version of the manuscript.

Funding: This work was partly sponsored by the Novo Nordisk Foundation in the frame of the "Accelerated Innovation in Manufacturing Biologics" (AIMBio) project (Grant number NNF19SA0035474).

Conflicts of Interest: The authors declare no conflict of interest.

References

1. Chatterjee, S. FDA Perspective on Continuous Manufacturing. In Proceedings of the IFPAC Annual Meeting, Baltimore, MD, USA, 22–25 January 2012.
2. ICH Expert Working Group. Q13 Continuous Manufacturing of Drug Substances and Drug Products. Available online: https://database.ich.org/sites/default/files/Q13_EWG_Concept_Paper.pdf (accessed on 11 December 2020).
3. Lee, S.L. Current FDA Perspective for Continuous Manufacturing. In Proceedings of the MIT-CMAC 2nd International Symposium on Continuous Manufacturing of Pharmaceuticals, Cambridge, MA, USA, 26–27 September 2016.
4. Nasr, M.; Krumme, M.; Matsuda, Y.; Trout, B.L.; Badman, C.; Mascia, S.; Cooney, C.; Jensen, K.D.; Florence, A.; Johnston, C.; et al. Regulatory Perspectives on Continuous Pharmaceutical Manufacturing: Moving From Theory to Practice. *J. Pharm. Sci.* **2017**, *106*, 3199–3206. [CrossRef] [PubMed]
5. Croughan, M.S.; Konstantinov, K.B.; Cooney, C. The future of industrial bioprocessing: Batch or continuous? *Biotechnol. Bioeng.* **2015**, *112*, 648–651. [CrossRef] [PubMed]
6. Konstantinov, K.B.; Cooney, C. White Paper on Continuous Bioprocessing 20–21 May 2014 Continuous Manufacturing Symposium. *J. Pharm. Sci.* **2015**, *104*, 813–820. [CrossRef] [PubMed]
7. Gernaey, K.V.; Gani, R. A model-based systems approach to pharmaceutical product-process design and analysis. *Chem. Eng. Sci.* **2010**, *65*, 5757–5769. [CrossRef]
8. Schaber, S.D.; Gerogiorgis, D.I.; Ramachandran, R.; Evans, J.M.B.; Barton, P.I.; Trout, B.L. Economic Analysis of Integrated Continuous and Batch Pharmaceutical Manufacturing: A Case Study. *Ind. Eng. Chem. Res.* **2011**, *50*, 10083–10092. [CrossRef]
9. Department of Health and Human Services, Food and Drug Administration. PAT Guidance for Industry-Framework for Innovative Pharmaceutical Development, Manufacturing and Quality Assurance. Available online: https://www.fda.gov/media/71012/download (accessed on 11 December 2020).
10. FDA. Quality Considerations for Continuous Manufacturing Guidance for Industry. *Food Drug Adm.* Available online: https://www.fda.gov/media/121314/download (accessed on 11 December 2020).
11. Langer, E. Biomanufacturing: Demand for Continuous Bioprocessing Increasing. *BioPharm Int.* **2020**, *33*, 5.
12. Godawat, R.; Konstantinov, K.; Rohani, M.; Warikoo, V. End-to-end integrated fully continuous production of recombinant monoclonal antibodies. *J. Biotechnol.* **2015**, *213*, 13–19. [CrossRef]
13. Aakesson, M.; Heitmann, M.; Tiainen, P. Integrated Continuous Biomanufacturing Process. U.S. Patent Application No. 15/306,938, 2 March 2017.
14. Desai, S.G. Continuous and semi-continuous cell culture for production of blood clotting factors. *J. Biotechnol.* **2015**, *213*, 20–27. [CrossRef]
15. Farid, S.S.; Thompson, B.; Davidson, A. Continuous bioprocessing: The real thing this time? In Proceedings of the10th Annual bioProcessUK Conference, London, UK, 3–4 December 2013.

16. Jones, S.; Castillo, F.; Levine, H. Advances in the development of therapeutic monoclonal antibodies. *BioPharm. Int.* **2007**, *20*, 96–114.
17. Agrawal, V.; Bal, M. Strategies for rapid production of therapeutic proteins in mammalian cells. *BioProcess Int.* **2012**, *10*, 32–48.
18. Ebeler, M.; Lind, O.; Norrman, N.; Palmgren, R.; Franzreb, M. One-step integrated clarification and purification of a monoclonal antibody using Protein A Mag Sepharose beads and a cGMP-compliant high-gradient magnetic separator. *New Biotechnol.* **2018**, *42*, 48–55. [CrossRef] [PubMed]
19. Käppler, T.; Cerff, M.; Ottow, K.E.; Hobley, T.J.; Posten, C. In situ magnetic separation for extracellular protein production. *Biotechnol. Bioeng.* **2009**, *102*, 535–545. [CrossRef] [PubMed]
20. Xu, J.; Xu, X.; Huang, C.; Angelo, J.; Oliveira, C.L.; Xu, M.; Xu, X.; Temel, D.; Ding, J.; Ghose, S.; et al. Biomanufacturing evolution from conventional to intensified processes for productivity improvement: A case study. *mAbs* **2020**, *12*, 1770669. [CrossRef] [PubMed]
21. Haringa, C.; Mudde, R.F.; Noorman, H.J. From industrial fermentor to CFD-guided downscaling: What have we learned? *Biochem. Eng. J.* **2018**, *140*, 57–71. [CrossRef]
22. Tajsoleiman, T.; Spann, R.; Bach, C.; Gernaey, K.V.; Huusom, J.K.; Krühne, U. A CFD based automatic method for compartment model development. *Comput. Chem. Eng.* **2019**, *123*, 236–245. [CrossRef]
23. Gargalo, C.L.; Heras, S.C.; de Las Jones, M.N.; Mansouri, S.S.; Krühne, U.; Gernaey, K.V. Towards the Development of Digital Twins for the Bio-Manufacturing Industry. Available online: https://doi.org/10.1007/10_2020_142 (accessed on 11 December 2020).
24. Busse, C.; Biechele, P.; de Vries, I.; Reardon, K.F.; Solle, D.; Scheper, T. Sensors for disposable bioreactors. *Eng. Life Sci.* **2017**, *17*, 940–952. [CrossRef]
25. Kornecki, M.; Schmidt, A.; Lohmann, L.; Huter, M.; Mestmäcker, F.; Klepzig, L.S.; Mouellef, M.; Zobel-Roos, S.; Strube, J. Accelerating Biomanufacturing by Modeling of Continuous Bioprocessing—Piloting Case Study of Monoclonal Antibody Manufacturing. *Processes* **2019**, *7*, 495. [CrossRef]
26. Huter, M.; Strube, J. Model-Based Design and Process Optimization of Continuous Single Pass Tangential Flow Filtration Focusing on Continuous Bioprocessing. *Processes* **2019**, *7*, 317. [CrossRef]
27. Carvalho, R.J.; Castilho, L.R.; Subramanian, G. Tools Enabling Continuous and Integrated Upstream and Downstream Processes in the Manufacturing of Biologicals. In *Continuous Biomanufacturing-Innovative Technologies and Methods*; Wiley: Hoboken, NJ, USA, 2017; pp. 31–68.
28. Castilho, L.; Anspach, F. CFD-aided design of a dynamic filter for mammalian cell separation. *Biotechnol. Bioeng.* **2003**, *83*, 514–524. [CrossRef]
29. Biomanufacturing Consortium (BioMAN)|MIT Center for Biomedical Innovation. Available online: http://cbi.mit.edu/research-overview/bioman/ (accessed on 31 August 2020).
30. BioPhorum. Available online: https://www.biophorum.com/ (accessed on 31 August 2020).
31. Udugama, I.A.; Feldman, H.; de las Heras, S.C.; Kizhedath, A.; Bryde-Jacobsen, J.; van den Berg, F.; Mansouri, S.S.; Gernaey, K.V. Biopro World Talent Campus: A week of real world challenge for biotechnology post-graduate students. *Educ. Chem. Eng.* **2018**, *25*, 1–8. [CrossRef]
32. DiCesare, C.; Yu, M.; Yin, J.; Zhou, W.; Hwang, C.; Tengtrakool, J.; Konstantinov, K. Development, qualification, and application of a bioreactor scale-down process: Modeling large-scale microcarrier perfusion cell culture. *Bioprocess. Int.* **2016**, *14*, 18–29.
33. Tajsoleiman, T.; Mears, L.; Krühne, U.; Gernaey, K.V.; Cornelissen, S. An Industrial Perspective on Scale-Down Challenges Using Miniaturized Bioreactors. *Trends Biotechnol.* **2019**, *37*, 697–706. [CrossRef] [PubMed]
34. Ram, R.J. Tools for Continuous Bioprocess Development. *BioPharm Int.* **2016**, *29*, 18–25.
35. Sartorius Ambr®15 Fermentation-High throughput Automated System|Sartorius. Available online: https://www.sartorius.com/us-en/products/fermentation-bioreactors/ambr-multi-parallel-bioreactors/ambr-15-fermentation (accessed on 9 October 2020).
36. Scott, C. Large-Scale Capacity Strategies: Single Use, Multiuse, or Both? *Bioprocess. Int.* **2019**. Available online: https://bioprocessintl.com/manufacturing/facility-design-engineering/large-scale-capacity-strategies-single-use-multiuse-or-both/ (accessed on 11 December 2020).
37. Goji, T.; Hayashi, Y.; Sakata, I. Evaluating "startup readiness" for researchers: Case studies of research-based startups with biopharmaceutical research topics. *Heliyon* **2020**, *6*, e04160. [CrossRef] [PubMed]

38. Udugama, I.A.; Taube, M.A.; Mansouri, S.S.; Kirkpatrick, R.; Gernaey, K.V.; Yu, W.; Young, B.R. A Systematic Methodology for Comprehensive Economic Assessment of Process Control Structures. *Ind. Eng. Chem. Res.* **2018**, *57*, 13116–13130. [CrossRef]
39. Thiess, H.; Gronemeyer, P.; Ditz, R.; Strube, J.; Zobel-Roos, S.; Subramanian, G. Engineering Challenges of Continuous Biomanufacturing Processes (CBP). In *Continuous Biomanufacturing-Innovative Technologies and Methods*; Wiley: Hoboken, NJ, USA, 2017; pp. 69–106.
40. Montes, F.; Gernaey, K.V.; Sin, G. Implementation of a Radial Basis Function control strategy for the crystallization of Ibuprofen under uncertainty. *Comput. Aided Chem. Eng.* **2018**, *44*, 565–570. [CrossRef]
41. O'Mahony, N.; Murphy, T.; Panduru, K.; Riordan, D.; Walsh, J. Adaptive process control and sensor fusion for process analytical technology. In Proceedings of the 2016 27th Irish Signals and Systems Conference (ISSC), Londonderry, UK, 21–22 June 2016; pp. 1–6.
42. Allmendinger, R.; Simaria, A.S.; Turner, R.; Farid, S.S. Closed-loop optimization of chromatography column sizing strategies in biopharmaceutical manufacture. *J. Chem. Technol. Biotechnol.* **2013**, *89*, 1481–1490. [CrossRef]
43. FDA. *PAT Guidance for Industry—A Framework for Innovative Pharmaceutical Development*; Manufacturing and Quality Assurance: Rockville, MD, USA, 2004.
44. Randek, J.; Mandenius, C.-F. On-line soft sensing in upstream bioprocessing. *Crit. Rev. Biotechnol.* **2017**, *38*, 106–121. [CrossRef]
45. Zimmermann, R.; Fiabane, L.; Gasteuil, Y.; Volk, R.; Pinton, J.-F. Measuring Lagrangian accelerations using an instrumented particle. *Phys. Scr.* **2013**, *2013*, 014063. [CrossRef]
46. Freesense. Available online: https://www.freesense.dk (accessed on 11 December 2020).
47. Pontius, K.; Junicke, H.; Gernaey, K.V.; Bevilacqua, M. Monitoring yeast fermentations by nonlinear infrared technology and chemometrics—Understanding process correlations and indirect predictions. *Appl. Microbiol. Biotechnol.* **2020**, *104*, 5315–5335. [CrossRef] [PubMed]
48. Landgrebe, D.; Haake, C.; Höpfner, T.; Beutel, S.; Hitzmann, B.; Scheper, T.; Rhiel, M.; Reardon, K.F. On-line infrared spectroscopy for bioprocess monitoring. *Appl. Microbiol. Biotechnol.* **2010**, *88*, 11–22. [CrossRef] [PubMed]
49. Gargalo, C.L.; Udugama, I.A.; Pontius, K.; Lopez, P.C.; Nielsen, R.F.; Hasanzadeh, A.; Mansouri, S.S.; Bayer, C.; Junicke, H.; Gernaey, K.V. Towards smart manufacturing: A perspective on recent developments in industrial measurement and monitoring technologies for bio-based production processes. *J. Ind. Microbiol. Biotechnol.* **2020**, *47*, 947–964. [CrossRef] [PubMed]
50. Oh, S.-K.; Yoo, S.J.; Jeong, D.H.; Lee, J.M. Real-time estimation of glucose concentration in algae cultivation system using Raman spectroscopy. *Bioresour. Technol.* **2013**, *142*, 131–137. [CrossRef]
51. De Assis, A.J.; Filho, R.M. Soft sensors development for on-line bioreactor state estimation. *Comput. Chem. Eng.* **2000**, *24*, 1099–1103. [CrossRef]
52. Veloso, A.C.A.; Rocha, I.; Ferreira, E.C. Monitoring of fed-batch E. coli fermentations with software sensors. *Bioprocess. Biosyst. Eng.* **2009**, *32*, 381–388. [CrossRef]
53. Sharma, S.; Tambe, S.S. Soft-sensor development for biochemical systems using genetic programming. *Biochem. Eng. J.* **2014**, *85*, 89–100. [CrossRef]
54. Krause, D.; Hussein, M.; Becker, T. Online monitoring of bioprocesses via multivariate sensor prediction within swarm intelligence decision making. *Chemom. Intell. Lab. Syst.* **2015**, *145*, 48–59. [CrossRef]
55. Luttmann, R.; Bracewell, D.G.; Cornelissen, G.; Gernaey, K.V.; Glassey, J.; Hass, V.C.; Kaiser, C.; Preusse, C.; Striedner, G.; Mandenius, C.-F. Soft sensors in bioprocessing: A status report and recommendations. *Biotechnol. J.* **2012**, *7*, 1040–1048. [CrossRef]
56. Ramin, P.; Mansouri, S.S.; Udugama, I.; Benyahia, B.; Gernaey, K.V. Modelling continuous pharmaceutical and bio-based processes at plant-wide level: A roadmap towards efficient decision-making. *Chim. Oggi Chem. Today* **2018**, *36*, 26–30.
57. Jacquemart, R.; VanderSluis, M.; Zhao, M.; Sukhija, K.; Sidhu, N.; Stout, J. A Single-use Strategy to Enable Manufacturing of Affordable Biologics. *Comput. Struct. Biotechnol. J.* **2016**, *14*, 309–318. [CrossRef] [PubMed]
58. Zürcher, P.; Shirahata, H.; Badr, S.; Sugiyama, H. Multi-stage and multi-objective decision-support tool for biopharmaceutical drug product manufacturing: Equipment technology evaluation. *Chem. Eng. Res. Des.* **2020**, *161*, 240–252. [CrossRef]

59. Shirahata, H.; Hirao, M.; Sugiyama, H. Multiobjective decision-support tools for the choice between single-use and multi-use technologies in sterile filling of biopharmaceuticals. *Comput. Chem. Eng.* **2019**, *122*, 114–128. [CrossRef]
60. Dallinger, D.; Kappe, C.O. Why flow means green–Evaluating the merits of continuous processing in the context of sustainability. *Curr. Opin. Green Sustain. Chem.* **2017**, *7*, 6–12. [CrossRef]
61. Jiménez-González, C.; Poechlauer, P.; Broxterman, Q.B.; Yang, B.-S.; Ende, D.A.; Baird, J.; Bertsch, C.; Hannah, R.E.; Dell'Orco, P.; Noorman, H.; et al. Key Green Engineering Research Areas for Sustainable Manufacturing: A Perspective from Pharmaceutical and Fine Chemicals Manufacturers. *Org. Process. Res. Dev.* **2011**, *15*, 900–911. [CrossRef]
62. May, M. Modular Bioprocessing Makes Adaptability a Snap: By swapping out and adding bioprocessing modules, biomanufacturers can modify functionality and adjust capacity quickly and economically. *Genet. Eng. Biotechnol. News* **2019**, *39*, 38–40. [CrossRef]
63. Scott, C. Sustainability in Bioprocessing. *BioProcess Int.* **2011**, 9. Available online: https://bioprocessintl.com/manufacturing/monoclonal-antibodies/sustainability-in-bioprocessing-323438/ (accessed on 11 December 2020).
64. Lonza Innovations in Pharma Biotech & Nutrition. Available online: https://annualreport.lonza.com/2019/segments/pharma-biotech-nutrition/innovations.html (accessed on 11 December 2020).
65. Udugama, I.A.; Gargalo, C.L.; Yamashita, Y.; Taube, M.A.; Palazoglu, A.; Young, B.R.; Gernaey, K.V.; Kulahci, M.; Bayer, C. The Role of Big Data in Industrial (Bio)chemical Process Operations. *Ind. Eng. Chem. Res.* **2020**, *59*, 15283–15297. [CrossRef]
66. Lopez, P.C.; Udugama, I.A.; Thomsen, S.T.; Roslander, C.; Junicke, H.; Mauricio-Iglesias, M.; Gernaey, K.V. Towards a digital twin: A hybrid data-driven and mechanistic digital shadow to forecast the evolution of lignocellulosic fermentation. *Biofuels, Bioprod. Biorefining* **2020**, *14*, 1046–1060. [CrossRef]

MDPI
St. Alban-Anlage 66
4052 Basel
Switzerland
Tel. +41 61 683 77 34
Fax +41 61 302 89 18
www.mdpi.com

Processes Editorial Office
E-mail: processes@mdpi.com
www.mdpi.com/journal/processes

www.ingramcontent.com/pod-product-compliance
Ingram Content Group UK Ltd.
Pitfield, Milton Keynes, MK11 3LW, UK
UKHW060121010826
14089UKWH00006B/99